Lecture Notes in Computer Science 2327

Edited by G. Goos, J. Hartmanis, and J. van Leeuwen

Springer
Berlin
Heidelberg
New York
Barcelona
Hong Kong
London
Milan
Paris
Tokyo

Hans P. Zima   Kazuki Joe Mitsuhisa Sato
Yoshiki Seo Masaaki Shimasaki (Eds.)

High Performance Computing

4th International Symposium, ISHPC 2002
Kansai Science City, Japan, May 15-17, 2002
Proceedings

 Springer

Volume Editors

Hans P. Zima
University of Vienna, Institute of Software Science
Liechtensteinstr. 22, 1090 Vienna, Austria
E-mail: zima@jpl.nasa.gov

Kazuki Joe
Nara Women's University, Department of Information and Computer Science
Kitauoyanishimachi, Nara City 630-8506, Japan
E-mail: joe@ics.nara-wu.ac.jp

Mitsuhisa Sato
University of Tsukuba, Institute of Information Science and Electronics
Tenno-dai 1-1-1, Tsukuba, Ibaraki 305-8577, Japan
E-mail: msato@is.tsukuba.ac.jp

Yoshiki Seo
NEC Corporation, Internet Systems Research Laboratories
4-1-1, Miyazaki, Miyamae, Kawasaki, Kanagawa 216-8555, Japan
E-mail: seo@ccm.cl.nec.cop.jp

Masaaki Shimasaki
Kyoto University
Yoshidahonmachi, Sakyo-ku, Kyoto 606-8501, Japan
E-mail: simasaki@kuee.kyoto-u.ac.jp

Cataloging-in-Publication Data applied for

Die Deutsche Bibliothek - CIP-Einheitsaufnahme

High performance computing : 4th international symposium ; proceedings /
ISHPC 2002, Kansai Science City, Japan, May 15 - 17, 2002. Hans P. Zima ...
(ed.). - Berlin ; Heidelberg ; New York ; Barcelona ; Hong Kong ; London ;
Milan ; Paris ; Tokyo : Springer, 2002
    (Lecture notes in computer science ; Vol. 2327)
    ISBN 3-540-43674-X

CR Subject Classification (1998): D.1-2, F.2, E.4, G.1-4, J.1-2, J.6, I.6

ISSN 0302-9743
ISBN 3-540-43674-X Springer-Verlag Berlin Heidelberg New York

Springer-Verlag Berlin Heidelberg New York
a member of BertelsmannSpringer Science+Business Media GmbH

http://www.springer.de

© Springer-Verlag Berlin Heidelberg 2002

Typesetting: Camera-ready by author, data conversion by Olgun Computergrafik
Printed on acid-free paper   SPIN 10846733   06/3142   5 4 3 2 1 0

Preface

I wish to welcome all of you to the International Symposium on High Performance Computing 2002 (ISHPC 2002) and to Kansai Science City, which is not far from the ancient capitals of Japan: Nara and Kyoto. ISHPC 2002 is the fourth in the ISHPC series, which consists, to date, of ISHPC '97 (Fukuoka, November 1997), ISHPC '99 (Kyoto, May 1999), and ISHPC 2000 (Tokyo, October 2000). The success of these symposia indicates the importance of this area and the strong interest of the research community. With all of the recent drastic changes in HPC technology trends, HPC has had and will continue to have a significant impact on computer science and technology. I am pleased to serve as General Chair at a time when HPC plays a crucial role in the era of the IT (Information Technology) revolution.

The objective of this symposium is to exchange the latest research results in software, architecture, and applications in HPC in a more informal and friendly atmosphere. I am delighted that the symposium is, like past successful ISHPCs, comprised of excellent invited talks, panels, workshops, as well as high-quality technical papers on various aspects of HPC. We hope that the symposium will provide an excellent opportunity for lively exchange and discussion about directions in HPC technologies and all the participants will enjoy not only the symposium but also their stay in Kansai Science City.

This symposium would not have been possible without the great help of many people who have devoted a tremendous amount of time and effort. I thank all those who have worked diligently to make ISHPC 2002 a great success. In particular I would like to thank Organizing Chair Takashi Arisawa of JAERI-KRE and the Organizing Committee members for their significant contribution to the planning and organization of ISHPC 2002. I would also like to thank the Program Chair Hans Zima of the University of Vienna/Jet Propulsion Laboratory/CalTech, Program Co-chair Mateo Valero of UPC (architecture track), William Gropp of Argonne National Laboratory (software track), Yoshitoshi Kunieda of Wakayama University (applications track), and the program committee members for their contribution to a technically excellent symposium program. Thanks are due to Workshop Chair Mitsuhisa Sato of the University of Tsukuba and Yoshiki Seo of NEC for organizing workshops on timely selected topics.

A last note of thanks goes to the Kayamori Foundation of Information Science Advancement, NEC, Fujitsu, Japan IBM, Japan SGI, KGT, Sumisho Electronics, and Mitsubishi Space Software for sponsoring the symposium.

May 2002   Masaaki Shimasaki

Foreword

The 4th International Symposium on High Performance Computing (ISHPC 2002, Kansai Science City, Japan, May 15–17, 2002), has been thoughtfully planned, organized, and supported by the ISHPC Organizing Committee and collaborative organizations.

The ISHPC 2002 program consists of three keynote speeches, several invited talks, workshops on OpenMP and HPF, two panel discussions, and several technical sessions covering theoretical and applied research topics on high performance computing which are representative of the current research activity in industry and academia. Participants and contributors to this symposium represent a cross section of the research community and major laboratories in this area, including the Kansai Research Establishment of the Japan Atomic Energy Research Institute, the Japan Society for Simulation Technology, SIGARCH and SIGHPC of the Information Processing Society Japan, and the Society for Massively Parallel Processing.

All of us on the program committee wish to thank the authors who submitted papers to ISHPC 2002. We received 57 technical contributions from 17 countries. Each paper received at least 3 peer reviews and, based on the evaluation process, the program committee selected 18 regular (12-page) papers. Since several additional papers received favorable reviews, the program committee recommended a poster session comprised of short papers. A total of 12 contributions were selected as short (8-page) papers for presentation in the poster session and inclusion in the proceedings.

The program committee also recommended two awards for regular papers: a distinguished paper award and a best student paper award. The distinguished paper award has been given to "Language and Compiler Support for Hybrid-Parallel Programming on SMP Clusters" by Siegfried Benkner and Viera Sipkova, and the best student paper award has been given to "Parallelizing Merge Sort onto Distributed Memory Parallel Computers" by Minsoo Jeon.

ISHPC 2002 has collaborated closely with two workshops: the second International Workshop on OpenMP: Experiences and Implementations (WOMPEI 2002) organized by Mitsuhisa Sato of the University of Tsukuba, and the first HPF International Workshop: Experiences and Progress (HiWEP 2002) organized by Yoshiki Seo of NEC. Invitation based submission was adopted by both workshops. The ISHPC 2002 program committee decided to include all papers of WOMPEI and HiWEP in the proceedings of ISHPC 2002.

We hope that the final program is of significant interest to the participants and serves as a launching pad for interaction and debate on technical issues among the attendees.

May 2002   Hans Zima

Foreword to WOMPEI

OpenMP is an emerging industry standard interface for shared memory programming of parallel computer applications. OpenMP allows applications written for the shared memory programming model to be portable to a wide range of parallel computers.

WOMPEI 2002 follows a series of workshops on OpenMP, such as WOMPAT 2001, EWOMP 2001, and WOMPEI 2000. This is the second OpenMP workshop held in Japan. It is part of the cOMPunity initiative to disseminate and exchange information about OpenMP.

The workshop consists of 2 invited talks, from SPEC HPG and OpenMP ARB, and 10 contributed papers. They report on some of the current research and development activities including tools and compilers for OpenMP, as well as experiences in the use of the language. We are also very pleased to have a joint panel discussion with HiWEP 2002 on "the parallel programming interface of the future."

We would like to thank the ISHPC Organizing Committee for giving us the opportunity to organize WOMPEI as part of the symposium. We would also like to thank the Program Committee, the cOMPunity, and the OpenMP ARB for their support. We hope that the program will be of interest to the OpenMP community and will serve as a forum for discussion on technical and practical issues related to OpenMP.

Mitsuhisa Sato
Eduard Ayguade

Foreword to HiWEP 2002

High Performance Fortran is a data parallel language that makes it possible to program efficient parallel codes for distributed memory parallel systems with minimal effort. Last year, several vendors started to provide long-awaited compilers that could be used for real parallelization with the help of JAHPF efforts. In the HUG 2000 meeting held in Tokyo in October 2000, many successful results using HPF were presented.

This workshop, HiWEP 2002, addresses recent progress in HPF software and experiences with programming in HPF and other distributed-parallel programming paradigms. HiWEP 2002 is organized as a workshop in association with ISHPC 2002 and consists of one keynote address, one invited talk, six contributed papers, and several short talks. We would like to thank the ISHPC 2002 Organizing Committee for giving us this opportunity. We are also very glad to have a joint panel discussion with WOMPEI on the future of parallel programming interfaces.

Kunihiko Watanabe
Yoshiki Seo
Yasuo Okabe

Organization

ISHPC 2002 is organized by the ISHPC Organizing Committee in cooperation
with the Kansai Research Establishment of the Japan Atomic Energy Research
Institute, the Japan Society for Simulation Technology, SIGARCH and SIGHPC
of the Information Processing Society Japan, and the Society for Massively Parallel Processing.

ISHPC 2002 Executive Committee

General Chair: Masaaki Shimasaki (Kyoto U, Japan)
Program Chair: Hans Zima (U Vienna, Austria)
Program Co-chair: Mateo Valero (UPC, Spain)
     William Gropp (Argonne, US)
     Yoshitoshi Kunieda (Wakayama U, Japan)
Organizing Chair: Takashi Arisawa (JAERI-KRE, Japan)
Publication & Treasury Chair: Kazuki Joe (NWU, Japan)
Local Arrangements Chair: Hayaru Shouno (NWU, Japan)
Workshop Chair: Mitsuhisa Sato (U Tsukuba, Japan)
     Kunihiko Watanabe (NIFS, Japan)

ISHPC 2002 Program Committee

Hideharu Amano (Keio U)
Taisuke Boku (U Tsukuba)
Claudia Dinapoli (CNR)
Shin-ichiro Mori (Kyoto U)
Hironori Nakajo (TUAT)
Olivier Teman (LRI)
Alex Veidenbaum (UCI)
Chuck Hansen (U Utah)
Chris Johnson (U Utah)
Yasunori Kimura (Fujitsu)
Mitsuhisa Sato (RWCP)
Valerie Taylor (Northwestern U)
Yutaka Akiyama (CBRC)
Ophir Frieder (IIT)
Stratis Gallopoulos (U Patras)
Mitsunori Miki (Doshisha U)
Hitoshi Oi (Florida Atlantic U)
Peter R.Taylor (UCSD)

Utpal Banerjee (Intel Corp.)
Doug Burger (U Texas Austin)
Michel Dubois (USC)
Andreas Moshovos (U Toronto)
Hiroshi Nakasima (TUT)
Stamatis Vassiliadis (U Delft)
Harvey Wasserman (Los Alamos)
Yasuhiro Inagami (Hitatchi)
Hironori Kasahara (Waseda U)
Allen Malony (U Oregon)
Yoshiki Seo (NEC)
Kathy Yelick (UCB)
Hamid Arabnia (Geogea U)
Mario Furnari (CNR)
Elias Houstis, (Purdue U)
Takashi Nakamura (NAL)
Mariko Sasakura (Okayama U)
Mitsuo Yokokawa (JAERI)

ISHPC 2002 Organizing Committee

Eduard Ayguade (UPC)
Hironori Nakajo (TUAT)
Toshinori Sato (Kyushu I)
Shinji Hioki (Tezukayama U)

Yutaka Ueshima (JAERI-KRE)
Steve Lumetta (UIUC)
Mariko Sasakura (Okayama U)
Hitoshi Oi (Florida Atlantic U)

WOMPEI 2002 Organization

General Chair:
Program Chair:

Mitsuhisa Sato (U Tsukuba, Japan)
Eduard Ayguade (UPC, Spain)

Program Committee:
Barbara Chapman (U Houston)
Hironori Kasahara (Waseda U)
Tim Mattson (Intel)

Rudolf Eigenmann (Purdue U)
Yoshiki Seo (NEC)
Matthijs van Waveren (Fujitsu)

HiWEP 2002 Organization

General Chair:

Kunihiko Watanabe
(National Institute of Fusion Science, Japan)

Program Chair:

Yoshiki Seo (NEC Corp.)

Program Committee:
Sigi Benkner (U Vienna)
Barbara Chapman (U Houston)
Hidetoshi Iwashita (Fujitsu)
Henk Sips (Delft U of Tech.)

Thomas Brandes (SCAI)
Masahiro Fukuda (JAERI)
Hitoshi Sakagami (Himeji Inst. of Tech.)

Local Organizing Chair:

Yasuo Okabe (Kyoto U)

Local Organizing Committee:
Mamiko Hata (JMSTEC)
Hiroshi Katayama (NEC)

Sachio Kamiya (Fujitsu)

Referees

A. Cohen	N. Naoyuki	H. Shouno
D. Crisu	N. Nide	P. Stathis
K. Itakura	E. Ogston	M. Takata
K. Joe	K. Okamura	W. Tang
H. Kamo	H. Okawara	T. Uehara
M. Koibuchi	S. Roos	A. Vakali
G. Kuzmanov	H. Saito	F. Vitobello
C. Lageweg	S. Saito	H. Wasserman
E. Lusk	F. Saito	S. Wong
M. Maeda	T. Sato	
M. Matsubara	J. Sebot	
T. Nakamura	K. Shimura	

Table of Contents

V. Architectures II

VI. HPC Systems

VII. Earth Simulator

VIII. Short Papers

IX. International Workshop on OpenMP: Experiences and Implementations (WOMPEI 2002)

X. HPF International Workshop: Experiences and Progress (HiWEP 2002)

The Gilgamesh MIND Processor-in-Memory Architecture for Petaflops-Scale Computing
(An Extended Abstract)

Thomas Sterling

California Institute of Technology
NASA Jet Propulsion Laboratory

1 Introduction

Implicit in the evolution of current technology and high-end system evolution
is the anticipated achievement of the implementation of computers capable of
a peak performance of 1 Petaflops by the year 2010. This is consistent with
both the semiconductor industry's roadmap of basic device technology development and an extrapolation of the TOP-500 list of the world's fastest computers
according to the Linpack benchmark. But if contemporary experiences with today's largest systems hold true for their descendents at the end of the decade,
then they will be very expensive (> \$100M), consume too much power (> 3
Mwatts), take up too much floor space (> 10,000 square feet), deliver very low
efficiency (< 10%), and are too difficult to program. Also important is the likely
degradation of reliability due to the multiplicative factors of MTBF and scale
of components. Even if these systems do manage to drag the community to the
edge of Petaflops, there is no basis of confidence to assume that they will provide
the foundation for systems across the next decade that will transition across the
trans-Petaflops performance regime. It has become increasingly clear that an
alternative model of system architecture may be required for future generation
high-end computers.

A number of critical factors determine the effectiveness of an architectural
strategy to achieving a practical high-end computing. At the strategic level
these include the ability to aggregate sufficient resources to achieve the targeted peak capabilities, the means to illicit efficient behavior from the available
resources, and characteristics that exhibit usability. Peak capabilities include the
total arithmetic and logic unit throughput, the aggregate storage capacities for
main memory and secondary storage, the interconnect bandwidth both within
the system and the external I/O, and the costs of these in terms of actual dol-
lars, power consumption, and floor space. System efficiency is to a significant
degree determined by the latency experienced by system resource accesses, the
overhead of performing critical time management functions, the contention for
shared resources such as network arbitration and memory bank conflicts, and
starvation due to insufficient task parallelism and load balancing. Usability is
more difficult to quantify but it of first importance to overall productivity that

H. Zima et al. (Eds.): ISHPC 2002, LNCS 2327, pp. 1–5, 2002.

can be accomplished with a system. Usability involves ease of programming, manageability, and reliability for system availability. Any system strategy to be effective and justify pursuit of ultimate implementation and application must address these three dominant factors and their constituent contributors.

Processor-in-Memory (PIM) technology and broad class of architectures present an innovative means and approach to address the challenges of high end computing. PIM fabrication processes permit the merging and co-location of both DRAM memory cells (SRAM as well) and CMOS logic on the same semiconductor die. PIM exposes enormous on-chip memory bandwidth, as well as significantly increasing available off-chip memory bandwidth. Compared to a conventional system with the same memory capacity, a PIM architecture may exhibit a usable memory bandwidth of a factor of a hundred to a thousand. Latency of access to nearest main memory can be reduced by as much as an order of magnitude. PIM greatly increases the number and availability of processing logic perhaps by as much as a factor of a hundred or more. And for a number of reasons, the energy required per operation and therefore the system power consumption can be dramatically reduced, at least as much as a factor of ten. Two general classes of PIM system architectures are of two types: 1) memory accelerators, and 2) those that treat data and tasks as named objects. Type 1 systems are very effective at exploiting streaming of blocked dense data through SIMD-like operations and can be very useful for applications that manipulate data within large arrays. Type 2 systems are general purpose and are well suited to the manipulation of sparse matrices, irregular data structures, and symbolic data representations. Work in both of these classes is being conducted both in architecture and software with some experimental systems having been implemented.

2   Goals of the Gilgamesh Project

The Gilgamesh project is developing a next generation PIM architecture of the Type 2 class for very high end computing for cost-effective Teraflops scale and practical Petaflops performance parallel computer systems. MIND is a PIM architecture under development to support two classes of high-end Type 2 system architectures. These are A) PIM arrays, and B) PIM as the smart memory layers in large distributed shared memory (DSM) systems. MIND chips can be employed alone in large multi-dimensional structures to perform all key functions of general purpose distributed computing with a global name space for data and tasks. They may also provide an effective basis for scalable embedded computing applications. MIND chips can be installed as the main memory and possibly a smart L3 cache (with SRAM instead of DRAM) to off-load the primary compute processors of the data oriented operations exhibiting low or no temporal locality and data reuse. These same chips can perform a number of overhead functions, especially those related to memory management responsibilities. As a result, much greater performance and efficiency may be achieved.

The goals of the MIND chip architecture design are:

- High performance through direct access to on-chip memory bandwidth.
- Enable high efficiency
- Low power by performing on-chip computing and performing on-the-fly power management
- Graceful degradation
- Real time response
- Scalability from standalone single chip to 1000 chips
- Support needs for smart memory in multiprocessor
- Low cost through simplicity of design
- Shared memory across system

3   Capabilities of the MIND Architecture

The single most important property of the MIND architecture is its ability to support virtual paged memory across an array of multiple MIND components including address translation and page migration. This is achieved through a distributed address translation table across the system and a multi-stage address translation process. The translation table is partitioned and distributed such that the responsibility for segments of the virtual address spaces is assigned to specific divisions of the physical address space. This permits source derived routing for any address translation for a message targeted to a virtual address. Preferential placing of pages puts a virtually addressed page at or near its entry in the corresponding page address translation entry. Within this narrow domain, the address translation entry specifies the exact location of the target page within the local node, chip, or collection of chips. In such cases, address translation is straightforward and efficient. But for a number of reasons, it may not be desirable or possible to conform to this restricted placement policy. In this case, the address translation entry points outside the local domain to a remote domain where the page is located, thus requiring multiple steps of address translation across the distributed system. This incurs some performance degradation but guarantees generality of page placement and address translation. To provide fast translation when locality permits, general purpose wide-registers are used to function as translation lookaside buffers (TLB) that enables a direct pointer to the valid physical space.

The MIND architecture supports message driven computation. A class of active messages called "parcels" (PARallel Communication ELements) is used to transmit data between MIND chips. But they also allow actions to be conveyed between MIND chips. This permits work to be conveyed to data as well as data to be transferred to work. An important aspect of parcel usage is that it provides a powerful latency hiding mechanism using decoupled and split transactions. Parcels are variable granularity packets that make possible efficient lightweight transfers and commands or large data transfers up to the size of an entire page (nominally 4K bytes). A parcel may invoke as simple an action as a remote load or store from memory or as complicated as instantiating an entire object. Parcels can even operate on physical resources for the purpose of state initialization and hardware diagnostics and reconfiguration. Parcels make it possible to conduct efficient operations across distributed sparse matrices and irregular data structures such as tree walking, parallel prefix operations, gather-scatters, and matrix transposition. Parcel packets include fields for target addresses (both physical and virtual), action specification, argument values, and the continuation). The action specification may be a simple basic operation (e.g. load, store), a lambda expression providing a sequence of instructions, or a call to a method of an object to start up a new process. The continuation is a specification of what happens after the work invoked by the parcel is completed. This can be as simple as the return address of a read operation or a point in a remote process definition at which to resume. Parcels are intended as very lightweight and hardware supported within the MIND chips for efficiency.

The MIND nodes (memory stack partition/processor logic pair) are multithreaded. While multithreading has historically been employed for latency hiding, MIND uses multithreading as a simplifying and unifying mechanism for local resource management, obviating the need for much of the processor hardware found on ordinary architectures. For example, MIND does not incorporate any branch prediction hardware. The multithreading is also used to cover for the latencies experienced within the MIND chip. The multithreading mechanism permits multiple on-chip resources to be operating at the same time, thus exploiting fine grain hardware parallelism.

4   Conclusions

The Gilgamesh MIND architecture exploits the high memory bandwidth and logic available through PIM technology to provide a truly virtualized memory and task management environment. It can be observed that computation tends to occur in one of two modes: high-speed compute-intensive register-to-register operations for high temporal locality, high data reuse, and high bandwidth data intensive memory operations for low or no temporal locality and data reuse. Conventional microprocessors excel at compute intensive operations but with low data reuse are relatively poor. This is a factor contributing to the relatively low efficiency observed on some of the largest multiprocessor systems. PIM provides the opportunity to address this second mode of computing and the Gilgamesh MIND architecture is designed to provide generalized mechanisms in support of memory oriented computation. It is found through analytical modeling that over this decade it will be possible to implement PIM chips with hundreds of memory-processor nodes. The MIND architecture departs from more typical PIM architectures by embracing and supporting virtual memory across system-wide PIM chip arrays and dynamically scheduling lightweight tasks efficiently across the system. For very large systems, reliability will become increasingly important and active fault tolerance techniques will be required. Gilgamesh systems incorporating sufficient number of MIND chips to realize Petaflops-scale performance will require millions of nodes. If any node constituted a single point failure, the uninterrupted operating time of such a system would unlikely exceed a few minutes. Therefore, new mechanisms and methods will be necessary to provide fault tolerant operation. The MIND architecture includes mechanisms for fault detection and fault isolation but neither guarantees total coverage nor deals with recovery or check-pointing. A prototype of the first generation MIND architecture is being developed on a specially designed FPGA based circuit board, permitting rapid implementation and iteration.

The UK e-Science Program and the Grid

Tony Hey

e-Science Core Programme, EPSRC / University of Southampton, UK
Tony.Hey@epsrc.ac.uk

Abstract. The talk describes the £ 120M UK 'e-Science' initiative and begins by defining what is meant by the term e-Science. The majority of the £ 120M, some £ 85M, is for the support of large-scale e-Science projects in many areas of science and engineering. The infrastructure needed to support such projects needs to allow sharing of distributed and heterogeneous computational and data resources and support effective collaboration between groups of scientists. Such an infrastructure is commonly referred to as the Grid. The remaining funds, some £ 35M, constitute the e-Science 'Core Program' and is intended to encourage the development of robust and generic Grid middleware in collaboration with industry. The key elements of the Core Program will be described including the construction of a UK e-Science Grid. In addition, the talk will include a description of the pilot projects that have so far received funding. These span a range of disciplines from particle physics and astronomy to engineering and healthcare. We conclude with some remarks about the need to develop a data architecture for the Grid that will allow federated access to relational databases as well as flat files.

H. Zima et al. (Eds.): ISHPC 2002, LNCS 2327, p. 6, 2002.

SPEC HPC2002: The Next High-Performance Computer Benchmark
(Extended Abstract)

Rudolf Eigenmann[1], Greg Gaertner[2], Wesley Jones[3],
Hideki Saito[4], and Brian Whitney[4]

[1] Purdue University
[2] Compaq Computer Corporation
[3] Silicon Graphics Incorporated
[4] Intel Corporation
[5] Sun Microsystems Incorporated

SPEC High-Performance Group

The High-Performance Group of the Standard Performance Evaluation Corporation (SPEC/HPG) [1] is developing a next release of its high-performance computer benchmark suite. The current suite, SPEChpc96 [2] is expected to be replaced by the SPEC HPC2002 suite in the second quarter of the year 2002. HPC2002 represents a minor update to SPEC's current high-performance computing benchmarks. In the last subsection of this document we describe SPEC's longer-term plans.

SPEC/HPG currently develops and maintains two benchmark suites, SPEC OMP2001 [3] and SPEC HPC2002. The OMP2001 suite is targeted at shared-memory platforms that support an OpenMP programming model. By contrast, the HPC2002 makes no restrictions about the target platform. In fact, facilitating cross-platform comparisons was an important motivation that had lead to SPEC's development of the HPC series of benchmark suites.

Compared to the previous suite, SPEC HPC2002 will provide the following changes.

- The current climate modeling benchmark will be replaced by the WRF weather modeling code.
- All HPC2002 benchmarks will include both an MPI and an OpenMP parallel version.
- The suite will make use of the standard *SPEC run tools* for compiling and running the codes, making it much easier to execute the benchmarks.
- New data sets will be defined, representing the small, medium, large, and extra-large benchmark input sets.

The following paragraphs describe these changes in more detail.

Benchmark Applications in SPEC HPC2002

SPEC HPC2002 will include three full-size applications in the areas of seismic processing, computational chemistry, and climate modeling, respectively. The

H. Zima et al. (Eds.): ISHPC 2002, LNCS 2327, pp. 7–10, 2002.

number of codes is the same as it was in SPEChpc96. However, the climate modeling code MM5 has been replaced by WRF. The other two applications are essentially the same as in SPEChpc96. They include SPECseis and SPECchem. The following paragraphs give a brief overview of the SPEC HPC2002 codes.

SPECclimate, the climate modeling code of the HPC96 suite will be replaced by the WRF weather model. The WRF weather model is part of a next-generation mesoscale forecast and assimilation modeling system. The system has a goal of advancing both the understanding and prediction of important mesoscale precipitation systems, and promote closer ties between the research and operational forecasting communities. The model is being developed as a collaborative effort among several NSF, Department of Commerce, and DoD sponsored institutions, together with the participation of a number of university scientists. Initially, two data sets will be used for the benchmark. Both data sets represent a 24-hour forecast of the weather over the continental United States. One is run at a resolution of 22km, while the larger data set is run at a resolution of 12km. WRF is written in C and Fortran90. It includes 143,000 lines of code.

SPECseis was developed by ARCO beginning in 1995 to gain an accurate measure of performance of computing systems as it relates to the seismic processing industry for procurement of new computing resources. It consists of a modeling phase which generates synthetic seismic traces for any size of data set, with a flexibility in the geometry of shots and receivers, ground structures, varying lateral velocity, and many other options. A subsequent phase stacks the traces into common midpoint stacks. There are two imaging phases which produce the valuable output seismologists use to locate resources of oil. The first of the two imaging phases is a Fourier method which is very efficient but which does not take into account variations in the velocity profile. Yet, it is widely used and remains the basis of many methods for acoustic imaging. The second imaging technique is a much slower finite-difference method, which can handle variations in the lateral velocity. This technique is used in many seismic migration codes today. The current SPECseis is missing Kirkoff and pre-stack migration techniques. SPECSeis contains some 20,000 lines of Fortran and C code.

SPECchem is used to simulate molecules ab initio, at the quantum level, and optimize atomic positions. It is a research interest under the name of GAMESS at the Gordon Research Group of Iowa State University and is of interest to the pharmaceutical industry. Like SPECseis, SPECchem is often used to exhibit performance of high-performance systems among the computer vendors. Portions of SPECchem codes date back to 1984. It comes with many built-in functionalities, such as various field molecular wave-functions, certain energy corrections for some of the wave-functions, and simulation of several different phenomena. Depending on what wave-functions you choose, SPECchem has the option to output energy gradients of these functions, find saddle points of the potential energy, compute the vibrational frequencies and IR intensities, and more. SPECchem contains over 100,000 lines of code written in Fortran and C.

MPI and OpenMP Parallelism

SPEChpc96 did not include shared-memory parallel versions of the benchmarks. Although the benchmark runrules allowed code modifications, so that shared-memory implementations could be created, developing such code versions was non-trivial. By contrast, SPEC HPC2002 includes OpenMP-based implementations, which will facilitate this task directly. The SPECseis benchmark can be made in either an MPI message passing or an OpenMP shared-memory version. Both versions exploit the same parallelism in the application. The other two benchmarks can also be made as combined MPI/OpenMP codes, with the two models exploiting different levels of parallelism.

Benchmark Make and Run Tools

A significant challenge, when obtaining SPEChpc96 benchmark numbers, was the process of making and executing the codes. The different benchmarks included separate make and run environments, which had to be learned individually by the benchmarker. By contrast, the HPC2002 suite will make use of the *SPEC run tools*, which have already proven their value in several other SPEC benchmark suites. These run tools facilitate the process of compiling, executing, validating, and even generating benchmark reports via an easy command line interface. We expect this change will make it significantly easier to generate benchmark results.

Data Sets

Today's high-performance computer systems are significantly more resourceful than at the time of release of SPEChpc96. Therefore, SPEC HPC2002 will include larger data sets. All three benchmark applications include four data sets, ranging from small to extra-large. A small data set will consume approximately an hour of runtime on a single CPU of a state-of-the-art computer. The extra-large data set will run two to three orders of magnitude longer. The memory requirement will follow from the criterion that each data set will represent a realistic workload for the problem being solved.

Future Development of SPEC HPC Benchmarks

While the described updates to the current SPEC high-performance benchmarks warranted the release of SPEC HPC2002 at this time, the SPEC High-Performance Group is actively working towards a next full release of a new HPC suite. The suite will include newer versions of the seismic processing and computational chemistry applications. It will also include applications in a number of additional application areas, such as computational fluid dynamics and computational biology.

References

1. Standard Performance Evaluation Corporation, "SPEC High-Performance Group," http://www.spec.org/hpg/.
2. Rudolf Eigenmann and Siamak Hassanzadeh, "Benchmarking with real industrial applications: The SPEC High-Performance Group," *IEEE Computational Science & Engineering*, vol. 3, no. 1, pp. 18–23, Spring 1996.
3. Vishal Aslot, Max Domeika, Rudolf Eigenmann, Greg Gaertner, Wesley B. Jones, and Bodo Parady, "SPEC OMP: A new benchmark suite for measuring parallel computer performance," in *OpenMP Shared-Memory Parallel Programming*, Springer Verlag, Heidelberg, Germany, July 2001, Lecture Notes in Computer Science #2104, pp. 1–10.

Language and Compiler Support for Hybrid-Parallel Programming on SMP Clusters*

Siegfried Benkner and Viera Sipkova

Institute for Software Science
University of Vienna, A-1090 Vienna, Austria
{sigi,sipka}@par.univie.ac.at

Abstract. In this paper we present HPF extensions for clusters of SMPs and their implementation within the VFC compiler. The main goal of these extensions is to optimize HPF for clusters of SMPs by enhancing the functionality of the mapping mechanisms and by providing the user with high-level means for controlling key aspects of distributed-memory and shared-memory parallelization. Based on the proposed language extensions, the VFC compiler adopts a hybrid parallelization strategy which closely reflects the hierarchical structure of SMP clusters by exploiting shared-memory parallelism based on OpenMP within nodes and distributed-memory parallelism utilizing MPI across nodes. We describe the language extensions, outline the hybrid parallelization strategy of VFC and present experimental results which show the effectiveness of these techniques.

Keywords: SMP clusters, HPF, OpenMP, MPI, hybrid parallelization

1 Introduction

Clusters of (symmetric) shared-memory multiprocessors (SMPs) are playing an increasingly important role in the high-performance computing arena. Examples of such systems are multiprocessor clusters from SUN, SGI, IBM, a variety of multi-processor PC clusters, supercomputers like the NEC SX-6 or the forthcoming Japanese Earth Simulator and the ASCI White machine. SMP clusters are hybrid-parallel architectures that consist of a number of nodes which are connected by a fast interconnection network. Each node contains multiple processors which have access to a shared memory, while the data on other nodes may usually be accessed only by means of explicit message-passing. Most application programs developed for SMP clusters are based on MPI [20], a standard API for message-passing which has been designed for distributed-memory parallel architectures. However, MPI programs which are executed on clusters of SMPs usually do not directly utilize the shared-memory available within nodes and thus may miss a number of optimization opportunities. A promising approach

---

* This work was supported by the Special Research Program SFB F011 "AURORA" of the Austrian Science Fund and by NEC Europe Ltd. as part of the NEC/Univ. Vienna ADVANCE project in co-operation with NEC C&C Research Laboratories.

H. Zima et al. (Eds.): ISHPC 2002, LNCS 2327, pp. 11–24, 2002.

for parallel programming attempts to combine MPI with OpenMP [24], a standardized shared-memory API, in a single application. This strategy aims to fully exploit SMP clusters by relying on data distribution and explicit message-passing between the nodes of a cluster, and on data sharing and multi-threading within nodes [9,12,18]. Although combining MPI and OpenMP allows optimizing parallel programs by taking into account the hybrid architecture of SMP clusters, applications written in this way tend to become extremely complex.

In contrast to MPI and OpenMP, High Performance Fortran (HPF) [14] is a high-level parallel programming language which can be employed on both distributed-memory and shared-memory machines. HPF programs can also be compiled for clusters of SMPs, but the language does not provide features for directly exploiting their hierarchical structure. Current HPF compilers usually ignore the shared-memory aspect of SMP clusters and treat such machines as pure distributed-memory systems.

In order to optimize HPF for clusters of SMPs, we have extended the mapping mechanisms of HPF by high-level means for controlling the key aspects of distributed and shared-memory parallelization. The concept of *processor mappings* enables the programmer to specify the hierarchical structure of SMP clusters by mapping abstract processor arrays onto abstract node arrays. The concept of *hierarchical data mappings* allows the separate specification of *inter-node data mappings* and *intra-node data mappings*. Furthermore, new intrinsic and library procedures and a new local extrinsic model have been developed. By using node-local extrinsic procedures, hybrid parallel programs may be constructed from OpenMP routines within an outer HPF layer. Based on these extensions, the VFC compiler [3] adopts a hybrid parallelization strategy which closely reflects the hierarchical structure of SMP clusters. VFC compiles an extended HPF program into a hybrid parallel program which exploits shared-memory parallelism within nodes relying on OpenMP and distributed-memory parallelism across nodes utilizing MPI.

This paper is organized as follows: In Section 2 we describe language extensions for optimizing HPF programs for SMP clusters. In Section 3 we give an overview of the main features of the VFC compiler and outline its parallelization strategy adopted for SMP clusters. Section 4 presents an experimental evaluation of the hybrid parallelization strategy. Section 5 discusses related work, followed by conclusions and a brief outline of future work in Section 6.

2   HPF Extensions for SMP Clusters

HPF has been primarily designed for distributed-memory machines, but can also be employed on shared-memory machines and on clusters. However, HPF lacks features for exploiting the hierarchical structure of SMP clusters. Available HPF compilers usually ignore the shared-memory aspect of SMP clusters and treat such machines as pure distributed-memory systems. These issues have been the main motivation for the development of cluster-specific extensions.

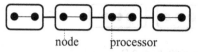

```
!hpf$ processors P(8)              !hpf$ processors Q(2,8)
!hpfC nodes N(4)                   !hpfC nodes N(4)
!hpfC distribute P(block) onto N   !hpfC distribute Q(*,block) onto N
```

Fig. 1. Examples of processor mappings: (left) 4x2 SMP cluster, (right) 4x4 cluster.

2.1 Abstract Node Arrays and Processor Mappings

HPF offers the concept of abstract processor arrays for defining an abstraction of the underlying parallel architecture. Processor arrays are used as the target of data distribution directives, which specify how data arrays are to be distributed to processor arrays. Although suitable for distributed-memory machines and shared-memory machines, processor arrays are not sufficient for describing the structure of SMP clusters. In order to specify the hierarchical topology of SMP clusters we introduce abstract *node arrays* and *processor mappings* (see Figure 1).

Processor mappings may be specified using the extended DISTRIBUTE directive with a processor array as distributee and a node array (declared by means of the NODES directive) as distribution target. Within processor mappings the HPF distribution formats BLOCK, GEN_BLOCK or "*" may be used. For example, the processor mapping directive DISTRIBUTE P(BLOCK) ONTO N, maps each processor of the abstract processor array P to a node of the abstract node array N according to the semantics of the HPF BLOCK distribution format.

The new intrinsic function NUMBER_OF_NODES is provided in order to support abstract node arrays whose sizes are determined upon start of a program. NUMBER_OF_NODES returns the actual number of nodes used to execute a program while the HPF intrinsic function NUMBER_OF_PROCESSORS returns the total number of processors in a cluster. Using these intrinsic functions, programs may be parallelized regardless of the actual number of nodes and processors per nodes.

2.1.1 Heterogeneous Clusters

While for homogeneous clusters the BLOCK distribution format is used in processor mappings, heterogeneous clusters, e.g. clusters where the number of processors per node varies, can be supported by means of the GEN_BLOCK distribution format of the Approved Extensions of HPF. Figure 2 shows a heterogeneous SMP cluster, consisting of 4 nodes with 2, 3, 4, and 3 processors, respectively. Here the GEN_BLOCK distribution format of HPF is utilized to specify that the number of processors within nodes varies.

2.1.2 Semantics of Processor Mappings

A processor mapping specifies for each processor array dimension whether distributed-memory parallelism, shared-memory parallelism or both may be exploited according to the following rules:

```
            integer, dimension(4):: SIZE = (/2,3,4,3/)
!hpf$ processors R(12)
!hpfC nodes N(4)
!hpfC distribute R(gen_block(SIZE)) onto N
```

Fig. 2. Specification of a heterogeneous SMP cluster using the GEN_BLOCK distribution format within a processor mapping directive.

(1) If a dimension of a processor array is distributed by BLOCK or GEN_BLOCK, contiguous blocks of processors are mapped to the nodes in the corresponding dimension of the specified node array. As a consequence, both distributed-memory parallelism and shared-memory parallelism may be exploited for all array dimensions that are mapped to a distributed processor array dimension.

(2) If for a dimension of a processor array a "*" is specified as distribution format, only shared-memory parallelism may be exploited across array dimensions that are mapped to that processor array dimension.

For example, in Figure 1 (b) only shared-memory parallelism may be exploited across the first dimension of Q, while both shared-memory and distributed-memory parallelism may be exploited across the second dimension.

By combining data distributions and processor mappings, *inter-node mappings* and *intra-node mappings* can be derived by the compiler.

An inter-node mapping determines for each node those parts of A that are owned by it. The implicit assumption is that those portions of an array owned by a node are allocated in an unpartitioned way in the shared memory of this node. Inter-node mappings are used by the compiler to control distributed-memory parallelization, i.e. data distribution and communication across nodes.

An intra-node mapping determines a mapping of the local part of an array assigned to a node of a cluster with respect to the processors within the node. Intra-node mappings are utilized by the compiler to organize shared-memory parallelization, i.e. work scheduling (work sharing) across concurrent threads within nodes. These issues are described in more detail in Section 3.

2.2 Hierarchical Data Mappings

In this section we describe additional extensions which allow users to specify hierarchical mappings for data arrays. A hierarchical data mapping comprises an *explicit inter-node mapping* and an *explicit intra-node mapping*, each specified by a separate directive. Compared to the basic concept of processor mappings as described previously, hierarchical data mappings provide a more flexible mechanism for the distribution of data on clusters of SMPs. However, in order to take advantage of this increased flexibility, it will be usually necessary to modify the mapping directives of existing HPF programs.

2.2.1 Explicit Inter-node Data Mappings

In order to specify a mapping of data arrays to the nodes of an SMP cluster, the DISTRIBUTE directive is extended by allowing node arrays to appear as distribution target. Such a mapping is referred to as *explicit inter-node mapping*. It maps data arrays to abstract nodes in exactly the same way as an original HPF distribute directive maps data arrays to abstract processors. Inter-node mappings are utilized by the compiler to organize distributed-memory parallelization, i.e. data distribution and communication across nodes. In the following example, array A is mapped to an abstract node array N.

```
!hpfC nodes N(2,2)
      real, dimension (8,8) :: A
!hpfC distribute A (block,block) onto N  ! inter-node mapping
```

As a consequence of the extended distribute directive, the section A(1:4,1:4) is mapped to node N(1,1), A(5:8,1:4) to N(2,1), A(1:4,5:8) to N(1,2), and A(5:8,5:8) is mapped to node N(2,2).

2.2.2 Explicit Intra-node Data Mappings

In order to specify a mapping of node-local data with respect to the processors within a node, the SHARE directive has been introduced. A mapping defined by the SHARE directive is referred to as *explicit intra-node data mapping*. As the name of the directive suggests, an intra-node mapping controls the work sharing (scheduling) of threads running within nodes. Besides the usual HPF distribution formats BLOCK, CYCLIC, and GEN_BLOCK, the OpenMP work-sharing formats DYNAMIC and GUIDED may be employed for this purpose.

The information provided by means of the share directive is propagated by the compiler to parallel loops. Various code transformations ensure that loops can be executed by multiple threads which are scheduled according to the specified work sharing strategy

Hierarchical data mappings may be specified regardless of whether a processor mapping has been specified or not. For example, in the code fragment

```
!hpfC nodes N(4)
      real, dimension (32,16) :: A
!hpfC distribute A(*, block) onto N   ! inter-node mapping
!hpfC share A (block,*)               ! intra-node mapping

!hpf$ independent
      do i = 1, 32
         a(i,:) = ...
      end do
```

the extended distribute directive specifies that the second dimension of A is distributed by BLOCK to the nodes of a cluster. The SHARE directive specifies that computations along the first dimension of A should be performed in parallel by multiple threads under a BLOCK work-scheduling strategy. If we assume that each node is equipped with four processors, the loop iteration space would be partitioned into four blocks of eight iterations and each block of iterations would be executed by a separate thread.

Note that although in some cases a hierarchical mapping can also be expressed by means of a usual HPF data distribution directive and a processor mapping, this is not true in general.

3 Hybrid Parallelization Strategy

In this section we outline how HPF programs that make use of the proposed extensions are compiled with the VFC compiler for clusters of SMPs according to a hybrid parallelization strategy that combines MPI and OpenMP.

3.1 Overview of VFC

VFC [3] is a source-to-source parallelization system which translates HPF+ programs into explicitly parallel programs for a variety of parallel target architectures. HPF+ [2] is an extension of HPF with special support for an efficient handling of irregular codes. In addition to the basic features of HPF, it includes generalized block distributions and indirect distributions, dynamic data redistribution, language features for *communication schedule reuse* [1] and the *halo concept* [4] for controlling irregular non-local data access patterns. VFC provides powerful parallelization strategies for a large class of non-perfectly nested loops with irregular runtime-dependent access patterns which are common in industrial codes. In this context, the concepts of communication schedule reuse and halos allow the user to minimize the potentially large overhead of associated runtime compilation strategies.

Initially, VFC has been developed for distributed-memory parallel machines. For such machines, VFC translates HPF programs into explicitly parallel, single-program multiple-data (SPMD) Fortran 90/MPI message-passing programs. Under the distributed-memory execution model, the generated SPMD program is executed by a set of processes, each executing the same program in its local address space. Usually there is a one-to-one mapping of abstract processors to MPI processes. Each processor only allocates those parts of distributed arrays that have been mapped to it according to the user-specified data distribution. Scalar data and data without mapping directives are allocated on each processor. Work distribution (i.e. distribution of computations) is mainly based on the owner-computes rule. Access to data located on other processors is realized by means of message-passing (MPI) communication. VFC ensures that all processors executing the target program follow the same control flow in a loosely synchronous style.

In the following we focus on the extensions of VFC for clusters of SMPs.

3.2 Hybrid Execution Model

The parallelization of HPF+ programs with cluster-specific extensions relies on a hybrid-parallel execution model. As opposed to the usual HPF compilation where a single-threaded SPMD node program is generated, a multi-threaded SPMD node program is generated under the hybrid execution model.

Under the hybrid model, an HPF program is executed on an SMP cluster by a set of parallel processes, each of which runs usually on a separate node. Each process allocates data it owns in shared memory, according to the derived (or explicitly specified) inter-node data mapping. Work distribution across node processes is usually realized by applying the owner computes strategy, which implies that each process performs only computations on data elements owned by it. Communication across node processes is realized by means of appropriate MPI message-passing primitives. In order to exploit additional shared-memory parallelism, each MPI node process generates a set of OpenMP threads which run concurrently in the shared address space of a node. Usually the number of threads spawned within node processes is equal to the number of processors available within nodes. Data mapped to a node is allocated in a non-partitioned way in shared memory, regardless of intra-node mappings. Intra-node data mappings are however utilized to organize parallel execution of threads by applying code transformations and inserting appropriate OpenMP directives.

3.3   Outline of the Hybrid Parallelization Strategy

The parallelization of extended HPF programs for clusters of SMPs can be conceptually divided into three main phases (1) inter-node and intra-node mapping analysis, (2) distributed-memory parallelization, and (3) shared-memory parallelization.

3.3.1   Deriving Inter-node and Intra-node Data Mappings

After the conventional front-end phases, the VFC compiler analyzes the distribution directives and processor mapping directives. As a result of this analysis, each dimension of a distributed array is classified as DM, SM, DM/SM or SEQ, depending on the type of parallelism that may be exploited. Then for each array dimension an inter-node data mapping and an intra-node data mapping is determined.

Assuming the following declarations

```
!hpf$ processors P(4)              ! usual HPF processor array
!hpfC nodes N(2)                   ! abstract node array
!hpfC distribute P(block) onto N   ! processor mapping
      real, dimension (100) :: A
!hpf$ distribute A(block) onto P   ! usual HPF distribution
```

the inter-node and intra-node mapping of A derived by the compiler are equivalent to those explicitly specified by the following directives:

```
!hpfC distribute A(block) onto N   ! "derived inter-node mapping"
!hpfC share A(block)               ! "derived intra-node mapping"
```

As a consequence, both distributed and shared-memory parallelism will be exploited for A.

On the basis of inter-node and intra-node mappings, the ownership of data both with respect to processes (nodes), and with respect to threads (processors within nodes) is derived and represented in symbolic form. Ownership information is then propagated to all executable statements and, at an intermediate code level, represented by means of ON_HOME clauses which are generated for assignment statements and loops accessing distributed arrays. Each loop is then classified as DM, SM, DM/SM or SEQ depending on the classification of the corresponding array dimension in the associated ON_HOME clause.

3.3.2   Distributed-Memory Parallelization

During the distributed-memory parallelization phase VFC generates an intermediate SPMD message-passing program based on inter-node data mappings. Array declarations are modified in such a way that each MPI process allocates only those parts of distributed arrays that are owned by it according to the inter-node mapping. Work distribution is realized by strip-mining all loops which have been classified as DM across multiple MPI processes. If access to non-local data is required, appropriate message-passing communication primitives are generated and temporary data objects for storing non-local data are introduced. Several communication optimizations are applied, including elimination of redundant communication, extraction of communication from loops, message vectorization, and the use of collective communication instead of point-to-point communication primitives.

The intermediate SPMD message-passing program generated after this phase could already be executed, however it would exploit only a single processor on each node of the cluster.

3.3.3   Shared-Memory Parallelization

The intermediate message-passing program is parallelized for shared memory according to the intra-node data mapping derived by VFC. The shared-memory parallelization phase makes use of OpenMP in order to distribute the work of a node among multiple threads. Work distribution of loops and array assignments is derived from the intra-node data mapping of the accessed arrays and realized by inserting corresponding OpenMP work-sharing directives and/or appropriate loop transformations (e.g. strip-mining). In this context, variables which have been specified as NEW in an HPF INDEPENDENT directive are specified as PRIVATE within the generated OpenMP directives.

Consistency of shared data objects is enforced by inserting OpenMP synchronization primitives (critical sections or atomic directives).

Furthermore, various optimizations are performed in order to avoid unnecessary synchronization. The potential overheads of spawning teams of parallel threads is reduced by merging parallel regions. Most of these optimization steps are conceptually similar to the communication optimizations performed during distributed-memory parallelization. Special optimizations are applied to loops performing irregular reductions on arrays for which a halo has been specified In order to minimize synchronization overheads for loops that perform irregular reductions on arrays for which a halo has been specified, special optimization techniques are applied [5].

Note that our current implementation of the hybrid parallelization strategy ensures that only the master thread performs MPI communication.

```
!hpf$ processors P(number_of_processors())
!hpfC nodes N(number_of_nodes())              ! abstract node array
!hpfC distribute P(block) onto N              ! processor mapping
      real, dimension (N) :: A
!hpf$ distribute A(block) onto P
      ...
!hpf$ independent
      do i = NL, NU
         a(i) = ... a(...) ...
      end do
```

(a) HPF program fragment with cluster-specific extensions (!hpfC)

```
      ...
!hpfC share A(block)                          ! derived inter-node mapping
      real, allocatable :: A(:)
      type(rt_dsc), pointer :: A_dsc          ! runtime descriptor of A
      ...                                     ! set up runtime descriptors
      allocate(A(vhpf_extent(A_dsc,1)))       ! allocate node-local part of A
      ...                                     ! compute node-local bounds
      call vhpf_loc_bounds_SM(A_dsc,NL,NU,lb_DM,ub_DM)
      call vhpf_comm(...)                     ! perform MPI communication
      do i = lb_DM, ub_DM                     ! process-local iterations
         a(i) = ...
      end do
      ...
```

(b) Intermediate (pseudo-)code after mapping analysis and DM parallelization.

```
      ...
      real, allocatable :: A(:)
      type(rt_dsc), pointer :: A_dsc          ! runtime descriptor of A
      ...                                     ! set up runtime descriptors
      allocate(A(vhpf_extent(A_dsc,1)))       ! allocate node-local part of A
      ...
      call vhpf_comm(...)                     ! perform MPI communication
      ...                                     ! spawn parallel threads
!$omp parallel, private(i,A_loc_lb_SM,A_loc_ub_SM)
                                              ! compute thread-local bounds
      call vhpf_loc_bounds_SM(A_dsc,lb_DM,ub_DM,lb_SM,ub_SM)
      do i = lb_SM, ub_SM                     ! thread-local iterations
         a(i) = ...
      end do
!$omp end parallel
      ...
```

(c) Final (pseudo-)code after SM parallelization using OpenMP.

Fig. 3. Sketch of hybrid parallelization strategy.

In Figure 3 the hybrid parallelization strategy as realized by VFC is sketched for a simple example. As shown in Figure 3 (b) during shared-memory parallelization the intermediate SPMD message-passing program is transformed by inserting OpenMP directives in order to exploit shared-memory parallelism within the nodes of the cluster. The OpenMP `parallel` directive ensures that multiple threads are generated on each node. Inside this parallel region the runtime routine `vhpf_loc_bounds_SM` computes from the iterations assigned to a node (i.e. the iterations from `lb_DM` to `ub_DM`) the chunk of iterations executed on each individual thread according to the derived intra-node data mapping.

3.3.4   Potential Advantages of Hybrid Parallelization

The hybrid parallelization strategy offers a number of advantages compared to the usual distributed-memory parallelization strategy as realized by most HPF compilers. The hybrid strategy reflects closely the topology of SMP clusters by exploiting distributed-memory parallelism across the nodes and shared-memory parallelism within nodes. It allows a direct utilization of the shared memory within nodes and usually requires less total memory than the DM parallelization strategy. For example, replicated data arrays, which are mostly accessed in a read only way, have to be allocated only once per node, while in the distributed-memory model replicated arrays have to be allocated by each process, resulting in multiple copies on each node.

Moreover, the hybrid model usually allows a more efficient handling of communication. Since data is distributed only across nodes and communication is performed only by the master thread, the hybrid model results in less messages as well as larger messages. However, on certain architectures, such a strategy, where only the master thread performs communication, may reduce the overall communication bandwidth.

Another important issue are unstructured computations, where data arrays are accessed indirectly by means of array subscripts. Under the distributed-memory model, an HPF compiler has to apply expensive runtime techniques to parallelize loops with indirect array accesses. However, if array dimensions which are accessed indirectly are mapped to the processors within a node, the overheads of runtime parallelization can be avoided due to the shared access.

Despite these differences both models usually result in the same degree of parallelism. Thus, for many applications only minor performance differences can be observed. In particular, this is true for codes which are characterized by a high degree of locality and independent computations.

In the following we present an experimental evaluation of the new language extensions and the hybrid parallelization strategy as provided by VFC using two benchmark codes on a Beowulf-type SMP PC cluster.

4   Experimental Results

For the performance experiments, we used a kernel from a crash-simulation code which originally has been developed for HPF+, and kernel from a numerical pricing module [21] developed in the context of the AURORA Financial Management

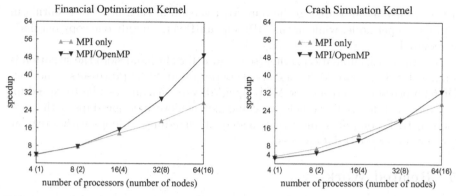

Fig. 4. Performance comparison of kernels parallelized with VFC under the distributed-memory parallelization strategy (MPI only) and under the hybrid parallelization strategy (MPI/OpenMP).

System [11]. Both kernels are based on an iterative computational scheme with an outer time-step loop. The main computational parts are performed in nested loops where large arrays are accessed by means of vector subscripts. In order to minimize the overheads that would be caused by usual runtime parallelization strategies (e.g. inspector/executor), non-local data access patterns are explicitly specified at runtime based on the concept of halos. Moreover, in both kernels the reuse of runtime generated communication schedules for indirectly accessed arrays is enforced by means of appropriate language features for communication schedule reuse [1,3].

Both kernels have been parallelized with the VFC compiler [3] and executed on a Beowulf cluster consisting of 16 nodes connected via Myrinet. Each node is equipped with four Pentium III Xeon processors (700MHz) and 2GB RAM.

For both kernels we compared a pure MPI version to a hybrid-parallel version based on MPI/OpenMP. Both versions have been generated from an HPF+ source program, in the latter case, with additional cluster-specific extensions. The `pgf90` compiler from Portland Group Inc. which also supports OpenMP has been used as a backend-compiler of VFC.

All kernels have been measured on up to 16 nodes, where on each node either four MPI processes (MPI only) or four OpenMP threads (MPI/OpenMP) were run. Speed-up numbers have been computed with respect to the sequential version of the code (i.e. HPF compiled with the Fortran 90 compiler).

In Figure 4 the speedup curves of the financial optimization kernel are shown on the left hand side. For up to 8 processors (2 nodes) the pure MPI version and the hybrid MPI/OpenMP version are almost identical. However, on more than 8 processors the MPI/OpenMP version becomes superior. The main reason for this performance difference is that the computation/communication ratio of the pure MPI version decreases faster than for the MPI/OpenMP version. In the pure MPI version the total number of messages is approximately four times larger than in the MPI/OpenMP version. Also the overall memory requirements of the pure MPI version are higher since the kernel contains several replicated arrays which are allocated in the pure MPI version in each process, resulting in four copies per node, while in the MPI/OpenMP version only one copy per node is required.

For the crash simulation kernel (see Figure 4, right hand side) the situation is similar. Here the pure MPI version is superior on up to 32 processors (8 nodes). On 64 processors, however, the MPI/OpenMP version achieves a better speedup.

Note that for both kernels we could not achieve any speedups with commercial HPF compilers due to an inadequate handling of loops with irregular (vector-subscripted) data accesses.

5   Related Work

Some of the extensions for SMP clusters described in this paper have been implemented also in the ADAPTOR compilation system [8,6]. ADAPTOR supports also the automatic generation of hybrid-parallel programs from HPF based on appropriate default conventions for processor mappings.

Several researchers have investigated the advantages of a hybrid programming model based on MPI and OpenMP against a unified MPI-only model. Cappelo [9] et al. investigated a hybrid-parallel programming strategy in comparison with a pure message-passing approach using the NAS benchmarks on IBM SP systems. In their experiments the MPI-only approach has provided better results than a hybrid strategy for most codes. They conclude that a hybrid-parallel strategy becomes superior when fast processors make the communication performance significant and the level of parallelization is sufficient. Henty [13] reports on experiments with a Discrete Element Modeling code on various SMP clusters. He concludes that current OpenMP implementations are not yet efficient enough for hybrid parallelism to outperform pure message-passing. Haan [12] performed experiments with a matrix-transpose showing that a hybrid-parallel approach can significantly outperform message-passing parallelization. On the Origin2000, the SGI data placement directives [23] form a vendor specific extension of OpenMP. Some of these extensions have similar functionality as the HPF directives, e.g. "affinity scheduling" of parallel loops is the counterpart to the ON clause of HPF. Compaq has also added a new set of directives to its Fortran for Tru64 UNIX that extend the OpenMP Fortran API to control the placement of data in memory and the placement of computations that operate on that data [7]. Chapman, Mehrotra and Zima [10] have proposed a set of OpenMP extensions, similar to HPF mapping directives, for locality control. PGI proposes a high-level programming model [17,19] that extends the OpenMP API with additional data mapping directives, library routines and environment variables. This model extends OpenMP in order to control data locality with respect to the nodes of SMP clusters. In contrast to this model, HPF, with the extensions proposed in this paper, supports locality control across nodes as well as within nodes.

All these other approaches introduce data mapping features into OpenMP in order to control data locality, but still utilize the explicit work distribution via the PARALLEL and PARDO directives of OpenMP. Our approach is based on HPF and relies on an implicit work distribution which is usually derived from the data mapping but which may be explicitly controlled by the user within nodes by means of OpenMP-like extensions.

A number of studies have addressed the issues of implementing OpenMP on clusters of SMPs relying on a distributed-shared memory (DSM) software layer. Hu et al. [15] describe the implementation of OpenMP on a network of shared-memory multiprocessors by means of a translating OpenMP directives into calls to a modified version of the TreadMarks software distributed-memory system. Sato et al. [22] describe the design of an OpenMP compiler for SMP clusters based on a compiler-directed DSM software layer.

6 Conclusions

Processor mappings provide a simple but convenient means for adapting existing HPF programs with minimal changes for clusters of SMPs. Usually only a node array declaration and a processor mapping directive have to be added to an HPF program. Based on a processor mapping, an HPF compiler can adopt a hybrid parallelization strategy that exploits distributed-memory parallelism across nodes, and shared-memory parallelism within nodes, closely reflecting the hierarchical structure of SMP clusters. Additional extensions are provided for the explicit specification of inter-node and intra-node data mappings. These features give users more control over the shared-memory parallelization within nodes, by using the SHARE directive.

As our experimental evaluation has shown, using these features performance improvements for parallel programs on clusters of SMPs can be achieved in comparison to a pure message-passing parallelization strategy. Another potential advantage is that with the hybrid-parallelization strategy shared memory within nodes can be exploited directly, often resulting in lower memory requirements.

For the future we plan to extend the hybrid compilation strategy of VFC by relaxing the current restriction that only the master threads on each process can perform MPI communication. This implies that a thread-safe MPI implementation supporting the features of MPI-2 for thread-based parallelization (cf. Section 8.7 of the MPI-2 Specification) must be available on the target SMP cluster. However, currently this is not the case on most clusters.

References

1. S. Benkner, P. Mehrotra, J. Van Rosendale, H. Zima. High-Level Management of Communication Schedules in HPF-like Languages, In *Proceedings of the International Conference on Supercomputing* (ICS'98), pp. 109-116, Melbourne, Australia, July 13-17, 1998, ACM Press.
2. S. Benkner. HPF+ – High Performance Fortran for Advanced Scientific and Engineering Applications, Future Generation Computer Systems, Vol 15 (3), 1999.
3. S. Benkner. VFC: The Vienna Fortran Compiler. *Scientific Programming*, 7(1):67–81, 1999.
4. S. Benkner. Optimizing Irregular HPF Applications Using Halos. *Concurrency: Practice and Experience*, John Wiley & Sons Ltd, 2000.

5. S. Benkner and T. Brandes. Exploiting Data Locality on Scalable Shared Memory Machines with Data Parallel Programs. In *Euro-Par 2000 Parallel Processing*, Lecture Notes in Computer Science 1900, Munich, Germany, September 2000.
6. S. Benkner, T. Brandes. High-Level Data Mapping for Clusters of SMPs, In *Proceedings 6th International Workshop on High-Level Parallel Programming Models and Supportive Environments*, San Francisco, USA, April 2001, Springer Verlag.
7. J. Bircsak, P. Craig, R. Crowell, Z. Cvetanovic, J. Harris, C. Nelson, and C. Offner. Extending OpenMP for NUMA Machines. In *Proceedings of SC 2000: High Performance Networking and Computing Conference*, Dallas, November 2000.
8. T. Brandes and F. Zimmermann. ADAPTOR - A Transformation Tool for HPF Programs. In K.M. Decker and R.M. Rehmann, editors, *Programming Environments for Massively Parallel Distributed Systems*, Birkhäuser Verlag, 1994.
9. F. Cappello and D. Etieble. MPI versus MPI+OpenMP on the IBM SP for the NAS Benchmarks. In *Proceedings of SC 2000: High Performance Networking and Computing Conference*, Dallas, November 2000.
10. B. Chapman, P. Mehrotra, and H. Zima. Enhancing OpenMP with Features for Locality Control. In *Proc. ECWMF Workshop "Towards Teracomputing - The Use of Parallel Processors in Meteorology"*, 1998.
11. E. Dockner, H. Moritsch, G. Ch. Pflug, and A. Swietanowski. AURORA financial management system: From Model Design to Implementation. Technical report AURORA TR1998-08, University of Vienna, June 1998.
12. O. Haan. Matrix Transpose with Hybrid OpenMP / MPI Parallelization. Technical Report, http://www.spscicomp.org/2000/userpres.html#haan, 2000.
13. D. S. Henty. Performance of Hybrid Message-Passing and Shared-Memory Parallelism for Discrete Element Modeling. In *Proceedings of SC 2000: High Performance Networking and Computing Conference*, Dallas, November 2000.
14. High Performance Fortran Forum. High Performance Fortran Language Specification. Version 2.0, Department of Computer Science, Rice University, 1997.
15. Y. Hu, H. Lu, A. Cox, and W. Zwaenepel. Openmp for networks of smps. In *Proceedings of IPPS.*, 1999.
16. P. V. Luong, C. P. Breshears, L.N. Ly. Costal Ocean Modeling of the U.S. West Coast with Multiblock Grid and Dual-Level Parallelism. In *Proceedings of SC2001*, Denver, USA, November 2001.
17. M. Leair, J. Merlin, S. Nakamoto, V. Schuster, and M. Wolfe. Distributed OMP – A Programming Model for SMP Clusters. In *Eighth International Workshop on Compilers for Parallel Computers*, pages 229–238, Aussois, France, January 2000.
18. R. D. Loft, S. J. Thomas, J. M. Dennis. Terascale spectral element dynamical core for atmospheric general circulation models. In *Proceedings SC2001*, Denver, USA, November 2001.
19. J. Merlin, D. Miles, V. Schuster. Extensions to OpenMP for SMP Clusters. In *Proceedings of the Second European Workshop on OpenMP*, EWOMP 2000.
20. Message Passing Interface Forum. MPI: A Message-Passing Interface Standard. Vers. 1.1, June 1995. MPI-2: Extensions to the Message-Passing Interface, 1997.
21. H. Moritsch and S. Benkner. High Performance Numerical Pricing Methods. In *4-th Intl. HPF Users Group Meeting*, Tokyo, October 2000.
22. M. Sato, S. Satoh, K. Kusano, and Y. Tanaka. Design of openmp compiler for an smp cluster. In *Proceedings EWOMP '99, pp.32-39.*, 1999.
23. Silicon Graphics Inc. MIPSpro Power Fortran 77 Programmer's Guide: OpenMP Multiprocessing Directives. Technical Report Document 007-2361-007, 1999.
24. The OpenMP Forum. OpenMP Fortran Application Program Interface. Version 1.1, November 1999. http://www.openmp.org.

Parallelizing Merge Sort
onto Distributed Memory Parallel Computers*

Minsoo Jeon and Dongseung Kim

Dept. of Electrical Engineering, Korea University
Seoul, 136-701, Korea
{msjeon,dkim}@classic.korea.ac.kr
Tel.: +822 3290 3232, Fax: +822 928 8909

Abstract. Merge sort is useful in sorting a great number of data progressively, especially when they can be partitioned and easily collected to a few processors. Merge sort can be parallelized, however, conventional algorithms using distributed memory computers have poor performance due to the successive reduction of the number of participating processors by a half, up to one in the last merging stage.

This paper presents load-balanced parallel merge sort where all processors do the merging throughout the computation. Data are evenly distributed to all processors, and every processor is forced to work in all merging phases. An analysis shows the upper bound of the speedup of the merge time as $(P - 1)/\log P$ where P is the number of processors. We have reached a speedup of 8.2 (upper bound is 10.5) on 32-processor Cray T3E in sorting of 4M 32-bit integers.

1 Introduction

Many comparison-based sequential sorts take $O(N \log N)$ time to sort N keys. To speedup the sorting multiprocessors are employed for parallel sorting. Several parallel sorting algorithms such as bitonic sort[1,6], sample sort[5], column sort[3] and parallel radix sort[7,8] have been devised. Parallel sorts usually need a fixed number of data exchange and merging operations. The computation time decreases as the number of processors grows. Since the time is dependent on the number of data each processor has to compute, good load balancing is important. In addition, if interprocessor communication is not fast such as in distributed memory computers, the amount of overall data to be exchanged and the frequency of communication have a great impact on the total execution time.

Merge sort is frequently used in many applications. Parallel merge sort on PRAM model was reported to have fast execution time of $O(\log N)$ for N input keys using N processors[2]. However, distributed-memory based parallel merge sort is slow because it needs local sort followed by a fixed number of iterations of merge that includes lengthy communication. The major drawback of the conventional parallel merge sort is in the fact that load balancing and processor

---

* This research was supported by KRF grant (no. KRF-99-041-E00287).

H. Zima et al. (Eds.): ISHPC 2002, LNCS 2327, pp. 25–34, 2002.

utilization get worse as it iterates; in the beginning every processor participates in merging of the list of N/P keys with its partner's, producing a sorted list of $2N/P$ keys, where N and P are the number of keys and processors, respectively; in the next step and on, only a half of the processors used in the previous stage participate in merging process. It results in low utilization of resource. Consequently, it lengthens the computing time. This paper introduces a new parallel merge sort scheme, called *load-balanced parallel merge sort*, that forces every processor to participate in merging at every iteration. Each processor deals with a list of size of about N/P at every iteration, thus the load of processors is kept balanced to reduce the execution time.

The paper is organized as follows. In sections 2 we present the conventional and improved parallel merge sort algorithms together with the idea how more parallelism is obtained. Section 3 reports experimental results performed on Cray T3E and PC cluster. Conclusion is given in the last section followed by performance analysis in the appendix.

2 Parallel Merge Sort

2.1 Simple Method

Parallel merge sort goes through two phases: local sort phase and merge phase. Local sort phase produces keys in each processor sorted locally, then in merging phase processors merge them in $\log P$ steps as explained below. In the first step, processors are paired as (sender, receiver). Each sender sends its list of N/P keys to its partner (receiver), then the two lists are merged by each receiver to make a sorted list of $2^1N/P$ keys. A half the processors work during the merge, and the other half sit idling. In the next step only the receivers in the previous step are paired as (sender, receiver), and the same communication and merge operations are performed by each pair to form a list of $2^2N/P$ keys. The process continues until a complete sort list of N keys is obtained. The detailed algorithm is given in Algorithm 1.

As mentioned earlier, the algorithm does not fully utilize all processors. Simple calculation reveals that only $P/\log P (= \{(P/2 + P/4 + P/8 + \cdots + 1)/(\log P \text{ steps})\})$ processors are used in average. It must have inferior performance to an algorithm that makes a full use of them, if any.

2.2 Load-Balanced Parallel Merge Sort

To keep each list of sorted data in one processor is simple and easy to deal with as long as the algorithm is concerned. However, as the size of the lists grows, sending them to other processors for merge is time consuming, and processors that no longer keep lists after transmission sit idling until the end of the sort. The key idea in our parallel sort is to distribute each (partially) sorted list onto multiple processors such that each processor stores an approximately equal number of keys, and all processors take part in merging throughout the execution.

Algorithm 1: Simple parallel merge sort
 P: the total number of processors (assume $P = 2^k$ for simplicity.)
 P_i: a processor with index i
 h: the number of active processors
begin
 $h := P$
 1. **forall** $0 \leq i \leq P - 1$
  P_i sorts a list of N/P keys locally.
 2. **for** $j = 0$ **to** $(\log P)\text{-}1$ **do**
  forall $0 \leq i \leq h - 1$
   if $(i < h/2)$ **then**
    2.1. P_i receives N/h keys from $P_{i+h/2}$
    2.2. P_i merges two lists of N/h keys into a sorted list of $2N/h$
   else
    2.3. P_i sends its list to $P_{i-h/2}$
  $h := h/2$
end

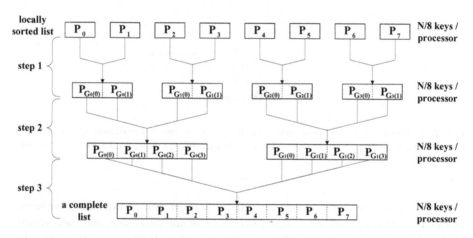

Fig. 1. Load-balanced parallel merge sort

Figure 1 illustrates the idea for the merging with 8 processors, where each rectangle represents a list of sorted keys, and processors are shown in the order that store and merge the corresponding list. It would invoke more parallelism, thus shorten the sort time. One difficulty in this method is to find a way how to merge two lists each of which is distributed in multiple processors, rather than stored in a single processor. Our design is described below that minimizes the key movement.

A *group* is a set of processors that are in charge of one sorted list. Each group stores a sorted list of keys by distributing them evenly to all processors. It also computes the *histogram* of its own keys. The histogram plays an important role in determining a minimum number of keys to be exchanged with others during

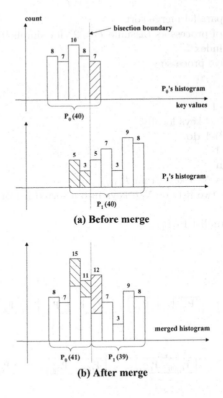

Fig. 2. Example of exchanging and merging keys by using histogram at first iteration (Processors P_0, P_1 both have 40 keys before merge, and 41 & 39 after merge, respectively.)

the merge. Processors keep nondecreasing (or nonincreasing) order for their keys. In the first merging step, all groups have a size of one processor, and each group is paired with another group called the *partner group*. In this step, there is only one communication partner per processor. Each pair exchanges its two boundary keys (a minimum and a maximum keys) and determines new order of the two processors according to the minimum key values. Now each pair exchanges group histograms and computes new one that covers the whole pair. Each processor then divides the intervals (histogram bins) of the merged histogram into two parts (i.e. *bisection*) so that the (half) lower indexed processor will keep the smaller half of the keys, the higher the upper. Now each processor sends out the keys that will belong to other processor(s) (for example, those keys in the shaded intervals are transmitted to the other processor in Figure 2). Each merges keys with those arriving from the paired processor. Now each processor holds $N/P \pm \Delta$ keys because the bisection of the histogram bins may not be perfect (we hope Δ is relatively small compared to N/P). The larger the number of histogram bins, the better the load balancing. In this process, only the keys in the overlapped intervals need to merge. It implies that keys in the non-overlapped interval(s) do not interleave with keys of the partner processor's during the merge. They are simply placed in a proper position in the merged list. Often there maybe no overlapped intervals at all, then no keys are exchanged.

From the second step and on, the group size (i.e. the number of processors per group) grows twice the previous one. Merging process is the same as before except that each processor may have multiple communication partners, up to the group size in the worst case. Now boundary values and group histograms are again exchanged between paired groups, then the order of processors is decided and histogram bins are divided into 2^i parts at the ith iteration. Keys are exchanged between partners, then each processor merges received keys. One cost saving method is used here called *index swapping*. Since merging two groups into one may require many processors to move keys, only ids of the corresponding processors are swapped to have a correct sort sequence if needed, in order not to sequentially propagate keys of processors to multiple processors. Index swapping minimizes the amount of keys exchanged among processors. The procedure of the parallel sort is summarized in Algorithm 2.

Algorithm 2: Load-balanced parallel merge sort
1. Each processor sorts a list of N/P keys locally and obtains local histogram.
2. Iterate $\log P$ times the following computation:
   2.1. Each group of processors exchanges boundary values between its partner group and determines the logical ids of processors for the merged list.
   2.2. Each group exchanges histograms with its paired group and computes a new histogram, then divides the bins into 2^{i-1} equal parts.
   /*At ith iteration, there are $P/2^{i-1}$ groups, each of which includes 2^{i-1} processors */
   2.3. Each processor sends keys to the designated processors that will belong to others due to the division.
   2.4. Each processor locally merges its keys with the received ones to obtain a new sorted list.
   2.5. Broadcast logical ids of processors for the next iterations.

Rather involved operations are added in the algorithm in order to minimize the key movement since the communication in distributed memory computers is costly. The scheme has to send boundary keys and histogram data at each step, and a broadcast for the logical processor ids is needed before a new merging iteration. If the size of the list is fine grained, the increased parallelism may not contribute to shortening the execution time. Thus, our scheme is effective when the number of keys is not too small to overcome the overhead.

3   Experimental Results

The new parallel merge sort has been implemented on two different parallel machines: Cray T3E and Pentium III PC cluster. T3E consists of 450 MHz Alpha 21164 processors and 3-D torus network. Pentium III PC cluster is a set

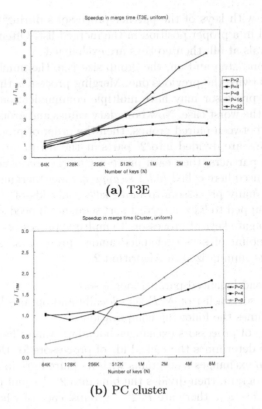

(a) T3E

(b) PC cluster

Fig. 3. Speedups of merge time on two machines with uniform distribution

Table 1. Machine parameters

	K_1 [msec/key]	K_2 [msec/key]	C
T3E	0.048	0.125	1.732
PC cluster	0.386	0.083	1.184

of 8 PCs with 1GHz Athlon CPUs interconnected by a 100Mbps Fast Ethernet switch. Maximum number of keys is limited by the capacity of the main memory of each machine. Keys are synthetically generated with two distribution functions (*uniform* and *gauss*) with 32-bit integers for PC cluster, and 64-bit integers for T3E. Code is written in C language with MPI communication library.

Parameters of the computation and communication performance of individual systems are measured as given Table 1. Notice that T3E is expected to achieve the highest performance enhancement due to having the biggest C introduced in Eq (9) in the appendix. The speedups in merge time of the load-balanced merge sort over the conventional merge sort are shown in Figure 3 and 4. The speedups with gauss distribution are smaller than those with uniform distribution since Δ in Eq. (7) is bigger in gauss distribution than in uniform distribution. The improvement gets better as the number of processors increases.

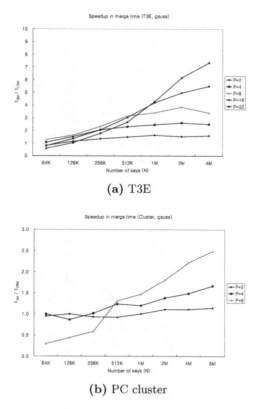

(a) T3E

(b) PC cluster

Fig. 4. Speedups of merge time on two machines with gauss distribution

The measured speedups are close to the predicted ones when the number of N/P is large. When N/P is small, the performance suffers due to the overhead such as in exchanging boundary values and histogram information, and broadcasting processor ids. Experimental results of T3E having the higher speedup supports the analytic result given in Eq. (8). The comparisons of the total sorting time and distribution of the load balanced merge sort with the conventional algorithm are shown in Figure 5. Local sort times of both methods remain same in one machine. The best speedup of 8.2 in merging phase is achieved on T3E with 32 processors.

4   Conclusion

We have improved the parallel merge sort by keeping and computing approximately equal number of keys in all processors through the entire merging phases. Using the histogram information, keys can be divided equally regardless of their distribution. We have achieved a maximal speedup of 8.2 in merging time for 4M keys on 32-processor Cray T3E, which is about 78% of the upper bound. This

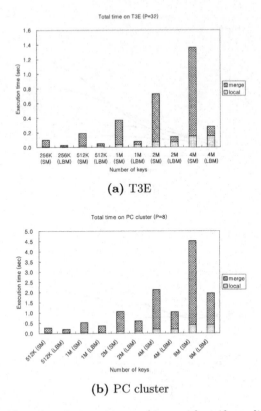

(a) T3E

(b) PC cluster

Fig. 5. Total sorting time on two machines with uniform distribution

scheme can be applied to parallel implementation of similar merging algorithms
such as parallel quick sort.

A   Appendix

The upper bound of the speedup of the new parallel merge sort is estimated
now. Let $T_{seq}(N/P)$ be the time for the initial local sort to make a sorted list.
$T_{comp}(N)$ represents the time for merging two lists, each with $N/2$ keys, and
$T_{comm}(M)$ is the interprocessor communication time to transmit M keys. For
the input of N keys, $T_{comm}(N)$ and $T_{comp}(N)$ are estimated as follows[4]:

$$T_{comm}(N) = S + K_1 \cdot N \qquad (1)$$

$$T_{comp}(N) = K_2 \cdot N \qquad (2)$$

where K_1 and K_2 are the average time to transmit one key and the average time
per key to merge N keys, respectively, and S is the startup time. The parameters
Ks and S are dependent on machine architecture.

For Algorithm1, step1 requires $T_{seq}(N/P)$. Step 2 repeats $\log P$ times, so execution time of the simple parallel merge sort (SM) is estimated as below:

$$T_{SM}(N, P) = T_{seq}\left(\frac{N}{P}\right) + \sum_{i=1}^{\log N}\left\{T_{comm}\left(\frac{2^{i-1}N}{P}\right) + T_{comp}\left(\frac{2^i N}{P}\right)\right\}$$

$$\approx T_{seq}\left(\frac{N}{P}\right) + \left\{T_{comm}\left(\frac{N}{P} + \frac{2N}{P} + \cdots + \frac{\frac{P}{2}N}{P}\right)\right.$$

$$\left. + T_{comp}\left(\frac{2N}{P} + \frac{4N}{P} + \cdots + \frac{PN}{P}\right)\right\}$$

$$= T_{seq}\left(\frac{N}{P}\right) + \left\{T_{comm}\left(\frac{N}{P}(P-1)\right) + T_{comp}\left(\frac{2N}{P}(P-1)\right)\right\} \quad (3)$$

In Eq.(3) the communication time was assumed proportional to the size of data by ignoring the startup time (Coarse-grained communication in most interprocessor communication networks reveal such characteristics).

For Algorithm 2, step 1 requires $T_{seq}(N/P)$. The time required in steps 2.1 and 2.2 is ignorable if the number of histogram bins is small compared to N/P. Since the maximum number of keys assigned to each processor is N/P, so at most N/P keys are exchanged among paired processors in step 2.3. Each processor merges $N/P + \Delta$ keys in step 2.4. Step 2.5 requires $O(\log P)$ time. The communication of steps 2.1 and 2.2 can be ignored since the time is relatively small compared to the communication time in step 2.3 if N/P is large (coarse grained). Since step 2 is repeated $\log P$ times, the execution time of the load-balanced parallel merge sort (LBM) can be estimated as below:

$$T_{LBM}(N, P) = T_{seq}\left(\frac{N}{P}\right) + \log P \cdot \left\{T_{comm}\left(\frac{N}{P}\right) + T_{comp}\left(\frac{N}{P} + \Delta\right)\right\} \quad (4)$$

To observe the enhancement in *merging phase only*, the first terms in Eqs. (3) and (4) will be removed. Using the relationship in Eqs. (1) and (2), merging times are rewritten as follows:

$$T_{SM}(N, P) = K_1 \cdot \frac{N}{P}(P-1) + K_2 \cdot \frac{2N}{P}(P-1) \quad (5)$$

$$T_{LBM}(N, P) = K_1 \cdot \frac{N}{P}\log P + K_2 \cdot \left(\frac{N}{P} + \Delta\right)\log P \quad (6)$$

A speedup of the load-balanced merge sort over the conventional merge sort, denoted as η, is defined as the ration of T_{CM} to T_{LBM}:

$$\eta = \frac{T_{SM}(N, P)}{T_{LBM}(N, P)}$$

$$= \frac{K_1 \cdot \frac{N}{P}(P-1) + K_2 \cdot \frac{2N}{P}(P-1)}{K_1 \cdot \frac{N}{P}\log P + K_2 \cdot \left(\frac{N}{P} + \Delta\right)\log P} \quad (7)$$

34   Minsoo Jeon and Dongseung Kim

If the load-balanced merge sort keeps load imbalance small enough to ignore Δ, and N/P is large, Eq. (7) can be simplified as follows:

$$
\begin{aligned}
\eta &= \frac{K_1 \cdot \frac{N}{P}(P-1) + K_2 \cdot \frac{2N}{P}(P-1)}{K_1 \cdot \frac{N}{P}\log P + K_2 \cdot \frac{N}{P}\log P} \\
&= \frac{K_1 + 2K_2}{K_1 + K_2} \cdot \frac{P-1}{\log P} \\
&= C \cdot \frac{P-1}{\log P}
\end{aligned}
\tag{8}
$$

where C is a value determined by the ratio of the interprocessor communication speed to computation speed of the machine as defined below

$$
C = \frac{K_1 + 2K_2}{K_1 + K_2} = 1 + \frac{K_2}{K_1 + K_2} = 1 + \frac{1}{\frac{K_1}{K_2}+1}
\tag{9}
$$

References

1. K. Batcher, "Sorting networks and their applications," *Proceedings of the AFIPS Spring Joint Computer Conference 32*, Reston, VA, 1968, pp.307-314.
2. R. Cole, "Parallel merge sort," *SIAM Journal of Computing*, vol. 17, no. 4, 1998, pp.770-785.
3. A. C. Dusseau, D. E. Culler, K. E. Schauser, and R. P. Martin, "Fast parallel sorting under LogP: experience with the CM-5", *IEEE Trans. Computers*, Vol. 7, Aug. 1996.
4. R. Hockney, "Performance parameters and benchmarking of supercomputers", *Parallel Computing*, Dec. 1991, Vol. 17, No. 10 & 11, pp. 1111-1130.
5. J. S. Huang and Y. C. Chow, "Parallel sorting and data partitioning by sampling", *Proc. 7th Computer Software and Applications Conf.*, Nov. 1983, pp. 627-631.
6. Y. Kim, M. Jeon, D. Kim, and A. Sohn, "Communication-Efficient Bitonic Sort on a Distributed Memory Parallel Computer", *Int'l Conf. Parallel and Distributed Systems (ICPADS'2001)*, June 26-29, 2001.
7. S. J. Lee, M. Jeon, D. Kim, and A. Sohn, "Partitioned Parallel Radix Sort," *J. of Parallel and Distributed Computing*, Academic Press, (to appear) 2002.
8. A. Sohn and Yuetsu Kodama, "Load Balanced Parallel Radix Sort," *Proceedings of the 12th ACM International Conference on Supercomputing*, July 1998.
9. R. Xiong and T. Brown, "Parallel Median Splitting and k-Splitting with Application to Merging and Sorting," *IEEE Transactions on Parallel and Distributed Systems*, Vol. 4, No. 5, May 1993, pp.559-565.

Avoiding Network Congestion
with Local Information*

E. Baydal, P. López, and J. Duato

Dept. of Computer Engineering
Universidad Politécnica de Valencia
Camino de Vera s/n, 46071 - Valencia, Spain
elvira@disca.upv.es

Abstract. Congestion leads to a severe performance degradation in
multiprocessor interconnection networks. Therefore, the use of techniques
that prevent network saturation are of crucial importance. Some recent
proposals use global network information, thus requiring that nodes exchange some control information, which consumes a far from negligible
bandwidth. As a consequence, the behavior of these techniques in practice is not as good as expected.
In this paper, we propose a mechanism that uses only local information to
avoid network saturation. Each node estimates traffic locally by using the
percentage of free virtual output channels that can be used to forward a
message towards its destination. When this number is below a threshold
value, network congestion is assumed to exist and message throttling is
applied. The main contributions of the proposed mechanism are two: i)
it is more selective than previous approaches, as it only prevents the
injection of messages when they are destined to congested areas; and ii)
it outperforms recent proposals that rely on global information.

1 Introduction

Massively parallel computers provide the performance that most scientific and
commercial applications require. Their interconnection networks offer the low
latencies and high bandwidth that is needed for different traffic kinds. Usually,
wormhole switching [7] with virtual channels [4] and adaptive routing [7] is used.
However, multiprocessor interconnection networks may suffer from severe saturation problems with high traffic loads, which may prevent reaching the wished
performance.

This problem can be stated as follows. With low and medium network loads,
the accepted traffic rate is the same as the injection rate. But if traffic increases
and reaches (or surpasses) certain level (the saturation point), accepted traffic
falls and message latency increases considerably. Notice that both latency and
accepted traffic are dependent variables on offered traffic. This problem appears
both in deadlock avoidance and recovery strategies [7]. Performance degradation appears because messages cyclically block in the network faster than they

---

* This work was supported by the Spanish CICYT under Grant TIC2000-1151-C07

H. Zima et al. (Eds.): ISHPC 2002, LNCS 2327, pp. 35–48, 2002.
© Springer-Verlag Berlin Heidelberg 2002

are drained by the escape paths in the deadlock avoidance strategies or by the deadlock recovery mechanism.

In order to solve this problem, the probability of cyclic waiting in the network has to be reduced. This can be accomplished by using different strategies. One approach is to avoid that network traffic reaches the saturation point. Although several mechanisms for multiprocessors and clusters of workstations (COWs) have already been proposed, they have important drawbacks that will be analyzed in Section 3.

In this paper, we propose and evaluate a mechanism to avoid network saturation that tries to overcome those drawbacks. It is intended for wormhole switching and is based on locally estimating network traffic by using the percentage of useful[1] free virtual output channels. Injection rate is reduced if this percentage is lower than a threshold (i.e. if the number of free channels is low). Indeed, if it is needed, message injection is completely stopped until the traffic decreases or a predefined interval time passes. This second condition avoids the starvation problem that some nodes may suffer otherwise. The mechanism has been tested in different situations, and compared with recent proposals [18], [19], achieving better results in all the evaluated conditions.

2   Related Work

Performance degradation in multiprocessor networks appears when network traffic reaches or surpasses the saturation point. From our point of view, there are two strategies to avoid this situation: congestion prevention and congestion recovery. Congestion prevention techniques require some kind of authorization from the message destination in order to inject it. The two most known techniques are based on opening connections, reserving the necessary resources before injecting the message, and on limiting the number of sent messages without receiving an acknowledgment. Both of them introduce overhead and, because of that, they will not be considered in this paper.

Congestion recovery techniques are based on monitoring the network and triggering some actions when congestion is detected. In this case, the solution has two steps: first, how to detect congestion, and second, what restrictions to apply.

About how network congestion is detected, there are two different kind of strategies. Ones are based on measuring the waiting time of blocked messages [10], [8], [17] while others use the number of busy resources in the network. In the first case, they may also detect deadlocked messages. Unfortunately, waiting time depends on message length. In the second group of strategies, busy resources may be channels [2], [5], [14] or buffers [19]. Information about occupied resources may be only local to the node or global from all the network nodes. Of course, global information may allow a more accurate network status knowledge, but with an important associated cost. In order to periodically update network status, a lot of control messages have to be sent across the network, thus wasting some amount of bandwidth. Mechanisms that use only local information are more limited, but

---

[1] the useful channels are the output channels returned by the routing function

they may obtain a better cost/performance tradeoff. As nodes do not need to exchange status information, more bandwidth is available to send data messages. As a consequence, mechanisms based on local information may outperform those that rely on global information.

Finally, the actions that are triggered to avoid network saturation can be divided in three groups. First, and more studied, is message throttling. We will refer to this later. The second group uses non-minimal routing to balance network load along alternate paths [5]. Finally, the third group [17] adds additional buffers ("congestion buffers") to slow down messages and reduce network traffic.

Message throttling [5] has been the most frequently used method. To stop, or at least, slow down injection is a classical solution for packet-switched networks [9]. Also, in multiprocessor and COWs interconnection networks, several strategies have been developed. Some proposals completely stops message injection when congestion is detected [12], [17], [19]. If several injection channels are available, a better way may consist of progressively disabling them. Also message transmissions may be delayed during increasing intervals [10]. Message throttling can be applied for a predefined time [17], [10], or until the traffic falls enough [2], [14], [19].

3   Features of a Good Congestion Control Mechanism

In order to design an efficient congestion control mechanism, in this section, we will describe the desirable features of such a mechanism. From our point of view, it should be robust, it should not penalize network behavior for low and medium loads and it should not generate new problems.

First, the mechanism should work properly for different conditions: different message destination distributions, message sizes and network topologies. However, saturation point depends on the distribution of message destinations. Because of that, a suitable mechanism for one distribution, may not work properly with another one. The same problems may appear when we change network topology. However, many of the previously proposed mechanisms have been analyzed with only one network size [2], [19] and the for uniform distribution of message destinations [5], [10], [8], [17]. Others do not achieve good results for each individual traffic pattern considered [18]. Also, in some cases performance strongly depends on message size [10], [17].

Second, the added mechanisms should not penalize the network when it is not saturated. Notice that, this situation is the most frequent [16]. Thus, when network traffic is not close to or beyond the saturation point, it would be desirable that the mechanism do not restrict or delay message injection. However, some of the previous proposals increase message latency before the saturation point [5], [18]. On the other hand, strategies based in non-minimal routing [8], [5] may also increase message latency.

Finally, the new mechanism should not generate new problems in the network. Some of the proposed mechanisms increase network complexity by adding new signals [17], [10] or even a sideband network [19], [18]. Others need to send extra information that increases traffic and makes worse the congestion situation [10].

4   An Injection Limitation Mechanism Based on Local Information

In this section, we will present another approach to prevent network saturation. It estimates network traffic locally at each network router by counting its number of free virtual channels. When network load is considered to be high, message throttling at the local node is applied, reducing the bandwidth available to inject messages into the network. Messages that can not be injected are buffered in a pending message queue.

The mechanism is based on the following idea. Before injecting a new message, a count of the number of free virtual output channels is performed. Only those channels that are useful for forwarding the message towards its destination are considered in the count. The idea behind this method is that although some network areas are congested, it does not matter if they are not going to be used by the message. Moreover, if the channels that may forward the message towards its destination are congested, the mechanism should prevent injection regardless if the other channels are free. Hence, despite the mechanism only uses local information to estimate network traffic, it only starts applying message throttling at the local node when it may contribute to increase network congestion.

Therefore, the number of useful free virtual output channels is used for estimating network traffic. Thus, injection of a message will be allowed only if this number exceeds a threshold. However, notice that this threshold value strongly depends on message destination distribution. To make the mechanism independent of message destination distribution, the number of useful free virtual output channels relative to the total number of virtual channels that may be used for forwarding the message towards its destination is used. Both this ratio and threshold value are lower than one. Therefore, a message that may use more virtual channels to reach its destination than another one will need more free virtual channels to be injected by the mechanism. The idea is to use a fixed threshold that is expected to work fine with all message destination distributions. However, the precise threshold value has to be tuned empirically. Figure 1 shows the above mentioned ratio for several message destination distributions. Although the throughput reached by different message destination distributions is not the same, the ratio value for the saturation points is very similar (around 0.3).

To illustrate the method, consider a bidirectional k-ary 3-cube network with 3 virtual channels per physical channel. Assume that optimal threshold has found to be equal to 0.3125 and that a message needs to cross two network dimensions. Then, the message may choose among 6 virtual output channels. Let F be the number of these that are free. If $\frac{F}{6}$ is greater than 0.3125, the message can be injected. In other words, at least two (0.3125*6) virtual channels must be free in order to allow message injection.

On the other hand, we have introduced another rule to avoid node starvation. After an interval detecting congestion and without be able to inject new messages, if a node has pending messages, it will inject the first of them regardless of the number of free virtual output channels. This rule avoids starvation. The best value of this time interval has to be empirically tuned.

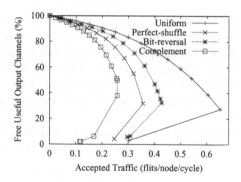

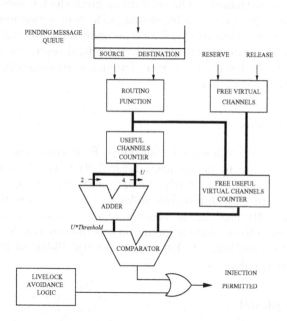

Fig. 1. Percentage of useful free output virtual channels versus traffic for a 8-ary 3-cube with a deadlock recovery fully adaptive routing algorithm and 3 virtual channels per physical channel for different distributions of message destinations. 16-flit messages.

Fig. 2. Hardware required for the new injection limitation mechanism.

Figure 2 shows the implementation of the mechanism. The routing function must be applied before injecting the newly generated message into the network. This requires replicating some hardware, but if a simple routing algorithm is used (see Section 5.2), this should not be a problem. The routing function returns the useful virtual output channels, for instance, in a bit-vector. Then, a circuit that performs a count on the number of "1" outputs the number of virtual channels returned by the routing function, U. On the other hand, the useful virtual output channels bit-vector is combined with virtual channel status, and the number of free useful virtual output channels, F, is counted. This can be easily done by a bitwise *and* operator followed by a circuit similar to the one that computes U. The comparison between $\frac{F}{U}$ and the threshold has been converted into a comparison between F and U multiplied by the threshold. This product is easily performed if the threshold is chosen as a sum of powers of two. For instance, if threshold is 0.3125 (as in Figure 2), it can be decomposed into $2^{-2} + 2^{-4}$. So, its product by U is as easy as adding U shifted right two positions and U shifted right three positions.

Finally, notice that although this hardware may add some delay to messages, it does not reduce clock frequency because it is not on the critical path.

5   Performance Evaluation

In this section, we will evaluate by simulation the behavior of the proposed congestion control mechanism. The evaluation methodology used is based on the one proposed in [6]. The most important performance measures are latency since generation[2] (time required to deliver a message, including the time spent at the source queue) and throughput (maximum traffic accepted by the network). Accepted traffic is the flit reception rate. Latency is measured in clock cycles, and traffic in flits per node per cycle.

5.1   Traffic Patterns

Message injection rate is the same for all nodes. Each node generates messages independently, according to an exponential distribution. Destinations are chosen according to the *Uniform, Butterfly, Complement, Bit-reversal,* and *Perfect-shuffle* traffic patterns, which illustrate different features. Uniform distribution is the most frequently used in the analysis of interconnection networks. The other patterns take into account the permutations that are usually performed in parallel numerical algorithms [11]. For message length, 16-flit, 64-flit and 128-flit messages are considered.

5.2   Network Model

The simulator models the network at the flit level. Each node has a router, a crossbar switch and several physical channels. Routing time and both transmission time across the crossbar and across a channel are all assumed to be equal to one clock cycle. On the other hand, each node has four injection/ejection channels. Although most commercial multiprocessors have only one injection/ejection channel, previous works [10], [7] have highlighted that the bandwidth available at the network interface may be a bottleneck that prevents achieving a high network throughput. As an example, Figure 3 shows the performance evaluation results for different number of injection/ejection ports for a 8-ary 3-cube network when messages are sent randomly to all nodes and with some locality. As can

---

[2] Notice that network latency only considers the time spent traversing the network

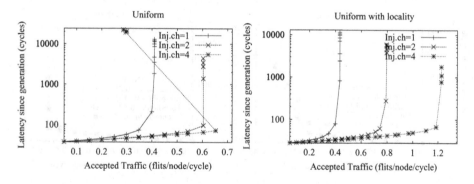

Fig. 3. Average message latency vs. accepted traffic for different injection/ejection ports. 8-ary 3-cube. 16-flit messages. TFAR routing with 3 virtual channels.

be seen, network throughput can be increased by 70% by using more than one injection port in the first case. When messages are sent more locally, throughput can be drastically increased (up to 2.9 times) using when four injection/ejection ports.

Concerning deadlock handling, we use software-based deadlock recovery [15] and a True Fully Adaptive routing algorithm (TFAR) [16], [15] with 3 and 4 (results not shown) virtual channels per physical channel. This routing algorithm allows the use of any virtual channel of those physical channels that forwards a message closer to its destination. TFAR routing outperforms deadlock-avoidance strategies [15], although when the number of virtual channels is large enough, both strategies achieve similar performance. In order to detect network deadlocks, we use the mechanism proposed in [13] with a deadlock detection threshold equal to 32 cycles.

We have evaluated the performance of the proposed congestion control mechanism on different bidirectional k-ary n-cubes. In particular, we have used the following network sizes: 256 nodes (n=2, k=16), 1024 nodes (n=2, k=32), 512 nodes (n=3, k=8), and 4096 nodes (n=3, k=16).

5.3   Performance Comparison

In this section, we will analyze the behavior of the mechanism proposed in section 4, which will be referred to as **U-channels** (Useful-channels). For comparison purposes, we will also evaluate the behavior of another recently proposed mechanism [19], which will be referred to as **Self-Tuned**. In this mechanism, nodes detect network congestion by using global information about the total number of full buffers in the network. If this number surpasses a threshold, all nodes apply message throttling. The use of global information requires to broadcast data between all the network nodes, which may generate a lot of traffic. A way of transmitting this control information is to use a sideband network with a far from negligible bandwidth [19]. To make a fair comparison, the bandwidth provided by the added sideband network should be considered as additional avail-

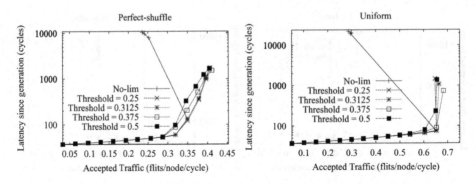

Fig. 4. Average message latency vs. accepted traffic for different threshold values for the U-channels mechanism. 8-ary 3-cube (512 nodes). 16-flit messages. 3 virtual channels.

able bandwidth in the main interconnection network, in mechanisms that do not need to exchange control messages. However, in the results that we present we do not consider this fact. If this additional bandwidth were considered, the differences, not only in throughput but also in latency, between **Self-Tuned** and the **U-channels** mechanism would be greater than the ones shown. Finally, results without any congestion control mechanism (**No-Lim**) are also shown.

Tuning the U-channels Mechanism. First, the **U-channels** mechanism must be properly tuned. Several thresholds have to be tested with different message destination distributions, message lengths and network topologies. We have tested the following threshold values: 0.25, 0.3125, 0.375, 0.5 and, in some cases, 0.125. Notice that all of them can be easily calculated as a sum of powers of two, as required in the implementation shown in Figure 2. For the sake of shortness, we will only show a subset of the results. Figure 4 shows the average message latency versus traffic for different threshold values for the uniform and perfect-shuffle traffic patterns for a 8-ary 3-cube (512 nodes). From the latency and throughput point of view, the highest threshold values lead to apply more injection limitation than necessary. As a consequence, message latency is increased due to the fact that messages are waiting at the source nodes. On the other hand, the lowest threshold value allows a more relaxed injection policy and trends saturating the network. For the message patterns shown, a good choice is 0.3125 for the 512-node network, but as you can see, a given topology may have more than one threshold that works properly.

In general, we have noticed that the optimal threshold mainly depends on the network topology (in particular, it depends on the number of nodes per dimension, k) and the number of virtual channels per physical channel, but it does not significantly depend on message destination distribution or message size. For the same radix k and number of dimensions n, the optimal threshold may decrease to the next value when the number of virtual channels per physical channel increases. On the other hand, the optimal threshold value may

Table 1. Optimal U-channels threshold values for several topologies and different number of virtual channels per physical channel.

Topology	Threshold (3 vc)	Threshold (4 vc)
256 (16 × 16), 1024 (32 × 32), 4096 (16 × 16 × 16)	0.375	0.3125
512 (8 × 8 × 8)	0.3125	0.25

increase for high k values. The explanation is simple. Increasing the number of virtual channels per physical channel decreases network congestion. Hence, we have to apply a more relaxed injection policy, with a lower threshold value. On the contrary, by increasing k, average traversed distance also increases and the maximum achievable network throughput is reduced [1]. Therefore, a higher threshold value is required. Table 1 shows the optimal thresholds for several topologies and different number of virtual channels.

We have also tested several interval time values (64, 128, 256 and 512 clock cycles) in order to prevent starvation. The lowest intervals (64 and 128) allows excessive message injection, nullifying the effect of the congestion control mechanism. Interval values of 256 and above work properly. Results shown use a value of 256 clock cycles.

Evaluation Results. Having tuned the new mechanism, we can compare all the analyzed congestion control mechanisms. Figures 5 through 10 show some of the results, for different distributions of message destinations, network and message sizes. As results for the different message lengths analyzed are qualitatively the same, usually we will only show the results for 16-flit messages.

All the simulations finish after receiving 500,000 messages, but only the last 300,000 ones are considered to calculate average latencies. As we can see, the **U-channels** congestion control mechanism avoids the performance degradation in all the analyzed cases, without increasing latency before the saturation point. Indeed, it usually improves network performance by increasing the throughput achieved when no congestion control mechanism is used. On the other hand, the **U-channels** mechanism outperforms the behavior of the **Self-Tuned** mechanism. **Self-Tuned** usually reaches lower throughput because due to its global nature, it applies injection restrictions in all nodes as soon as congestion is detected in some area network. Indeed, in some cases, the performance degradation is not fully removed. This behavior of the **Self-Tuned** mechanism was already described in [18]. It must be noticed that unlike [18], [19], we use four injection/ejection ports. Indeed, if we use the bandwidth provided by the sideband network used in **Self-Tuned** as additional data bandwidth in the **U-channels** and **No-Lim** mechanisms the differences between the mechanisms would be even greater, in terms of both traffic and message latency.

We have also used bursty loads that alternates periods of high message injection rates, corresponding to saturation intervals, with periods of low traffic under the saturation. In this case, we inject a given number of messages into the net-

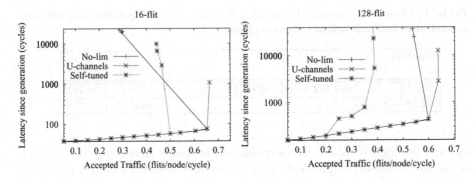

Fig. 5. Average message latency versus accepted traffic. Uniform distribution of message destination. 8-ary 3-cube (512 nodes). 3 virtual channels. U-channels threshold=0.3125.

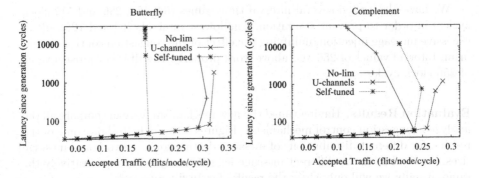

Fig. 6. Average message latency versus accepted traffic. 8-ary 3-cube (512 nodes). 16-flit messages. 3 virtual channels. U-channels threshold=0.3125.

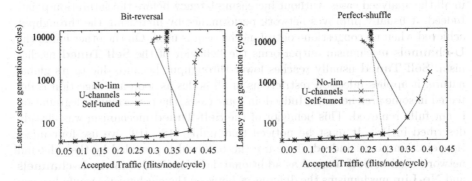

Fig. 7. Average message latency versus accepted traffic. 8-ary 3-cube (512 nodes). 16-flit messages. 3 virtual channels. U-channels threshold=0.3125.

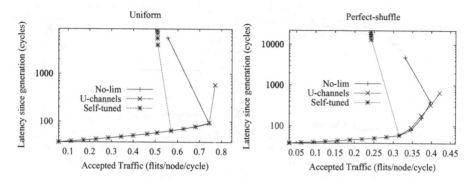

Fig. 8. Average message latency versus accepted traffic. 8-ary 3-cube (512 nodes). 16-flit messages. 4 virtual channels. U-channels threshold=0.25.

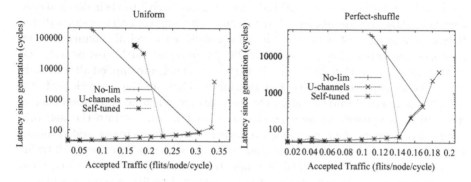

Fig. 9. Average message latency versus accepted traffic. 16-ary 2-cube (256 nodes). 16-flit messages. 3 virtual channels. U-channels threshold=0.375.

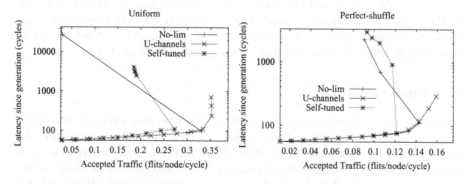

Fig. 10. Average message latency versus accepted traffic. 16-ary 3-cube (4096 nodes). 16-flit messages. 3 virtual channels. U-channels threshold=0.375.

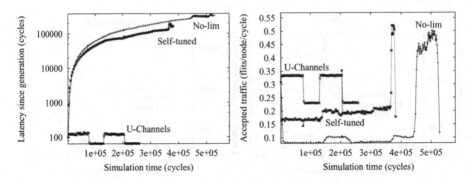

Fig. 11. Performance with variable injection rates. Uniform distribution of message destinations. 16-flits. 16-ary 2-cube, 3 virtual channels. U-channels threshold=0.375.

work and simulation goes on until all the messages arrive to their destinations. Figure 11 shows the results for a 2-ary 16-cube (256 nodes) with a uniform distribution of message destinations, 3 virtual channels per physical channel, 400,000 messages generated with a rate of 0.34 flits/node/cycle (high load period) and 200,000 messages with 0.23 flits/node/cycle. This loads are applied alternatively twice. As we can see, with the **U-channels** mechanism the network perfectly accepts the injected bursty traffic. On the contrary, when no congestion control mechanism is applied, as soon as the first burst is applied into the network, congestion appears. As a consequence, latency strongly increases and throughput falls down. Lately, after some time injecting with low rate, network traffic starts recovering but the arrival of a new traffic burst prevents it. Congestion only disappears in the last period of time (accepted traffic increases), when no new messages are generated. Concerning **Self-Tuned** mechanism, we can see that it excessively limits injection rate, significantly reducing the highest value of accepted traffic and increasing the time required to deliver all the injected messages. This time is another performance measure that is strongly affected by the presence of network saturation. Figure 11 shows that **U-channels** delivers the required number of messages in half the time than **No-Lim**, while **Self-Tuned** achieves an intermediate value between both of them.

6   Conclusions

In this paper, we have proposed a mechanism to prevent network saturation based on message throttling. This mechanism uses the number of useful (the ones that may be selected by the routing algorithm) free virtual output channels relative to the useful virtual output channels to estimate locally network traffic. Message injection is allowed only if this number exceeds a given threshold. Although the threshold has to be empirically tuned, we have found that it does not strongly depend on message destination distribution or message length. It only slightly depends on network topology, specially on radix k and the number of virtual channels per physical channel. Moreover, it guarantees that no node suffers from starvation by periodically allowing the injection of queued messages. The proposed mechanism needs to be implemented in the routing control unit but it does not affect clock frequency, because it is not in the critical path. Moreover, it does not require extra signaling.

From the presented evaluation results, we can conclude that the mechanism proposed in this paper verifies the three desirable properties of a good congestion control method (see Section 3). Thus, its behavior outperforms other previous proposals. First, it works well with all tested message destination distributions, message lengths and network topologies. Second, it does not penalize normal network behavior, without increasing latency before the saturation point. Moreover, it increases the maximum throughput achieved by the network with respect to the results without applying any congestion control mechanism. Finally, it does neither require extra signals nor sending control messages to be implemented. Although some extra hardware is required to implement the mechanism, it is not very complex.

The mechanism has been also compared -in a quantitative form- with a recently proposed mechanism that uses global network information in order to detect network congestion, outperforming its behavior in all the analyzed cases.

References

1. A. Agarwal, "Limits on interconnection network performance", *IEEE Trans. on Parallel and Distributed Systems*, vol. 2, no. 4, pp. 398–412, Oct. 1991.
2. E. Baydal, P. López and J. Duato, "A Simple and Efficient Mechanism to Prevent Saturation in Wormhole Networks", in *14th. Int. Parallel & Distributed Processing Symposium*, May 2000.
3. W. J. Dally and C. L. Seitz, "Deadlock-free message routing in multiprocessor interconnection networks," *IEEE Trans. on Computers*, vol. C–36, no. 5, pp. 547–553, May 1987.
4. W. J. Dally, "Virtual-channel flow control," *IEEE Trans. on Parallel and Distributed Systems*, vol. 3, no. 2, pp. 194–205, March 1992.
5. W. J. Dally and H. Aoki, "Deadlock-Free Adaptive Routing in Multicomputer Networks Using Virtual Channels", *IEEE Trans. on Parallel and Distributed Systems*, vol. 4, no. 4, pp. 466–475, April 1993.
6. J. Duato, "A new theory of deadlock-free adaptive routing in wormhole networks," *IEEE Trans. on Parallel and Distributed Systems*, vol. 4, no. 12, pp. 1320–1331, Dec. 1993.
7. J. Duato, S. Yalamanchili and L.M. Ni, *Interconnection Networks: An Engineering Approach*, IEEE Computer Society Press, 1997.
8. C. Hyatt and D. P. Agrawal, "Congestion Control in the Wormhole-Routed Torus With Clustering and Delayed Deflection" Workshop on Parallel Computing, Routing, and Communication (PCRCW'97), June 1997, Atlanta, GA.
9. V. Jacobson, "Congestion Avoidance and Control", *Proc. ACM SIGCOMM'88*, Stanford, CA, August 1988.
10. J. H. Kim, Z. Liu and A. A. Chien, "Compressionless routing: A framework for Adaptive and Fault-Tolerant Routing," in *IEEE Trans. on Parallel and Distributed Systems*, Vol. 8, No. 3, 1997.

11. F. T. Leighton, *Introduction to Parallel Algorithms and Architectures: Arrays, Trees, Hypercubes*. San Mateo, CA, USA, Morgan Kaufmann Publishers, 1992.
12. P. López and J. Duato, "Deadlock-free adaptive routing algorithms for the 3D-torus: Limitations and solutions," in *Proc. of Parallel Architectures and Languages Europe 93*, June 1993.
13. P. López, J.M. Martínez and J. Duato, "A Very Efficient Distributed Deadlock Detection Mechanism for Wormhole Networks," in *Proc. of High Performance Computer Architecture Workshop*, Feb. 1998.
14. P. López, J.M. Martínez and J. Duato, "DRIL: Dynamically Reduced Message Injection Limitation Mechanism for Wormhole Networks," *1998 Int. Conference Parallel Processing*, August 1998.
15. J.M. Martínez, P. López, J. Duato and T.M. Pinkston, "Software-Based Deadlock Recovery Technique for True Fully Adaptive Routing in Wormhole Networks," *1997 Int. Conference Parallel Processing*, August 1997.
16. T.M. Pinkston and S. Warnakulasuriya, "On Deadlocks in Interconnection Networks", in *the 24th Int. Symposium on Computer Architecture*, June 1997.
17. A. Smai and L. Thorelli , "Global Reactive Congestion Control in Multicomputer Networks", *5th Int. Conference on High Performance Computing*, 1998.
18. M. Thottetodi, A.R. Lebeck, S.S. Mukherjee, "Self-Tuned Congestion Control for Multiprocessor Networks", *Technical Report CS-2000-15, Duke University*, 2000.
19. M. Thottetodi, A.R. Lebeck, S.S. Mukherjee, "Self-Tuned Congestion Control for Multiprocessor Networks", *Proc. of 7th. Int. Symposium on High Performance Computer Architecture*, 2001.

Improving InfiniBand Routing
through Multiple Virtual Networks*

J. Flich, P. López, J.C. Sancho, A. Robles, and J. Duato

Department of Computer Engineering
Universidad Politécnica de Valencia, Spain
{jflich,plopez,jcsancho,arobles,jduato@gap.upv.es}

Abstract. InfiniBand is very likely to become the de facto standard for communication between nodes and I/O devices (SANs) as well as for interprocessor communication (NOWs). The InfiniBand Architecture (IBA) defines a switch-based network with point-to-point links whose topology is arbitrarily established by the customer. Often, the interconnection pattern is irregular. Up*/down* is the most popular routing scheme currently used in NOWs with irregular topologies. However, the main drawbacks of up*/down* routing are the unbalanced channel utilization and the difficulties to route most packets through minimal paths, which negatively affects network performance. Using additional virtual lanes can improve up*/down* routing performance by reducing the head-of-line blocking effect, but its use is not aimed to remove its main drawbacks. In this paper, we propose a new methodology that uses a reduced number of virtual lanes in an efficient way to achieve a better traffic balance and a higher number of minimal paths. This methodology is based on routing packets simultaneously through several properly selected up*/down* trees. To guarantee deadlock freedom, each up*/down* tree is built over a different virtual network. Simulation results, show that the proposed methodology increases throughput up to an average factor ranging from 1.18 to 2.18 for 8, 16, and 32-switch networks by using only two virtual lanes. For larger networks with an additional virtual lane, network throughput is tripled, on average.

1 Introduction

InfiniBand [8] has been recently proposed as an standard for communication between processing nodes and I/O devices as well as for interprocessor communication. Several companies, including the leading computer manufacturers, support the InfiniBand initiative. InfiniBand is designed to solve the lack of high bandwidth, concurrency and reliability of existing technologies (PCI bus) for system area networks. Moreover, InfiniBand can be used as a platform to build networks of workstations (NOWs) or clusters of PCs [13] which are becoming a cost-effective alternative to parallel computers. Currently, clusters are based on different available network technologies (Fast or Gigabit Ethernet [21],

---

* This work was supported by the Spanish CICYT under Grant TIC2000-1151 and by Generalitat Valenciana under Grant GV00-131-14.

H. Zima et al. (Eds.): ISHPC 2002, LNCS 2327, pp. 49–63, 2002.

Myrinet [1], ServerNet II [7], Autonet [20], etc...). However, they may not provide the protection, isolation, deterministic behavior, and quality of service required in some environments.

The InfiniBand Architecture (IBA) is designed around a switch-based interconnect technology with high-speed point-to-point links. Nodes are directly attached to a switch through a Channel Adapter (CA). An IBA network is composed of several subnets interconnected by routers, each subnet consisting of one or more switches, processing nodes and I/O devices.

Routing in IBA subnets is distributed, based on forwarding tables (routing tables) stored in each switch, which only consider the packet destination node for routing [9]. IBA routing is deterministic, as the forwarding tables only store one output link per every destination ID. Moreover, virtual cut-through switching is used [10]. IBA switches support a maximum of 16 virtual lanes (VL). VL15 is reserved exclusively for subnet management, whereas the remaining VLs are used for normal traffic. Virtual lanes provide a mean to implement multiple logical flows over a single physical link [3]. Each virtual lane has associated a separate buffer. However, IBA virtual lanes cannot be directly selected by the routing algorithm. In IBA, to route packets through a given virtual lane, packets are marked with a certain Service Level (SL), and SLtoVL mapping tables are used at each switch to determine the virtual lane to be used. According to the IBA specifications, service levels are primarily intended to provide quality of service (QoS). However, virtual lanes can be also used to provide deadlock avoidance and traffic prioritization.

2   Motivation

Usually, NOWs are arranged as switch-based networks whose topology is defined by the customer in order to provide wiring flexibility and incremental expansion capability. Often, due to building constraints, the connections between switches do not follow any regular pattern leading to an irregular topology. The irregularity in the topology makes the routing and deadlock avoidance quite complicate. In particular, a generic routing algorithm suitable for any topology is required. However, the IBA specification does not establish any specific routing algorithm to be used.

Up*/down* [20,1] is the most popular routing algorithm used in the NOW environment. Other routing schemes have been proposed for NOWs, such as the adaptive-trail [14], the minimal adaptive [19], the smart-routing [2] and some improved methodologies to compute the up*/down* routing tables (DFS) [17].

The adaptive-trail and the minimal adaptive routing algorithms are adaptive and allow the existence of cyclic dependencies between channels provided that there exist some channels (escape paths) without cyclic dependencies to avoid deadlocks. However, these escape paths must be selected at run-time, when the other channels are busy. Hence, both routings are not supported by IBA because routing in IBA is deterministic and it cannot select among virtual channels. On the other hand, the smart-routing algorithm is deterministic and it has been shown that it is able to obtain a very high performance [6]. However, its main drawback is the high time required to compute the routing tables. In fact, it has not been possible yet to compute the routing tables for medium and large network sizes. Therefore, in this paper, we will focus on the up*/down* routing.

The main advantage of using *up*/down** routing is the fact that it is simple and easy to implement (see Section 3). However, there exist several drawbacks that may noticeably reduce network performance. First of all, this routing scheme does not guarantee all the packets to be routed through minimal paths. This problem becomes more important as network size increases. In general, *up*/down** concentrates traffic near the root switch, often providing minimal paths only between switches that are allocated near the root switch [16,6]. Additionally, the concentration of traffic in the vicinity of the root switch causes a premature saturation of the network, thus obtaining a low network throughput and leading to an uneven channel utilization.

As stated earlier, IBA allows the use of several virtual lanes, which are closely related to service levels. Although these virtual lanes are primarily intended to provide QoS, they can also be used to improve network performance. In fact, we have recently proposed [18] the use of virtual lanes to reduce the head-of-line blocking effect when using the up*/down* routing algorithm in IBA. In particular, we were using virtual lanes as if were different virtual networks, because packets remained in a particular virtual network (virtual lane) while crossing the network. However, as we were using the same up*/down* tree on each particular virtual network, we were not addressing the main drawbacks of up*/down*. Notice that, if several virtual networks are available, we can use a different routing algorithm on each particular virtual network. We can take advantage of this fact to obtain more minimal paths and a better traffic balance.

In this paper, we take on this challenge by proposing an easy way to improve the performance of InfiniBand networks by using virtual lanes in an efficient way. In particular, we propose the use of different up*/down* trees on each virtual network.

Notice that according to IBA specifications, we need a number of service levels equal to the number of virtual lanes for our purposes. However, service levels are primarily intended for QoS. Therefore, any proposal aimed to improve performance in InfiniBand by using virtual lanes and service levels should limit as much as possible the number of resources devoted to this purpose.

The rest of the paper is organized as follows. In section 3, the up*/down* routing scheme and its implementation on IBA is described. Section 4 describes the proposed methodology to improve network performance. In section 5 the IBA simulation model is described, together with the discussion on the performance of the proposal. Finally, in section 6 some conclusions are drawn.

3   Up*/Down* Routing on IBA

Up*/down* routing is the most popular routing scheme currently used in commercial networks, such as Myrinet [1]. It is a generic deadlock-free routing algorithm valid for any network topology.

Routing is based on an assignment of direction labels ("up" or "down") to the operational links in the network by building a BFS spanning tree. To compute a BFS spanning tree a switch must be chosen as the root. Starting from the root, the rest of the switches in the network are arranged on a single spanning tree [20].

After computing the BFS spanning tree, the "up" end of each link is defined as: 1) the end whose switch is closer to the root in the spanning tree; 2) the end whose switch has the lowest identifier, if both ends are at switches at the same tree level. The result of this assignment is that each cycle in the network has at least one link in the "up" direction and one link in the "down" direction. To avoid deadlocks while still allowing all links to be used, this routing scheme uses the following up*/down* rule: a legal route must traverse zero or more links in the "up" direction followed by zero or more links in the "down" direction. Thus, cyclic channel dependencies [1] [4] are avoided because a packet cannot traverse a link in the "up" direction after having traversed one in the "down" direction.

Unfortunately, $up^*/down^*$ routing cannot be applied to InfiniBand networks in a straightforward manner because it does not conform to IBA specifications. The reason for this is the fact that the $up^*/down^*$ scheme takes into account the input port together with the destination ID for routing, whereas IBA switches only consider the destination ID. Recently, we have proposed two simple and effective strategies to solve this problem [15,12]. In particular, we will use the destination renaming technique proposed in [12].

4   New Strategy for IBA Routing

The basic idea proposed in this paper is the following. For a given network, several up*/down* trees can be easily computed by considering different root nodes. Moreover, for a given source-destination pair, some trees may allow shorter paths than others. Indeed, some trees may offer minimal paths while others not. At first, the idea is to use two of these trees to forward packets through the network. To avoid conflicts between them, every tree will use a different virtual lane (i.e., there will be two virtual networks). Hence, for a given source-destination pair, the tree which offers the shortest path can be used. This will mitigate the non-minimal path problem of basic up*/down* routing. Additionally, as there are now two different trees with two different roots and the paths are distributed between them, network traffic is better balanced.

In order to implement this scheme on IBA, we need to use one virtual lane for each of the virtual networks that correspond to each up*/down* tree. This is easily done by assigning a different SL to each virtual network and mapping each SL to a given virtual lane with the SLtoVL mapping table. Hence, we need only two SLs to implement the proposed mechanism with two up*/down* trees on IBA. Notice that there are up to 15 available SLs, so the remaining ones can be used for other purposes. Additionally, for every source-destination pair,

---

[1] There is a channel dependency from a channel c_i to a channel c_j if a packet can hold c_i and request c_j. In other words, the routing algorithm allows the use of c_j after reserving c_i.

a path from one of the two up*/down* trees will be used. Hence, all the packets sent from the source host to the destination host will use the SL assigned to the selected tree. Thus, every source node needs a table that returns the SL that must be used to forward a packet to a given destination host.

It must be noticed that this routing scheme based on the use of multiple virtual networks is deadlock-free, as packets are injected into a given virtual network and remain on it until they arrive at their destinations (i.e., packets do not cross from one virtual network to another).

Let us explain now how both trees are selected. Ideally, both up*/down* trees should complement each other, in such a way that those paths that are non-minimal in one tree are minimal in the another tree. Another idea is to select the two trees that better balance network links, following some heuristic. In practice, though, this problem translates into how to select the two root nodes. We propose a simple and intuitive strategy. The first root node is the node which has the lowest distance to the rest of nodes (as Myrinet does [1]). The second root is the node that is farthest from the first one. In this way, if the whole set of paths are equally distributed between both trees, as the root nodes are very far from each other, we could expect that many paths from one tree do not interfere with the paths from the another tree (i.e., they do not use the same physical links). As a consequence, a better traffic balance should be expected by using this straightforward strategy. More complex strategies can also be used. However, this issue is beyond the scope of this paper.

Once we have computed the full set of paths for each of the up*/down* trees, the final set of paths needs to be selected. In this set, only one path for every source-destination host has to be included[2]. Again, several strategies can be used. We propose here the straightforward one of randomly selecting one path from the set of all shortest paths (this set is the union of the shortest paths of both trees). Notice that as up*/down* is a partially adaptive routing algorithm, one of the trees may supply more paths than the another one. Hence, with the proposed strategy to select the final paths, this tree will have more chance of being selected than the another tree. Other more complex strategies could select paths trying to better balance link utilization.

The proposed routing scheme based on the use of multiple virtual networks can be easily extended to use more than two (say n) different up*/down* trees. The possibility of having more trees may lead to reduce even more average distance (by supplying shorter paths or even minimal ones) and to achieve a better link utilization. This may be specially noticeable in large networks. The only drawback of using more than two trees is that every new tree needs a different virtual network, and thus, consumes a new SL/VL in InfiniBand. While this may not be a problem for some applications, it can be a limiting factor for other applications that require traffic prioritization and quality of service.

If we have to compute n up*/down* trees, our criteria is to select the n root nodes that satisfy the following two conditions (and in this order): i) the distance

---

[2] Notice, though, that IBA allows different paths from a given source-destination pair with the virtual addressing feature, but this is beyond the scope of this paper.

between all pair of roots is roughly the same (i.e., the roots are equidistant and far away from each other), and ii) the distance between any pair of the n roots is the longest one. Again, we use a simple algorithm that tries to balance network traffic by randomly selecting the final path to be used.

Finally, notice that the proposal of routing through multiple virtual networks can be generalized to use not only up*/down* and other tree-based routing algorithms but also any deadlock-free routing algorithm. As once a packet enters a virtual network it does not switch to another virtual network, the resulting routing algorithm is deadlock-free.

5 Performance Evaluation

In this section we will evaluate the proposed strategy. For this purpose, we have developed a detailed simulator that allows us to model the network at the register transfer level following the IBA specifications [9]. First, we will describe the main simulator parameters and the modeling considerations we have used in all the evaluations. Then, we will evaluate the use of multiple up*/down* trees under different topologies and different traffic patterns.

5.1 Simulation Model

In the simulator, we will use a non-multiplexed crossbar on each switch. This crossbar supplies separate ports for each VL. We will use a simple crossbar arbiter based on FIFO request queues per output crossbar port, that will select the first request that has enough space in the corresponding output crossbar port. If there is sufficient buffer capacity in the output buffer, the packet is forwarded. Otherwise, the packet must wait at the input buffer. A round-robin arbiter is used to select the next VL that contains a packet to be transmitted over the physical link. The crossbar bandwidth will be set accordingly to the value of the injection rate of the links. Buffers will be used both at the input and the output side of the crossbar. Buffer size will be fixed in both cases to 1 KB.

The routing time at each switch will be set to 100 ns. This time includes the time to access the forwarding and SLtoVL tables, the crossbar arbiter time, and the time to set up the crossbar connection. Links in InfiniBand are serial. The link speed is fixed to 2.5 Gbps. Therefore, a bit can be injected every 0.4 ns. With 10/8 coding [9] a new byte can be injected into the link every 4 ns. We will model 20 m copper cables with a propagation delay of 5 ns/m. Therefore, the fly time (time required by a bit to reach the opposite link side) will be set to 100 ns.

For each simulation run, we assume that the packet generation rate is constant and the same for all the end-nodes. Once the network has reached a steady state, the packet generation rate is equal to the packet reception rate. We will evaluate the full range of traffic, from low load to saturation. Moreover, we will use a uniform distribution of packet destinations in all the evaluations. For some particular evaluations, we will also use the hotspot traffic pattern. Packet length is 32-bytes.

Every packet will be marked with the SL that corresponds to the up*/down* tree that the packet must use. In each switch, the SLtoVL table is configured to assign the same VL to each SL. Packets will continue through the same VL (same tree) until they reach the destination end-node. In all the presented results, we will plot the average packet latency[3] measured in nanoseconds versus the average accepted traffic[4] measured in bytes/ns/switch.

We will analyze irregular networks of 8, 16, 32 and 64 switches randomly generated. These network topologies will be generated taking into account some restrictions. First, we will assume that every switch in the network has 8 ports, using 4 ports to connect to other switches and leaving 4 ports to connect to hosts. Second, each switch will be connected to one switch by exactly one link. Ten different topologies will be generated for each network size. Results plotted in this paper will correspond to the most representative topology for every network size.

5.2   Simulation Results

In this section, we will study in detail the possible benefits of using different up*/down* trees through different virtual networks to increase network performance. In a first study, we will evaluate how network performance can be increased when using two up*/down* trees and selecting the farthest root nodes. This study will take into account different traffic patterns. In a second study, we will consider the use of three up*/down* trees and will evaluate its impact on network performance. In particular, we are interested in detecting under which circumstances it is worth using an additional up*/down* tree through a new virtual network. Finally, in the last study, we will evaluate how effective is the strategy used to select the root nodes (farthest nodes) and how the selection of an alternative pair of root nodes could affect to network performance

5.3   Using Two Virtual Networks

Figures 1.a, 1.b, and 1.c show evaluation results when using one and two up*/down* trees for 16, 32, and 64-switch networks, respectively. Packet size is 32 bytes and a uniform distribution of packet destinations is used. For comparison reasons, when using one up*/down* tree, two virtual lanes are equally used.

We can observe that for the 16-switch network the use of two up*/down* trees increases network throughput by a factor of 1.63. Moreover, when network size increases, benefits are much more noticeable. Network throughput is increased by a factor of 3 for the 32-switch network.

The benefits of using an additional up*/down* tree are mainly due to the better traffic balance achieved over the network. In Figures 2.a and 2.b we can

---

[3] Latency is the elapsed time between the generation of a packet at the source host until it is delivered at the destination end-node.

[4] Accepted traffic is the amount of information delivered by the network per time unit.

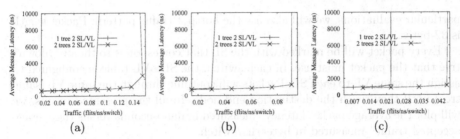

Fig. 1. Average packet latency vs. traffic. Network size is (a) 16, (b) 32, and (c) 64 switches. Packet size is 32 bytes. Uniform distribution of packet destinations.

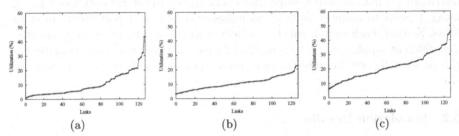

Fig. 2. Link utilization. (a) One up*/down* tree and (b, c) two up*/down* trees. Traffic is (a, b) 0.03 flits/ns/switch and (c) 0.09 flits/ns/switch. Network size is 32 switches. Packet size is 32 bytes. Uniform distribution of packet destinations.

observe the utilization of all the links connecting switches for the 32-switch network when using one and two up*/down* trees, respectively. Links are sorted by link utilization. Network traffic is 0.03 bytes/ns/switch (the routing algorithm with one up*/down* tree is reaching saturation). We can see that when using one up*/down* tree (Figure 2.a) there is a high unbalance of the link utilization. Most of the links have a low link utilization (50% of links with a link utilization lower than 10%), whereas few links exhibit a high link utilization (link utilization higher than 30%). The links with a high link utilization correspond to the links connected to the root node. These links cause a premature saturation of the network.

However, when using two up*/down* trees (Figure 2.b) we can observe that, for the same traffic point, the traffic is better distributed over the links. Almost all the links exhibit a link utilization lower than 20%. This is due to the fact that there are now two root nodes in the network and, on average, the traffic is equally distributed over both root nodes. Figure 2.c shows the link utilization of the routing based on two up*/down* trees at its saturation point (traffic is 0.09 bytes/ns/switch). We can observe that even at this traffic point the routing exhibits a good traffic distribution over the links.

For other random topologies, Table 1 shows minimum, maximum, and average factor of throughput increases for 10 different topologies generated for each network size. We can see how, on average, using two up*/down* trees helps to obtain a higher network throughput. In particular, 8-switch networks increase,

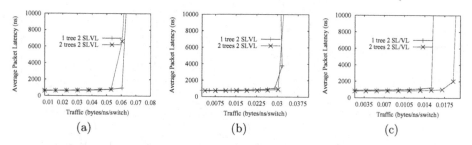

Fig. 3. Average packet latency vs. traffic. Network size is (a) 16, (b) 32, and (c) 64 switches. Packet size is 32 bytes. Hotspot distribution. One hotspot. 20% of hotspot traffic.

Table 1. Factor of throughput increase when using two up*/down* trees. Packet size is 32 bytes. Uniform distribution of packet destinations.

Sw	Min	Max	Avg
8	0.96	1.47	1.18
16	1.01	2.01	1.55
32	1.59	3.00	2.18
64	1.81	2.83	2.14

on average, their throughput by a factor of 1.18. For larger networks, average factor of network throughput increases significantly.

We can see also that for large networks of 64 switches (Figure 1.c) the increase in network throughput when using two up*/down* trees is 1.95 and, on average, is 2.14 (Table 1). This improvement is significantly lower than the improvement obtained with a 32-switch network (Figure 1.b). The reason for this lower improvement will be explained in the next section and basically will be related to the need of using an additional up*/down* tree.

Previous results have been obtained with a uniform distribution of packet destinations. Although this traffic pattern is commonly used, other traffic patterns could be more realistic in these kind of networks. Indeed, in these networks is typical to find some disk devices (TCA) attached to switches that are accessed by hosts (HCA). We model this behavior with a hotspot traffic distribution. With this distribution, all the hosts in the network will send a percentage of the traffic to a particular host (the hotspot). The rest of the traffic will be sent by using a uniform distribution of packet destinations. The hotspot will be randomly selected for each network. Different percentages of traffic sent to the hotspot will be evaluated. We will refer to this traffic as hotspot traffic. Also, different number of hotspots will be used in order to model networks with several attached disks. When dealing with more than one hotspot, the hotspot traffic will be uniformly distributed among all the hotspots.

In Figures 3.a, 3.b, and 3.c, we can observe the performance results when there is one hotspot in 16, 32, and 64-switch networks, respectively. Hotspot traffic is 20%. We can observe that for the 16-switch network, the use of two

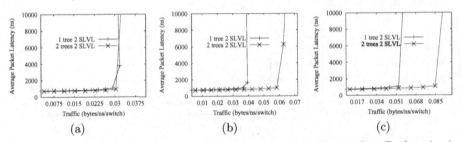

(a)   (b)   (c)

Fig. 4. Average packet latency vs. traffic. Network size is 32 switches. Packet size is 32 bytes. Hotspot distribution. One hotspot. Hotspot traffic is (a) 20%, (b) 10%, and (c) 5%.

Table 2. Factor of throughput increase when using two up*/down* trees. Packet size is 32 bytes. Hotspot traffic. One hotspot. Different percentages of hotspot traffic.

	5%			10%			20%		
Sw	Min	Max	Avg	Min	Max	Avg	Min	Max	Avg
8	0.84	1.63	1.16	0.91	1.59	1.16	0.77	1.31	1.00
16	1.13	1.94	1.48	0.93	1.53	1.18	0.70	1.50	0.99
32	1.52	2.82	1.95	1.30	1.93	1.67	0.78	1.67	1.25
64	1.36	2.84	1.86	1.01	2.10	1.36	0.81	1.86	1.28

up*/down* trees does not help in increasing network throughput. Even, the use of two up*/down* trees may behave slightly worse than when using one up*/down* tree. This is due to the fact that the hotspot may affect differently both up*/down* trees (in the routing with two up*/down* trees), dealing to a sudden saturation of one tree. In particular, one of the up*/down* trees may behave worse than the another one simply because it handles much more traffic due to the hotspot location. This up*/down* tree enters saturation, prematurely leading to the overall network saturation. However, for 64-switch networks the use of two up*/down* trees increases network throughput by a factor of 1.27. In this case, the hotspot does not affect in a severe way any of both up*/down* trees.

On the other hand, if the hotspot traffic is reduced, the benefits of using two up*/down* trees become significant again. Figures 4.a, 4.b, and 4.c show the performance results for the 32-switch network when the hotspot traffic is 20%, 10%, and 5%, respectively. We can observe that for a 10% hotspot traffic the use of two up*/down* trees obtains an increase of network throughput by a factor of 1.5. If the hotspot traffic is much lower (5%), the factor improvement is 1.8.

Table 2 shows the average, minimum, and maximum factors of throughput increase when there is one hotspot in the network with different hotspot traffics and different network sizes. We can observe that, for severe hotspots (20% of hotspot traffic), only large networks of 32 and 64 switches obtain better results with the use of two up*/down* trees. When the hotspot is less severe (10% and 5% of hotspot traffic), the routing that uses two up*/down* trees behaves, on

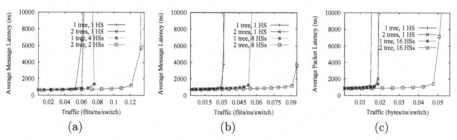

Fig. 5. Average packet latency vs. traffic. Network size is (a) 16, (b) 32, and (c) 64 switches. Packet size is 32 bytes. Hotspot traffic pattern. 20% of hotspot traffic.

Table 3. Factor of throughput increase when using two up*/down* trees. Packet size is 32 bytes. Hotspot traffic. Several hotspots. 20% of hotspot traffic.

Sw	HS	Min	Max	Avg
8	2	0.90	1.19	1.03
16	4	1.20	1.75	1.49
32	8	1.41	2.57	1.93
64	16	1.60	2.61	2.02

average, much better than the routing with only one up*/down* tree for every network size.

However, as more hotspots appear in the network (modeling more disks), the negative effect is diminished. We can see in Figures 5.a, 5.b, and 5.c the network performance when there is more than one hotspot in the network, for 16, 32, and 64-switch networks, respectively. We can observe that, as more hotspots appear in the network, higher throughput values are obtained when using the routing with two up*/down* trees. In particular, for 16-switch networks with 4 hotspots (Figure 4.a) network throughput is increased by a factor of 1.65. For 32 and 64-switch networks with 8 and 16 hotspots, respectively, the factor of throughput improvement ranges from 1.64 to 2.5.

Table 3 shows average, minimum, and maximum factors of throughput increase for different number of hotspots. We can observe that when there is more than one hotspot in the network, the routing with two up*/down* trees significantly increases network throughput.

5.4 Using an Additional Virtual Network

In this section we focus on the evaluation of the routing algorithm that uses an additional virtual network. This virtual network will be used by a new up*/down* tree. The selection of the root node will be done in the same way as the previous case. That is, the three chosen root nodes will be at the same distance each other and at the farthest positions. We are interested in obtaining the trade-off between throughput improvement and the use of an additional virtual network (SL/VL).

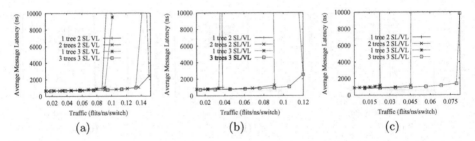

Fig. 6. Average packet latency vs. traffic. Network size is (a) 16, (b) 32, and (c) 64 switches. Packet size is 32 bytes. Uniform distribution of packet destinations.

Table 4. Factor of throughput increase when using three up*/down* trees. Packet size is 32 bytes. Uniform distribution of packet destinations.

	3 trees vs 2 trees			3 trees vs 1 tree		
Sw	Min	Max	Avg	Min	Max	Avg
8	1.00	1.20	1.12	1.03	1.47	1.22
16	0.95	1.23	1.08	1.09	2.17	1.61
32	0.96	1.60	1.24	1.70	3.42	2.58
64	1.21	1.78	1.52	2.63	3.61	3.22

Figures 6.a, 6.b, and 6.c show the performance results when one, two, and three up*/down* trees are used for 16, 32, and 64-switch networks, respectively. Packet size is 32 bytes and a uniform distribution of packet destinations is used. When using one up*/down* tree, and for comparison reasons, two and three virtual lanes are equally used.

We can observe that, when using an additional up*/down* tree, improvements are lower except for the 32 and 64-switch networks. In particular, network throughput is only increased by a factor of 1.15 for the 16-switch network. The use of an additional up*/down* tree does not help in small networks because the use of two up*/down* trees is enough to balance the network in an efficient way.

However, in medium-sized and large networks (32 and 64 switches), the use of an additional up*/down* tree helps to increase significantly network throughput. In particular, network throughput is increased by factors of 1.33 and 1.78 for 32 and 64-switch networks, respectively.

Table 4 shows minimum, maximum, and average factors of throughput increase when using three up*/down* trees. We can observe how, on average, the benefits are noticeable only for 32 and 64-switch networks. Throughput is increased, on average, by a factor of 2.58 and 3.22 for 32 and 64-switch networks, respectively. We can observe that it is not worth for small and medium networks the use of an additional virtual network (SL/VL) to increase network throughput. However, for larger networks is highly recommended as network throughput is even tripled, on average.

To sum up, the use of an additional up*/down* tree is highly related to the network size and it is worth using it only with large networks at the expense of consuming an additional SL/VL resource.

5.5   Selection of the Root Nodes

In this last study we evaluate how the way we choose the root nodes for each up*/down* tree may affect the achieved network performance. In particular, we are interested in foreseeing the throughput that could be achieved by using an alternative pair of root nodes and knowing if it is worth making a deeper study about the selection of root nodes.

For this, we have computed all the possible routings for every possible pair of root nodes for a particular 16-switch network. For each routing (240 routings) we have simulated its behavior and obtained its throughput for a uniform distribution of packet destinations and with 32-byte packets. Figure 7 shows all the achieved throughput values. Every pair of possible roots are shown in the X-bar and they are sorted by throughput. We have also plotted the throughput achieved when considering the farthest root nodes (0.0145 bytes/ns/switch).

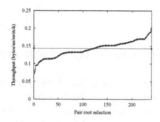

Fig. 7. Average packet latency vs. traffic. Network size is 32 switches. Packet size is 32 bytes. Uniform distribution of packet destinations.

As we can see, there is a high variation of network throughput depending on the selection of the root nodes. We can obtain a lower throughput (0.07 bytes/ns/switch) with some particular root nodes as well as a higher throughput (up to 0.2 bytes/ns/switch).

Notice that we are in the middle way when using the farthest root nodes in the topology. Therefore, with a clever selection of the root nodes we could even increase the network throughput by an additional factor of 1.40. As a conclusion, it is worth undertaking a deeper study of the selection of the root nodes.

6   Conclusions

In this paper, we have proposed a new routing methodology to improve IBA network performance by using virtual lanes in an efficient way. This routing methodology does not use virtual lanes just to reduce the head-of-line blocking effect on the input buffers. Rather, virtual lanes are viewed as separate virtual networks and different routing algorithms are used at each virtual network. In particular, different up*/down* trees are used at each virtual network. With this we try to achieve a better traffic balance and to allow most packets to follow minimal paths.

By using just two virtual networks (2 SL/VL) it is possible to increase network throughput by average factors of improvement of 1.55 and 2.18 for 16 and 32-switch networks, respectively. Only for very large networks (64 switches) it is worth using an additional virtual network (3 SL/VL). In particular, 64-switch network throughput is tripled. All of these results have been obtained assuming a straightforward strategy to select the root nodes and the final set of paths.

We have also evaluated the selection of alternative pairs of root nodes. We have seen that a better selection of the root nodes could lead to a better network performance. Therefore, it is worth deepening more in the selection of the root nodes.

As future work, we plan to study better criteria to select the appropriate number of root nodes and their locations in the network. Also, we plan to evaluate the use of better routing algorithms and the combination of different routing algorithms.

References

1. N. J. Boden et al., Myrinet - A gigabit per second local area network, *IEEE Micro*, vol. 15, Feb. 1995.
2. L. Cherkasova, V. Kotov, and T. Rokicki, "Fibre channel fabrics: Evaluation and design," in *Proc. of 29th Int. Conf. on System Sciences*, Feb. 1995.
3. W. J. Dally, Virtual-channel flow control, *IEEE Transactions on Parallel and Distributed Systems*, vol. 3, no. 2, pp. 194-205, March 1992.
4. W. J. Dally and C. L. Seitz, Deadlock-free message routing in multiprocessors interconnection networks, *IEEE Transactions on Computers*, vol. C-36, no. 5, pp. 547-553, May. 1987.
5. J. Duato, A. Robles, F. Silla, and R. Beivide, A comparison of router architectures for virtual cut-through and wormhole switching in a NOW environment in Proc. of *13th International Parallel Processing Symposium*, April 1998.
6. J. Flich, P. Lopez, M.P. Malumbres, J. Duato, and T. Rokicki, "Combining In-Transit Buffers with Optimized Routing Schemes to Boost the Performance of Networks with Source Routing," *Proc. of Int. Symp. on High Performance Computing*, Oct. 2000.
7. D. García and W. Watson, Servernet II, in *Proceedings of the 1997 Parallel Computer, Routing, and Communication Workshop*, Jun 1997.
8. InfiniBand^{TM} Trade Association, http://www.infinibandta.com.
9. InfiniBand^{TM} Trade Association, *InfiniBand^{TM} architecture. Specification Volumen 1. Release 1.0.a.* Available at http://www.infinibandta.com.
10. P. Kermani and L. Kleinrock, Virtual cut-through: A new computer communication switching technique, *Computer Networks*, vol. 3, pp. 267-286,1979.
11. C. Minkenberg and T. Engbersen, A Combined Input and Output Queued Packet-Switched System Based on PRIZMA Switch-on-a-Chip Technology, in *IEEE Communication Magazine*, Dec. 2000.

12. P. López, J. Flich, and J. Duato, Deadlock-free Routing in InfiniBand^{TM} through Destination Renaming, in *Proc. of 2001 International Conference on Parallel Processing (ICPP'01)*, Sept. 2001.
13. G. Pfister, *In search of clusters*, Prentice Hall, 1995.
14. W. Qiao and L. M. Ni, "Adaptive routing in irregular networks using cut-through switches," in *Proc. of the 1996 International Conference on Parallel Processing*, Aug. 1996.
15. J.C. Sancho, A. Robles, and J. Duato, Effective Strategy to Compute Forwarding Tables for InfiniBand Networks, in *Proc. of 2001 International Conference on Parallel Processing (ICPP'01)*, Sept. 2001.
16. J.C. Sancho and A. Robles, Improving the Up*/down* routing scheme for networks of workstations in *Proc. of Euro-Par 2000*, Aug. 2000.
17. J.C. Sancho, A. Robles, and J. Duato, New methodology to compute deadlock-free routing tables for irregular networks, in *Proc. of CANPC'2000*, Jan. 2000.
18. J.C. Sancho, J. Flich, A. Robles, P. López and J. Duato, "Analyzing the Influence of Virtual Lanes on InfiniBand Networks," submitted for publication.
19. F. Silla and J. Duato, Tuning the Number of Virtual Channels in Networks of Workstations, in *Proc. of the 10th International Conference on Parallel and Distributed Computing Systems (PDCS'97)*, Oct. 1997.
20. M. D. Schroeder et al., Autonet: A high-speed, self-configuring local area network using point-to-point links, *SRC research report 59*, DEC, Apr. 1990.
21. R. Sheifert, *Gigabit Ethernet*, Addison-Wesley, April 1998.

Minerva: An Adaptive Subblock Coherence Protocol for Improved SMP Performance[*]

Jeffrey B. Rothman[1] and Alan Jay Smith[2]

[1] Lyris Technologies, Berkeley, CA 94704, USA
jeffrey@lyris.com
[2] Computer Science Division, University of California, Berkeley, CA 94720, USA
smith@cs.berkeley.edu

Abstract. We present a new cache protocol, **Minerva**, which allows the effective cache block size to vary dynamically. **Minerva** works using sector caches (also known as block/subblock caches). Cache consistency attributes (from the MESI set of states) are associated with each 4-byte word in the cache. Each block can itself have one of the attributes invalid, exclusive or shared. Each block also has a current subblock size, of 2^k words and a confidence value for hysteresis. The subblock size is reevaluated every time there is an external access (read or invalidate) to the block. When a fetch miss occurs within a block, a subblock equal to the current subblock size is fetched. Note that the fetch may involve a gather operation, with various words coming from different sources; some of the words may already be present.

Depending on the assumed cache sizes, block sizes, bus width, and bus timings, we find that **Minerva** reduces execution times by 19–40%, averaged over 12 test parallel programs. For a 64-bit wide bus, we find a consistent execution time reduction of around 30%. Our evaluation considers the utility of various other optimizations and considers the extra state bits required.

1 Introduction

Symmetric multiprocessor (SMP) systems with a small number of processors are becoming increasingly popular. These bus-based systems suffer performance loss due to "traffic jams" on the single interconnect between the processors and memory. This excessive traffic is caused by two problems, false sharing and dead sharing. False sharing results from the enforcement of cache consistency (coherence) between shared blocks, causing coherence operations to take place, even though there are no individual words that are actually shared. Dead sharing is a result of block granularity cache coherence enforcement, in that most of the data transferred between caches is not referenced before being invalidated by another cache.

Our examination of multiprocessor workloads in [31] showed that although most cache blocks behave well, most of the misses (and thus traffic) are caused by a few badly behaving blocks. The reference patterns of the frequently accessed blocks had some interesting patterns (for 64-byte blocks): (1) only a single word has been referenced in a block between the time the block was fetched and by the time it was invalidated

---

[*] Funding for this research has been provided by the State of California under the MICRO program, and by AT&T Laboratories, Cisco Corporation, Fujitsu Microelectronics, IBM, Intel Corporation, Maxtor Corporation, Microsoft Corporation, Sun Microsystems, Toshiba Corporation, and Veritas Software Corporation.

H. Zima et al. (Eds.): ISHPC 2002, LNCS 2327, pp. 64–77, 2002.

in 40 percent of blocks; (2) around 60 percent of the invalidated blocks that are dirty have only a single dirty (modified) word. Of the blocks that had more than 2 words written, the next largest number of words written was 16 (the maximum number per block).

To address this problem, we present our new coherence protocol **Minerva**, which employs dynamically sized subblocks to reduce the effects of false and dead sharing. **Minerva** is evaluated by comparing it to schemes with fixed (sub)block invalidation and fetch sizes using traces of 12 parallel programs. In addition, we also examine read-broadcasting (snarfing) and bus read-sharing (sharing the results of pending read transactions between multiple caches).

Minerva functions by tracking consistency on a word basis using MESI states, and a block state to determine whether the block is invalid, present in only one cache (exclusive) or possibly in multiple caches (shared). Each block also maintains a *current* subblock size, of 2^k words and a confidence value for hysteresis. The current subblock size (and confidence level) is reevaluated every time there is an external access (read or invalidate) to the block. When a fetch miss occurs within a block, a subblock equal to the current subblock size is fetched. Note that the fetch may involve a gather operation, with various words coming from different sources; some of the words may already be present.

Despite the apparent complexity of the protocol, it can be implemented with fairly simple combinational logic. The nature of **Minerva** makes it easy and convenient to also implement optimizations such as read-sharing, read-broadcast (snarfing), and write-validate. For non-parallel workloads, **Minerva** just converges to the Illinois protocol on 64-byte blocks, so performance for conventional workloads is also high. Using the **Minerva** protocol in combination with other simple cache optimizations, we find that the execution time is reduced by approximately 30% for a 64-bit wide shared bus, and beyond 40% for a 32-bit wide bus.

The remainder of this paper is organized as follows: Section 2 provides some background on cache memories and the issues involved in cache coherency for multiprocessor systems and describes previous published subblock coherence schemes. Section 3 describes our workloads and our simulation methodology. Section 4 introduces our new protocol **Minerva**. The results and comparisons of the simulations are provided in Section 5. We present our conclusions in Section 6. An extended version of this paper is available as [28], where we provide additional consideration of issues such as bus utilization, the effect of other performance improvements, a breakdown of misses by type, a breakdown of delays by cause, the performance of **Minerva** for various cache sizes, a logic implementation of Minerva, and additional information on cache implementations in commercial products.

2   Background

Much research on cache coherence protocols for SMP systems has focussed on determining the best protocol for fixed block sizes [3, 11, 14, 22, 24]. It has been found that invalidation protocols, such as Illinois [26] typically outperform update protocols, but protocols that can dynamically switch between update and invalidate can do slightly better [14], and other protocol extensions, such as sequential prefetching, can provide significant performance improvement [6]. None of the fixed-block protocols, however, have measured the impact of reducing false sharing on execution time, although [8] analyzes the effect of several protocols on the number of shared memory misses.

Some research attacking the false sharing problem has focussed on associating several coherence/transfer units (subblocks) with each address tag. This type of design has been studied extensively for uniprocessor systems [18, 23, 27, 30, 32, 33]. Several designs have been proposed for multiprocessor systems using a fixed size subblock [1, 5, 16], showing good performance improvement over systems using block size coherence granularity. Variable size block coherence was investigated in [7], showing better performance than fixed size blocks for most workloads, but they were unable to simulate the effect of their protocol on execution time and no implementation details were provided. One dynamically adjustable subblock protocol allowed a block to be divided into two (possibly unequal) pieces for coherence purposes, with a slight improvement in performance [21].

2.1   Sharing Issues

False sharing occurs when blocks are shared due to block granularity coherence, but the individual words are not shared. This was identified as a problem in [12]. It has been investigated in a number of papers such as [4]. [9, 37] found that false sharing is generally not the major source of cache misses. Attempts to restructure data to avoid false sharing were found to have some success in [10, 19, 31, 36].

Our analysis of sharing patterns in [31] using an invalidation-based protocol with an infinite cache showed that false sharing is the largest source of misses for some programs; in those cases false sharing becomes a larger source of misses than true sharing and cold start misses once the block size is sufficiently large (16 bytes and greater). We found also that approximately 80 percent of bus traffic consists of data that is unused before being invalidated, but is required to be transferred because of block granularity coherence. This effect, which we call *dead sharing*, is mostly closely associated with false sharing, although it can be caused by true sharing (e.g., one shared word and 15 inactive words in a block).

3   Methodology

Our work is based on TDS (trace-driven simulation). Initially our research used execution-driven simulation (EDS), which is slightly more accurate [15]. However we found that results could be generated much more quickly using modern PCs and workstations (using compressed traces generated by our EDS tool **Cerberus** [29]) than on the slower DEC5000 workstations on which our EDS system depends. To keep our simulations as accurate as possible, synchronization objects (barriers and locks) are simulated at run-time. Examination of some of the key data points simulated by both EDS and TDS showed extremely similar results.

3.1   Workload Characteristics

We examined a variety of parallel programs (12) to provide the results for this paper. Ten of the programs (barnes, cholesky, fmm, locus, mp3d, ocean, pthor, raytrace, volrend, water) come from the SPLASH 1 and 2 suites from Stanford, which have been available to the research community as a *de facto* benchmark for comparing parallel program execution. These programs have all been used in a number of papers analyzing parallel code performance, and are described and characterized in more detail in

[31, 34, 38]. The other two programs (**topopt** and **pverify**) were created by the CAD group at U.C. Berkeley, and used for measurements at Berkeley and the University of Washington [2, 10, 12, 13]. These workloads may considered to be dated; however, (1) they are still popular for researching parallel computation; and (2) they represent the more typical effort of regular programmers who have not optimized their code to minimize coherence traffic.

Program Characteristics								
Programs	References (Millions)		Data Space (KBytes)		Fraction of Data References				
			Shared	Private	Shared			Private	
	Inst	Data	(KBytes)		Reads	Writes	Locks	Reads	Writes
barnes	114.23	42.31	33.02	34.64	0.16	0.00	0.00	0.46	0.38
cholesky	90.67	34.32	970.02	783.38	0.50	0.07	0.01	0.29	0.13
fmm	288.30	166.82	380.41	460.14	0.14	0.01	0.00	0.35	0.51
locus	805.62	164.45	1405.70	1151.79	0.56	0.02	0.00	0.26	0.16
mp3d	174.88	60.82	701.91	181.53	0.32	0.22	0.00	0.30	0.16
ocean	234.09	92.37	140.16	984.45	0.26	0.03	0.01	0.56	0.13
pthor	275.86	97.76	1233.09	1026.75	0.38	0.05	0.04	0.35	0.18
pverify	181.32	55.24	23.08	149.67	0.47	0.02	0.01	0.32	0.19
raytrace	471.08	196.94	667.30	2144.09	0.32	0.00	0.00	0.43	0.25
topopt	655.75	141.60	19.22	38.76	0.81	0.09	0.00	0.08	0.02
volrend	351.62	79.92	395.61	2340.98	0.48	0.01	0.00	0.29	0.23
water	366.23	127.67	44.51	102.37	0.18	0.02	0.00	0.58	0.23
	Total				Average				
Overall	4009.6	1260.2	6014	9399	0.38	0.04	0.01	0.35	0.21

Table 1. Reference characteristics of programs for 16-processor simulation.

The characteristics of the workload are summarized in Table 1. The fraction of shared accesses (a shared access being a reference to a block that is accessed by multiple processors) has quite a large variation; it ranges from 0.15 in **fmm** to 0.90 in **topopt** with an average value of 0.43. Due to memory space limitations in our trace generation machines, 16 processors was the maximum number we could simulate for all workloads. The runs represent full execution of a problem, to capture the entire behavior of the program. We picked 16 processors for our experiments to provide an example of high sharing contention for (hotspot) blocks and the common bus.

3.2   Bus Design and Timings

Our simulation testbed consists of 16 processors (MIPS architecture) communicating with an interleaved shared memory over a split-transaction 32- (and 64-bit) wide common bus. Each processor has its own cache, which snoops the bus for coherency information. The caches are dual-ported, to support snoopy cache (bus watching) operations without conflicting with the processors' access to their caches. We simulate bus transactions cycle-by-cycle to provide the highest level of accuracy and to properly model contention. We model simple processors: they are single issue, and single cycle instruction execution. Because much of the processors' time is spent stalled waiting for data or synchronization, the effect of increasing the instruction-level parallelism would show better relative performance for the **Minerva** protocol [25].

Our bus design incorporates the major features of modern high-performance busses. Table 2 shows the timing of the various bus operations. Times shown are in bus cycles, which are 4 processor cycles in duration. The timings we use are typical of a contemporary system with 400MHz processors, a 100 MHz shared bus, and 60ns SDRAM (6 bus cycle latency for the first word(s) with data available every bus cycle after that). The following paragraphs contain some of the operational details of our simulated bus.

To perform a read transaction on the bus, the processor must first arbitrate for the bus, except when it was the last agent to use the bus (bus parking). If the bus is busy

Transaction	Number of Bus Cycles
Bus Arbitration	0/2/4 Cycles
Initiate Read	1 Cycle
Cache-to-Cache Read	1 Cycle per Bus-Width
Memory Latency	6 Cycles
Transfer Time	1 Cycle per Bus-Width
Write-back	1 Cycle per Dirty Bus-Width
Invalidation	1 Cycle
Lock operation	Read + 1 Cycle

Table 2. Duration of bus transactions. All transactions except memory response are preceded by bus arbitration.

when a bus request is made, the arbitration is pipelined and the bus requires only two cycles to grant bus ownership once the current bus transaction has ended. Four bus cycles are required to gain ownership when the bus is empty and the processor does not currently have ownership.

If one or more of the other caches has the requested data (which can be as large as a block or as small as a word), the data is immediately transferred between the caches, gathering the data from multiple caches if necessary (possibly with other caches passively receiving the data if *snarfing* is enabled). When cache-to-cache transactions occur, the request to main memory is automatically aborted. If the particular word that caused the cache miss is not available in any cache, the main memory responds to the request, after a delay of six bus cycles. The final phase of data transfer requires 1 bus cycle for each bus-width of data transferred. In the case of a read-modify-write instruction (LOCK), the bus is held for an additional bus cycle to indicate whether the lock was successful and the transferred subblock should be invalidated in other caches. Some additional bus transactions specific to **Minerva** are described in Section 4.

3.3   Cache Structures

The simulated caches are incorporated on-chip with the processor, leading to single (processor) cycle access times. Two-level caches were not simulated in the system because the workload would not significantly exercise the second level of cache; a single level is sufficient for these programs. Much of our focus is on bus transactions; the levels of cache are irrelevant for these comparisons. Instruction caches were ignored (i.e., instruction references always hit in the cache) for these simulations as the instruction miss ratios for these programs are very small and would have little effect on the execution times, while simulating the instruction cache would have significantly increased the time to evaluate each data point. The simulations used fully-associative caches to eliminate conflict misses caused by data-mapping artifacts in the data space, so the misses reported only contain cold-start, capacity and (mostly) coherence misses.

For the standard or base case, our experiments used the Illinois [26] write-invalidate-based protocol with fixed block size in a normal (non-sectored) cache. The Illinois protocol also forms the basis of the state changes of **Minerva** at the word level. We chose the Illinois protocol for two reasons: (1) it is an invalidation-based protocol, which generates less bus traffic than update-based protocols [22] and generally outperform update-based protocols [14]; (2) write-invalidate protocols like the Illinois protocol are the most popular class of protocols that are actually implemented in real systems [17, 35], which makes them a more attractive target for performance improvement.

In addition to our dynamic subblock size invalidation protocol **Minerva**, fixed size subblock (sector cache) and full block protocols were simulated. The fixed size subblocks that were simulated ranged from 4 bytes up to the (64-byte) block size. The results we present show 64-byte blocks using the Illinois protocol (full block invalidation), 8- and 16-byte subblocks and the **Minerva** adaptive protocol. The cache sizes we tested were 4 to 128 Kbytes; for the most part we present the results for the 64 Kbyte caches. All the protocols measured here include the simple cache transaction optimizations (read-sharing and read-broadcast) we evaluated in [28].

4   Minerva Protocol Description

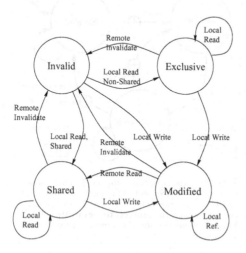

Fig. 1. State transitions for coherence units using the Illinois protocol. These states are used in all the protocols evaluated.

Minerva is an invalidation-based protocol that maintains coherence state for each word in a block, which allows it to perform variable size fetch and invalidation transactions, but correctly track the state of each word. The size of the transactions are based on each block's estimate of the most appropriate subblock size. Each word can independently be present or absent in the cache, as in sector caches for uniprocessors [32].

Figure 1 shows the word state transitions, which are the same as the block transitions for the Illinois protocol [26]. Each word (or subblock for fixed size subblock protocols we evaluated) within a block has one of the MESI states associated with it (**modified (m)**, **exclusive (e)**, **shared (s)**, and **invalid (i)**), depending on whether the data is different than the main memory value, only in one cache but consistent with memory, potentially shared among several caches, or not a valid word, respectively.

A block state is associated with each block as a whole, to track if copies of the block exist in other caches (Figure 2). If the block is in the **exclusive** state, the block (i.e., words from the block) is only in one cache. This allows a write-validate operation, by which words can be overwritten without first requiring the block be fetched from main memory [20]. If the block is in the **shared** state, then write-validate is disabled.

When a cache observes an external event (other than write-backs) associated with a block it contains, it evaluates the minimum size subblock that contains all the most recently written words for that block (transitions in Figure 2 with the ⊕ symbol). Associated with each block is the current subblock size and the confidence level. The current subblock size consists of 2^k words, aligned on a 2^k word boundary. The confidence level is incremented (up to a saturation level, 3 for our experiments) if the newly calculated subblock is the same size as the old subblock. Otherwise, the confidence level is decremented. Once the confidence level is 0, the new subblock size replaces the old subblock size. The confidence level aids in settling the subblock size on the most commonly measured value.

The initial subblock size is 4 words (16 bytes), which was determined to be the best size by simulation. However, when a miss occurs to a block that is not present in any cache, the entire block is fetched. This is the only case in which the subblock size is ignored; it speeds-up fetching of non-shared data. Misses to a block that is present in other caches fetch only the requested subblock, whether the data must come from the caches or from main memory. It is possible to determine the amount of data that must be retrieved from main memory based on the responses of the caches in the cycle after the request address and request vector is presented to the bus.

One important issue for the **Minerva** protocol is that the data for fetch misses may come from multiple bus agents, which is not an issue for fixed-size subblock protocols. When a fetch miss occurs, the processor issuing the fetch provides the address of the missing word and a bit vector of the particular subblock pieces it wants over the bus. If a requested (target) word is available from another cache, that word and as much of the subblock as is available is immediately supplied by other caches; since word state is maintained by the caches, unavailable words are just marked invalid. If the target missing word is not in any of the other caches, the main memory responds somewhat later with the missing data; since the caches and the memory "know" which words are available from the caches, the memory supplies only the words unavailable in any of the caches.

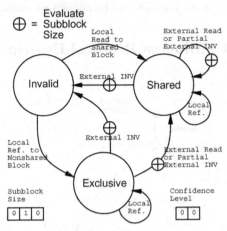

Fig. 2. State transitions for blocks in the adaptive **Minerva** protocol. Any observed read or invalidate bus transactions to the block triggered remotely cause the protocol to reevaluate the subblock size based upon the number of words written locally.

Caches supply the particular words they have in a cooperative manner, so the entire request is fulfilled with the latest values. It is possible that multiple agents will containing the same word will drive the bus at the same time; however, only shared data can be available in multiple caches, so the values driven must be identical. Since the data for a block is transferred in a fixed order, it is not a significant issue for caches to synchronize and contribute their parts of the block over the bus at the appropriate time.

The protocol is aided by letting blocks that have no remaining valid subblocks stay in the cache until removed by the natural LRU block eviction process, to keep behavioral information intact. For frequently accessed shared blocks, there is a reasonably good possibility that the block will be fetched back into the cache again and the learned behavioral information will still be available. Experiments with smaller (16K) caches [28] show that some invalid blocks remain in the cache long enough to be reused by blocks with the same address tag (thus having access to behavioral data); however, under more strenuous workloads the invalid blocks may not survive long enough to be useful. Note that this type of behavior is a peculiarity of the particular workload we study - parallel shared memory programs. Such programs frequently have different processors operating on the same data over short time intervals. Other workloads would likely show very different behavior.

4.1   Implementation Details

Figure 3 provides a schematic of the information required to maintain coherence using (a) block granularity; (b) subblock granularity; and (c) the extra requirements for the

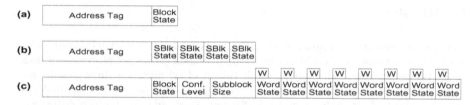

Fig. 3. Information required for each block for (a) normal Illinois, (b) Illinois subblock coherence, and (c) the adaptive protocol. For case (c), the W bit tracks recent writes to the words.

Minerva protocol. The Illinois protocol (Figure 3a) requires the least number of bits, needing only the address tag (required by all protocols) and 2 bits to track the block state. For the pure subblock protocols (Figure 3b), 2 state bits are required for each subblock. For the adaptive protocol, a fair amount of additional information is required to provide the ability to dynamically change the subblock size (Figure 3c). In addition to the 2 bits to track the state for each word, the state of the block as a whole is tracked (whether it is exclusive, invalid, or shared), $\lceil log_2(1 + log_2 \frac{block\ size}{4}) \rceil$ bits are used to track the subblock size (e.g., 3 bits for 64-byte blocks with five possible subblock sizes: 4, 8, 16, 32, and 64 bytes), and 2 bits as the confidence level (maximum value 3).

Associated with each word in each cache block (in our adaptive protocol) is a W (write) bit (Figure 3c) that tracks the words that have been written since the last external event to the block was observed over the bus (as defined in Figure 2). The intent is to provide the most accurate idea of what the adaptive subblock size should be. The W bit is set whenever a write occurs to that word, and cleared after the subblock size has been evaluated.

A significant advantage the 64-byte block Illinois protocol has over the others presented here is the cost of implementation, in terms of tag bits and consistency and replacement logic. For a 64K 4-way set-associative data cache with 64-byte blocks and 32-bit addresses, full-block Illinois requires 21.5 bits for each block (18 address tag, 2 state bits, and 6 LRU (least recently used) bits per set). 64-byte block **Minerva** requires 74.5 bits (18 address tag bits, 34 state bits, 16 W bits, 5 adaptive bits and 6 LRU bits for each set). This increases the number of bits associated with the data cache by 9.9 percent or about 6.78K bytes over the Illinois implementation. Since modern caches are likely to use longer addresses as well as set-associative cache organizations, the extra overhead per cache block will actually be less than the 9.9 percent estimated here. Based on the performance improvement the adaptive cache provides by reducing execution time and data traffic, it is well worth the investment in the extra bits.

An important feature of our dynamic subblock evaluation algorithm (and any useful cache coherence protocol) is that it is implementable in hardware. Determining the subblock size requires two gate levels of logic; a priority encoder then finds the minimum subblock size that spans all the written bits. The circuitry to compare the new subblock size with the old subblock size and determine whether to increment (or decrement) the confidence level requires another 4 gate levels of logic. Additionally some logic is required to increment or decrement the confidence level and to select which subblock size. So the entire logic to evaluate this algorithm is relatively small and can easily be performed in parallel with other actions occurring to the block (such as state changes, invalidation of data, etc.).

5   Results

This section presents the results of our multiprocessor simulations. We evaluate 4 different protocols using 64-byte blocks with 64K byte data caches per processor for each of the twelve programs. The four protocols consist of (1) normal 64-byte Illinois protocol (labeled **N** in the figures); (2) a subblock protocol using eight-byte subblocks (**E**); (3) a sixteen-byte subblock protocol (**S**); and (4) our adaptive protocol **Minerva**, which has an initial subblock size of sixteen bytes (**A**). These labels are consistently used for the figures in the following sections. The subblock protocols fetch and invalidate data strictly using fixed subblock size operations. The subblock states and state transitions are identical to the Illinois protocol (Figure 1). Each of the protocols evaluated in this section incorporate the simple performance enhancements described in [28]. These enhancements consisted of read-broadcast (a.k.a. *snarfing*) and bus read-sharing, which allows multiple outstanding read requests to combine and share results of a single request. Bus read-sharing reduced execution time by 4.4 percent on average, snarfing reduced it by 11.9 percent; together they reduced execution time by 14.5 percent (arithmetic average) for 64 Kbyte caches with 64-byte blocks.

Our simulations model all the events which occur between the processors and main memory (excluding instruction fetches). We simulated 16 processors for each workload, which is the maximum number that our trace generator could provide for all workloads.

5.1   Execution Time

Figure 4 shows the breakdown of processor execution time for 64K byte (per processor) data caches. The time is broken down into uniprocessor execution (actually performing calculations), uniprocessor stall (waiting for the main memory to respond), multiprocessor time: execution, read stall, write stall, and invalidation stall. Synchronization stall is also measured, which is the time the average processor spends waiting to acquire locks and waiting at barriers. From this figure, we can see that neither the 8- nor the 16-byte subblock consistently are the best fixed subblock size, nor do they always outperform the normal (no subblock) case. **Minerva** outperforms the full-block Illinois protocol in 11 out of 12 cases (slower by 1.4 per-

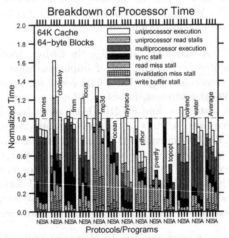

Fig. 4. Execution time, normalized to full-block coherence, 64-byte blocks, 64K cache.

cent for **cholesky**) and beats the fixed subblock sizes for all of the programs. On average, **Minerva** reduces execution time by 25.2 percent over the 64-byte full-block protocol for 64K byte caches (using arithmetic averages). It does this by reducing the stall time for both invalidations and read fetches by about half. If we consider the snarfing and bus read-sharing discussed in [28], the improvement over the original unoptimized 64-byte block Illinois coherence protocol exceeds 33 percent on average. This is a tremendous amount of improvement that can take place without having to recode any of the programs. Other improvements such as weaker consistency models (which may require recoding the programs) can also be done in addition to these hardware changes, potentially providing a greater performance boost.

Synchronization stall time is fairly prominent in some of the programs. Much of this time is due to stalling at barriers when the workload is not well distributed. For example, **cholesky** uses a set of barriers to allow the master processor (processor 0) to execute in single processor mode in the middle phase of the calculation. This has the effect of wasting in excess of 15 percent of total execution time, which for that workload, exceeds the time that the processors spend actually performing calculations in multiprocessor mode.

Note that we use arithmetic averages in Figure 4 and other similar figures. As we will explain later in Section 5.3, the geometric average is better for calculating the effect of various improvements in the execution time, but the arithmetic average is more suitable for these figures. However, this tends to underestimate the real performance improvement using the **Minerva** protocol, which reduces the average execution time by more than 40 percent when the proper measure of central tendency (geometric) is used. Due to space limitations, additional performance metrics have been removed, but can be found in [28].

5.2   Performance Costs

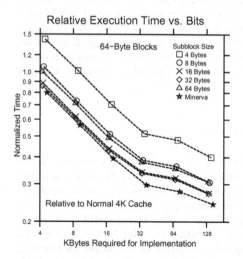

Fig. 5. Relative execution time for a given number of bits, 64-byte blocks (1.0=4 Kbyte cache, 64-byte block, 64-byte subblock).

Figure 5 shows the performance of various cache configurations, showing the number of bits (data, tag and other overhead) for 4 to 128 Kbyte caches, 64-byte blocks, and 4- to 64-byte subblocks. For each (approximately) vertical grouping, the cache sizes are the same; the variation in the horizontal direction is caused by differences in the number bits required for address tags and state bits. 4-byte subblocks generally have the worst performance for a given block size, 16- and 32-byte subblocks perform the best of the fixed subblock sizes, 8-byte and 64-byte subblocks have similar performance. The adaptive caches (indicated with the ★ symbols) show the best performance overall. The performance improvement using **Minerva** protocol over the normal Illinois protocol is sufficient to outperform caches two to four times as large (for 16K caches and beyond). Even though the **Minerva** protocol requires more bits to manage a given amount of data space, it manages the cache so well that it is worth devoting the extra few percent of bits required.

5.3   Sensitivity Analysis

To project this design to future architectures, it is necessary to determine how sensitive these results are to different timing parameters. For example, future designs will use

Sensitivity of Performance to System Parameters								
CS	BW	BFreq	TpC	PFreq	DD	SOR	MR	Comments
64K	32	100	1	400	60	16.0%	41.2%	**Standard Configuration**
16K	32	100	1	400	60	11.5%	38.3%	Small Cache
8K	32	100	1	400	60	6.5%	19.4%	Tiny Cache
64K	64	100	1	400	60	13.8%	29.6%	Wider Bus
16K	64	100	1	400	60	10.7%	27.0%	Smaller Cache and Wider Bus
64K	64	200	1	1000	50	14.5%	31.5%	1GHz Processor With Fast Memory
64K	64	300	2	1200	50	13.0%	28.6%	1.2GHz Processor With RAMBUS

Table 3. Execution time reduction vs. unoptimized Illinois with variation of simulation parameters (CS: cache size, BW: bus width in bits, BFreq: bus frequency in MHz, TpC: transfers per cycle, PFreq: processor frequency in MHz, DD: DRAM delay in ns, SOR: simple optimization reduction, MR: **Minerva** reduction).

more advanced DRAMs that have higher bandwidth, wider busses, and faster processors. Table 3 shows the effects of changing various memory system parameters. Unlike the figures and tables in in the previous sections which use arithmetic averages because of the desire to show such features as the breakdown of execution time and bus utilization, Table 3 shows the geometric mean of the speedups achieved using various system parameters. Regressions we performed using the geometric average show a much better fit than those using the arithmetic average (the R^2 measurement of fit with value 0.76 for geometric average vs. 0.30 for arithmetic average).

The results here show that the simple optimizations (snarfing and bus read-sharing) lead to a reduction of execution time (compared to unoptimized 64-byte block Illinois) of 6.5 to 16.0 percent, and the simple optimizations combined with the **Minerva** protocol lead to a reduction of execution time of 19.4 to 41.2 percent, depending on the configuration parameters. Generally the wider the bus, the less the **Minerva** protocol will outperform the Illinois protocol, but it is generally a useful performance improvement.

5.4   Effects of Software Restructuring

In [31], we investigated the impact of certain data structure improvements on the execution time. We found that changes in the data layout due to information obtained by profiling the worst behaving 10 (64-byte) blocks in four of our programs could lead to an average reduction of 51 percent of the false sharing misses, 20 percent of the total cache fetch misses (both measured using an infinite cache). For reasonable size caches, this results in a 15 percent reduction in the execution time for both 16K and 64K byte caches (using 64-byte blocks). As a further evaluation of the **Minerva** protocol, we examine the effect of the protocol on code that has been restructured to eliminate many of the false sharing problems.

Table 4 shows the reduction of execution time for the four workloads restructured to reduce false sharing in [31], when the simple optimizations from [28] are applied and with the **Minerva** protocol. *Original* refers to the unoptimized source code; *restructured* refers to the modified code. For these four workloads, **Minerva** reduces execution time by 55.9 percent for the original code and 44.8 percent for the optimized code. A direct comparison of the execution times using **Minerva** for the original and optimized times shows only a 2.9 percent improvement for the optimized code (vs. the 15 percent reduction using the Illinois protocol reported in [31]). So **Minerva** complements the effects of data restructuring, but **Minerva** alone provides most of the improvement without requiring modification of the original source code.

Execution Time Reduction with Restructured Code			
	Program	Simple Opts	Minerva
Original	Barnes	27.4%	36.6%
	Pthor	19.7%	53.0%
	Topopt	51.4%	85.1%
	Water	2.2%	15.4%
	Average	27.4%	55.9%
Modified	Barnes	25.4%	31.8%
	Pthor	19.0%	49.6%
	Topopt	50.3%	71.0%
	Water	1.9%	7.16%
	Average	26.4%	44.8%

Table 4. Execution time reduction of **Minerva** on restructured code. Geometric averages are used for the averages.

5.5 Observations

Despite the reasonably good improvement in execution time and other statistics that the **Minerva** protocol provides, it is obvious that much more needs to be done in software to improve some of these programs. An examination of Figure 4 shows that only a tiny fraction of execution time in the parallel phase of programs is used performing calculations (labelled **multiprocessor execution**) for some of these programs. Such programs are **cholesky, mp3d, pthor, pverify**, and **topopt**. There is very little point in trying to speed up the execution time by adding more processors to solving these problems, as the read and invalidation stall times heavily dominate the calculation time. There are also other programs in which the initialization time in uniprocessor mode also dominates the multiprocessor calculation time, also leading to poor speed-ups with more processors. Programs that fall into this category are **cholesky, locus, raytrace**, and **volrend**. This leads us to two general principles that should be considered when writing parallel programs: (1) uniprocessor initialization time should be minimized; methods must be explored to initialize the data space in parallel; (2) data objects should be constrained as much as possible to single processors to reduce communications overhead. In addition, processors should be given tasks that are as independent as possible; barriers should be very rarely used (unlike **cholesky** and **fmm**). Using these principles, programs should be designed to have behavior like **water**, where multiprocessor execution (calculation) time still heavily dominates the total run time, even with 16 processors on the job.

6 Conclusion

In this paper we have presented and evaluated our new cache coherency protocol **Minerva**. **Minerva** is a MESI-based protocol that allows each block in each processor's cache to determine the most appropriate subblock size for conducting coherence and fetching operations. Our new protocol in combination with other simpler and more general optimizations result in up to a 40 percent average reduction of execution time on a realistic shared memory machine with a 32-bit wide bus, and around 30 percent for a 64-bit wide bus. Other statistics show positive virtues, such as a reduction of bus utilization, and only a small increase in the total number of fetch and invalidation transactions, much less than for fixed size subblock protocols.

Minerva achieves its success by reducing the invalidation size when it is wasteful to invalidate whole cache blocks. Smaller invalidation and fetch sizes aid in reducing the dead sharing traffic caused predominantly by false sharing behavior. By reducing bus traffic, processors spend much less time stalled waiting for bus transactions to complete and proportionally more time performing calculations.

An important feature of this protocol is that it can simply and efficiently be implemented in hardware, using a few levels of combinational logic to calculate the smallest subblock size every time certain external events affecting that cache block are observed on the bus. It also takes advantage of its ability to track word level coherence to implement write-validate, allowing writes to invalid words in order to reduce data fetches.

Our experiments with restructured code show that **Minerva** is able to make further improvements in performance, to such an extent that false sharing caused unintentionally by programmers can be largely eliminated in hardware. This frees the programmer from worrying about data layout and aids in fully realizing the simplicity of programming using the shared memory model.

References

1. Craig Anderson and Jean-Loup Baer. Design and Evaluation of a Subblock Cache Coherence Protocol for Bus-Based Multiprocessors. Technical Report TR-94-05-02, University of Washington, May 1994.
2. Craig Anderson and Jean-Loup Baer. Two Techniques for Improving Performance on Bus-based Multiprocessors. In *Proc. First IEEE Symposium on High-Performance Computer Architecture*, pages 264–275, Raleigh, NC, January 22–25 1995.
3. J. K. Archibald. A cache coherence approach for large multiprocessor systems. *Proc. 2nd International Conference on Supercomputing*, pages 337–345, July 1988.
4. William J. Bolosky and Michael L. Scott. False Sharing and its Effect on Shared Memory Performance. In *Proc. USENIX Symposium on Experiences with Distributed and Multiprocessor Systems*, pages 57–71, San Diego, CA, September 22–23 1993.
5. Yung-Syau Chen and Michel Dubois. Cache Protocols with Partial Block Invalidations. In *Proc. Seventh International Parallel Processing Symposium*, pages 16–23, Newport, CA, April 13–16 1993.
6. Fredrik Dahlgren, Michel Dubois, and Per Stenström. Combined Performance Gains of Simple Cache Protocol Extensions. In *Proc. 21st Annual International Symposium on Computer Architecture*, pages 187–197, Chicago, IL, April 18–21 1994.
7. Czarek Dubnicki and Thomas J. LeBlanc. Adjustable Block Size Coherent Caches. In *Proc. 19th Annual International Symposium on Computer Architecture*, pages 170–180, Gold Coast, Queensland, Australia, May 19–21 1992.
8. Michel Dubois, Jonas Skeppstedt, Livio Ricciulli, Krishnan Ramamurthy, and Per Stenström. The Detection and Elimination of Useless Misses in Multiprocessors. In *Proc. 20th Annual International Symposium on Computer Architecture*, pages 88–97, San Diego, CA, May 16–19 1993.
9. Michel Dubois, Jonas Skeppstedt, and Per Strenström. Essential Misses and Data Traffic in Coherence Protocols. *Journal of Parallel and Distributed Computing*, pages 108–125, September 1995.
10. Susan J. Eggers and Tor E. Jeremiassen. Eliminating False Sharing. In *Proc. 1991 International Conference on Parallel Processing*, pages I–377–I–381, St. Charles, IL, August 12–17 1991.
11. Susan J. Eggers and Randy H. Katz. A Characterization of Sharing in Parallel Programs And Its Applicibility to Coherency Protocol Evaluation. In *Proc. 15th Annual International Symposium on Computer Architecture*, pages 373–382, Honolulu, HI, May 30–June 2 1988.
12. Susan J. Eggers and Randy H. Katz. Evaluating the Performance of Four Snooping Cache Coherency Protocols. In *Proc. 16th Annual International Symposium on Computer Architecture*, pages 2–15, Jerusalem, Israel, May 28–June 1 1989.
13. Susan J. Eggers and Randy H. Katz. The Effect of Sharing on the Cache and Bus Performance of Parallel Programs. In *Proc. Third International Conference on Architectural Support for Programming Languages and Operating Systems*, pages 257–270, Boston, MA, April 3–6 1989. ACM.
14. Jeffrey D. Gee and Alay Jay Smith. Evaluation of Cache Consistency Algorithm Performance. In *Proc. Fourth International Workshop on Modeling, Analysis, and Simulation of Computer and Telecommunication Systems*, pages 236–248, San Jose, CA, February 1–3 1996.
15. Stephen R. Goldschmidt and John L. Hennessy. The Accuracy of Trace-Driven Simulation of Multiprocessors. In *Proc. 1993 ACM SIGMETRICS Conference on Measurement and Modeling of Computer Systems*, pages 146–157, Santa Clara, CA, May 10–14 1993.
16. James R. Goodman. Coherency For Multiprocessor Virtual Address Caches. In *Proc. 2nd International Conference on Architectural Support for Programming Languages and Operating Systems*, pages 72–81, Palo Alto, CA, October 5–8 1987.

17. John Hennessy and David A. Patterson. *Computer Architecture, A Quantitative Approach.* Morgan-Kaufmann, 2nd edition, 1996.
18. Mark D. Hill and Alan Jay Smith. Experimental Evaluation of On-Chip Microprocessor Cache Memories. In *Proc. 11th Annual International Symposium on Computer Architecture*, pages 158–166, Ann Arbor, MI, June 5–7 1984.
19. Tor E. Jeremiassen and Susan J. Eggers. Reducing False Sharing on Shared Memory Multiprocessors through Compile Time Data Transformations. In *Proc. Fifth ACM SIGPLAN Symposium on Principles and Practice of Parallel Programming*, pages 179–188, Santa Barbara, CA, July 19–21 1995.
20. Norman P. Jouppi. Cache Write Policies and Performance. In *Proc. 20th Annual International Symposium on Computer Architecture*, pages 191–201, San Diego, California, May 16–19 1993.
21. Murali Kadiyala and Laxmi N. Bhuyan. A Dynamic Cache Sub-block Design to Reduce False Sharing. In *Proc. International Conference on Computer Design: VLSI in Computers and Processors*, pages 313–318, Austin, TX, October 2–4 1995.
22. David J. Lilja. Cache Coherence in Large-Scale Shared-Memory Multiprocessors: Issues and Comparisons. *ACM Computing Surveys*, 25(3):303–338, September 1993.
23. J. S. Liptay. Structural Aspects of the System/360 Model 85, Part II: The Cache. *IBM Systems Journal*, 7(1):15–21, 1968.
24. Håkan Nilsson and Per Stenström. An Adaptive Update-Based Cache Coherence Protocol for Reduction of Miss Rate and Traffic. In C. Halatsis, D. Maritsas, G. Philokyprou, and S. Theodoridis, editors, *Proc. PARLE '94 (Parallel Architectures and Languages Europe)*, pages 363–374, Athens, Greece, July 4–8 1994.
25. Vijay S. Pai, Parthasarathy Ranganathan, and Sarita V. Adve. The Impact of Instruction-Level Parallelism on Multiprocessor Performance and Simulation Methodology. In *Proc. The third International Symposium on High-Performance Computer Architecture*, pages 72–83, San Antonio, TX, February 1–5 1997.
26. Mark S. Papamarcos and Janak H. Patel. A Low-Overhead Coherence Solution for Multiprocessors with Private Cache Memories. In *Proc. 11th Annual International Symposium on Computer Architecture*, pages 348–354, Ann Arbor, MI, June 5–7 1984.
27. Steven Przybylski. The Performance Impact of Block Sizes and Fetch Strategies. In *Proc. 17th Annual International Symposium on Computer Architecture*, pages 160–169, Seattle, WA, May 28–31 1990.
28. Jeffrey B. Rothman and Alan Jay Smith. **Minerva**: An Adaptive Subblock Coherence Protocol for Improved SMP Performance. Technical Report UCB/CSD-99-1087, Computer Science Division, University of California, Berkeley, December 1999.
29. Jeffrey B. Rothman and Alan Jay Smith. Multiprocessor Memory Reference Generation Using **Cerberus**. In *Seventh International Symposium on Modeling, Analysis and Simulation of Computer and Telecommunication Systems (MASCOTS '99)*, pages 278–287, College Park, MD, October 24-28 1999.
30. Jeffrey B. Rothman and Alan Jay Smith. The Pool of Subsectors Cache Design. In *Proc. International Conference on Supercomputing*, pages 31–42, Rhodes, Greece, June 20–25 1999.
31. Jeffrey B. Rothman and Alan Jay Smith. Analysis of Shared Memory Misses and Reference Patterns. In *Proc. International Conference on Computer Design (ICCD2000)*, September 17–20 2000.
32. Jeffrey B. Rothman and Alan Jay Smith. Sector Cache Design and Performance. In *Eighth International Symposium on Modeling, Analysis and Simulation of Computer and Telecommunication Systems (MASCOTS 2000)*, August 29–September 1 2000.
33. André Seznec. Decoupled Sectored Caches: conciliating low tag implementation cost and low miss ratio. In *Proc. 21st Annual International Symposium on Computer Architecture*, pages 384–393, Chicago, IL, April 18–21 1994.
34. Jaswinder Pal Singh, Wolf-Dietrich Weber, and Anoop Gupta. SPLASH: Stanford Parallel Applications for Shared-Memory. Technical report, Stanford University, June 1992. Report No. CSL-TR-92-526.
35. Per Stenström. A Survey of Cache Coherence Schemes for Multiprocessors. *IEEE Computer*, 23(6):12–25, June 1990.
36. Josep Torrellas, Monica S. Lam, and John L. Hennessy. Measurement, Analysis, and Improvement of the Cache Behavior of Shared Data in Cache Coherent Multiprocessors. Technical report, Stanford University, February 1990. Report No. CSL-TR-90-412.
37. Josep Torrellas, Monica S. Lam, and John L. Hennessy. False Sharing and Spatial Locality in Multiprocessor Caches. *IEEE Transactions on Computers*, 43(6):651–663, June 1994.
38. Steven Cameron Woo, Moriyoshi Ohara, Evan Torrie, Jaswinder Pal Singh, and Anoop Gupta. The SPLASH-2 Programs: Characterization and Methodological Considerations. In *Proc. 22nd Annual International Symposium on Computer Architecture*, pages 24–36, Santa Margherita Ligure, Italy, June 22–24 1995.

Active Memory Clusters: Efficient Multiprocessing on Commodity Clusters

Mark Heinrich, Evan Speight, and Mainak Chaudhuri

Cornell University, Computer Systems Lab, Ithaca, NY 14853, USA
{heinrich,espeight,mainak}@csl.cornell.edu
http://www.csl.cornell.edu/

Abstract. We show how novel active memory system research and system networking trends can be combined to realize hardware distributed shared memory on clusters of industry-standard workstations. Our active memory controller extends the cache coherence protocol to support transparent use of address re-mapping techniques that dramatically improve single-node performance, and also contains the necessary functionality for building a hardware DSM machine. Simultaneously, commodity network technology is becoming more tightly-integrated with the memory controller. We call our design of active memory commodity nodes interconnected by a next-generation network *active memory clusters*. We present a detailed design of the AMC architecture, focusing on the active memory controller and the network characteristics necessary to support AMC. We show simulation results for a range of parallel applications, showing that AMC performance is comparable to that of custom hardware DSM systems and far exceeds that of the fastest software DSM solutions.

1 Introduction

With the advent of low-cost, high-performance commodity components, networks of industry-standard workstations have captured the interest of both industry and research institutions over the past several years. Referred to as NOWs (network of workstations), COWs (collection of workstations), COPs (collection of PCs), etc., these *clusters* are typically comprised of commodity high-performance uniprocessor or small-scale symmetric multiprocessor (SMP) nodes containing two or four processors, large amounts of local memory (on the order of 1-4 GB of RAM), and a high-speed network such as Myrinet, cLAN, or ServerNet that provides zero-byte inter-node latencies less than $10\mu s$.

Clusters. As a whole, the machines comprising a cluster make up a distributed-memory parallel machine, with each node containing a complete version of the operating system, its own memory hierarchy, and I/O subsystem. This "virtual parallel machine" lends itself naturally to message passing APIs such as MPI [16]. However, many clusters are comprised of SMP nodes that already contain cache-coherent hardware, presenting the application programmer with two distinct memory interfaces: shared memory between the processors local to each machine,

H. Zima et al. (Eds.): ISHPC 2002, LNCS 2327, pp. 78–92, 2002.

and distributed memory between processors that are not co-located. Because of the high overhead incurred when sending and receiving messages, there are clear performance advantages to using the native SMP load/store interface for communication between co-located processors, but accesses to remote memory then require an entirely different communication model. Utilizing two different communication models in the same program is problematic, particularly when the number of threads and processors per machine is not known at compile time.

Software DSM. In an attempt to address the programming concerns of cluster-based parallel computing, several *software DSM* systems have been built that provide shared-memory parallel programming on the cluster as a whole without requiring specialized hardware (e.g., Ivy [14], TreadMarks [8], Munin [2], Brazos [22], and Shasta [20]). These systems typically use page-level granularity to enforce coherence, manipulating the virtual memory page protections to trap protection violations and executing the coherence protocol in software on the main microprocessor. Utilizing page-level granularity also allows the high cost of communication to be amortized over a large coherence unit (4-16 KB for a typical page). Although multiple-writer protocols [2] and relaxed consistency models [2,8] significantly reduce the amount of communication necessary to maintain coherence between nodes, high communication costs often result in poor performance.

Other software DSM systems instrument application code to check for coherence actions that need to be performed before each access to shared memory (e.g., the Shasta system [20]). These systems can then implement coherence at any granularity desired, but the high handler overhead and the fact that the network is typically integrated on the I/O bus rather than the memory bus still results in the choice of pages for data transfer. Subsequently, such systems also incur large software overhead compared to hardware DSM systems. Despite this overhead, and other factors such as high synchronization rates, frequent sharing, or large amounts of false sharing that can severely hinder the performance of software DSM systems, the cost advantages of software DSM clusters make them a viable alternative in many situations.

Hardware DSM. Alternatively, some systems provide direct hardware support for scalable shared memory, adding substantial cost to each individual node in exchange for higher performance on parallel applications. Such *hardware DSM* machines include DASH [13], the SGI Origin 2000 [12], and the Sun S3.mp [17]. Most high-performance hardware DSM machines have tightly-integrated node or memory controllers that connect the microprocessor both to the memory system and to a proprietary high-speed switching network. The scalable coherence protocols used in such machines are implemented either in hardware finite-state machines or in software running on an embedded programmable device in the controller. Despite the resulting high performance of these systems, and efforts to show that the necessary additional hardware to support hardware DSM in commodity workstations and servers is small [13], high-end PC servers and engineering workstations have yet to integrate the additional functionality needed to build seamless hardware DSM from COTS (commodity off-the-shelf) components. As discussed in Section 2, the primary reason for this is that this additional hardware does nothing to improve single-node performance.

In this paper, we show that our research in active memory systems and our subsequent active memory controller, combined with emerging network technology trends, can create the realization of hardware DSM performance on commodity clusters at the cost of previous software DSM efforts, with a concomitant improvement in single-node performance. We call the result of this convergence *active memory clusters* (AMC).

In Section 2 we explain the key differences between hardware and software DSM systems in more detail, and how our research in active memory systems narrows the gap between the two. Section 3 presents the design of active memory clusters and discusses the architectural, network, and operating systems issues in implementing AMC. Section 4 discusses the parallel performance of AMC relative to that of hardware DSM solutions, and presents simulated AMC speedup results for several parallel applications. In addition, we explore the performance impact of varying the latency and bandwidth of the commodity network interconnect. Section 5 concludes the paper.

2   Differences between HW DSM and SW DSM

To the naïve eye, a physical comparison of a hardware DSM machine with a modern software DSM system based on clusters reveals few differences. Both machines are constructed out of individual commercial boxes connected together with proprietary high-speed networks. Closer examination reveals three main differences between the two systems:

- hardware DSM networks are faster and more tightly-integrated
- nodes in software DSM systems run separate versions of the operating system
- hardware DSM requires a specialized node controller

The first difference is the speed and integration level of the network. Typically the communication latency in software DSM networks is an order of magnitude more than in hardware DSM networks ($\sim 10\mu s$ versus under $1\mu s$). In addition, commodity motherboards integrate the network on the I/O bus versus the tighter integration on the node or memory controller in hardware DSMs. These are real differences, but they are rapidly disappearing. The computing industry's new InfiniBand network (discussed further in Section 3.2) has latencies on the order of $1\mu s$ (similar to hardware DSM latencies) and will be connected directly to the memory controller on future commodity motherboards [7].

Unlike hardware DSM systems in which every node is under the control of a single operating system, software DSM systems run in an environment where each node executes its own version of the host operating system. The critical aspect of this distinction with respect to hardware DSM is the lack of a central page table accessible by all nodes in the system. Instead, each operating system maintains its own set of virtual-to-physical mappings. As discussed in Section 3.3, we can remove this distinction by making a "distributed page table" that acts like the centralized page table in a hardware DSM-capable operating system. Alternately, a NUMA-aware OS providing a single-system image across the entire cluster can be used in the context of an AMC system.

The only remaining architectural difference is the last one listed above: the specialized node controller. In hardware DSM machines, this node controller implements the directory-based coherence protocol at a fine granularity and offloads the overhead from the main microprocessor. If these functions were integrated into the memory controller of a commodity server, there would be essentially no remaining difference between the two architectures.

While the integration of the specialized hardware DSM functions into commodity servers and workstations is possible, the economic arguments have not been compelling enough to include this functionality for two main reasons. Most importantly, this additional functionality does not improve single-node performance, meaning it is unlikely it would ever be included in commodity servers. Second, the size of the high-performance hardware DSM market has never been large enough to warrant true commodity nodes with hardware DSM support.

The debate about whether the necessary controller functionality will ever become commodity would be left at that, except for two factors. First, there is a trend toward placing more CPUs per machine in today's SMP boxes (e.g., the 32-processor SMP solution currently offered by Unisys) that is naturally accompanied by a higher cost per box. The desire to keep the cost of individual boxes low, while retaining the ability to program an entire cluster as if it were a single SMP, will be a powerful economic argument for including support for hardware shared-memory in commodity cluster components. Second, we will show how recent research into the single-node performance benefits of *active memory systems* argues for their inclusion in high-end servers, providing hardware DSM support "for free" if the active memory support is implemented in a flexible manner.

2.1 Active Memory Systems

One of the biggest challenges facing modern computer architects is overcoming the *memory wall* [19]. Technology trends dictate that the gap between processor and memory performance is widening. Though good cache behavior mitigates this problem to some extent, memory latency remains a critical performance bottleneck in modern high-performance processors. Heavily-pipelined clocked memory systems have improved memory bandwidth, but do nothing to address memory latency or reduce the number of cache misses incurred by the processor.

One approach to reducing the gap between processor and memory performance is to move processing into the memory system by using active memories [4,5,9,15,18,23] or active memory controllers [3,10]. In many such systems, parts of a program that have poor cache behavior are executed in the memory system, thereby reducing cache misses and memory bandwidth requirements. Other work employs address re-mapping techniques to re-structure data (like linked lists or non-unit-stride accesses) so that the processor can access them in a more cache-efficient manner. Recently, we made the observation that active memory techniques can be treated as an extension of the cache coherence

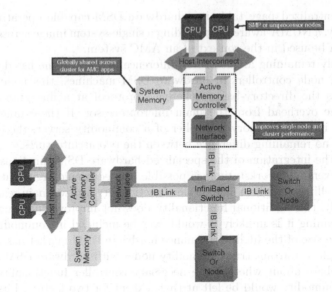

Fig. 1. Active Memory Cluster Configuration.

protocol [10] because of the coherence problem created by address re-mapping techniques. Our active memory controller implements the coherence protocol and the extensions necessary to support active memory operations on data in the memory system, allowing the memory system to transparently and coherently access any data in the system. In this paper we detail our active memory controller architecture, its integration with emerging commodity network technology, and the resulting high-performance, low-cost, coherent DSM architecture of AMC. Our particular active memory controller approach is described in more detail in Section 3.1.

Convergence. We return now to our discussion of hardware DSM versus software DSM systems. Our active memory controller that is designed to improve uniprocessor and single-node performance contains the same functionality needed to enable cluster-based hardware DSM systems. This fact – that active memory controllers and hardware DSM controllers share much of the same functionality – lies at the heart of the AMC design. It strengthens the cases of both the active memory and hardware DSM advocates for inclusion in commodity servers. Even if the individual arguments for including specialized controller functionality fall short, their combined benefits may be enough to finally produce commodity nodes with active memory controllers.

3   Active Memory Clusters Implementation

This section discusses the implementation details of our active memory cluster system (see Figure 1), focusing on the three architectural differences between current hardware and software DSM machines that we presented in Section 2. We first describe our active memory controller functionality, present our active memory controller design, and show how it can dramatically improve uniprocessor and single-node performance. We further explain how this functionality enables a hardware DSM machine to be constructed from cluster components. We then discuss the ramifications of upcoming tightly-integrated commodity networks and the details relevant to AMC. Finally, we end with a discussion of the system software issues related to our active memory clusters implementation.

3.1   Active Memory Controller Functionality

Recent active memory proposals have advocated the technique of re-mapping the address space of a process in an application-specific manner. Accesses to this space are then used as a signal to the memory controller to perform "active" operations rather than satisfying these accesses from physical memory [3,10,15]. For example, when performing matrix operations that require row and column traversals, one traversal uses the cache effectively whereas the other does not. By providing multiple memory viewpoints of the same matrix using shadow address spaces, row traversals are unchanged, whereas column traversals are treated as row traversals of a matrix at a different (shadow) address. The active memory controller fetches individual double-words from a column and returns them in a single cache line. Data is therefore provided in blocks that can be cached efficiently by the main processor. The result is good cache behavior for both row and column traversals of the matrix. Such an approach speeds up many scientific applications by using the processing capability in the memory system. Similar re-mapping techniques are used to speedup sparse matrix codes, scatter/gather operations, reductions, and linked-list-intensive programs. These, and other active memory operations are described in more detail in [10].

The main challenge with this active memory approach is solving the cache coherence problem it creates. For example, if columns of a matrix are being written via a different address space during column traversals, the next row traversal via the normal address space will return incorrect or stale data unless care is taken or costly cache flushes are performed. The key insight into solving the coherence problem in active memory systems is that the active memory controller controls both the coherence protocol *and* the fetching of the data requested by the processor. In architectures like the FLASH multiprocessor [11] and the S3.mp [17], the coherence protocol itself is programmable or extensible. Thus, active memory support is, in effect, an extension of the cache coherence protocol. In this case, the active memory controller enforces coherence between the original and shadow address spaces.

Our work with active memory controllers indicates that the occupancy of an active memory controller is significantly reduced by the introduction of a hardware unit to support the fast assembly of cache lines from individual double words (and their disassembly as well). Adding this special data unit to our flexible coherence engine reduces the occupancy of active memory operations by up to an order of magnitude, thereby improving overall system performance. At

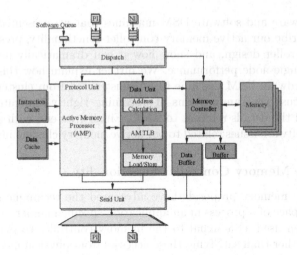

Fig. 2. Active Memory Controller Microarchitecture.

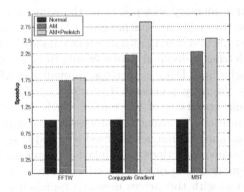

Fig. 3. Uniprocessor Active Memory Speedup.

the same time, a flexible engine need not slow down non-active requests, as the processing on the AMP in Figure 2 occurs in parallel with the memory lookup needed to satisfy the cache miss [10].

Our active memory controller microarchitecture (see Figure 2) combines a flexible coherence engine with the aforementioned special data unit to support fast address re-mapping and dynamic cache line assembly. Detailed simulations of this microarchitecture can be found in [10], but we repeat representative results here. Figure 3 shows uniprocessor speedup between 1.75 and 2.8 when using our active memory controller versus systems with normal memory controllers. In these applications the number of application cache misses is reduced by over a factor of two, and prefetching the transformed shadow address space offers room for additional speedup not available in the normal application. The source code of these applications was changed minimally (on the order of 10 lines of code) to achieve these results. As shown in [10], the same architecture can also improve the performance of single-node multiprocessors via the same methods. In this paper we will further show how to extend this architecture to create multi-node DSM systems.

Because we leverage cache coherence, our active memory controller design achieves this performance without requiring cache flushes on the processor. Similarly, our design is also compatible with the SMP nodes commonly used in clusters (where, because of process migration, architectures that rely on cache flushes must flush *all* the caches on the node, even for uniprocessor applications). In the remainder of this section we describe how to use this same architecture to build active memory clusters and highlight the relevant features of our microarchitecture to AMC.

Active Memory Controller Requirements. Our active memory controller manages the cache coherence protocol. Consequently, our flexible active memory controller already tracks sharing information in the system and invalidates and retrieves cache lines from the processor caches. For active memory support the processor can communicate with the memory controller through uncached writes to a portion of the address space where these writes are interpreted as commands by the memory controller (e.g. notifying the controller about shadow address spaces during initialization). An active memory controller must also have the ability to dispatch both normal and active coherence handlers by looking at certain bits in the request and address on the processor bus.

Additional support required for AMC. To support AMC, the memory controller must have a network interface as shown in Figure 1 in addition to the processor interface. In forthcoming network architectures such as InfiniBand, the network interface will be moved closer to the main CPU. We discuss the effect of the memory controller being directly connected to the network interface in more detail in Section 3.2. The additional features needed for AMC beyond those already present in the active memory controller are the ability to dispatch handlers from the network interface (including deadlock avoidance in the scheduling mechanism) and the existence of the corresponding coherence handlers for network messages. In flexible active memory controllers with a direct interface to the network, the additional coherence handlers that are needed consist of protocol code running on the memory controller (AMP in Figure 2) and do not necessitate architectural changes to the active memory controller.

3.2   Network Integration

Commodity architectures are witnessing evolutionary changes in network integration as the network connection moves from a plug-in card on the distant I/O bus to a routing chip directly connected to the memory controller (see Figure 1). Although these networks are designed primarily for use in storage networks and support user-level heavyweight protocols, our active memory controller only uses the physical routing capabilities of these directly-connected network switches. In addition, coherence messages travel between memory controllers only and are not forwarded up to the processor for handling via interrupts. Instead, an active memory cluster handles the messages entirely in the memory controller, similar to a hardware DSM machine.

Our AMC controller design uses a system area network (SAN), such as InfiniBand, as its network interface. However, the active memory controller does not use the high-level (and higher overhead) user-level protocols that run over the SAN network link when sending and receiving the messages that comprise the coherence protocol. Instead, our controller uses the fast underlying link-level performance to synthesize and handle messages to support DSM. InfiniBand supports these low-level messages via its *raw packet format*, which has an overhead of 14 bytes on top of the data payload of the packet [7]. Although this message overhead is more than that present in many hardware DSM machines, it is still small compared to the data payload (typically a cache line, or 128 bytes in our system).

The latency and bandwidth characteristics of system area networks such as InfiniBand, as well as the physical routing capabilities, are comparable to networks used in today's hardware DSM machines. InfiniBand supports virtual lanes, a critical feature for supporting deadlock-free request/reply networks in DSM machines. The "hop time" through an InfiniBand switch is currently on the order of 150 ns, and will likely drop in the near future with later generations of the hardware. The bandwidth of a single link in the InfiniBand network is 250 MB/s (2.0 Gb/s). Although this is less bandwidth than the network in an Origin 2000 (800 MB/s), it is still 2.5 times higher than that used in the most advanced current software DSM machines. Coupled with the its low latency, this network will provide an excellent base on which to build a high-performance DSM machine. We show simulation results of a prototype active memory cluster in Section 4 with varying network latencies and bandwidth parameters.

3.3 Operating System Issues

As mentioned in Section 2, software DSM runtime layers are typically used on clusters of workstations, each running their own version of the host operating system. The majority of these clusters consist of the same types of machines running the same version of the same operating system. The reasons for this include the cost advantage of buying in bulk, ease of cluster-wide system management, and the avoidance of the overhead associated with translating between different data representations. Clusters providing high-performance parallel programming environments are not groups of machines sitting on desks in an office environment, but rather sets of rack-mounted machines sequestered in machine rooms. Thus, while companies such as *Entropia* seek to use heterogenous, widely-spread machines for large-scale independent computations, realistically we do not expect high-performance cluster machines to be constructed of different components.

In AMC, as in other hardware and software DSM implementations, the virtual address of the shared global memory region must be the same on all nodes in the cluster. Furthermore, unlike software DSM, the virtual-to-physical mapping must be the same across the cluster because the AMC memory controller main-

Table 1. AMC hardware configuration.

Parameter	Value
Number of Nodes	1-32
Processors per Node	1
Processor Clock Speed	2.0 GHz
System Clock Speed	400 MHz (2.5 ns)
Primary Data Cache	32 KB, 32 B line size
Primary Instruction Cache	32 KB, 64 B line size
Secondary Cache	1 MB, 128 B line size
Max. Outstanding Misses	4
Network Interface Delay	Inbound: 40 ns; Outbound: 20 ns
Memory Latency from AMC Controller	125 ns to first double-word

tains coherence based on the physical addresses presented. With a single system image, the OS running on the cluster provides this shared page table directly for AMC. With separate operating systems, a small AMC runtime library is used to ensure the VA-to-PA mappings are consistent across the cluster during program initialization. Thus, AMC requires that the host OS support the ability to request that a specific virtual address map to a physical address within a certain range of the physical memory address space, a feature commonly used to map I/O devices into a process' virtual space. With this ability, we can create a "distributed page table" in which the shared region of memory resides at the same virtual and physical location across all OS instances. This procedure, as well as issues regarding pinning of shared pages and shared pointer distribution, are detailed in [6]. For operating systems with readily-available source (i.e., Linux or FreeBSD), we can modify the virtual memory manager to provide the necessary functionality required by AMC. Operating systems such as Windows 2000 can also provide this functionality through the use of device drivers.

Several operating systems are currently available that provide NUMA support, including Irix from SGI, Windows XP from Microsoft, and a NUMA-aware version of Linux. Any of these operating systems will run on active memory clusters with little or no modification.

4   AMC Performance Results

We simulated several parallel applications from the SPLASH-2 application suite [24] running on an AMC system connected via a SAN with InfiniBand-like characteristics: FFT with 1M points, LU decomposition with a 512x512 matrix, Ocean with a 514x514 ocean, Radix-Sort with 2M keys and a radix of 32, and Water with 1024 molecules.

These execution-driven simulations use the parameters described in Table 1. Our AMC simulator models our flexible active memory controller from Figure 2 connected to a network with the latency, bandwidth, and overhead characteristics of a system area network such as InfiniBand. The AMC results assume a network comprised of 16-port switches connected in a fat-tree configuration for

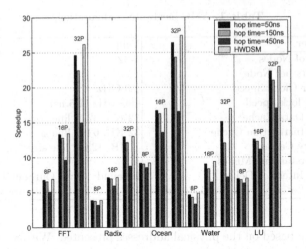

Fig. 4. AMC SPLASH-2 speedup.

varying network latencies. We present results for 3 different network "hop times" corresponding to current and future SAN architecture generations. Results for 450 ns show the performance with first-generation SAN parameters, which have been available for some time. The 150 ns performance numbers reflect the current state-of-the-art in system area networks, and the 50 ns results show the performance of AMC as system area networks mature. For comparison, results for a custom hardware DSM system are shown as well. The HWDSM system modeled in detail here has 12.5 ns hop times and is connected in a mesh configuration. Note that uniprocessor execution times for the HWDSM and AMC simulations are identical, indicating that speedup comparisons between the architectures is a valid metric for performance. The coherence protocol running on the AMC controller is dynamic pointer allocation [21], a linked-list protocol that scales well to the processor counts used here. Figure 4 reports speedup on 8, 16, and 32 processor systems for the parallel execution times of the applications.

AMC performance. As shown in Figure 4, the performance of the SPLASH-2 benchmarks on AMC is remarkably good. When the interconnection network has a 50 ns hop time, all applications perform nearly as well as the custom hardware DSM implementation, despite the fact that the HWDSM architecture has a hop-time over four times faster than this AMC configuration. At the other end of the spectrum, when the network latency is 450 ns we find that AMC performs significantly worse than HWDSM. For 32-processors, HWDSM performs an average of 72% better than AMC across all 5 applications.

The bars representing 150 ns hop times most closely represent the current InfiniBand architecture, which we are using as our representative SAN. At 150 ns, HWDSM outperforms AMC by 16% for FFT, 7% for Radix-Sort, 12% for Ocean, 40% for Water, and 9.4% for LU. Even if system area networks do not ultimately achieve the 50 ns latency represented in our experiments, the bandwidth of SANs is sure to increase in subsequent generations due to the importance of I/O. For

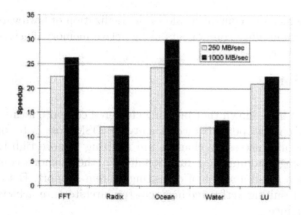

Fig. 5. AMC SPLASH-2 speedup at different network bandwidth (32 P, 150 ns hops).

example, in the InfiniBand architecture it is possible to add additional links to increase bandwidth substantially over the 250 MB/s shown here.

To evaluate the impact of increasing the bandwidth of AMC's network, we also simulated our five applications with a network bandwidth of 1 GB/s. The results for 32 processors are shown in Figure 5. This quadrupling of the network bandwidth improved the performance of all 5 applications, most notably that of Radix, FFT, and Ocean, all of which have much higher network requirements than Water or LU. Radix shows a 46% improvement in speedup with the additional bandwidth, Ocean improves by 18%, and FFT is 14% faster.

We also expect the performance of a traditional software DSM system to improve when using future system area network architectures due to three features of emerging network architectures: a significant reduction in memory latency, the capability of remote DMA operations, and reliable communication mechanisms. Nevertheless, we still expect AMC to outperform software DSM due to software DSM's remaining high kernel overhead involved with page exception handling, the large granularity of sharing required, and the processing overhead for runtime structures such as `diffs` and `twins`. By way of comparison, the best reported software DSM speedup on 32 processors for these applications was achieved by running a software DSM system on the Origin 2000, resulting in speedups of 5.5 for FFT, 16.5 for LU, 5.9 for Ocean, and 14.1 for Water [1]. AMC achieves substantially better speedup than even this optimistic software DSM scenario.

Finally, we emphasize that despite the impressive results, the parallel performance of AMC does little toward strengthening the argument for putting the required AMC functionality in commodity servers. The only way that active memory controllers will be included in commodity servers is if the industry embraces the research results showing that active memory techniques like those discussed in Section 3.1 will significantly improve the performance of a standalone machine. When this happens, the insight of this paper is that designing a *flexible* active memory controller allows the realization of hardware DSM performance with commodity components using active memory clusters.

5   Conclusions

In this paper, we have shown that building clustered compute farms that deliver hardware DSM parallel performance at software DSM cost is becoming a real possibility. The only necessary changes are utilizing a more tightly-integrated network technology, adding new functionality in the memory controller, and perhaps adding a small amount of operating system support. Detailed simulations show the resulting active memory clusters architecture achieves excellent parallel performance.

Of these three enabling mechanisms, the most problematic is the additional memory controller functionality. Historically, the additional cost of changing the memory controller to support hardware DSM was not justified by the expected performance gain for two main reasons. First, the required support did nothing to help single-node performance, which is the metric by which the computer industry measures any technology for inclusion in "commodity" boxes. Second, the number of users that would benefit from such a costly change to the memory controller architecture was relatively small. However, research in active memory systems has argued for enhanced memory controller functionality to improve *single-node* performance. By treating active memory support as an extension of the cache coherence protocol, we showed that the controller support needed for active memory and hardware DSM is almost identical, provided the active memory mechanisms are implemented in a flexible manner.

Industry has already begun addressing the issue of network/system integration to support the low communication latency required by the cluster-based systems in use today. We used InfiniBand as an example of such a network architecture, but others exist that would work equally well with active memory clusters. The only remaining issue is one of system software. Software DSM machines have had a distinct advantage in this area, because the use of commodity operating systems is an important factor in keeping both initial system cost and subsequent upgrade costs low. The active memory cluster architecture can certainly be used with an operating system that natively supports hardware DSM. Alternately, simple functionality provided by device drivers can be used with commodity operating systems to achieve hardware DSM performance, even in the absence of a specialized DSM operating system.

In summary, our research in the area of active memory systems shows that active memory techniques substantially improve single-node performance in cases where caching behavior is poor. The results presented in this paper also show that this additional memory controller functionality, if implemented in a flexible manner, results in clusters of servers with the necessary components to achieve the parallel performance of a hardware distributed shared memory system. These two performance benefits are a strong argument for the inclusion of active memory systems in commodity nodes, resulting in a parallel computing platform comprised of industry-standard servers that exhibit parallel performance far surpassing that of traditional cluster and software DSM efforts.

Acknowledgments

This research was supported in part by Cornell's Intelligent Information Systems Institute, Microsoft Corporation, and NSF CAREER Award CCR-9984314.

References

1. Bilas, A., Liao, C., Singh, J.P.: Using Network Interface Support to Avoid Asynchronous Protocol Processing in Shared Virtual Memory Systems. In *Proceedings of the 26th International Symposium on Computer Architecture*, May 1999.
2. Carter, J. B., et al.: Design of the Munin Distributed Shared Memory System. *Journal of Parallel and Distributed Computing*, **29**(2):219–227, September 1995.
3. Carter, J. B., et al.: Impulse: Building a Smarter Memory Controller. In *Proceedings of the Fifth International Symposium on High Performance Computer Architecture* January 1999.
4. Gokhale, M., Holmes, B., Iobst, K.: Processing in Memory: the Terasys Massively Parallel PIM Array. *Computer*, **28**(3):23–31, April 1995.
5. Hall, M., et al.: Mapping Irregular Applications to DIVA, A PIM-based Data-Intensive Architecture. *Supercomputing*, Portland, OR, Nov. 1999.
6. Heinrich, M., Speight, E.: Active Memory Clusters: Efficient Multiprocessing on Next-Generation Servers. Technical Report CSL-TR-2001-1014, Computer Systems Lab, Cornell University, August, 2001.
7. InfiniBand Architecture Specification, Volume 1.0, Release 1.0. InfiniBand Trade Association, October 24, 2000.
8. Keleher, P., et al.: TreadMarks: Distributed Shared Memory on Standard Workstations and Operating Systems. In *Proceedings of the Winter 1994 USENIX Conference*, pages 115–132, January 1994.
9. Kang, Y., et al.: FlexRAM: Toward an Advanced Intelligent Memory System. *International Conference on Computer Design*, October 1999.
10. Kim, D., Chaudhuri, M., Heinrich, M.: Leveraging Cache Coherence in Active Memory Systems. Technical Report CSL-TR-2001-1018, Computer Systems Laboratory, Cornell University, November 2001.
11. Kuskin, J., et al.: The Stanford FLASH Multiprocessor. In *Proceedings of the 21st International Symposium on Computer Architecture*, pages 302–313, April 1994.
12. Laudon, J., Lenoski, D.: The SGI Origin: A ccNUMA Highly Scalable Server. In *Proceedings of the 24th International Symposium on Computer Architecture*, pages 241–251, June 1997.
13. Lenoski, D., et al.: The Stanford DASH Multiprocessor. *IEEE Computer*, **25**(3):63–79, March 1992.
14. Li, K., Hudak, P.: Memory Coherence in Shared Virtual Memory Systems. In *ACM Transactions on Computer Systems*,7(4):321–359, November 1989.
15. Manohar, R., Heinrich, M.: A Case for Asynchronous Active Memories. In *ISCA 2000 Solving the Memory Wall Problem Workshop*, June 2000.
16. Message Passing Interface Forum. MPI: A Message-Passing Interface Standard, Version 1.0, 1994.

17. Nowatzyk, A., et al.: The S3.mp Scalable Shared Memory Multiprocessor. In *Proceedings of the 24th International Conference on Parallel Processing*, 1995.
18. Oskin, M., Chong, F. T., Sherwood, T.: Active Pages: A Computation Model for Intelligent Memory. In *Proceedings of the 25th International Symposium on Computer Architecture*, 1998.
19. Saulsbury, A., Pong, F., Nowatzyk, A.: Missing the Memory Wall: The Case for Processor/Memory Integration. In *Proceedings of the 23rd International Symposium on Computer Architecture*, pages 90–101, May 1996.
20. Scales, D. J., Gharachorloo, K., Thekkath, C. A.: Shasta: A Low-Overhead Software-Only Approach for Supporting Fine-Grain Shared Memory. In *Proceedings of the Seventh International Conference on Architectural Support for Programming Languages and Operating Systems*, pages 174–185, October 1996.
21. Soundararajan, R., et al.: Flexible Use of Memory for Replication/Migration in Cache-Coherent DSM Multiprocessors. In *Proceedings of the 25th International Symposium on Computer Architecture*, pages 342–355, June 1998.
22. Speight, E., Bennett, J. K.: Brazos: A Third Generation DSM System. In *Proceedings of the First Usenix Windows NT Symposium*, August 1997.
23. Torrellas, J., Yang, L., Nguyen, A.-T.: Toward a Cost-Effective DSM Organization that Exploits Processor-Memory Integration In *Proceedings of the 6th International Symposium on High-Performance Computer Architecture*, January 2000.
24. Woo, S. C., et al.: The SPLASH-2 Programs: Characterization and Methodological Considerations. In *Proceedings of the 22nd International Symposium on Computer Architecture*, pages 24–36, June 1995.

The Impact of Alias Analysis
on VLIW Scheduling

Marco Garatti[1,2], Roberto Costa[1,3],
Stefano Crespi Reghizzi[1,4], and Erven Rohou[2]

[1] Politecnico di Milano, {garatti,crespi}@elet.polimi.it
[2] STMicroelectronics, erven.rohoue@st.com
[3] Harvard University, roberto.costa@ensta.org
[4] CNR – CESTIA

Abstract. This experiment studies the speed-up increase that alias analysis (AA) produces on code for very long instruction word machines. AA is done on-demand when requested by the scheduler, in order to eliminate critical arcs of the data dependence graph. Different heuristic criteria are investigated for deciding when to compute alias information,and they show that only a fraction of the alias relation really contributes to the program speed-up. A qualitative study shows that the quality of the initial code affects the speedup alias analysis can give. The results should help compiler designers for VLIW machines in making cost effective AA decisions.

1   Introduction

A VLIW (very long instruction word) processor relies on the compiler for code scheduling. But high parallelism cannot be achieved unless the compiler accurately determines dependencies between instructions, which in turn requires some form of alias analysis (AA). An alias occurs at some program point during execution when two or more names exist for the same memory location. Without AA conservative assumptions force the compiler to introduce dependencies which may not exist in reality and hinder code movement and parallelization. Compiler designers have raised the questions of how vital is AA and how accurate it should be. Educated guesses claim that the final performance of compiled VLIW code is very sensitive to the extent and accuracy of AA. But precise, exhaustive AA is CPU and memory intensive, and would increase the already substantial complexity of compilers for VLIW machines; on the other hand cheap AA may not be sufficient. Moreover, exhaustive computation may be an overkill, since the scheduler does not need alias relations for instructions which are already constrained by other dependencies or by resource unavailability.

To our knowledge no experimental data were available to assess the direct benefits of AA on code performance, when we started this investigation on the impact of AA on scheduling performances. Actually a few quantitative studies (for example [11]) reports the impact of AA on the accuracy of static dataflow computation (such as liveness) which indirectly affects scheduling. In our study

H. Zima et al. (Eds.): ISHPC 2002, LNCS 2327, pp. 93–105, 2002.

the scheduler is a client of a server performing *AA on-demand*. Using various heuristics the scheduler decides when a certain piece of alias information is needed and how important it is for improving performances. Several heuristics are considered to reduce the length of the critical path and the number of queries to the AA server. Clearly one could take the same approach for other phases of compilation which need alias information, such as register promotion.

Building on a research compiler for C (MachineSUIF[13] and SUIF[6]) we developed new AA on-demand algorithms. Using an existing scheduling library we built a superblock scheduler, we measured code performances for a VLIW-like processor and several benchmarks. Then we related code speed-ups to two aspects of AA: heuristic criteria for alias querying and number of queries to the AA server. Finally we started to investigate the relationship between initial code quality and the overall benefit AA can provide. As we found that poorly optimized code had little speedup from AA, we run a second experiment after introducing some code optimization, confirming the direct relation between level of optimization and productivity of AA.

The results provide new quantitative indications that should be of interest to compiler designers, for choosing a cost-effective AA policy.

Section 2 introduces the basic concepts, a running example and develops the concept of "on-demand" AA. Section 3 focuses on scheduling and its interaction with AA. Section 4 discusses the mutual effect between AA and other optimizations. Section 5 describes the experimental setting and the results. The conclusion mentions related work and future developments.

2   Alias Analysis On-Demand

We give enough information to understand the algorithm, but we do not present its novel aspects in detail. We assume familiarity with the concepts of ordered sequence of instructions, basic block and superblock, control flow graph (CFG), data dependence graph (DDG), static analysis, VLIW and ILP architectures, and related ones (see e.g. [1], [9], and [10]).

Two memory locations belong to the *may alias relation* in a program statement s if there is at least one path that reaches s along which the two variables can be alias. An analysis is *flow-sensitive* if it takes into account the order of program statements. Flow-insensitive analysis assumes the statements can be executed in any order [9]. Let p be a pointer and s a statement. The *point-to set* of p in s is the set of memory locations p may point to in s.

An analysis is *structure-wise* (resp. *array-wise*) if it can distinguish different structure fields (resp. different array cells).

AA on-demand is performed only where and when optimizations and transformations need alias information. The queries AA on-demand can answer are: *Are p_1 in s_p and q_1 in s_q possible alias?* Here p_1, q_1 are pointers, s_p, s_q are statements, and both pointers are alive in both statements. The on-demand approach can be advocated on several grounds. First, only a fraction of alias relations matters for a specific optimization. For example an instruction scheduler is interested only on aliases that can shorten the schedule. Second, since code optimization phases modify the program, AA should be redone after each transformation, which is not practical unless the analysis is incremental, or very fast and necessarily inaccurate algorithms are used. AA on-demand does not incur in the same penalty because it stores little or no information about the program.

With respect to the usual classification of AA algorithms our approach operates at assembly language level, because the alias information computed by a front-end would be invalidated by the transformations to machine instruction level. Our analysis is flow-sensitive, structure wise, array unwise, intraprocedural. We justify the two latter restrictions.

Truly in relevant applications (like multi-media) arrays are often used in critical loops, therefore the ability to disambiguate array elements is important. But, in agreement with most other AA projects [2,3,5,7,14], we assumed this task is performed by other compiler phases at a higher level, when the information about array accesses is usually easier to retrieve.

The current experiment does not use interprocedural analysis, but an upper bound on the gain it can provide is presented.

The analysis uses a program representation based on RISC-like instructions. All the instructions can operate on registers, variables and immediates. Figure 1 introduces our running example.

```
INT global  ....
FUNCTION f1
RECORD REC{int f1,int f2}
RECORD REC *p1, **p2
INT arr[10], a, b;
....
B0 (10)  lda r1 <- x
   (11)  add r1 <- r1,4
   (12)  lda r5 <- p1 # r5 is p2
   (13)  lda p1 <- x
   (14)  cmp r10,0
   (15)  cbr 18 # cbr to 16
B1 (16)  ldc r11 <- 0
   (17)  jmp 19 # branch to 19
B2 (18)  ldc r11 <- 1
B3 (19)  lda r4 <- y
   (20)  add r4 <- r4,8
   (21)  lda r6 <- y
```

```
        (22)  str (r5) <- r6
        (23)  cmp r11,0
        (24)  cbr 27
B4 (25)  lda r2 <- a
   (26)  jmp 28
B5 (27)  lda r2 <- b
B6 (28)  mov r3 <- r1
   (29)  str (r3) <- r11
   (30)  str (r1) <- r12
   (31)  lda r13 <- arr
   (32)  ldc r10 <- 0
B7 (33)  mul r14 <- r10,4
   (34)  add r13 <- r13,r14
   (35)  str (r13) <- 0
   (36)  add r10,1
   (36)  cmp r10,10
   (37)  cbr 33
B8 (38)  ret
```

Fig. 1. Running Example

Memory Representation

A memory location is denoted by a *symbolic address*, which can be either a real address or the position in the code where the address is computed (e.g. by a

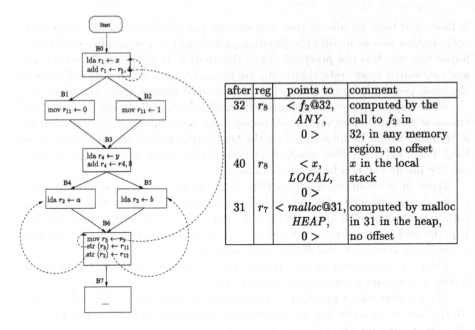

after	reg	points to	comment
32	r_8	< f_2@32, ANY, 0 >	computed by the call to f_2 in 32, in any memory region, no offset
40	r_8	< x, LOCAL, 0 >	x in the local stack
31	r_7	< malloc@31, HEAP, 0 >	computed by malloc in 31 in the heap, no offset

Fig. 2. Point-to set computation examples (from fig 1. Use-def chains are dotted.

malloc invocation). Memory locations, also called *final objects*, are described by a 3-tuple (examples in Figure 2):

1. basic object: either the address of a known symbol or the statement where the value is computed. More precisely a basic object can be:
   - address of a local or a global variable, or address of a function
   - address returned by a memory allocation function (e.g. *malloc*)
   - address returned by a non-memory allocation function
   - address computed by a load statement
   In the first case the basic object is a symbol, while in the last three it is a program statement
2. region: the memory region where the object is allocated. The possible regions are: local stack (LOCAL), global area (GLOBAL), code area (CODE), heap space (HEAP), caller space (CALLER). The caller space is the memory visible to any procedure that calls the current one, i.e. all the memory except the local stack. This is useful for pointers passed as parameters to a procedure, because they cannot (at least with their initial value) point to the local stack.
3. offset: a constant indicating a fixed distance from the basic object.

An object allocated by a dynamic function is identified by the program point where the allocation routine is invoked. Therefore all the objects allocated by the same *malloc* statement are assumed to be alias, a frequent hypothesis in AA projects [2,3,5,7,14].

Combinations of regions are allowed, e.g. a pointer to elements within local stack or global area. (The abbreviation ANY stands for any region).

The AA Algorithm

AA analysis is requested by a driver (the instruction scheduler in this work) which submits a query of the form ¡$(p_1,s_p)(q_1,s_q)$¿. Two steps are executed:

1. Computation of point-to sets: the point-to sets of p_1 in s_p and of q_1 in s_q.
2. The extended intersection (see below) of the two sets is returned.

The on-demand analysis needs some initial data from data-flow analysis, namely: reaching definitions and possible predecessors. The possible predecessors of a basic block B are the blocks in the same procedure that can be executed before B. If the program representation is in SSA [9] and allows registers and no variables, then the analysis becomes completely on-demand.

Step 1. A backwards walk on the use-definition chains finds the definitions of the requested object reaching the given point. Suppose the scheduler wishes to anticipate instruction 30 knowing its immediate dependencies are satisfied after instruction 27. But swapping the order of two *store* is in general not safe by the conservative assumption that alias may be present. The scheduler issues the query: are $(r_3,29)$ and $(r_2,30)$ alias? Figure 2 shows the relevant part of the CFG to determine the points-to sets of r_3 and r_1. Step 1 searches for any definition of r_3 that reaches the first *str* in B6, and it finds the *mov* in B6. Since this is a copy it searches for the definition of r_1 and it finds the *add* in B0. In this statement the constant operand is disregarded, while for the other operand r_1 the algorithm searches again for reaching definitions, finding the *lda* in B0. Combining all this information the point-to set for $(r_3,29)$ is: $< \&x, LOCAL, 4 >$. The computation of the point-to set of r_2 in B6 is similar. There are two reaching definitions: the *lda* in B4 and the *lda* in B5. In neither case there is a need to follow other chains and the point-to set of $(r_2,30)$, after the merge of the two paths, is: $< \&a, LOCAL, 0 >, < \&b, LOCAL, 0 >$

The analysis also needs to perform a safety check. If the pointer p under investigation has its address taken (this means the program contains at least one statement like $p_2 = \&p$), the algorithm must look for the presence of *implicit definitions*. A definition is implicit if the object *obj* content is changed using a name different from *obj*. The algorithm checks that statements executable between the definition and the use, cannot implicitly define the object under investigation. The information on the possible predecessors is used to compute the set of instructions that can be executed between the definition and the use. Suppose AA is computing the point-to set of p_1 in 23. The only explicit definition of p_1 is in 13, but the analysis cannot conclude that p_1 points to x because between 23 and the definition there is code that potentially contains implicit definitions (in particular 22). In general, this check is not very frequent because variables whose address is taken are a minority.

During point-to set computation the algorithm must also pay attention to cyclic and unknown definitions. If a pointer value cannot be determined because of a recursive definition (like $add\ p_1 \leftarrow p_1, 4$) then the recursive chain is simply skipped because the algorithm assumes this is code to access a particular array cell. Since our AA is array unwise there is no need to follow this chain. Similarly if a symbol is added to a pointer, and the symbol can have an unknown value, then it is not considered. Suppose the AA is computing the point-to set of r_{13} in 35. The definition in 34 is skipped and the only one considered is 31. The analysis correctly concludes that r_{13} points inside arr.

Step 2. It examines the two point-to sets computed by step 1 to determine if the pointers can be alias. The basic test is essentially intersection emptyness, but we call it *extended intersection* because, if the intersection is empty, some special cases must be considered before concluding that the pointers are not aliases. In particular if an address is computed by a load or by a non-allocation function, some particular conditions must be imposed.

Program Restrictions

This AA algorithm works on almost any standard C programs with the following restrictions:

- record fields are always accessed using statically and intraprocedurally computable constants. In particular, if a value is added to a structure pointer, it is a field selector if and only if it is smaller than the structure size;
- record field offsets must be positive numbers;
- structure pointer casting is allowed only for homomorphic types;
- longjmp/setjmp/signals are only used to capture events that cause program termination.

A common viewpoint is that AA should be able to analyze any kind of code. As an extreme case [15] claims that it is important, at least in some cases, to correctly handle out of boundaries array accesses. Our philosophy is different: we support only standard C programs. Moreover if elimination of a rarely used feature greatly simplifies analysis, then we think it worth to restrict the class of analyzable programs. In particular the programmer could insert a compilation flag specifying if a certain feature (e.g. the hypothesis on the structure field selection) is absent.

3 Scheduling with AA Information

Each time a memory accessing instruction is encountered while the data dependence graph (DDG) is being built, a conservative dependence arc is added to other instructions accessing memory. This provision is necessary if the scheduler does not have enough information to rule out interference between those instructions.

The availability of on-demand AA permits to discard some dependencies added in a conservative way because of possible memory aliases. One could envision, during the construction of the DDG, to query the AA every time a new arc is about to be added because of possible memory alias. But this approach is not convenient because AA queries are time consuming and should be only performed when they are likely to pay off. But this is not known until the DDG has been entirely built. Therefore DDG construction remains unchanged, we only mark memory dependencies arcs. We assume that all the arcs that can be added by a transitive closure are omitted. This is a common choice to decrease the size of the DDG. The scheduler must select which of the arcs are worth trying to delete through AA. If an arc can be deleted because AA reports no alias, the DDG must be updated. But removing the arc that represents the non existing memory dependency is not enough, because the DDG contains implicit (because initially redundant) arcs.

A complete description of the DDG update algorithm is presented in [4]

Alias Query Heuristics

The scheduler can issue AA queries to try to shorten the scheduler, using different heuristics next described.

Static Criterion. The criterion considers, given a DDG, which queries could decrease the lengths of the longest paths in the DDG or, in equivalent terms, the heights of the highest nodes in the graph. All the nodes belonging to a critical path are inspected for memory-induced arcs and such dependencies are queried; this phase is repeated as long as the length of the critical path is decreased.

We call this criterion *static* because all the queries are performed before issuing any instructions, and, for this reason, are not influenced at all by the evolution of the state of the scheduling algorithm while it is being executed. Its main strengths are:

- it is completely independent of the scheduling algorithm chosen, because it is performed entirely before any instruction has been issued;
- it takes into account the whole DDG in order to shorten its longest paths and not only the local situation that can occur while issuing instructions.

On the other hand, the main drawback is that it does not take into account the resource constraints of the target machine. But it may happen that, even if the longest paths of the DDG have been shortened, the schedule produced for a given target processor is not shorter, if resource constraints prevent exploiting the relaxation of dependencies.

Dynamic Criterion. We have also used a *dynamic* criterion that forces the scheduler to call AA while issuing an instruction, or determining the instructions to issue. This criterion is not independent of the algorithm; it is specific to cycle-based schedulers like top-down or bottom-up ones. The idea is that, every time the scheduler is choosing which instruction to issue, it looks not only for ready instructions, but also for those that would become ready if their memory-induced dependencies were removed. We call such instructions *maybe-ready.*

Under the dynamic criterion, AA is invoked every time that, among all the instructions that could be issued, the priority of the most urgent maybe-ready instruction is higher than that of the most urgent ready instruction. Now, the scheduler has to decide at each iteration whether to issue a ready instruction, to advance to the next cycle, or to query about a memory-induced dependence.

The advantage of the dynamic criterion over the static one is that it takes into account the resources of the target machine (since a resource checker is used to determine which instructions can be issued), hence a query is performed only if the processor can take advantage of it. However, the dynamic criterion is less good than the static for understanding the global worthiness of the queries in the region being scheduled. For instance, it might happen that the dynamic criterion queries about instructions that look urgent (perhaps because they lie on a critical path) and that it discovers, after a later query and the modifications it caused to the DDG, that they were not.

In conclusion, both criteria have pros and cons that make them incomparable from a theoretical point of view. Examples in [4] show cases in which each approach outperforms the other. In Section 5 the two approaches are experimentally compared.

The previous criteria can be combined to form a *static+dynamic* criterion. Such a criterion combines the advantages of both, but the price to pay may be an excessive number of queries.

Finally, in order to compare the efficiency of the criteria, we introduce a *full* criterion, which calls for querying the AA for every memory-induced dependence existing in the DDG.

The full criterion, like the static one, is completely independent of the algorithm chosen for scheduling and of the target machine. This criterion is equivalent to using the AA library in a global and not on-demand way.

4   Effect of Optimizations

Alias information improves the precision of static analysis and the quality of many code transformations. For example [11] shows that reaching definitions can be more accurate if alias information is available. In the case of instruction scheduling there is a very interesting relationship between AA and other optimizations. In particular we experimentally found that if a program is optimized, higher speed-up from AA can be obtained.

The main reason of this behavior is simple. If the original code is not optimized it frequently happens that critical paths are dominated by non-memory related instructions. In these cases the scheduler does not need alias information, hence no benefit comes from its knowledge. The more optimized the code, the more likely it is that memory related instructions are on a critical path. Figure 3 shows a simple example.

```
(x) lda r₃ <- array b
loop:
(0) lda r₂ <- array a
(1) add r₁ <- r₁,4
(2) add r₂ <- r₁,r₂
(3) str 0(r₂) <- 2
(4) str 0(r₃) <- 4
(5) mul r₄ <- r₁,4
(6) add r₄ <- r₄,1
(7) cmple r₅ <- r₁,10000
(8) cbr r₅,loop
```

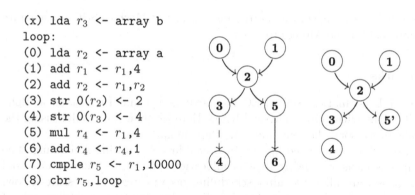

Fig. 3. Impact of optimizations on AA effect. DDG before and after optimization

Suppose the processor has an instruction *mulby4andadd*. In this case instructions 5 and 6 can be merged into 5': *mul4add* $r_4 < -r_1, 1$. After this transformation the memory dependence is on the critical path, and the AA can prove the two *str* instructions are independent. The scheduler (assuming no resource conflicts) can generate a shorter schedule.

5   Experimental Setting and Measurements

Ideally the experiments should be performed using an optimizing VLIW compiler. Since none was available to us for research, we developed a compilation line based on SUIF and Machine SUIF, version 2 [6] [13]. The exact extent to which our results carry to optimizing compilers is unknown, but we conjecture that both speed-up and number of queries would increase.

Target Machines

Our work was oriented towards VLIW processors as target machines, but unfortunately none of the targets supported by Machine SUIF is a VLIW processor. To overcome this problem we specified an Alpha processor that resembles a VLIW machine. We refer to this processor as alpha-vliw. Its instruction is a subset of Alpha, as many instructions are never generated. The number of integer registers has been increased from 32 to 64. alpha-vliw is able to issue up to 4 instructions per cycle, but only one can be memory related (load or store). Latencies of instructions have been decreased. Most non-branch instructions have a latency of 1, except multiply, whose latency is 2. Any hardware for out-of-order execution of instructions or detection of resource violation has been removed.

Since no simulator exists for alpha-vliw, it is not possible to execute the code generated and scheduled for this fictitious machine. We used the profiling information about the execution frequency of basic blocks to compute the execution time of the scheduled program. The execution frequencies were obtained producing code for a real Alpha machine. In order to partially check the correctness of code, we took advantage of the retargeting features of Machine SUIF, to generate code for a real 2-issue Alpha processor.

Benchmarks

We run 27 benchmarks from SPEC 95, Mediabench and PtrDist on a Alpha 21164 700MHz, with 1Gb of physical RAM. Here we show few results (using the alpha-vliw processor), the remaining ones are similar.

Figure 4 shows the speed-up obtained enabling AA during scheduling. The speed-up was measured before register allocation, but very similar results hold when registers are allocated after scheduling because register pressure is low. We did not consider the impact of cache misses, but assuming a similar ratio of misses holds for both codes, with/without AA, the results should be unaffected. Results in the left part of the picture are for completely unoptimized code, while in the right part the initial code had been generated using an optimizing code generator. The experiments are identified by the following labels:

label	Meaning	label	Meaning
1	Static Criterion	2	Dynamic Criterion
3	Static+Dynamic Criterion	4	Full Criterion
4 Fake	Fake Full Criterion	SB	Superblocks Scheduling

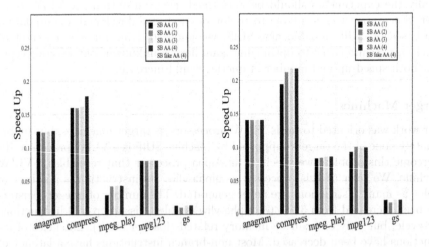

Fig. 4. Speed-up with different AA heuristics. ($Speed - up = \frac{\Delta cycles}{original\ cycles}$)

By *fake* analysis we mean the complete removal of all memory dependencies. This operation does not preserve the program semantics, but it provides an upper bound of the benefit AA can give: at most 19% for unoptimized and 28% for optimized code. For unoptimized code the speedup brought by our AA ranges

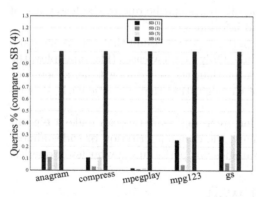

benchmark	$SB(3)$	$SB(2)$	$SB(ad)$
anagram	1.722	1.637	1.13
compress	4.756	3.872	2.4
mpeg_play	437.1	395.6	337.4
mpg123	695.5	692.4	385.9
gs	835.4	575.21	137.6
benchmark	A_t/S_t	A_t/S_t	A_t/S_t
anagram	12%	11.6%	8.4%
compress	20%	17.2%	11.8%
mpeg_play	66.8%	65%	42%
mpg123	50%	49%	42%
gs	38%	30.5%	9.7%

Fig. 5. Number of queries (percentage), AA execution times and (bottom) ratio AA time/scheduling time (sec)

between 1% and 17%. The static+dynamic query heuristics is slightly superior to the static and dynamic ones, and quite close to the speedup obtainable by full AA. For optimized code the speedup of AA (static+dynamic query heuristic) is consistently higher, from 1% to 23%.

We also computed the time spent to perform AA. The results have been obtained on a software prototype where little effort has been devoted to optimize the analysis algorithm and the compiler infrastructure. The times shown in Figure 5 include computation of reaching definitions, possible predecessors and translating the code representation into the format used by the AA algorithm. The AA time is reported also as percentage of the total time needed by scheduling with AA. Figure 5 also shows the number of AA queries performed by different strategies. These numbers are reported as percentage of the queries made by the full criterion.

Main Results Summary

The experimental findings show that AA can significantly speed-up the scheduled code. The benefit varies from benchmark to benchmark and it is also dependent on the quality of the initial code. Benefits between 10% and 20% are frequently achievable. Occasionally for some benchmarks the speed-up was poor: such cases present a code without pointers and with few arrays. The type and quality of the initial code may greatly affect the benefit provided by AA.

Comparing the benefits of on-demand versus full alias analysis we found that most elements of the alias relation are irrelevant for instruction scheduling. In particular no more than 20% of the possible queries provide a benefit.

Our analysis (though not interprocedural) removes on the average about [40−50]% of the memory dependencies. In general the analysis performs quite close to the *fake* analysis. The worst cases are mpegplay and mpg123. In mpegplay few queries are performed and this suggests that the code is not optimized enough. This fact is supported also by the increment of AA benefit deriving from the use of optimized code. In mpg123 the difference can be due to the heavy use of arrays and the lack of interprocedural information.

Concerning the time necessary to perform AA we found that it is in general a fraction of the time required to schedule. Only in a few cases (like mpegplay, the worst case we found) it becomes greater than the scheduling time, a behavior due to the current implementation of the algorithm that is particularly slow on long definition chains.

Experimental results also show that querying heuristics can make a big difference. In particular, quite surprisingly, the dynamic criterion provides almost the same speed-up given by other strategies but performing by far fewer queries.

6   Conclusion and Related Work

This work presents to the best of our knowledge, four original contributions:

- A new algorithm for on-demand AA performed at machine code level.
- An upper bound to the benefit alias information can provide to instructions scheduling. More sophisticated AA algorithms could further improve speed-up but never more than the fake analysis.
- A quantitative study of the impact of alias information on VLIW instructions scheduling and on the percentage of alias relation that is important to the instruction scheduler. Speed-ups of 10-20% can be expected with little increase in compilation time.
- A qualitative study of the interaction between original code quality and benefit the AA can provide.

Although AA is a widely studied topic ([8,15,3,14,12]) and some papers have considered analysis of assembly code ([2] and [5]) and on-demand analysis [7] for higher level code, no study is known to us on the impact of AA on code performance.

In particular the sensitivity of scheduling performances on alias information is a subject much talked about but never quantitatively studied. Our experiments show the speed-up that a non cyclic instruction scheduler can obtain exploiting alias information. It is also clear that schedulers can achieve 90% of the benefit from AA reducing the number of queries to 10%.

The last point shows that the gain from AA is strictly related to the original code quality.

Future work: we intend to refine the study of the relationship of alias analysis benefits and level of initial program optimization. It is already in an advanced stage of development a Machine SUIF support for the LX (ST200) processor, a VLIW design by ST Microelectronics. Preliminary results confirm what we have already obtained with the alpha-vliw. Finally it would be interesting to redo our experiments using different instruction schedulers, in particular a loop scheduler.

Acknowledgement

This research was made possible by the contribution of persons of Harvard University (Michael Smith, Glenn Holloway, and Gang Chen), Politecnico di Milano (Giampaolo Agosta and Vincenzo Martena), and ST Microelectronics (Marco Cornero, Andrea Ornstein and Julien Zory). Thanks to Stefan Freudenberger and HP people at the Cambridge lab for suggestions.

References

1. A. V. Aho, R. Sethi, and J. D. Ullman. *Compilers Principles, Techniques, and Tools*. Addison Wesley, 1986.
2. W. Amme, P. Braun, E. Zehendner, and F. Thomasset. Data dependence analysis of assembly code. In *Proceedings of the 1998 PACT*, pages 340–347, 1998.
3. B. Cheng and W. W. Hwu. A practical interprocedural pointer analysis framework. Technical Report IMPACT-99-01, University of Illinois, Urbana, IL, USA, 1999.
4. R. Costa. Integrating on-demand alias analysis into schedulers for advanced microprocessors. Master's thesis, Politecnico di Milano, December 2001.
5. S. Debray, R. Muth, and M. Weippert. Alias analysis of executable code. In *Conference Record of POPL '98*, pages 12–24, San Diego, California, 1998.
6. G. Aigner et al. *An Overview of the SUIF2 Compiler Infrastructure*. Computer Systems Laboratory, Stanford University.
7. N. Heintze and O. Tardieu. Demand-driven pointer analysis. *ACM SIGPLAN Notices*, 36(5):24–34, May 2001.
8. W. Landi and B. G. Ryder. A safe approximate algorithm for interprocedural pointer aliasing. In *Proceedings PLDI*, volume 27, pages 235–248, New York, NY, July 1992. ACM Press.
9. S. S. Muchnick. *Advanced compiler design and implementation*. Morgan Kaufmann, 1997.
10. D. A. Patterson and J. L. Hennessy. *Computer Architecture: A Quantitative Approach*. Morgan Kaufmann, San Mateo, CA, 1990.
11. M. Shapiro and S. Horwitz. The effects of the precision of pointer analysis. *Lecture Notes in Computer Science*, 1302, 1997.
12. M. Shapiro and S. Horwitz. Fast and accurate flow-insensitive points-to analysis. In ACM, editor, *Conference record of POPL '97*, pages 1–14, 1997.
13. M. D. Smith and G. Holloway. *An Introduction to Machine SUIF and Its portable Libraries for Analysis and Optimization*. DEAS, Harvard University, 2001.
14. B. Steensgaard. Points-to analysis in almost linear time. In *Conference Record of POPL'96)*, pages 32–41. ACM Press, January 21–24, 1996.
15. R. P. Wilson and M. S. Lam. Efficient context-sensitive pointer analysis for C programs. In *Proceedings of the PLDI '95*, pages 1–12, 18–21 June 1995.

Low-Cost Value Predictors
Using Frequent Value Locality

Toshinori Sato[1,2] and Itsujiro Arita[1]

[1] Department of Artificial Intelligence
[2] Center for Microelectronic Systems
Kyushu Institute of Technology
tsato@ai.kyutech.ac.jp

Abstract. The practice of speculation in resolving data dependences has been recently studied as a means of extracting more instruction level parallelism (ILP). Each instruction's outcome is predicted by value predictors. The instruction and its dependent instructions can be executed in parallel, thereby exploiting ILP aggressively. One of the serious hurdles for realizing data speculation is the huge hardware budget required by the predictors. In this paper, we propose techniques that exploit frequent value locality, resulting in a significant budget reduction. Based on these proposals, we evaluate two value predictors, named the *zero-value predictor* and the *0/1-value predictor*. The zero-value predictor generates only value 0. Similarly, the 0/1-value predictor generates only values 0 and 1. Simulation results show that the proposed predictors have greater performance than does the last-value predictor which requires a hardware budget twice as large as that of the predictors. Therefore, the zero- and the 0/1-value predictors are promising candidates for cost-effective and practical value predictors which can be implemented in real microprocessors.

1 Introduction

Modern microprocessors boost performance by exploiting instruction level parallelism (ILP). However, several obstacles limit ILP. These include dependences between instructions and are classified into three classes – control, name, and data dependences. This paper focuses on reducing data dependences. Recent studies have shown that these data dependences can be speculatively resolved by value prediction. An outcome of an instruction is predicted by means of value predictors, and the instruction and its dependent instructions can then be executed in parallel, thereby exploiting ILP aggressively.

However, there is a serious hurdle for exploiting data speculative execution in real microprocessors. In order to efficiently utilize value prediction without incurring misspeculation penalties, value predictors must operate at high predictability. Some predictors such as hybrid[21] and context-based[18] predictors achieve significantly high predictability rates at considerable hardware cost. This hardware cost is one of the serious hurdles for realizing data speculation, and

H. Zima et al. (Eds.): ISHPC 2002, LNCS 2327, pp. 106–119, 2002.

thus it should be alleviated. Furthermore, even embedded microprocessors support out-of-order execution[10]. Thus, low-cost value predictors are required.

Many studies[3–5, 12–15] have been devoted to reducing the hardware cost of value predictors. One way of reducing this cost is through the use of simple circuits. For example, direct-mapped tables are smaller than highly associative ones. Reducing the number of entries in the value prediction table by carefully selecting predicting instructions[4, 5, 12, 13] is also a simple technique to reduce cost. Tag array size can be saved by employing partial resolution, using fewer tag address bits than necessary to uniquely identify every instruction[15]. Full resolution of value predictors is not necessary since they do not have to be correct all the time. Data array size can be reduced if the value predictor keeps only low-order bits of data values, exploiting narrow width values. It has been found that across SPECint95 benchmark programs, over 50% of integer operands are 16 bits or less[1]. Therefore, it is possible to keep only low-order bits of data values in the value predictor in order to reduce its data array size[3, 14]. In this paper, we propose an alternative technique to obviate the data array cost.

The organization of the rest of this paper is as follows: Section 2 surveys related work. Section 3 describes our evaluation environment. Section 4 presents how frequent value locality is found and proposes cost-effective value predictors which are evaluated in Section 5. Finally, Section 6 concludes the paper.

2 Related Work

Value prediction[8, 11] is a speculative technique which executes instructions using predicted data values. Data dependences are speculatively resolved and thus ILP is increased. Many studies have proposed value prediction mechanisms, some of which achieve predictability rates as high as 80%[18, 21]. However, these predictors such as 2-level, hybrid, and context-based predictors require considerable hardware cost for realizing their high predictabilities.

Morancho et al.[12], Rychlik et al.[13], Del Pino et al.[5], and Calder et al.[4] have examined the capacity constraints of value predictors. Morancho et al.[12], Rychlik et al.[13], and Del Pino et al.[5] proposed to reduce hardware cost by classifying instructions based on their value predictability. Easily predictable instructions use simpler predictors such as last-value and stride predictors, whose hardware cost is low. High-cost predictors such as 2-level and the context-based predictors are used only for hard-to-predict instructions. Calder et al.[4] proposed filtering instructions based on their level of criticality. Only instructions regarding critical paths are held in the value predictor, resulting in capacity saving. On the other hand, Fu et al.[7] and Tullsen et al.[20] remove value prediction hardware completely with the aid of the compiler management of values in registers.

Partial resolution is first evaluated in the research area of branch prediction. Fagin[6] proposed to reduce the tag array bitwidth in branch target buffers (BTBs). He investigated the tradeoff between bitwidth and branch prediction accuracy, and found that it is possible to restrict the bitwidth of the tag array to 2 bits without severe performance loss. We evaluated the partial resolution of several value predictors[15] and found that no tag bits are required for the last-value predictor and that only 2 bits are required for the hybrid predictor.

Recently, we examined a technique for reducing hardware cost by exploiting narrow width values[14]. Brooks et al.[1] found that across SPECint95 benchmark programs, over 50% of integer operands are 16 bits or less. That is, only low-order bits of the value prediction table are utilized. Therefore, we propose to keep only low-order bits of data values in the prediction table in order to reduce its hardware cost. Simulation results showed that performance improvement due to data speculation is maintained accompanied by the hardware cost savings of 45.1%[14]. A similar technique was proposed by Burtscher et al.[3], where infrequently changing high-order bits are shared by several values held in the prediction table.

3   Evaluation Environment

In this section, we describe our evaluation environment by explaining a processor model and benchmark programs.

3.1   Processor Model

We use two types of simulators for this study. One is a functional simulator for evaluating predictability and the other is a timing simulator for evaluating processor performance. The timing simulator models wrong path execution caused by misspeculations. We implemented the simulators using the SimpleScalar tool set (ver.3.0a)[2]. SimpleScalar/PISA instruction set architecture (ISA) is based on MIPS ISA.

The timing simulator models a realistic 8-way out-of-order execution superscalar processor based on a register update unit[19] which has 128 entries. Each functional unit can execute any operation. The latency for execution is 1 cycle except in the cases of multiplication (4 cycles) and division (12 cycles). A 4-port, non-blocking, 128KB, 32B block, 2-way set-associative L1 data cache is used for data supply. It has a load latency of 1 cycle after the data address is calculated and a miss latency of 6 cycles. It has a backup of an 8MB, 64B block, direct-mapped L2 cache which has a miss latency of 18 cycles for the first word plus 2 cycles for each additional word. A memory operation that follows a store whose data address is unknown cannot be executed. A 128KB, 32B block, 2-way set-associative L1 instruction cache is used for instruction supply and also has the backup of the L2 cache which is shared with the L1 data cache. For control prediction, a 1K-entry 4-way set associative BTB, a 4K-entry gshare-type 2-level adaptive branch predictor, and an 8-entry return address stack are used. The branch predictor is updated at the instruction commit stage.

For the purpose of comparison with our proposal, we also evaluate the last-value predictor[11]. Its main structure is the Value History Table (VHT). The VHT has Tag, Data Value, and Conf fields. The Tag field is for identifying each instruction. As determined by a previous study[15], the Tag field does not always contribute to predictor performance and thus we omit it in this study. The Data Value field stores the last result of the associated instruction. When the same instruction is encountered, the Data Value is used for its predicted value. The Conf field is a saturating up-down counter that determines whether oa not the instruction should be predicted. The counter is incremented or decremented whenever a prediction is correct or incorrect, respectively. In our study, 2-bit saturating counters are used. The threshold value necessary to trigger a prediction is 2. We use a direct-mapped table for the predictor, since we are interested in cost-effective predictors.

When a misspeculation occurs, it is necessary to revert the processor state to a safe point where the speculation is initiated. We use an instruction reissue mechanism which selectively flushes and reissues misspeculated instructions. We have already proposed a practical instruction reissue mechanism[17]. When comparing an actual value with its associated predicted value, a one cycle penalty of the comparison stage is included when every mispredicted instruction is reissued.

3.2 Benchmark Programs

This study uses the SPECint95 benchmark suite. We focus on improving the performance of only integer programs because it tends to be more difficult to obtain high levels of parallelism from these types of programs than from floating-point programs. The input files are modified to achieve a practical evaluation time. We use the object files provided by the University of Wisconsin-Madison[2]. Each program is executed to completion. The candidate instructions for value prediction are register-writing ones, and did not include branch and store instructions. We counted only committed instructions.

4 Cost-Effective Value Predictors

Frequent value locality[22] is a new kind of locality, and is distinct from traditional value locality[11] in that the former is observed across several instructions and the latter is defined in each instruction. This section investigates how frequent values are observed and proposes cost-effective value predictors.

4.1 Frequently Occurring Values

Zhang et al.[22] observed that at any given point in the program's execution, a small number of distinct values occupy a vast fraction of referenced locations, when the values involved in memory accesses are traced. Furthermore, values 0 and 1 are the most frequently accessed values in the case of SPECint95 programs. We extend their study to trace all register-writing instructions. The simulation results are summarized in Figure 1. Each bar is divided into two parts. The bottom part (black) indicates the percentage of dynamic instructions which generate the value 0. The top one (gray) is for the value 1. On average, 15.3% of dynamic

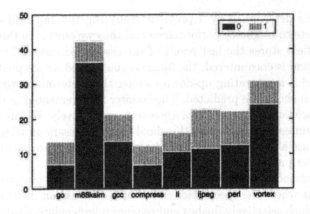

instructions generate the value 0. While this is much smaller than the frequency in memory access, which is nearly 50%, this frequency is still large enough to exploit these characteristics. In addition, the percentage of dynamic instructions which generates the value 1 is 7.3% on average. Tables 1 and 2 summarize instructions which frequently generate the values 0 and 1, respectively, for each benchmark program. The first column is the program name. The second column presents the top five instructions which frequently produce the value 0 or 1. The next one indicates the percent occupancy that is defined as the number of the associated instructions over that of all dynamic instructions which generate the value. The last column is percent frequency, meaning how frequently the associated instruction generates the value. This can be explained by using the case of 099.go. In this case, 34.79% of all dynamic instructions which produce the value 0 is instruction lw. In addition, 9.07% of all dynamic instructions lw generates the value 0. As can be easily seen, the instruction lw most frequently produces the value 0. The next frequent instruction is instruction addu. It is interesting that any subtract instructions do not appear in Table 1. It is also interesting that logical instructions occupy small parts. In contrast, logical instructions frequently generate value 1 as can be seen in Table 2. In the following subsection, we propose cost-effective value predictors utilizing the localities of values 0 and 1.

4.2   Zero-Value Predictor

Since the value 0 is the most frequent value generated, processors may benefit from a value predictor which provides only value 0. We call this predictor *zero-value predictor*[16]. It can be implemented by removing the Data Value field completely from the VHT. That is, the zero-value predictor consists of only Conf and Tag[1] fields. This significantly reduces the hardware cost of the predictor.

---

[1] As we will see in Section 5.1, the Tag field is optional.

Table 1. Instructions frequently generating value of zero

program	inst.	%occu.	%freq.	program	inst.	%occu.	%freq.
099.go	lw	34.79	9.07	130.li	lw	49.86	11.58
	slt	27.05	51.92		addu	23.23	15.99
	addu	18.15	5.42		andi	12.04	63.56
	sll	5.70	2.93		lbu	7.61	26.76
	alti	5.46	37.62		addiu	1.81	0.85
124.m88ksim	lw	16.71	29.69	132.ijpeg	lw	22.64	9.84
	addu	16.67	45.55		addu	15.86	5.64
	sll	13.83	98.82		slt	15.15	57.81
	and	11.07	56.89		sll	11.22	11.33
	andi	8.36	37.48		sltu	10.72	47.78
126.gcc	lw	32.93	14.30	134.perl	lw	30.98	11.02
	addu	23.79	19.27		addu	18.41	12.75
	sltiu	6.23	41.83		slt	8.48	74.71
	sll	6.23	14.96		andi	6.79	50.39
	slt	5.21	45.91		sltu	5.17	25.32
129.compress	lw	22.20	5.84	147.vortex	lw	53.04	27.54
	slti	18.10	45.61		addu	37.06	41.26
	slt	17.46	31.19		sltu	2.58	37.42
	sltu	7.54	42.00		andi	1.39	16.19
	and	7.23	31.55		lbu	1.13	41.54

4.3   0/1-Value Predictor

Figure 1 shows that the value 1 also occurs frequently. Hence, it is expected that applying the frequent one-value locality by building the *0/1-value predictor* improves the value predictabilities. In order to distinguish the values 0 and 1, the 0/1-value predictor should have a 1-bit Data Value field in the VHT. In addition, its replacement policy becomes a complex issue. When an instruction produces the value 0 or 1, the data value field simply holds the value. However, if the generated value is neither 0 nor 1, we can determine a replacement policy according to many choices. In this paper, we use a simple replacement policy. Only when the generated value is 1 does the data value field in the VHT keep the value 1. Otherwise, it keeps the value 0. In other words, the 0/1-value predictor has a priority on the value 0. This decision is proper since the value 0 is the most frequently generated value, as we have already seen in Figure 1.

5   Evaluation

In this section, we present simulation results. First, we evaluate predictability. We define predictability as the number of instructions that are (correctly and incorrectly) predicted by a value predictor over that of all register-writing instructions. We also define prediction coverage as the number of instructions correctly predicted over that of all register-writing instructions. On the other

Table 2. Instructions frequently generating value of one

program	inst.	%occu.	%freq.	program	inst.	%occu.	%freq.
099.go	lw	30.57	7.56	130.li	sltu	22.18	90.51
	slt	26.40	48.08		sltiu	16.63	95.05
	addiu	25.04	8.76		lbu	16.62	30.36
	slti	9.54	62.38		slti	16.13	100.0
	addu	3.20	0.91		andi	8.56	23.48
124.m88ksim	sltu	33.32	66.15	132.ijpeg	slti	26.17	87.34
	slti	33.27	99.93		lw	19.94	10.86
	addiu	33.18	16.34		addiu	19.65	12.78
	sra	0.07	0.39		sltu	9.35	52.22
	lw	0.03	0.01		slt	8.82	42.19
126.gcc	addiu	20.93	7.17	134.perl	sltiu	20.01	88.54
	lw	18.69	4.54		sltu	19.08	74.68
	sltiu	15.50	58.17		addiu	18.84	9.30
	slti	15.15	71.08		lw	10.69	3.04
	slt	10.99	54.09		lbu	6.59	31.95
129.compress	slt	45.45	68.81	147.vortex	addiu	52.04	18.03
	slti	25.47	54.39		sltu	15.49	62.58
	sltu	12.30	58.00		lw	11.26	1.61
	addiu	4.95	1.14		lhu	6.51	35.90
	sra	2.57	8.61		addu	5.50	1.71

hand, prediction accuracy, which is the percentage of instructions correctly predicted over all predicted instructions, can be easily obtained using the following equation.

$$(Prediction\ accuracy) = \frac{(Prediction\ coverage)}{(Predictability)}$$

After that, processor performance is evaluated.

5.1 Predictability

There is a tradeoff between predictability and prediction accuracy. If every instruction initiates value prediction in order to increase predictability, hard-to-predict instructions must be included and thus the prediction accuracy is degraded. Therefore, only easily predictable instructions should be carefully selected. In the cases of the zero- and the 0/1-value predictors, their predictabilities are severely limited since they can predict only the value 0 or 1. And thus, the tradeoff point in the predictors is more difficult to find than in any conventional value predictors. From these considerations, we investigate how a confidence mechanism affects predictabilities. After the confidence mechanism is determined, we evaluate the relationship between the table capacity and predictabilities. Please remember that we use the functional simulator for evaluating predictabilities.

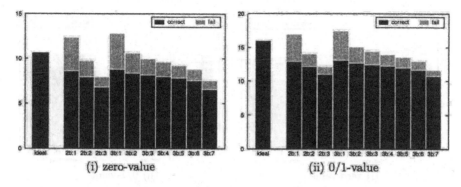

Fig. 2. (%)Predictability (infinite capacity): 130.li

Confidence. We evaluate 2-bit and 3-bit saturating up-down counters for the confidence mechanism. The threshold value for triggering every prediction varies between 1 and 3 for the 2-bit counter and between 1 and 7 for the 3-bit counter. Figure 2 summarizes the results for the zero- and the 0/1-value predictors. Due to space limitation, only results for 130.li are shown. It should be noted that the VHT has infinite capacity. The left-most bar indicates the upper boundary for the predictability which is determined by the frequent value localities observed in Section 4.1. The remaining bars indicate the predictabilities and the prediction coverages for all confidence mechanisms evaluated. Each simulation result is indicated by a group consisting of the bit length of the counter and the threshold value. For example, if a simulation result is denoted as 3b:5, it presents the confidence mechanism as the 3-bit counter, and the threshold value is 5. Each bar is divided into two parts. The lower part (black) indicates the percentage of the instructions whose data value is correctly predicted. The upper part (gray) indicates the percentage that is mispredicted. That is, the lower part is the prediction coverage, while the sum of the two parts is the predictability.

We can find the following. Figure 2 indicates that, first, there is no significant difference between the 2-bit and the 3-bit counters. Thus, in regard to hardware cost efficiency, we select the 2-bit counter for the confidence mechanism. Second, in the case of a threshold value of 1, the percentage of misprediction is considerably high. On the other hand, a threshold value of 3 limits predictability. These observations indicate that the threshold value is 2. Throughout the rest of this paper, the 2-bit saturating up-down counter with a threshold value of 2 is used.

Capacity. Next, we present how table capacity affects predictabilities. Figure 3 summarized the results. The left-most bar indicates the case of infinite capacity, which we have seen above. The remaining bars indicates the cases of finite capacity tables. For each group of six bars, the bars from left to right are for the results of 1K-, 2K-, 4K-, 8K-, 16K-, and 32K-entry VHTs, respectively. Each bar is divided into two parts as in Figure 2. The left group presents the results of the VHT which has the Tag field, and the right one presents that of the VHT whose

114   Toshinori Sato and Itsujiro Arita

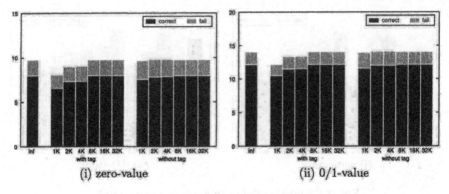

(i) zero-value   (ii) 0/1-value

Fig. 3. (%)Predictability (finite capacity): 130.li

Tag field is removed. We would like to consider the VHT without tag because
the evaluated predictors are special kinds of last-value predictor and hence it
is expected that the tag contributes little to the predictabilities and processor
performance improvement as we have already seen in [15]. It is also desirable
that removing the tag further significantly reduces the hardware cost. However,
the tagless VHT increases aliasing. Different instructions share a single predictor
entry. In general, an aliasing may be constructive, destructive, or neutral. However, in the case of the zero-value predictor, the predicted value is only 0 and
hence destructive aliasing rarely occurs. In order to increase constructive aliasing, the zero-value predictor can consist of two individual tables, one of which is
for high confidence instructions while the other is for low confidence ones, such
as the bi-mode branch predictor[9]. On the other hand, the 0/1-value predictor
suffers destructive aliasing more often than does the zero-value predictor since
it holds both values 0 and 1 in each entry. This problem also can be mitigated
by the using bi-mode predictor, which has two dedicated tables for predicting
the values 0 and 1. We leave consideration of the bi-mode prediction for future
study.

The following are observed. First, the VHT without tag marks slightly larger
predictability than does the one with tag at the risk of large misspredictions,
as we expected. Second, for most programs, the predictability of the VHT with
8K-entry is comparable to that of the VHT with infinite capacity. Based on these
observations, we determine that the table capacity is 8K. This results in an 8K-
entry tagless zero-value predictor whose hardware cost is only 2K bytes. This
is approximately equal to the costs of 256- and 512-entry last-value predictors
with and without tag, respectively, and is less than 2% in the area of on-chip
caches in modern microprocessors. On the other hand, the cost for the 8K-entry
0/1-value predictor is 3K bytes since it has a 1-bit Data Value field.

5.2 Processor Performance

In this subsection, we evaluate how much the zero- and the 0/1-value predictors
contribute to processor performance. Each predictor has an 8K-entry VHT whose

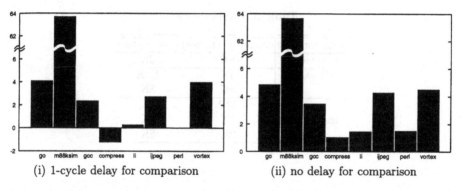

Fig. 4. (%)Processor performance improvement (zero-value)

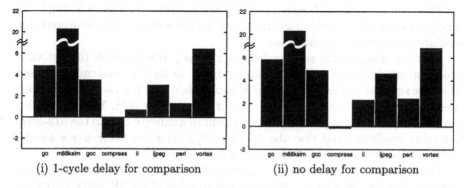

Fig. 5. (%)Processor performance improvement (0/1-value)

Tag field is removed. For measuring performance, we use committed instructions per cycle (IPC). Only useful instructions are considered for counting the IPC. We do not count nop instructions. We define percent speedup as the increased IPC over the IPC of the baseline model. After that, we compare the proposed value predictors with the last-value predictor. It should be noted that a timing simulator is used for evaluating processor performance.

Figure 4(i) presents the percent speedup when the 8K-entry zero-value predictor is utilized. Except for 129.compress, substantial speedup is observed. Especially in the case of 124.m88ksim, a 63.8% speedup is achieved. We can see that the speedup figure closely follows the pattern of the frequent value locality depicted in Figure 1. While the zero-value predictor can generate only value 0, data speculation contributes considerably to processor performance. From these observations, it is confirmed that utilizing the frequent zero-value locality is an effective way of exploiting ILP.

Since the zero-value predictor generates only value 0, the comparison between a predicted value (always 0) and its actual value is very simple. Therefore, it may be possible to verify each prediction during the execution stage, and thus the one cycle penalty of each mispredicted instruction can be removed. The results of

Table 3. Hardware cost

predictor type	capacity (entries)	cost (bytes)
zero-value	8K	2K
0/1-value	4K	1.5K
	8K	3K
last-value	512	2K
	1K	4K
	2K	8K

this scenario are summarized in Figure 4(ii). Processor performance is improved for all programs.

The results for the 0/1-value predictor are presented in Figure 5. While the absolute values of the speedup are different from those of the zero-value predictor, similar tendencies can be observed.

Next, we compare the zero- and the 0/1-value predictors with the last-value predictor which is the simplest value predictor so far. We evaluate six cases. They are summarized in Table 3, which also explains the hardware cost of every predictor evaluated. Note that no predictor has the Tag field. We can see that the 8K-entry zero-value predictor is equivalent in hardware cost to the 512-entry last-value predictor, and that the 4K-entry 0/1-value predictor has a smaller hardware cost than does the 512-entry last-value predictor.

The percent of performance improvement is shown in Figure 6. For each group of six bars, the first one indicates the speedup of the 8K-entry zero-value predictor. The second and the third bars indicate those of the 4K- and 8K-entry 0/1-value predictors. The remaining bars from left to right are for the 512-, 1K-, and 2K-entry last-value predictors, respectively. Note that the last-value predictor suffers a one cycle penalty from every mispredicted instruction but that the zero- and the 0/1- value predictors do not. Primary observation is that the zero-value predictor contributes more to processor performance than does the 1K-entry last-value predictor, which is twice as large in hardware cost as the 8K-entry zero-value predictor. Moreover, for most programs, the zero-value predictor is comparable in speedup to the 2K-entry last-value predictor. In the case of 124.m88ksim, the zero-value predictor overwhelms the last-value predictor. This is a good example for explaining the benefit gained from the frequent zero-value locality. Because the last-value predictor evaluated has a quite small capacity, each entry holding the value 0 is replaced by the other values. This diminishes the benefit. On the other hand, the 8K-entry zero-value predictor has sufficient capacity for capturing the frequent zero-value locality. From these observations, we can confirm the cost-efficiency of the zero-value predictor.

The 0/1-value predictor has different characteristics from the zero-value predictor. In the cases of 099.go, 126.gcc, 134.perl, and 147.vortex, even the 4K-entry 0/1 value predictor attains a speedup comparable to the 2K-entry last-value predictor, which requires a hardware cost five times as large. In the cases

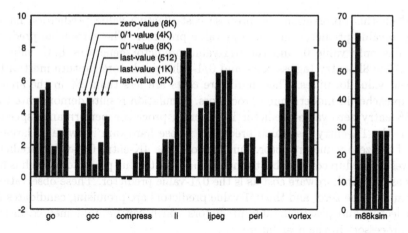

Fig. 6. (%)Comparison with last-value predictor

of 132.ijpeg and 124.m88ksim, the discrepancy of speedup between the 0/1-value and the last-value predictors is modest. These observations confirm that the 0/1-value predictor is also a cost-effective alternative to the zero-value predictor. However, in the cases of 129.compress and 130.li, the discrepancy is significant. When we compare the 0/1-value predictor with the zero-value predictor, the most interesting result is found in the case of 124.m88ksim. While the predictability of the 0/1-value predictor is larger than that of the zero-value predictor and the prediction accuracies of both predictors are almost perfect, the speedup in the former is less than one third of the latter. This is due to the serious destructive aliasing between the values 0 and 1. If the 0/1-value predictor cannot predict instructions producing the value 0 on critical paths while the zero-value predictor can, its speedup is severely diminished. Nonetheless, it is confirmed that the 0/1-value predictor is as cost-effective as the zero-value predictor, since the former is faster in speedup than the latter except for the cases of 129.compress and 124.m88ksim.

6   Conclusions

Recently, the practice of speculation in resolving data dependences has been studied as a means of aggressively extracting ILP. One of the serious hurdles for realizing data speculation is the huge hardware budget required by the value predictors. This is because the predictors which achieve high predictability for alleviating misspeculation penalties require considerable hardware cost. This paper uses these observations to explore ways to reduce the hardware budget of data value predictors.

Detailed simulation results show the existence of is frequent value locality in programs. We find that across SPECint95 benchmark programs, on average 15.3% and 7.3% of dynamic instructions generate the value 0 and 1, respectively.

Based on this observation, we propose cost-effective value predictors, named the zero-value predictor and the 0/1-value predictor. The zero-value predictor generates only value 0, and the 0/1-value predictor generates both values 0 and 1. The 8K-entry tagless zero- and 0/1-value predictors capture most of the frequent value localities. Their hardware costs are less than 2% in the area of on-chip caches in modern microprocessors. Simulation results demonstrate that the 8K-entry zero-value predictor is faster in processor performance speedup than is the 1K-entry last-value predictor, whose hardware is twice as large as that of the zero-value predictor. Furthermore, the 4K-entry 0/1-value predictor achieves a speedup comparable to the 2K-entry last-value predictor, which is five times as large in hardware cost as is the 0/1-value predictor. These observations confirm that the zero- and the 0/1-value predictors are promising candidates for cost-effective and practical value predictors which can be implemented in real microprocessors in the near future.

Future study includes investigation and evaluation of the bi-mode zero- and 0/1-value predictors for reducing mispredictions caused by the destructive aliasing.

Acknowledgments

This work is supported in part by a Grant-in-Aid for Encouragement of Young Science (#12780273) and a Grant-in-Aid for Scientific Research (B) (#12780273) from the Japan Society for the Promotion of Science.

References

1. Brooks D., Martonosi M.: Dynamically exploiting narrow width operands to improve processor power and performance. 5th Int. Symp. on High Performance Computer Architecture (1999)
2. Burger D., Austin T.M.: The SimpleScalar tool set, version 2.0. ACM SIGARCH Computer Architecture News, 25(3) (1997)
3. Burtscher M., Zorn B.G.: Hybridizing and coalescing load value predictors. Int. Conf. on Computer Design (2000)
4. Calder B., Reinman G., Tullsen D.M.: Selective value prediction. 26th Int. Symp. on Computer Architecture (1999)
5. Del Pino S., Pinuel L., Moreno R.A., Tirado F.: Value prediction as a cost-effective solution to improve embedded processor performance. 3rd Int. Meeting on Vector and Parallel Processing (2000)
6. Fagin B.: Partial resolution in branch target buffers. IEEE Trans. Comput., 46(10) (1997)
7. Fu C-y., Jennings M.D., Larin S.Y., Conte T.M.: Software-only value speculation scheduling. Tech. Rep., Dept of Electr. and Comp. Eng., North Carolina State University (1998)
8. Gabbay F.: Speculative execution based on value prediction. Tech. Rep., Dept of Electr. Eng., Technion (1996)
9. Lee C-C., Chen I-C.K., Mudge T.N.: The bi-mode branch predictor. 30th Int. Symp. on Microarchitecture (1997)

10. Levy M. NEC processor goes out of order. Microprocessor Report, 15(9) (2001)
11. Lipasti M.H., Wilkerson C.B., Shen J.P.: Value locality and load value prediction. Int. Conf. on Architectural Support for Programming Languages and Operation Systems VII (1996)
12. Morancho E., Llaberia J.M., Olive A.: Split last-address predictor. Int. Conf. on Parallel Architectures and Compilation Techniques (1998)
13. Rychlik B., Faistl J.W., Krug B.P., Kurland A.Y., Sung J.J., Velev M.N., Shen J.P.: Efficient and accurate value prediction using dynamic classification. Tech. Rep., Dept of Electr. Comp. Eng., Carnegie Mellon University (1998)
14. Sato T., Arita I.: Table size reduction for data value predictors by exploiting narrow width values. 14th Int. Conf. on Supercomputing (2000)
15. Sato T., Arita I.: Partial resolution in data value predictors. 29th Int. Conf. on Parallel Processing (2000)
16. Sato T., Arita I.: Reducing hardware budget of data value predictors by exploiting frequent value locality. IEICE Tech. Rep. CPSY2000-62 (2000)
17. Sato T.: Evaluating the impact of reissued instructions on data speculative processor performance. Microprocessors and Microsystems, 25(9) (2002)
18. Sazeides Y., Smith J.E.: Implementations of context based value predictors. Tech. Rep., Dept of Electr. Comp. Eng., University of Wisconsin-Madison (1997)
19. Sohi G.S.: Instruction issue logic for high-performance, interruptible, multiple functional unit, pipelined computers. IEEE Trans. Comput., 39(3) (1990)
20. Tullsen D.M., Seng J.S.: Storageless value prediction using prior register values. 26th Int. Symp. on Computer Architecture (1999)
21. Wang K., Franklin M.: Highly accurate data value prediction using hybrid predictors. 30th Int. Symp. on Microarchitecture (1997)
22. Zhang Y., Yang J., Gupta R.: Frequent value locality and value-centric data cache design. Int. Conf. on Architectural Support for Programming Languages and Operation Systems IX (2000)

Integrated I-cache Way Predictor and Branch Target Buffer to Reduce Energy Consumption

Weiyu Tang, Alexander Veidenbaum, Alexandru Nicolau, and Rajesh Gupta

University of California at Irvine, Irvine CA 92697-3425, USA

Abstract. In this paper, we present a Branch Target Buffer (BTB) design for energy savings in set-associative instruction caches. We extend the functionality of a BTB by caching way predictions in addition to branch target addresses. Way prediction and branch target prediction are done in parallel. Instruction cache energy savings are achieved by accessing one cache way if the way prediction for a fetch is available.

To increase the number of way predictions for higher energy savings, we modify the BTB management policy to allocate entries for non-branch instructions. Furthermore, we propose to partition a BTB into ways for branch instructions and ways for non-branch instructions to reduce the BTB energy as well.

We evaluate the effectiveness of our BTB design and management policies with SPEC95 benchmarks. The best BTB configuration shows a 74% energy savings on average in a 4-way set-associative instruction cache and the performance degradation is only 0.1%. When the instruction cache energy and the BTB energy are considered together, the average energy-delay product reduction is 65%.

1 Introduction

To exploit instruction level parallelism, modern processors support out of order execution engines. This requires instruction fetch across basic block boundaries to keep the execution units busy. Many instructions will be fetched and issued before a branch target address can be resolved.

A BTB [16] is used for branch target address prediction. There are two flavors of BTB design, coupled and decoupled where the tradeoff in performance and energy is fixed at the design time.

The coupled BTB design is used in Alpha 21264 [12] and UltraSparc-II [1], where the BTB is integrated with the instruction cache. Each cache line has two additional fields, a "line-prediction" field and a "way-prediction" field, to predict a cache line for the next fetch.

The decoupled design is used in Pentium 4 [8], Athlon [2] and G5 [7], where a BTB is organized as a separate cache. The number of BTB entries in Pentium 4, Athlon and G5 is 4K, 2K and 4K, respectively. The branch instruction address is used to index a BTB. If a matching entry is found and the branch is predicted taken, the BTB will provide the target address for the next fetch.

In a processor with a coupled BTB design, one instruction cache way is accessed when the way prediction is available. Therefore the instruction cache

H. Zima et al. (Eds.): ISHPC 2002, LNCS 2327, pp. 120–132, 2002.
© Springer-Verlag Berlin Heidelberg 2002

energy is lower than that in a processor with a decoupled BTB design where all the ways are accessed in parallel for high performance.

However, the accuracy of target address prediction in a coupled BTB is often lower than that in a decoupled design with same number of entries and associativity. "Way-prediction" and "line-prediction" fields in a coupled BTB can only point to one cache line. Prediction misses often occur if there are multiple branches in a cache line or branches change direction. A decoupled BTB can still provide accurate prediction in such cases. In modern processors with long pipelines, high target address prediction accuracy by the BTB is crucial for high performance.

In this paper, we propose a **Way-Predicting BTB** (WP-BTB) for energy savings in set-associative instruction caches. A WP-BTB makes modifications to a conventional decoupled BTB design. When a branch is allocated an entry in the BTB, in addition to the branch target address, way predictions for both the cache line with the target address and the cache line with the fall-through address are saved in the WP-BTB. When a branch accesses the WP-BTB for next fetch address and the branch has an entry in the WP-BTB, both the next fetch address and the corresponding way prediction are provided. Hence the next fetch needs only to access one cache way.

To enable more way predictions for higher instruction cache energy savings, we also modify the WP-BTB allocation policy. Non-branch instructions are also allocated entries in the WP-BTB. This increases the number of BTB accesses and hence overall BTB energy consumption. To keep the BTB energy low, a set-associative WP-BTB can be partitioned into ways for branch instructions and ways for non-branch instructions. As a consequence, the overall BTB energy is reduced due to lower per BTB access energy as not all the ways are accessed.

The rest of this paper is organized as follows. In Section 2, we briefly describe related work on cache and BTB energy savings. We present in Section 3 the design of WP-BTB. The experimental results are given in Section 4. This paper is concluded with future work in Section 5.

2   Related Work

2.1   Cache Energy Savings

High utilization of the instruction memory hierarchy is needed to exploit instruction level parallelism. As a consequence, energy consumption by on-chip instruction caches can comprise as high as 27% of the CPU energy [11].

Cache way partitioning is often used for energy savings in set-associative caches. The energy consumption by one cache way in a n-way set-associative cache is approximately $1/n$ the total cache energy consumption when all the ways are accessed.

Way prediction technique as used in Alpha 21264 can reduce instruction cache energy as only one cache way is accessed when the prediction is correct. To minimize way prediction miss penalty in a coupled BTB design, tags for all the ways are accessed in parallel with predicted way data access. The percentage of energy consumption by tag access increases with the number of cache

ways. Based on energy parameters generated by Cacti [17], the relative energy
consumption by a 4-way tag access and a one-way data access in a 32KB 4-way
set-associative cache is 44% and 56% respectively. In an 8-way cache, the energy
consumption by an 8-way tag access and a one-way data access is 66% and 34%
respectively.

Another method of way prediction is proposed in [10]. A table is used to
save for each cache set the most recently used way. On a cache access, the
table is accessed first to retrieve the way prediction. Then the predicted way
is speculatively accessed. On a way prediction miss, the remaining ways are
accessed in the next cycle. This approach suffers from serialization, which may
affect the processor cycle time if cache access is on a critical path of the processor.

Alpha 21164 uses another kind of access serialization for L2 cache energy
savings. Tags for all ways are accessed first to determine which way holds the
data. Then that particular data way is accessed. This kind of access serialization
nearly doubles cache access time and cannot apply to the L1 cache where short
cache access time is critical to performance.

[3] recognizes that not all the cache ways are needed in an application and
proposes to turn off some ways for energy savings based on application demands.
All the remaining ways are accessed in parallel. Hence energy savings are much
smaller than those in way prediction based techniques where only one way is
accessed most of the time. This technique also cannot apply to applications with
large working set because the number of energy expensive L1 cache misses may
increase dramatically.

Cache subbanking and line buffers [9] are used for cache energy savings. These
techniques are orthogonal to way prediction based energy saving techniques.
For example, when cache subbanking is used in a n-way set-associative cache,
n subbanks, one from each cache way, have to be accessed in parallel. If the
cache way holding the data is known before the access, then only one subbank
needs to be accessed. As a consequence, the percentage of energy savings by way
prediction based techniques is unchanged.

Filter cache [13] and L-cache [4] place a small and hence low-energy cache
in front of the L1 cache. This design has high performance penalty and is only
effective for applications with a small working set.

2.2   BTB Energy Savings

A large BTB is required for high performance. The number of BTB accesses is
large and its energy consumption is also high. Rather small 512-entry BTB of
the Pentium Pro consumes about 5% of the processor energy [14].

BTB is often organized as a set-associative cache and several ways can be
turned off for energy savings if application demands are low [3]. [15] uses a BTB
predictor for BTB energy savings. The assumption is that BTB misses occur in
bursts. If a miss in the BTB is predicted for the next access, then BTB is not
accessed for the next branch instruction.

Decreasing the BTB entry size can also reduce the BTB energy. In G5 [7],
instruction size is 32 bits. Instead of using 32 bits each for branch address and
target address, the number of bits for each of them is 13 and 19 respectively. This
approach trades branch target address prediction accuracy for energy savings.

3   WP-BTB Design

3.1   Way Prediction for Branch Instructions

Each entry of the WP-BTB is augmented with two fields for way prediction:

- *twp*: **T**arget **W**ay **P**rediction for the cache line with the branch target address;
- *fwp*: **F**all-through **W**ay **P**rediction for the cache line with the fall-through address.

Each entry in the WP-BTB is in the following format:

$$(\text{branch\_addr}, \text{target\_addr}, \text{lru}, \text{twp}, \text{fwp}).$$

Branch_addr and *target_addr* are used for target address prediction and *lru* is used to manage the replacement in a set-associative BTB.

When a branch address hits in the WP-BTB, the predicted target address and the corresponding way prediction are provided simultaneously. Using way prediction, the next fetch needs only to access one cache way.

The following hardware support is needed for way prediction in the WP-BTB:

- A queue, called **W**ay **Q**ueue (WQ), holds recently hit ways;
- A queue, called **B**TB **U**pdate **R**equest **Q**ueue (BURQ), holds pending WP-BTB update requests;
- Additional fields for each branch instruction in the pipeline:
  - *wq_ptr*: pointer to the next entry in the WQ;
  - *fwp, twp*: way predictions of the hit WP-BTB entry of this branch.

For a branch, the next fetch line is either the fall-through cache line or the target cache line depending on the direction of the branch. WQ and *wq_ptr* are used to track the way for either the target cache line or the fall-through cache line. BURQ is needed because the WP-BTB update requests cannot be processed immediately if the next fetch address misses from the cache or the target address prediction is not correct. Fields *fwp* and *twp* are needed for way prediction update.

Figure 1 shows modifications (labels "1", "2", "3", "4" and "5") to the conventional design when an instruction fetch finishes. First, the hit way is added to the WQ. Second, if the head entry in the BURQ is waiting for the finish of this fetch, then that entry is ready to commit changes to the WP-BTB. Third, if a branch instruction is found, the *wq_ptr* of this branch is set to the next entry in the WQ. When a branch instruction commits and the branch prediction is correct, *wq_ptr* is used to retrieve the way for either the fall-through cache line or the target cache line. Fourth, if the way prediction for the next fetch is found in the WP-BTB, the next fetch needs only to access one way. Otherwise, all cache ways are enabled for the next fetch as indicated by label "5".

Modifications to the branch instruction commit are shown in Figure 2. In a conventional BTB, if a branch has an entry in the BTB, changes to the BTB can be updated immediately because the target address is already resolved. In the

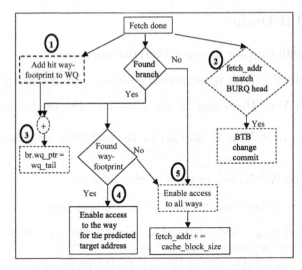

Fig. 1. Modifications when instruction fetch finishes.

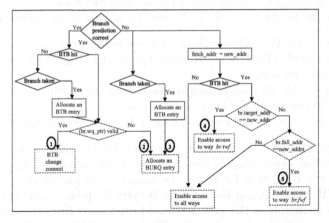

Fig. 2. Modifications for branch instruction commit.

WP-BTB, the way prediction is also needed. Thus changes to the WP-BTB can
be updated only after the way prediction is available. There are two scenarios
when way prediction is not available (label "2" and "3"). In the first scenario,
the branch target prediction is correct but the next fetch address is missing from
the cache. In the second scenario, the predicted address is not correct and the
cache line with the correct address is to be fetched in the next cycle. In either
scenario, the way prediction will be available once next fetch finishes. Therefore
the WP-BTB update request will be added to BURQ first. Then once the next
fetch finishes, the head entry of the BURQ will be ready to update the WP-BTB
(label "2" in Figure 1). On branch prediction misses, if the way prediction for
the cache line with the correct fetch address is found in the WP-BTB (label "4"
and "5"), then only one way needs to be accessed for the next fetch.

3.2   Way Prediction for Non-branch Instructions

As only branch instructions access the WP-BTB, way prediction is limited to speculative fetch addresses. As a consequence, instruction cache energy savings are only available to speculative instruction fetches. To enable more instruction cache energy savings, non-branch instructions are allowed to access the WP-BTB for way prediction. For this purpose, every non-branch instruction in the pipeline needs the following additional fields, which are similar to the fields added for branch instructions:

- wq_ptr: pointer to the next entry in the WQ;
- fwp : way prediction for the fall-through cache line.

Wq_ptr is needed to track the way prediction for the next fetch line and fwp is needed for WP-BTB way prediction update.

If there is no branch in a cache line that is under fetch, the following changes are needed:

- the address of the first instruction is used to index the WP-BTB to find the way prediction for the fall-through cache line;
- the first instruction in the cache line will set wq_ptr to the next entry in the WQ;
- if the way prediction is found, then only one way will be enabled for the next fetch.

When a non-branch instruction commits, if its wq_ptr is valid and it doesn't have an entry in the WP-BTB, an entry will be allocated. Then, if the way prediction is available at this time, the WP-BTB can be updated immediately. In case the way prediction is not available, an entry will be added to the WQ to wait for the finish of the next fetch.

3.3   WP-BTB Partitioning

As a WP-BTB is organized as a cache, way-based cache energy saving techniques can also apply to a set-associative WP-BTB. A WP-BTB can be partitioned into ways for branch instructions and ways for non-branch instructions. We assume that there is predecoding or other mechanisms to determine whether a WP-BTB access is by a branch instruction or by a non-branch instruction. Therefore only the corresponding WP-BTB ways need to be enabled. Reduced switching activities per WP-BTB access result in BTB energy savings.

4   Experimental Results

4.1   Simulation Setup

We use the SimpleScalar toolset [6] to model an out-of-order superscalar processor. The processor parameters shown in Table 1 roughly correspond to those in

Table 1. System configuration.

Parameter	Value
branch pred.	combined, 4K 2-bit chooser, 4k-entry bimodal, 12-bit, 4K-entry global
BTB	2K-entry, 4-way
RUU	64
LSQ	32
fetch queue	16
fetch width	4
int. ALUs	4
flt. ALUs	2
int. Mult/Div	2
flt. Mult/Div	2
L1 Icache	64KB, 4-way, 32B block
L1 Dcache	64KB, 4-way, 32B block
L2 cache	512KB, 4-way, 64B block

a current high-end microprocessor. Wattch [5] is used for energy estimation. A set of SPEC95 benchmarks are simulated. The test input set is used.

We have evaluated the performance and energy of five WP-BTB configurations:

- *base*: only branch instructions access the WP-BTB;
- *share*: both branch and non-branch instructions can access all ways in the WP-BTB;
- 1_3: the WP-BTB is partitioned into 1 way for branch instructions and 3 ways for non-branch instructions;
- 2_2: the WP-BTB is partitioned into 2 ways for branch instructions and 2 ways for non-branch instructions;
- 3_1: the WP-BTB is partitioned into 3 ways for branch instructions and 1 way for non-branch instructions.

4.2 Results

Figure 3 shows branch hit rate in the WP-BTB. As branch prediction is not available for a branch missing from the WP-BTB, low hit rate means low prediction accuracy. The hit rate decreases with effective BTB capacity, the number of entries available for branch instructions. *Share* almost always has the highest hit rate. The rate by 3_1 is almost same as that by *share*. For some benchmarks, the difference in hit rate by different configurations is very small. These benchmarks either don't have many branches like *applu*, or have many branches with high access locality like *ijpeg* and *li*. BTB capacity decrease won't impact hit rate much for these benchmark. On the other hand, *hydro2d*, *su2cor*, *tomcatv* and *gcc* are affected the most by the decrease in BTB capacity and associativity. They have many branches that cannot fit in a small and low associativity BTB.

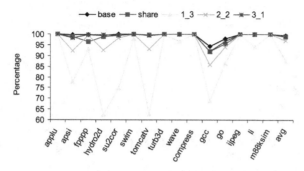

Fig. 3. Branch hit rate in the WP-BTB.

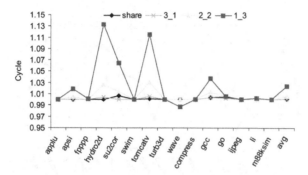

Fig. 4. Normalized execution time.

Figure 4 shows normalized execution time with regard to *base*. For 2_2 and 1_3, five benchmarks–*apsi*, *hydro2d*, *su2cor*, *tomcatv* and *gcc* show high performance degradation. Referring to Figure 3, we notice that execution time increases whenever the branch hit rate in BTB decreases. Hit rate decrease results in performance degradation. There is virtually no performance degradation for *share* and 3_1 as the branch hit in the BTB is unchanged. The average performance degradation for *share*, 3_1, 2_2 and 1_3 is 0.1%, 0.1%, 0.6% and 2.4% respectively.

Figure 5 shows the non-branch access hit rate in the WP-BTB. *Share* has the highest hit rate, followed in order by 1_3, 2_2 and 3_1 as the number of entries available for non-branch access decreases. Most benchmarks have small working set and 1K entries for non-branch instructions are enough for way prediction by most instructions. Hence the difference in hit rate among 2_2 , 1_3 and *share* is very small and the hit rate for *share* and 1_3 is close to 100%.

Noticeable differences in hit rate can be found in half of the benchmarks between 2_2 and 3_1. For these benchmarks, 512 entries for non-branch instructions are not enough for way prediction by instructions in the working set. For example, *fpppp* has a large working set, which is evident as instruction cache miss rate in *fpppp* increases from 0.5% to 7.5% when the cache size decreases from 64KB to 32KB. Therefore the differences in hit rate among different configurations are large.

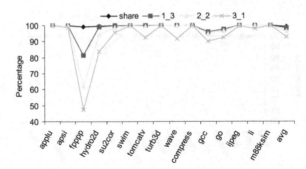

Fig. 5. Non-branch access hit rate in the WP-BTB.

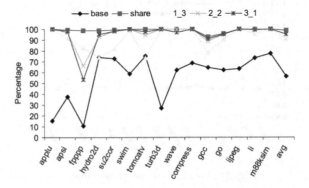

Fig. 6. BTB hit rate for branch and non-branch accesses.

Figure 6 shows BTB hit rate for branch and non-branch accesses. High hit rate indicates more number of way predictions. *Share* has the highest hit rate and *base* has the lowest. The hit rate for partitioned BTB configurations varies. 1_3 generally has lower hit rate than 2_2 and 3_1. For all benchmarks except *fpppp*, there is at least one partitioned configuration that has a hit rate as high as *share*.

Figure 7 shows instruction cache energy savings. The percentage in savings follows the same trend as the BTB hit rate shown in Figure 6. When there is a hit in the BTB, the way prediction for the next fetch line can be retrieved and the next fetch needs only to access one cache way. High hit rate results in high instruction cache energy savings. The average percentage in energy savings for *share*, 1_3, 2_2 and 3_1 is 74.2%, 68.3%, 71.6% and 71.3% respectively. *Share* results in the best instruction cache energy savings.

Figure 8 shows processor energy contribution (clocking energy excluded) by the instruction cache and the BTB for configuration *base*. The instruction cache contributes a much larger percentage to the processor energy than the BTB. The contribution by the instruction cache ranges from 17% in *fpppp* to 31% in *gcc* with an average of 23%. The contribution by the BTB ranges from 0.25% in *fpppp* to 4.55% in *ijpeg* with an average of 2.97%.

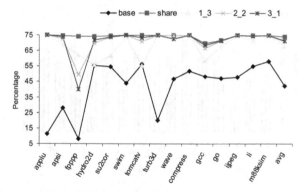

Fig. 7. Instruction cache energy savings.

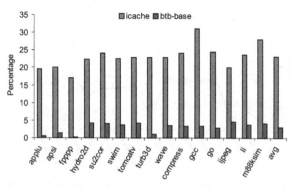

Fig. 8. % processor energy contribution by the instruction cache and the BTB for configuration *base*.

Figure 9 shows processor energy contribution by the instruction cache and the BTB for configuration *share*. Comparing with Figure 8, we notice that the instruction cache energy decreases dramatically. On the other hand, average BTB energy consumption nearly doubles with additional BTB activity by non-branch instruction accesses. For some benchmarks such as *swim* and *ijpeg*, the energy consumption by the BTB is larger than that by the instruction cache. The average BTB energy is roughly equal to the average instruction cache energy. This figure shows the need to keep BTB energy under control as well.

Figure 10 shows processor energy contribution by the BTB with different WP-BTB configurations. The BTB energy by *share* is much higher than that by other configurations because the number of BTB accesses increases dramatically. Although the number of accesses in partitioned BTB configurations also increases, per access energy decreases because not all the ways are accessed. For 5 benchmarks, *base* achieves the minimal BTB energy. For the remaining 10 benchmarks, 1_3 achieves the minimal BTB energy. 1_3 has the lowest average BTB energy, followed in order by 2_2, 3_1 and *base*. The differences among them

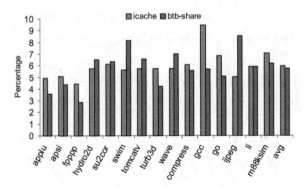

Fig. 9. % processor energy contribution by the instruction cache and the BTB for configuration *share*.

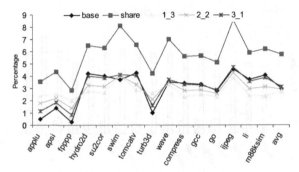

Fig. 10. % processor energy contribution by the BTB with different WP-BTB configurations.

are small. We conclude that partitioned BTB configurations don't increase BTB energy much even with more number of accesses.

Figure 11 shows total energy contribution by the instruction cache and the BTB. Partitioned BTB configurations achieve the minimal energy for all benchmarks except *fpppp*. Which partitioned configuration is the best for a given benchmark depends on the working set size and the ratio of branch instruction accesses to the non-branch instruction accesses. For example, 3_1 is the best for *applu* and *turb3d*. In these benchmarks, the working set size is small and most of the BTB accesses are by non-branch instructions. Single BTB way for non-branch instruction accesses results in minimal overall energy. As either compiler or profiling techniques may determine the branch to non-branch access ratio and the working set size, the optimal BTB partition may be determined statically based on application demands for the best tradeoff in performance and energy.

Figure 12 shows energy-delay product for the instruction cache and the BTB. For each benchmark, point *min* represents the minimal energy-delay product for this benchmark among all the configurations. The minimal energy-delay product ranges from 0.29 for *m88ksim* to 0.42 for *fpppp* and the average is 0.35. An

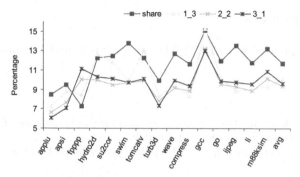

Fig. 11. % processor energy contribution by the instruction cache and the BTB with different WP-BTB configurations.

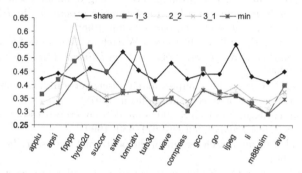

Fig. 12. Energy-delay product for the instruction cache and the BTB.

average 65 percent energy-delay product reduction for the instruction cache and the BTB is achieved by the WP-BTB.

5   Conclusion

In this paper, we have presented a way-predicting BTB design. It enables way prediction by both branch and non-branch instructions for energy savings in set-associative instruction caches. In addition, a set-associative WP-BTB is partitioned into ways for branch instruction and ways for non-branch instructions to keep the WP-BTB energy under control.

Configuration *share* is the best in terms of instruction cache energy. It achieves an average 74% of energy saving in a 4-way set-associative instruction cache with only 0.1% performance degradation. When the BTB energy and the instruction cache energy are considered together, partitioned BTB configurations are better than the *share* configuration and can achieve 65% reduction in energy-delay product.

We are currently investigating other techniques such as compiler/profiling techniques and dynamic mechanisms to determine application-specific BTB configuration with the best tradeoff in energy and performance.

Acknowledgments

This work is supported by DARPA/ITO under PACC and DIS programs.

References

1. K.B. Normoyle et al. UltraSparc-IIi: expanding the boundaries of system on a chip. *IEEE Trans. Micro*, 18(2):14–24, 1998.
2. Advanced Micro Devices, Inc. *AMD athlon processor architecture*, 2000. White paper.
3. D. H. Albonesi. Selective cache ways: on-demand cache resource allocation. In *Int'l Symp. Microarchitecture*, pages 248–259, 1999.
4. N. Bellas, I. Hajj, and C. Polychronopoulos. Using dynamic cache management techniques to reduce energy in a high-performance processor. In *Int'l Symp. on Low Power Electronics and Design*, pages 64–69, 1999.
5. D. Brooks, V. Tiwari, and M. Martonosi. Wattch: a framework for architectural-level power analysis and optimizations. In *Int'l Symp. Computer Architecture*, pages 83–94, 2000.
6. D. Burger and T.Austin. The simplescalar toolset, version 2.0. Technical Report TR-97-1342, University of Wisconsin-Madison, 1997.
7. M. Check and T. Slegel. Custom S/390 G5 and G6 microprocessors. *IBM Journal of Research and Development*, 43(5/6):671–680, 1999.
8. G. Hinton et al. The microarchitecture of the pentium 4 processor. *Intel Technology Journal*, Q1, 2001.
9. K. Ghose and M. Kamble. Reducing power in superscalar processor caches using subbanking, multiple line buffers and bit-line segmentation. In *Int'l Symp. on Low Power Electronics and Design*, pages 70–75, 1999.
10. K. Inoue, T. Ishihara, and K. Murakami. Way-predicting set-associative cache for high performance and low energy consumption. In *Int'l Symp. on Low Power Electronics and Design*, pages 273–275, 1999.
11. J. Montanaro et al. A 160-MHz, 32-b, 0.5-W CMOS RISC microprocessor. *IEEE Journal of Solid-State Circuits*, 32(11):1703–14, 1996.
12. R. Kessler. The Alpha 21264 microprocessor. *IEEE Micro*, 19(2):24–36, 1999.
13. J. Kin, M. Gupta, and W. Mangione-Smith. The filter cache: An energy efficient memory structure. In *Int'l Symp. Microarchitecture*, pages 184–193, 1997.
14. S. Manne, A. Klauser, and D. Grunwald. Pipeline gating: speculation control for energy reduction. In *Int'l Symp. Computer Architecture*, pages 132–141, 1998.
15. E. Musoll. Predicting the usefulness of a block result: a micro-architectural technique for high-performance low-power processors. In *Int'l Symp. Microarchitecture*, pages 238–247, 1999.
16. C. Perleberg and A. Smith. Branch target buffer design and optimization. *IEEE Trans. Computers*, 42(4):396–412, 1993.
17. S. Wilton and N. Jouppi. An enhanced access and cycle time model for on-chip caches. Technical Report 93/5, Digital Western Research Laboratory, 1994.

A Comprehensive Analysis
of Indirect Branch Prediction

Oliverio J. Santana[1], Ayose Falcón[1], Enrique Fernández[2], Pedro Medina[2],
Alex Ramírez[1], and Mateo Valero[1]

[1] Dpto. de Arquitectura de Computadores, Universidad Politécnica de Cataluña
{osantana,afalcon,aramirez,mateo}@ac.upc.es
[2] Dpto. de Informática y Sistemas, Universidad de Las Palmas de Gran Canaria
{efernandez,pmedina}@dis.ulpgc.es

Abstract. Indirect branch prediction is a performance limiting factor
for current computer systems, preventing superscalar processors from
exploiting the available ILP. Indirect branches are responsible for 55.7%
of mispredictions in our benchmark set, although they only stand for
15.5% of dynamic branches. Moreover, a 10.8% average IPC speedup is
achievable by perfectly predicting all indirect branches.

The Multi-Stage Cascaded Predictor (MSCP) is a mechanism proposed
for improving indirect branch prediction. In this paper, we show that a
MSCP can replace a BTB and accurately predict the target address of
both indirect and non-indirect branches. We do a detailed analysis of
MSCP behavior and evaluate it in a realistic setup, showing that a 5.7%
average IPC speedup is achievable.

Keywords: microarchitecture, branch prediction, Branch Target Buffer,
indirect branch, Multi-Stage Cascaded Predictor

1 Introduction

Current computer systems rely on instruction level parallelism (ILP) to achieve
high performance. Superscalar processors are designed to exploit ILP, but control
hazards prevent them from taking advantage of all the available ILP. One of
the most used mechanisms to overcome control hazards is branch prediction. A
decoupled branch prediction architecture [2] uses a direction predictor to predict
if conditional branches will be taken or not taken and a target address predictor
to predict where a taken branch will go. There is a wide amount of research
about improving the accuracy of branch direction predictors [15,9,10], but the
use of a branch target buffer (BTB) is commonly accepted as a good way of
predicting target addresses [8,16,12]. However, some authors [7,1,3,4] claim that
the target addresses of indirect branches are not accurately predicted by a simple
BTB.

In this paper, we show that indirect branch prediction is a hard problem for
computer systems and that there is room for improvement. In our benchmark
set, branch instructions are correctly predicted 91.2% of the time on average.

H. Zima et al. (Eds.): ISHPC 2002, LNCS 2327, pp. 133–145, 2002.

If we do not take into account return instructions, indirect branches are only correctly predicted 45.3% of the time. Overall, indirect branches are responsible for 55.7% of all branch mispredictions, although they only stand for 15.5% of dynamic branch instructions. A 10.8% average IPC speedup can be obtained by perfectly predicting all indirect branches.

The Multi-Stage Cascaded Predictor (MSCP) is a mechanism described by Driesen et al. [6] for improving the prediction of indirect branches. In their previous study, they evaluate MSCP behavior using only indirect branch traces. We contribute to the study of this technique by showing that MSCP can work well with complete programs, replacing the BTB. Using a MSCP to predict the target address of branch instructions provides an increase in target address prediction accuracy, as well as an improvement in processor performance. This improvement indicates that MSCP can predict accurately target addresses of both indirect and non-indirect branches.

We do a detailed analysis of MSCP behavior in order to understand how it works. MSCP improves processor performance by providing an important reduction in mispredictions due to indirect branches. We show that, in a three level configuration, the intermediate level is poorly used. Most branches, whose target addresses are easy to predict, use the first level, while difficult to predict branches are better predicted using the last level.

This deep analysis of MSCP allows us to find better configurations and to optimize its performance in a realistic setup. Using a *gskewed* branch direction predictor [10] and replacing the BTB with a MSCP of an equivalent hardware budget we achieve 5.7% average IPC speedup. We show that this speedup cannot be achieved by a larger BTB. Therefore, problems in target address prediction in studied benchmarks are not due to conflict misses, but to the unpredictability of indirect branches using a simple BTB.

We also analyze the behavior of some static indirect branches in our integer benchmark suite. There is a small amount of static indirect branches that are responsible for a great number of mispredictions in some benchmarks. These branches are indirect jumps in high level switch structures and function calls using pointers. Therefore, not only object oriented programs will find beneficial an improvement in indirect branch prediction. In addition, we show that a difficult to predict indirect branch not necessarily should have a lot of different dynamic target addresses. The predictability of an indirect branch in MSCP depends on whether or not there is correlation of its target address with the target address of previously executed branches.

The remainder of this paper is organized as follows. Section 2 exposes previous related work. In section 3 we describe our simulation environment and the selected benchmark set. In section 4 we show the relative importance of indirect branches with regards to the rest of branch instructions. In section 5 we describe the MSCP and analyze its behavior, proposing a realistic configuration. In section 6 we study the behavior of some static indirect branches. Finally, section 7 exposes our concluding remarks.

2   Related Work

One of the best known approaches to indirect branch prediction is the return address stack (RAS) proposed by Kaeli et al. [7]. In [1] Calder et al. show that indirect branches will be a limiting factor to performance with the increase of popularity of object oriented programming. They analyze some static and dynamic techniques for improving the prediction of indirect function calls. We show in this paper that better predicting indirect branches is important not only for object oriented programs.

In [11] Nair proposes to record the path of target addresses of recent branches in the history register instead of a pattern of bits meaning taken or not taken branches. This technique is used for indexing the prediction tables of MSCP. In [3] Chang et al. propose a target cache to predict the target address of indirect branches. This mechanism has a prediction table indexed by a value obtained hashing the branch instruction address with the contents of a history register. Stark et al. present in [14] a mechanism which index its prediction table using different history lengths for each branch, according to previously collected profile information.

The main source of inspiration of this paper is the previous work of K. Driesen and U. Hölzle in indirect branch prediction. In [4] they explore a variety of two-level predictor configurations dedicated to predict the target address of indirect branches. In [5] Driesen et al. propose the cascaded predictor as a new way of combining two prediction tables. This predictor keeps easy to predict branches in a first level table and allows difficult to predict branches, i.e. indirect branches, to use a second level table, which takes advantage of correlation for keeping more information. The Multi-Stage Cascaded Predictor is a generalization of this technique proposed in [6]. We contribute to the study of this mechanism by showing that it is able to replace a BTB and accurately predict the target address of both indirect and non-indirect branches. We do a detailed analysis of its internal behavior and evaluate its impact on processor performance presenting IPC measures.

3   Experimental Methodology

Our data has been obtained using the *sim-outorder* simulator from the *SimpleScalar 3.0 tool set*. This simulator, configured to use the Alpha instruction set architecture, models a six stage pipeline with a register update unit (RUU). We have collected detailed information about static branch instructions from *sim-outorder*. The baseline configuration is a 4-way issue processor described in table 1. This configuration uses a 2048 entry 4-way associative BTB for target address prediction and a 64 entry RAS for predicting return instructions. A *gskewed* predictor [10] is used to predict the direction of conditional branches. It is a 12 KB predictor with three 16K entry tables and a history length of 14 bits.

We have selected the nine programs from the SPECint 95 and 2000 benchmarks where indirect branch prediction is a harder problem. These benchmarks

Table 1. Configuration of the baseline processor

fetch width	4
issue width	4
RUU entries	64
load/store queue entries	32
integer ALUs	4
floating point ALUs	4
integer multiplier/dividers	1
floating point multiplier/dividers	1
first level instruction cache	32 KB, 2-way associative, block size: 32 B
first level data cache	32 KB, 2-way associative, block size: 32 B
second level unified i+d cache	1 MB, 4-way associative, block size: 64 B

allow us to measure MSCP performance with complete programs instead of using only indirect branch traces. They also allow us to show that not only object
oriented programs would find MSCP beneficial. The selected benchmarks were
compiled with the Compaq C V5.8-015 compiler on Compaq UNIX V4.0 with
options *-O2*, *-g3* and *-non_shared*. SPECint 95 benchmarks were executed to
completion using a modified version of the *test* or *train* input set. SPECint 2000
benchmarks were executed to completion or until 500 million instructions were
executed. SPECint 2000 used the *test* input set, excluding *253.perlbmk*, which
used the *train* input set.

4   The Importance of Indirect Branches

In this section we show the importance of indirect branches with regards to
the rest of branch instructions. By perfectly predicting all indirect branches, we
get a 10.8% average IPC speedup over our baseline configuration, getting 20% in
253.perlbmk or even 31% in *134.perl*. This data shows the potential improvement
a mechanism for better predicting indirect branches can achieve and that such
a mechanism will be worthwhile.

Figure 1 shows the prediction accuracy of branch instructions. The first
bar represents prediction accuracy for indirect branches excluding return instructions, which are already correctly predicted by a RAS [7]. The second bar
represents prediction accuracy for all branch instructions. While the average
branch prediction accuracy is 91.2%, indirect branches are correctly predicted
only 45.3% of the time. This low prediction accuracy is caused by the BTB,
since indirect branches are always taken in the Alpha instruction set architecture. Therefore, a BTB cannot predict accurately the target address of indirect
branches.

In figure 2 we show the percentage of indirect branches over the total number
of dynamic branch instructions and the percentage of all branch mispredictions
due to indirect branches. The first bar indicates that indirect branches are a small
percentage of branch instructions in our benchmark set. However, the second bar

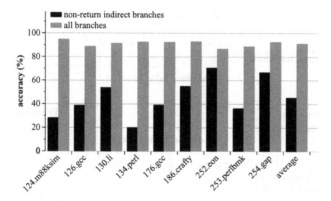

Fig. 1. Comparison of prediction accuracy for indirect branches, excluding returns, and all branches

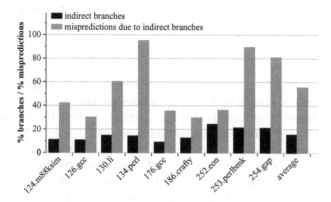

Fig. 2. Comparison of the percentage of indirect branches over the total number of dynamic branch instructions and the percentage of all branch mispredictions due to indirect branches

points out that indirect branches are responsible for most branch mispredictions. On average, indirect branches represent only 15.5% of dynamic branch instructions, although they are responsible for 55.7% of all mispredictions. This makes clear that, although research has been focused on branch direction predictors, better ways of predicting indirect branches should be developed.

5   Analysis of Multi-stage Cascaded Predictor

In this section we analyze and evaluate the Multi-Stage Cascaded Predictor. A more detailed study can be found in [13]. First of all, we do a description of MSCP in section 5.1. In section 5.2 we present an analysis of MSCP behavior and performance measures using IPC. We show the influence of MSCP in indirect branch prediction and how different table levels are used. Finally, we evaluate

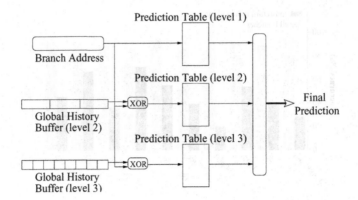

Fig. 3. Multi-Stage Cascaded Predictor

MSCP in section 5.3, showing that this mechanism can outperform a BTB with the same hardware budget in a realistic setup.

5.1 Description

The Multi-Stage Cascaded Predictor, shown in figure 3, is designed to predict the target address of branch instructions. Therefore, this predictor replaces the BTB in a decoupled branch prediction architecture [2]. MSCP consists on three prediction tables that make their prediction in parallel. Each prediction table is indexed by *xoring* the branch instruction PC with the contents of a history register. This history is built using the target address of recently executed branches.

The number of branches used to build the history is called path length. The higher the table level is, the larger its path length should be. First level table uses a zero path length, so it is indexed only with the branch instruction PC and behaves like a BTB. MSCP chooses the prediction from the higher level which has information about the branch being predicted. Higher levels are supposed to be more accurate because they keep more information about each branch by using correlation with a larger path length.

To work well, MSCP should keep easy to predict branches in the first level table while only difficult ones should be allowed to use higher level tables. A new branch, that is, a branch which cannot be predicted because MSCP has no information about it, is always introduced in the first level table. Only if the prediction for later instances of that branch fails repeatedly it will promote to higher level tables.

5.2 Behavior Analysis

In order to study the behavior of MSCP we evaluate a configuration with three prediction tables. Each one is a 2048 entry 4-way associative table. Therefore, this configuration has 6K entries, three times more than the baseline BTB. The path length is three branches for the second level and eight branches for the third level, as in [6].

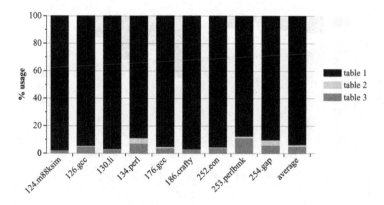

Fig. 4. Percentage of usage of each table prediction in MSCP over the whole number of predictions done

The size of this MSCP configuration is large enough to analyze its behavior without resource limitations. In this section we use a perfect branch direction predictor to ensure that all branch mispredictions are actually target address mispredictions. With this approach, all non-indirect branches are easily predictable, so we can focus on the predictability problems of indirect branches.

We obtain a 5.7% average IPC speedup along our benchmark set by replacing the baseline BTB with the analyzed setup of MSCP. The benchmark *134.perl* even achieves a 21.5% IPC speedup. MSCP gets an average reduction of 55% of this kind of mispredictions, achieving even 87% in *134.perl* or 93% in *254.gap*. This data shows that MSCP improves processor performance by better predicting indirect branches.

Behavior of Each MSCP Prediction Table. To better understand how MSCP works, we analyze the use of each table. In figure 4 we show the percentage of usage of each table prediction over the total number of predictions done. Most predictions have been done by the first level table. This means that a majority of branches can be easily predicted by a simple BTB without correlation and do not need to be upgraded toward higher levels. The third level table is the one that predicts difficult branches, while the second is rarely used.

Almost all target mispredictions are caused by predictions done by the third level table. This happens because predictable branches remain in the first level while only difficult branches are promoted to the third level table. Therefore, the first and the second level tables will only predict branches that they are able to predict, while the third level table will predict not only the branches it is able to predict, but also those branches which cannot be predicted by any table.

Figure 5 shows the percentage of indirect branch predictions done by each table over the total number of indirect branch predictions. Most indirect branch predictions are done by the third level table because indirect branches are supposed to be too difficult for the first level table to predict them. However, there is a small amount of indirect branches in some benchmarks that can be correctly

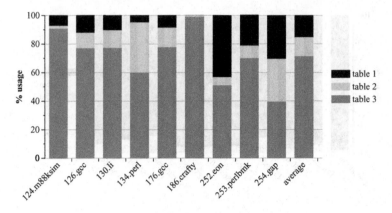

Fig. 5. Percentage of indirect branch predictions done by each table

predicted by the first level table. This happens to indirect branches having only a single target address or a target address that repeats in consecutive instances of the branch. In addition, we can observe that the second level table is poorly used in comparison with the third level one in the majority of benchmarks. This data points out that MSCP would work well using only two levels of prediction tables.

5.3   Evaluation

In the previous section we analyzed a MSCP three times larger than the baseline BTB. We explore now some possible configurations of MSCP using approximately the same hardware budget than the baseline BTB, that is, 2048 entries. We are not taking into account the little cost of global history registers and associated logic.

Figure 6 shows IPC speedup of some MSCP configurations over the baseline using a perfect branch direction predictor. We have simulated a three level MSCP, labeled as *3t* in figure 6. This configuration has a 1024 entry 2-way associative table in the first level and a 512 entry direct mapped table in both the second and the third level. Path length three is used in the second level and path length eight is used in the third level. We have also simulated three two level MSCP configurations, labeled as *2t3*, *2t8* and *2t9*. These configurations have a 1536 entry 3-way associative table in the first level and a 512 entry direct mapped table in the second level. The second level is indexed using path length three (*2t3*), eight (*2t8*) and nine (*2t9*) branches.

The best MSCP configuration shown in figure 6 is the one with two levels using a path length of nine branches to index second level prediction table. This configuration achieves a 6.4% average IPC speedup using 2048 entries while a 6K entry one achieves only a 5.7% average IPC speedup, as we said in section 5.2. This happens because 2048 entries is a size big enough and a larger path length is more beneficial than bigger prediction tables. We can also see that the three

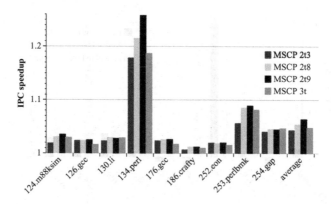

Fig. 6. Some 2048 entry MSCP configurations IPC speedup over baseline configuration

level MSCP performance is worse than the two level configurations with path length eight and nine. The first level table is smaller in the three level MSCP than in the two level one. This fact causes more conflict misses in the three level MSCP.

We have also evaluated a 4096 entry 4-way associative BTB, twice the size of the baseline BTB. This larger BTB achieves very little speedup over the baseline, since a 2048 entry BTB is big enough to avoid the majority of conflict misses. Therefore, using MSCP is a better option than increasing the size of the BTB.

Realistic Branch Direction Predictor. All previous data in this section has been obtained using a perfect branch direction predictor. Next, we show that MSCP outperforms the 2048 entry 4-way associative baseline BTB using a realistic branch direction predictor. Figure 7 shows IPC speedup for the best 2048 entry MSCP configuration described above using a 12 KB *gskewed* branch direction predictor. This speedup is measured over the baseline BTB using the same *gskewed* predictor. This data is compared with speedup obtained perfectly predicting all indirect branches. As we can see, this MSCP configuration achieves 5.7% average IPC speedup with regards to 10.8% average potential IPC speedup. We should highlight that *134.perl* benchmark is achieving almost all its potential speedup, but there is still room for improvement in the remaining benchmarks.

Besides, we have simulated a path correlated target address predictor, that is, a 2048 entry 4-way associative table managed in the same way as the second level of MSCP and using a path length of nine branches. This mechanism is similar to the target cache proposed by Chang et al. [3] but applied to all branches instead of only indirect branches. Figure 7 also shows that the path correlated BTB only achieves a 4.3% average IPC speedup and even loses performance with regards to the baseline configuration in three benchmarks. This effect is clear in *126.gcc* and *176.gcc* due to conflict misses caused by their high number of static branches. The first level table of MSCP prevents easy to predict branches from using path correlation, avoiding aliasing; this not happens in the path correlated BTB.

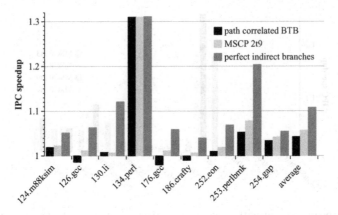

Fig. 7. Comparison of the performance of the best MSCP 2048 entry configuration, using a *gskewed* branch direction predictor, with a perfect indirect branch predictor and a path correlated BTB

6 Analysis of Static Branches

In this section we analyze the behavior of some static indirect branches. We have found [13] that a little number of static branches are responsible for almost all indirect jump and indirect function call mispredictions in the majority of benchmarks. For example, only three branches are responsible for 99% of mispredictions due to indirect jumps in *124.m88ksim*. Only one branch is responsible for 99% of mispredictions due to indirect function calls in *254.gap*.

We have selected 10 static indirect branches for a more detailed analysis. Table 2 shows information collected about these branches; each of the 10 entries in the table corresponds to one of them. The first column identifies the branch and the second one shows if it is an indirect jump or a function call. All studied indirect jumps correspond to high level switch structures and all studied indirect function calls correspond to calls using a pointer to the called function [13]. The next four columns have the percentage of all mispredictions caused by the branch, the number of different dynamic targets found for the branch, the percentage of time the branch went to a different target address than in its latest execution and the reduction in mispredictions achieved using MSCP.

Data of the branch selected from the benchmark *252.eon*, which is written in C++, points out that we can expect a good performance of MSCP with object oriented programs. The number of mispredictions for this branch is reduced in a 62.5% by using MSCP. Besides, a reduction of almost all mispredictions in the three indirect branches studied in *134.perl* is achieved, so MSCP can also be beneficial for not object oriented programs.

The number of different targets where an indirect branch can go after execution is not necessarily related with its predictability or unpredictability. Although some difficult to predict branches have a lot of different dynamic targets, as the branch selected from *253.perlbmk* which has 43 targets, there are other difficult branches with only three or four different targets. However, these branches have

Table 2. Data of the 10 selected static indirect branches

	type	misp	tgt	chn	red
124.m88ksim, file dpath.c, line 444	jump	8.8%	28	81.2%	24.7%
124.m88ksim, file stats.c, line 83	jump	8.5%	9	78.2%	75.6%
130.li, file xldmem.c, line 446	jump	4.0%	3	36.5%	40.3%
130.li, file xleval.c, line 106	call	3.3%	22	83.5%	21.5%
134.perl, file eval.c, line 137	jump	36.1%	7	80%	99.9%
134.perl, file eval.c, line 450	jump	27.1%	12	99.9%	99.9%
134.perl, file cmd.c, line 224	jump	9.0%	5	66.6%	99.9%
186.crafty, file swap.c, line 165	jump	4.0%	8	92.9%	48.3%
252.eon, file mrMaterial.cc, line 50	call	5.49%	4	79.0%	62.5%
253.perlbmk, file run.c, line 30	call	41.2%	43	94.2%	55.4%

a high percentage of executions in which their target address is different than the one observed in their latest execution. Therefore, the predictability of an indirect branch depends not on the number of different targets it has but on the predictability of the sequence of changes in its target address. More research should be done in order to analyze the path of different targets that an indirect branch follows and how we can better predict it.

7 Conclusions

In this paper we have shown that indirect branches are an important limiting factor for computer systems. Indirect branches are responsible for 55.7% of all mispredictions on average along our benchmark set, although they represent only 15.5% of dynamic branch instructions. Besides, indirect branches, excluding return instructions, have a prediction accuracy of 45.3% while the average prediction accuracy for all branch instructions is 91.2%. Taking into account this data, it is not surprising that a 10.8% IPC speedup is achievable by perfectly predicting all indirect branches.

We contribute to the study of Multi-Stage Cascaded Predictor [6] by showing that it can replace a BTB and accurately predict the target address of all branch instructions, improving processor performance. This improvement is due to an important reduction in indirect branch mispredictions achieved by MSCP. We have done a detailed analysis of MSCP behavior. We show that the majority of MSCP predictions are done by first level table, since most branches are easy to predict by a simple BTB. Difficult to predict branches are frequently upgraded up to the third level table, so second level one is poorly used.

We have found that the best MSCP configuration is one with only two prediction tables. This configuration has a 1536 entry 3-way associative first level table and a 512 entry direct mapped second level table. Second level is indexed using correlation with the target address of the latest nine executed branches. In a realistic setup, using a *gskewed* conditional branch direction predictor, a 5.7% IPC speedup is obtained over a BTB with the same hardware budget.

We have also shown that integer benchmarks can take advantage of better predicting indirect branches. These benchmarks do not have as much indirect branches as object oriented programs. However, only a little number of static indirect branches can harm the processor performance in such a way that techniques to improve their prediction will be worthwhile. Improving the prediction of indirect branches will be beneficial for programs that call functions using pointers, like *253.perlbmk*, or which have some high level switch structures depending on input data, like *gcc* compiler. Nevertheless, indirect branch prediction would be even a more important topic in computer architecture since object oriented programs, like *252.eon*, are becoming more popular.

Acknowledgements

This research has been supported by CICYT grant TIC-2001-0995-C02-01, CEPBA and an Intel scholarship grant. O. J. Santana is also supported by Generalitat de Catalunya grant 2001FI-00724-APTIND. A. Falcón is also supported by Ministry of Education of Spain grant AP2000-3923. E. Fernández is supported by Gobierno de Canarias. The authors would like to thank Jesús Corbal and Fernando Latorre for their valuable comments, as well as Carlos Navarro for his previous work in the simulation tool.

References

1. B. Calder and D. Grunwald. Reducing indirect function call overhead in C++ programs. *21st Symp. on Principles of Programming Languages*, 1994.
2. B. Calder and D. Grunwald. Fast & accurate instruction fetch and branch prediction. *21st Intl. Symp. on Computer Architecture*, 1994.
3. P. Y. Chang, E. Hao and Y. Patt. Target prediction for indirect jumps. *24th Intl. Symp. on Computer Architecture*, 1997.
4. K. Driesen and U. Hölzle. Accurate indirect branch prediction. *25th Intl. Symp. on Computer Architecture*, 1998.
5. K. Driesen and U. Hölzle. The cascaded predictor: economical and adaptive branch target prediction. *31st Intl. Symp. on Microarchitecture*, 1998.
6. K. Driesen and U. Hölzle. Multi-Stage Cascaded Prediction. *5th Intl. Euro-Par Conf.*, 1999
7. D. Kaeli and P. Emma. Branch history table prediction of moving target branches due to subroutine returns. *18th Intl. Symp. on Computer Architecture*, 1991.
8. J. Lee and A. Smith. Branch prediction strategies and branch target buffer design. *IEEE Computer Magazine, 17(1)*, 1984.
9. S. McFarling Combining branch predictors. *Digital Equipment Corporation, WRL Technical Note TN-36*, 1993.
10. P. Michaud, A. Seznec and R. Uhlig. Trading conflict and capacity aliasing in conditional branch predictors. *24th Intl. Symp. on Computer Architecture*, 1997.
11. R. Nair. Dynamic path-based branch correlation. *28th Intl. Symp. on Microarchitecture*, 1995.
12. C. Perleberg and A. Smith. Branch target buffer design and optimization. *IEEE Transactions on Computers, 42(4)*, 1993.

13. O. J. Santana, A. Falcón, E. Fernández, P. Medina, A. Ramírez and M. Valero. Analysis and evaluation of the Multi-Stage Cascaded Predictor. *Departamento de Arquitectura de Computadores, UPC, Technical Report DAC-UPC-2001-24*, 2001.
14. J. Stark, M. Evers and Y. Patt. Variable length path branch prediction. *8th Intl. Conf. on Architectural Support for Programming Languages and Operating Systems*, 1998.
15. T. Y. Yeh and Y. Patt. Two level adaptive training branch prediction. *24th Intl. Symp. on Microarchitecture*, 1991.
16. T. Y. Yeh and Y. Patt. A comprehensive instruction fetch mechanism for a processor supporting speculative execution. *25th Intl. Symp. on Microarchitecture*, 1995

High Performance and Energy Efficient Serial Prefetch Architecture

Glenn Reinman[1], Brad Calder[2], and Todd Austin[3]

[1] Computer Science Department, University of California, Los Angeles
[2] Department of Computer Science and Engineering,
University of California, San Diego
[3] Electrical Engineering and Computer Science Department, University of Michigan

Abstract. Energy efficient architecture research has flourished recently, in an attempt to address packaging and cooling concerns of current microprocessor designs, as well as battery life for mobile computers. Moreover, architects have become increasingly concerned with the complexity of their designs in the face of scalability, verification, and manufacturing concerns.

In this paper, we propose and evaluate a high performance, energy and complexity efficient front-end prefetch architecture. This design, called *Serial Prefetching*, combines a high fetch bandwidth branch prediction and efficient instruction prefetching architecture with a low-energy instruction cache. Serial Prefetching explores the benefit of decoupling the tag component of the cache from the data component. Cache blocks are first verified by the tag component of the cache, and then the accesses are put into a queue to be consumed by the data component of the instruction cache. Energy is saved by only accessing the correct way of the data component specified by the tag lookup in a previous cycle. The tag component does not stall on a I-cache miss, only the data component. The accesses that miss in the tag component are speculatively brought in from lower levels of the memory hierarchy. This in effect performs a prefetch, while the access migrates through the queue to be consumed by the data component.

1 Introduction

At a high-level, a modern high-performance processor is composed of two processing engines: the *front-end processor* and the *execution core*. This producer and consumer relationship between the front-end and execution core creates a fundamental bottleneck in computing, *i.e.*, execution performance is strictly limited by fetch performance.

An energy efficient fetch design that still achieves high performance is important because overall chip energy consumption may limit not only what can be integrated onto a chip, but also how fast the chip can be clocked [6]. Brooks et al. [2] report that instruction fetch and the branch target buffer are responsible for 22.2% and 4.7% respectively of power consumed by the Intel Pentium Pro. Brooks also reports that caches comprise 16.1% of the power consumed by Alpha

H. Zima et al. (Eds.): ISHPC 2002, LNCS 2327, pp. 146–159, 2002.

21264. Montanaro et al. [5] found that the instruction cache consumes 27% of power in their StrongARM 110 processor.

The goal of the research presented in this paper is to create a fetch engine that provides:

1. high fetch bandwidth for wide issue processors
2. a complexity effective design that will scale to future processor technologies
3. efficient instruction prefetching to better tolerate memory latencies
4. a design that is as energy efficient as possible, while still achieving the above three goals

In [12,13], we proposed a fetch architecture that provides high fetch bandwidth for wide issue architectures and can scale to future processor technologies by using a multi-level branch predictor decoupled from the instruction cache. The decoupled branch predictor enabled a highly accurate instruction prefetch architecture.

In this paper, we propose a new prefetch architecture that builds upon our prior design. We examine integrating an energy efficient serial access instruction cache into a high performance instruction prefetch architecture to achieve the above design goals. This design features an intelligent cache replacement and consistency mechanism that reduces the complexity of our prior instruction prefetching scheme from [13].

2   An Energy Efficient Multi-component Cache

Instruction cache performance is vital to the processor pipeline. Associativity is a useful technique to improve cache performance by reducing conflict misses in the cache. The conventional set-associative cache design probes the tag and data components of the cache in parallel to reduce the cache access time. We refer to this design as the *parallel* cache. This approach wastes energy in the bitlines and sense amps of the cache as it must drive all associative ways of the data component.

A *serial* cache design breaks up the instruction cache lookup into two components – the tag comparison and the data lookup. The data component is responsible for the majority of the power consumed in the access. If the way is known or predicted before the data component is accessed, we will avoid unnecessarily driving the bitlines of other ways of the cache and decrease the number of necessary sense amps. This design has been used for L2 caches, and recently for data caches on graphics cards [9, 7]. The Alpha 21264 [8] splits the tag and data component of its direct-mapped second level cache, effectively creating a serial L2 cache design. Solomon et al. [17] examined the use of a serial cache along with their Micro-Operation Cache.

The energy efficient cache architecture we use in this paper is the *multi-component cache (MC)*. This cache has the same tag component arrangement as a regular set-associative instruction cache, but rather than a single set-associative data component, there are a number of direct mapped data components. The

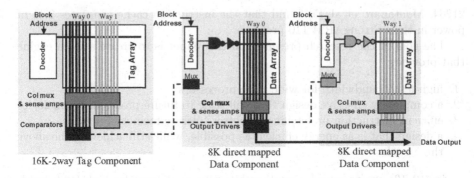

Fig. 1. A 16KB 2-way set-associative multi-component (MC) cache. This has the same tag component as the instruction cache, but with multiple data components. Each data component is a direct mapped cache. For a cache of size C that is A-way set associative, there are A direct mapped caches of size $\frac{C}{A}$ forming the data components.

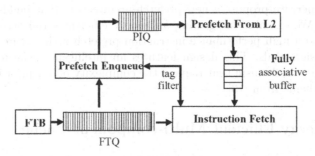

Fig. 2. The fetch directed prefetching architecture.

16KB 2-way associative configuration shown in Figure 1 has two direct mapped data components, each only 8KB in size. A 16KB 4-way set associative MC cache would have four 4KB direct mapped data components. Each direct mapped data component has its own decoder, sense amps, and other auxiliary structures. At most one data component is enabled at each access, depending on tag information. In this paper, we examine incorporating the MC cache design into our high bandwidth fetch directed prefetching architecture.

3   High Bandwidth Fetch Architecture

In our prior work (shown in Figure 2), we explored an architecture that decoupled the branch prediction architecture from the instruction fetch unit (including the instruction cache) to provide latency tolerance and fetch stream look-ahead. The branch predictor and instruction fetch unit are separated by a queue of fetch addresses (branch predictions) called the *Fetch Target Queue* (FTQ) [14]. The FTQ has two primary functions, it provides latency tolerance between the branch prediction architecture and the instruction fetch unit, and it provides a glimpse at the future fetch stream of the processor.

The ability of the FTQ to tolerate latency between the branch prediction architecture and instruction cache enables a multilevel branch predictor hierarchy called the fetch target buffer (FTB) [12]. The FTB combines a small first level predictor that scales well to future technology sizes with a larger, pipelined second level structure, which provides the capacity needed for accurate branch prediction. With sufficient branch predictions stored in the FTQ, the architecture is able to tolerate the latency of the second level branch predictor access while the instruction fetch unit continues consuming predictions already stored in the FTQ.

The fetch stream look-ahead provided by the FTQ provides a set of addresses that can be used to perform other optimizations, such as *Fetch-Directed Prefetching* (FDP) [13]. The stream of fetch PCs stored in the FTQ represents the path of execution that the execution core will be following. In that study, we used the stream of fetch PCs in the FTQ to guide instruction cache prefetching.

To increase the accuracy of the prefetches, we examined using a variety of filtering techniques to reduce the number of instruction cache prefetches performed. We used what we call *Cache Probe Filtering*, which uses the instruction cache tag array to verify potential instruction cache prefetch requests. Prefetches are only performed if the cache block is not already in the instruction cache. As it is relatively inexpensive to replicate ports on the instruction cache tag array, a separate port on the tag array can be used to verify FTQ entries for prefetching.

These techniques provide a significant boost to performance, and in this paper, we examine a new architecture that improves the energy efficiency and reduces the complexity of the fetch directed prefetching architecture.

3.1 Complexity Concerns

In Figure 2, the branch predictor feeds fetch blocks into the FTQ where they are consumed by the instruction fetch unit, which contains the instruction cache. Each FTQ entry contains a fetch block PC, a fetch distance, and a branch target. The fetch block prediction stored in a given FTQ entry may span up to five instruction cache blocks.

With FDP, *any* entry in the FTQ can potentially initiate a prefetch. This would require a connection from *each* FTQ entry to the prefetch engine via a multiplexor. This could have substantial design and performance implications. Rather than allowing a prediction to proceed from any entry in the FTQ, we could restrict the number of FTQ entries the architecture is allowed to initiate prefetches from. However, we found that having those restrictions resulted in a significant performance loss. If we restrict prefetching to only the entries at the head of the FTQ, the prefetcher is not able to look far enough ahead to provide a timely prefetch. Restricting prefetching to FTQ entries towards the tail of the FTQ (near the branch predictor) results in reduced coverage, since those FTQ entries are not occupied with fetch blocks a sufficient fraction of the time to provide maximum benefit. Therefore, it is beneficial to have a large window of cache blocks to prefetch from in the FTQ in order to achieve the maximum performance.

For each potential prefetch address, the fetch-directed prefetch architecture in [13] uses a tag port in the instruction cache, to first see if the cache block is in the cache. This is called *Cache Probe Filtering*, and significantly increases the accuracy of the prefetcher. It filters L2 prefetch requests, thereby reducing energy dissipation and saving bus bandwidth. Therefore, the FDP architecture can require a given instance of a cache block to access the instruction cache tag array twice (first for prefetch filtering, and secondly for the actual cache access).

The complexity of processing any FTQ entry to perform a prefetch, and the fact that the FDP architecture requires two tag lookups for every cache block motivated us to examine a design where we effectively move part of the FTQ between the tag and data components of the instruction cache lookup. We call this architecture the *Serial Prefetch* architecture, and describe it in detail in the next section.

4   Serial Prefetch Architecture

To address the issues described in the previous section, we start by decoupling the tag component from the data components of the instruction cache. Figure 3 shows the *Serial Prefetch* (SP) architecture. There are three pipeline stages: branch prediction, tag check, and data lookup. An MC instruction cache is used, and split into tag and data components separated by a queue of cache block requests called the *cache block queue (CBQ)*. The tag component does not block, but the data component blocks on a lookup if the data is not ready, as explained below.

The tag component of the instruction cache consumes fetch block addresses from the FTQ and verifies whether or not the cache blocks within the fetch block are already in the instruction cache. If a cache block is not found by the tag lookup, the prefetch engine can speculatively fetch the block from the L2 cache. The tag component inserts an entry into the CBQ that corresponds to a single cache block request. The data component then consumes this entry and if the block was found to be in the cache (i.e. a known cache hit), it uses the additional state in the CBQ entry to drive the appropriate associative way. We can also use the entries stored in the CBQ to guide cache replacement.

The FTB can provide up to five cache blocks in a single fetch block, but it is expensive (both in terms of power and access time) to add additional ports to the data component of the instruction cache to handle multiple cache blocks. Simply adding extra ports to the tag component of the instruction cache is cheaper than multiporting the data component. This allows the tag component to run ahead of the data component to examine the future cache block request stream of the processor. With the CBQ, the tag component can also run ahead of the data component if the data component has stalled due to a full instruction window or if it is waiting on data from a cache miss.

However, there is a serious consistency issue which must be addressed. If the tag component is allowed to run ahead of the data component, it is possible that a cache data block may be in contention for replacement *after* the tag compo-

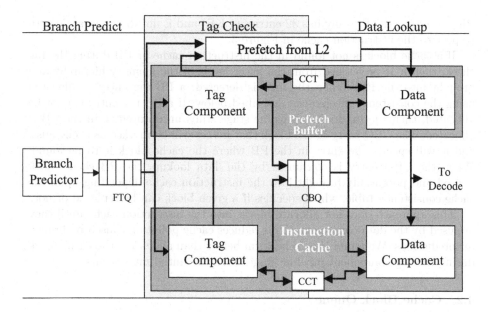

Fig. 3. The pipeline for the serial prefetch architecture.

nent has already verified that it is in a particular way of the instruction cache. When the cache block address is sitting in the CBQ waiting to be consumed, its corresponding data block must not be evicted to ensure correctness. To provide for this, we propose the use of a cache consistency mechanism called the *Cache Consistency Table* (CCT), which will be explained in Section 4.3.

We will now examine the structures of the serial prefetch architecture in more detail.

4.1   Prefetch Buffer and Cache Misses

As branch mispredictions do occur, it is not always desirable to bring cache blocks directly into the instruction cache, as they may be on a mispredicted path and could potentially replace a useful block. Therefore, we make use of a separate fully associative structure to hold cache blocks, the *Prefetch Buffer* (PB), which holds cache blocks (analogous to the prefetch buffer from [13]).

The PB works in parallel with the instruction cache for both tag checks and data lookups. The tag components of the instruction cache and PB consume an entry from the FTQ and check each cache block address in the fetch block to determine whether it is a hit in the instruction cache or PB — or if it missed in both and must be retrieved from the lower levels of the memory hierarchy. A new entry is inserted in the CBQ for each cache block that is verified from the FTQ. Then, when the entry is consumed from the CBQ, extra bits contained in the entry will indicate which directed mapped data component (to save energy) should be accessed, and if that component is part of the instruction cache or

PB. Note that the PB only has 32 entries (1 KB) and is not drawn to scale with respect to the instruction cache in Figure 3.

If a cache block is not found in the instruction cache or PB during the tag check, then it is prefetched from the lower levels of the memory hierarchy and brought into the PB. When the miss is detected, a PB tag entry is allocated using the consistency mechanism described below. If a PB tag entry cannot be allocated, then the tag lookup pipeline stage stalls until an entry in the PB is allocated. The CBQ entry of the block that misses will be marked as cache miss, and it will specify the entry in the PB where the cache block is to be found. When the CBQ entry is consumed by the data lookup, the PB cache block is used and potentially brought into the instruction cache depending upon the cache consistency table, which specifies if a given block can be replaced or not.

This approach does not allocate blocks into the instruction cache until they are used by the data component. This reduces cache pollution caused by branch mispredictions. We found this to perform better than allocating the cache block during the tag component pipeline stage when the initial miss occurs.

4.2   Cache Block Queue

The CBQ holds a cache block address, *block location* bits, *instruction cache way* bits, and *PB index* bits. The block location bits represent whether the cache block that the entry represents is in the instruction cache, in the PB, or is to be brought into the PB from another level of the memory hierarchy. The way bits are fed into the data component of the instruction cache on an instruction cache hit. These bits indicate which direct mapped component should be activated (i.e. what data way the tag component found the data in – the output from the comparators of the tag array). When the cache block hits in the PB, the PB index is used to keep track of the location of the cache block in the PB. If the cache block is prefetched from lower levels of the memory hierarchy, the PB index holds the location where the prefetched cache block will be stored.

The size of the CBQ can be used to control the amount of prefetching that occurs. The larger the queue, the more the tag comparator can be allowed to run ahead of the data output components, and the more cache blocks (not found in the instruction cache by the tag component) that can potentially be brought into the PB. The further ahead the tag component runs of the data components, the earlier the prefetch can occur and the more memory latency that can be hidden. However, there is also a greater chance of speculating down a mispredicted control path. Therefore the size of the CBQ trades the benefit obtainable from prefetching with the amount of power potentially wasted on mispredicted control paths (similar to the tradeoffs inherent in FTQ size). The CBQ is flushed on a branch misprediction.

4.3   Consistency Mechanism

Because the tag component can verify cache blocks far in advance of the data component and we perform replacement in the data lookup pipeline stage, we need some consistency mechanism to guarantee that cache blocks verified by the tag component are not evicted during cache replacements before they can be accessed in the data lookup stage. We maintain an extra table, called the *cache consistency table (CCT)*, to guarantee this. The CCT is a tagless buffer, with one entry for every cache block in the instruction cache, but with much smaller blocks – each CCT block only holds an N-bit counter. For example, assuming the use of a 3-bit counter, a 16KB 2-way associative instruction cache would only need a 320 byte table (a structure roughly 1% of the size of the instruction cache).

The counter stored in each CCT entry represents the *number of outstanding verified cache block requests* sitting in the CBQ for the corresponding data block in the instruction cache. When the tag component verifies a cache block in the tag check stage, the N-bit counter in the CCT corresponding to that cache block is incremented. When a data component accesses an instruction cache block, the N-bit counter in the CCT corresponding to that cache block is decremented. On a misprediction or misfetch, the CCT is flushed (set to zero), just as the CBQ is also flushed. The mapping from instruction cache to CCT is implicit and does not require tags – since both structures have the same number of entries and associativity, they have identical decoders.

This consistency mechanism also extends to the prefetch buffer. The PB has its own dedicated CCT as shown in Figure 3. If the desired cache block is not found in the instruction cache, but is in the PB, a N-bit counter associated with this block in the PB is incremented each time that block is placed in the CBQ. This is necessary to preserve cache consistency so that a cache block in the PB that a CBQ entry points to is not replaced.

The CBQ provides a means to look ahead at the behavior of the instruction cache. In addition to reducing energy by only accessing the data component known to contain the desired cache block, the CBQ can also *help guide instruction cache or PB replacement*, with the help of the CCT. When a cache block is brought into the instruction cache from the PB, a block is chosen for replacement from the set that has a zero CCT entry. This ensures that a cache block, which a later CBQ entry wants to use, does not get removed from the cache. This policy overrides the standard LRU replacement policy of the instruction cache. If all cache blocks in a particular cache set are marked in the CCT (meaning no replacement for the new cache block exists), then the block is not put into the instruction cache, and instead just stays in the PB until used again or is replaced. Similarly, the replacement policy ensures that entries in the PB with non-zero N-bit counters are not replaced by new prefetches until they are have been using by the CBQ entry that incremented their N-bit counter. In this manner, the PB acts as a flexible depository of instruction cache blocks for contended cache sets – directed by the CBQ and CCT.

On a branch misprediction, all entries in the PB have their CCT entries cleared (i.e. set to zero) – but the PB is *not* flushed. New PB entries are allocated based on a LRU replacement policy of the entries that have cleared N-bit counters. If an entry in the PB with a cleared N-bit counter matches a desired cache block in the tag check stage, the N-bit counter is incremented, and any second level cache access is avoided. This way, when a mispredicted short forward jump or other wrong-path prefetch is encountered, the prefetching performed on the mispredicted path can be reused. When the processor encounters a miss in both the PB and instruction cache, and there are no entries in the PB with cleared N-bit counters, the tag components of both the PB and instruction cache stall until the N-bit counter of a PB entry has been decremented to provide a space for the desired cache block. This will happen either once the CBQ drains far enough to clear an N-bit counter or if a branch misprediction is detected.

In this paper, we use 5-bit CCT counters when using a 32 entry CBQ, and 3-bit CCT counters when using a 12 entry CBQ.

5   Methodology

The simulator used in this study was derived from the SimpleScalar/Alpha 3.0 tool set [3], a suite of functional and timing simulation tools for the Alpha AXP ISA. The timing simulator executes only user-level instructions, performing a detailed timing simulation of an aggressive 8-way dynamically scheduled microprocessor with two levels of instruction and data cache memory. Simulation is execution-driven, including execution down any speculative path until the detection of a fault, TLB miss, or branch misprediction.

To perform our evaluation, we collected results for 5 of the SPEC95 C benchmarks plus 2 of the SPEC2000 C benchmarks. We only selected benchmarks that exhibited adequate instruction cache miss behavior in presenting our results – but there was no performance degradation for benchmarks without instruction cache pressure. The programs were compiled on a DEC Alpha AXP-21164 processor using the DEC C and C++ compilers under OSF/1 V4.0 operating system using full compiler optimization (-O4 -ifo). Each benchmark was fast forwarded before actual simulation as described by Sherwood et. al. [16]. Their approach uses basic block fingerprinting to determine how far to fast forward in order to have accurate simulation points.

5.1   Baseline Architecture

Our baseline simulation configuration models a next generation out-of-order processor microarchitecture. We've selected parameters to capture underlying trends in microarchitectural design. The processor has a large window of execution; it can fetch up to 8 instructions per cycle. It has a 128 entry re-order buffer with a 32 entry load/store buffer. We simulated perfect memory disambiguation (perfect Store Sets [4]). Therefore, a load only waits on a store it is really data dependent upon. To compensate for the added complexity of disambiguating loads and stores in a large execution window, we increased the store forward latency to 3 cycles.

There is an 8 cycle minimum branch misprediction penalty. The processor has 8 integer ALU units, 4-load/store units, 2-FP adders, 2-integer MULT/DIV, and 2-FP MULT/DIV. The latencies are: Int ALU 1 cycle, Int MULT 7 cycles, Int DIV 12 cycles, FP ALU 4 cycles, FP MULT 4 cycles, and FP DIV 12 cycles. All functional units are fully pipelined allowing a new instruction to initiate execution each cycle.

We use a 128 entry 4-way associative FTB with a 2K entry 4-way associative second level FTB. Each fetch block stored in the FTB can span up to five sequential cache blocks. We use the McFarling bi-modal gshare predictor [10], with an 8K entry gshare table and a 64 entry return address stack in combination with the FTB. We use a 32 entry FTQ in conjunction with the FTB.

5.2 Memory Hierarchy

We rewrote the memory hierarchy in SimpleScalar to model bus occupancy, bandwidth, and pipelining of the second level cache and main memory. We provide results for instruction caches with a single ported data component and dual ported tag component. We found that this port configuration provides the best tradeoff between performance and energy dissipation. Additional tag or data ports impacted energy dissipation more than they helped performance. The data cache for each configuration is a 4-way set associative 32KB cache with 32 byte lines. Results are gathered using a 16KB 2-way associative instruction cache.

The second level cache is a unified 1 MB 4-way set associative pipelined L2 cache with 64-byte lines. The L2 hit latency is 12 cycles, and the round-trip cost to memory is 100 cycles. The L2 cache has only a single port. The L2 cache is pipelined to allow a new request every 4 cycles, so the L2 bus can transfer 8 bytes/cycle. The L2 bus is shared between instruction cache block requests and data cache block requests.

5.3 Energy Model

The amount of energy consumed by a circuit influences layout issues, power-supply requirements, thermal considerations, and even reliability [11]. We are interested in building an architecture that combines high performance with energy efficiency.

The energy data we need to generate results is gathered using the new CACTI cache model version 2.0 developed by Reinman and Jouppi [15]. CACTI 2.0 contains a detailed model of the wire and transistor structure of on-chip memories, verified by hspice. We modified CACTI 2.0 to model the timing and energy consumption of the front-end structures of our architecture. CACTI 2.0 uses data from $0.80\mu m$ process technology and can then scale timing data by a constant factor to generate timings for other process technology sizes. We examine timings for the $0.10\mu m$ process technology size, which makes use of a $1.1V$ Vdd.

CACTI 2.0 reports energy data for successful cache accesses. We modified CACTI 2.0 to report energy data for successful accesses, misses, tag probes, and writes. Since we are concerned with instruction caches, we only examine cache writes as replacements from lower levels of the memory hierarchy. Also, we modified CACTI 2.0 to support extra ports on just the tag array of the cache.

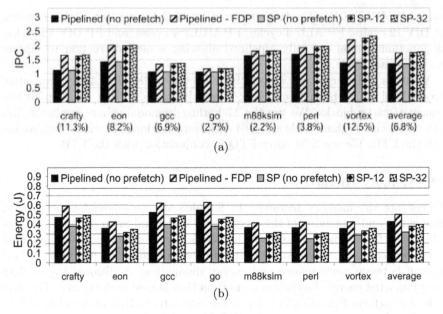

Fig. 4. IPC and Energy results

We further modified CACTI 2.0 to estimate the power consumption of all front-end structures, including the FTB, FTQ, instruction cache, L2 cache (a unified cache – but we only counted power from front-end requests, not from data cache misses), and other auxiliary structures that have been introduced (CBQ, CCT, and PB). For each, we modified the BITOUT, ADDRESS BITS, and block size parameters appropriately. When we report energy dissipation results in Joules, this includes the power dissipated by all the above listed front-end structures.

6   Results

Figures 4(a) and (b) present IPC and Energy results for five architectures. The first two bars represent the pipelined parallel instruction cache from prior work – with and without FDP using cache probe filtering: Pipelined (no prefetch) and Pipelined-FDP. The remaining three bars represent the serial prefetch architecture with the MC cache. SP (no prefetch) is the serial prefetch architecture with a 0 entry CBQ. The other two bars represent the serial prefetch architecture with a 12-entry CBQ (SP-12) and a 32-entry CBQ (SP-32). All instruction caches are 16KB 2-way set associative, with two tag array ports and a single data array port. Instruction cache miss rates are shown under each benchmark in (a).

Because both the pipelined parallel cache and the serial prefetch architecture feature instruction cache accesses across 2 pipeline stages, the performance of these architectures without any form of prefetching (i.e. without a CBQ) is

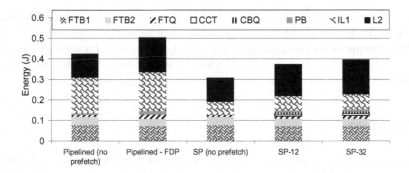

Fig. 5. Energy breakdown for 16KB 2-way set associative cache.

identical. However, the use of the MC cache in the serial prefetch architecture provides a 24% reduction in energy dissipation on average over the pipelined parallel cache. The use of a 32-entry CBQ provides a 31% improvement in IPC over the architecture with a 0-entry CBQ (no prefetch).

For all benchmarks, the serial prefetch architecture with a 12 or 32 entry CBQ is able to perform as well or better than the pipelined parallel cache with fetch directed prefetching. The SP-32 architecture is able to use 21% less energy on average than the parallel pipelined cache with FDP. The SP-12 architecture is able to use 26% less energy.

The difference between the SP-12 and SP-32 architectures is relatively small for most benchmarks, with the exception of vortex. This benchmark is plagued by a larger instruction cache miss rate than other benchmarks and can therefore take full advantage of the intelligent replacement policy provided by the CCT in the serial prefetch architecture.

Finally, we have traded complexity in the design of both the FTQ and the FDP prefetch mechanism for the addition of the CBQ and CCT to the Serial Prefetch architecture. By consolidating prefetch verification and enqueue to a single site (the tag component of the MC cache), we have reduced the amount of wire needed around the FTQ. This could have a large impact at future technology sizes. Moreover, we have reduced the number of accesses to the tag array of the instruction cache and the number of times that an FTQ entry needs to access an address generator. The CBQ, CCT, and PB are all small structures which will scale well to future technology sizes [1]. Also, by integrating prefetching into the regular operation of the instruction cache, we have eliminated some of the scheduling complications that might arise from having three distinct sources of memory requests (now the instruction cache demand misses and instruction prefetching requests originate from the same source).

Figure 4(b) showed that the Pipelined cache architectures dissipate considerably more energy than the Serial Prefetch architectures. Figure 5 explores this more closely by breaking down the average energy consumption of the 7 benchmarks by front-end component. The instruction cache, L2 cache, and first level FTB dissipate the most energy of the front-end structures shown in this

Figure. The FTQ, CBQ, and CCT dissipate very little energy relative to these structures. Prefetching adds to the energy dissipation of the front-end – but, the SP-32 technique in Figure 4(b) dissipate slightly less energy on average than the Pipelined (no prefetch) architecture, which does *not* even include any form of prefetching. This is due to the energy efficient MC cache design.

7   Summary

In this paper we examined integrating an energy efficient instruction cache into a high fetch bandwidth architecture with prefetching. The goal was to not sacrifice any performance, but at the same time to reduce the energy footprint and the complexity of the high fetch bandwidth architectures.

We examined the complexity inherent in the design of a fetch-directed prefetching architecture, and proposed a novel serial prefetch architecture that decouples the tag and data components of the instruction cache. This architecture reduces the complexity of the FTQ and energy dissipation over our prior fetch-directed prefetching architecture. This is made possible by use of a new cache consistency mechanism (called the CCT) that coordinates the tag and data components of the instruction cache and provides an improved form of cache replacement.

The best performing serial prefetch architecture provides a 31% improvement in IPC over the parallel cache with no form of prefetching, and can reduce energy dissipation by 21% over a parallel instruction cache with fetch directed prefetching. This design is able to scale well to more highly associative caches. The intelligent replacement mechanism can even provide high performance in applications with severe instruction cache thrashing. Finally, the serial prefetch architecture exhibits less complexity along the critical timing path by reducing wire congestion around the FTQ, and only requiring one tag lookup for a cache block when it is prefetched.

Acknowledgments

We would like to thank Keith Farkas, Norm Jouppi, and the anonymous reviewers for providing useful feedback on this paper. This work was funded in part by Semiconductor Research Corporation contract number 2001-HJ-897, by DARPA/ITO under contract number DABT63-98-C-0045, and a grant from Intel Corporation.

References

1. V. Agarwal, M. Hrishikesh, S. Keckler, and D. Burger. Clock rate versus ipc: The end of the road for conventional microarchitectures. In *27th Annual International Symposium on Computer Architecture*, 2000.

2. D. Brooks, V. Tiwari, and M. Martonosi. Wattch: A framework for architectural-level power analysis and optimizations. In *27th Annual International Symposium on Computer Architecture*, 2000.
3. D. C. Burger and T. M. Austin. The simplescalar tool set, version 2.0. Technical Report CS-TR-97-1342, University of Wisconsin, Madison, June 1997.
4. G. Chrysos and J. Emer. Memory dependence prediction using store sets. In *25th Annual International Symposium on Computer Architecture*, June 1998.
5. J. Montanaro et al. A 160 mhz 32b 0.5w cmos risc microprocessor. In *Digital Technical Journal*, August 1997.
6. L. Gwennap. Power issues may limit future cpus. Microprocessor Report, August 1996.
7. H. Igehy, M. Eldridge, and K. Proudfoot. Prefetching in a texture cache architecture. In *Proceedings of the 1998 Eurographics/SIGGRAPH Workshop on Graphics Hardware*, 1999.
8. R. Kessler. The alpha 21264 microprocessor. In *IEEE Micro*, April 1999.
9. J. McCormack, R. McNamara, C. Gianos, L. Seiler, N. Jouppi, and K. Correll. Neon: a single-chip 3d workstation graphics accelerator. In *Proceedings of the 1998 EUROGRAPHICS/SIGGRAPH workshop on Graphics Hardware*, 1999.
10. S. McFarling. Combining branch predictors. Technical Report TN-36, Digital Equipment Corporation, Western Research Lab, June 1993.
11. J. Rabaey. *Digital Integrated Circuits*. Prentice Hall Electronics and VLSI Series., 1996.
12. G. Reinman, T. Austin, and B. Calder. A scalable front-end architecture for fast instruction delivery. In *26th Annual International Symposium on Computer Architecture*, May 1999.
13. G. Reinman, B. Calder, and T. Austin. Fetch directed instruction prefetching. In *32st International Symposium on Microarchitecture*, November 1999.
14. G. Reinman, B. Calder, and T. Austin. Optimizations enabled by a decoupled front-end architecture. In *IEEE Transactions on Computers*, April 2001.
15. G. Reinman and N. Jouppi. Cacti version 2.0. http://www.research.digital.com/wrl/people/jouppi/CACTI.html, June 1999.
16. T. Sherwood, E. Perelman, and B. Calder. Basic block distribution analysis to find periodic behavior and simulation points in applications. In *International Conference on Parallel Architectures and Compilation Techniques*, September 2001.
17. B. Solomon, A. Mendelson, D. Orenstien, Y. Almog, and R. Ronen. Micro-operation cache: A power aware frontend for variable instruction length isa. In *International Symposium on Low Power Electronics and Design*, 2001.

A Programmable Memory Hierarchy
for Prefetching Linked Data Structures

Chia-Lin Yang[1] and Alvin Lebeck[2]

[1] National Taiwan University, 1 Roosevelt Rd. Sec. 4, Taipei, Taiwan
yangc@csie.ntu.edu.tw
[2] Duke University, Durham, NC 27708, USA
alvy@cs.duke.edu

Abstract. Prefetching is often used to overlap memory latency with computation for array-based applications. However, prefetching for pointer-intensive applications remains a challenge because of the irregular memory access pattern and pointer-chasing problem. In this paper, we use a programmable processor, a prefetch engine (PFE), at each level of the memory hierarchy to cooperatively execute instructions that traverse a linked data structure. Cache blocks accessed by the processors at the L2 and memory levels are proactively pushed up to the CPU.
We look at several design issues to support this programmable memory hierarchy. We establish a general interaction scheme among three PFEs and design a mechanism to synchronize the PFE execution with the CPU. Our simulation results show that the proposed prefetching scheme can reduce up to 100% of memory stall time on a suite of pointer-intensive applications, reducing overall execution time by an average 19%.

1 Introduction

The widening gap between processor cycle time and main memory access time makes techniques to alleviate this disparity essential for building high performance computer systems. Caches are recognized as a cost-effective method to bridge this gap. However, the effectiveness of caches is limited for programs with poor locality. Programs with regular access patterns can often employ prefetching techniques to hide memory latency. Unfortunately, these techniques are difficult to apply to pointer-intensive data structures because the address stream lacks regularity and there are data dependencies between elements in the structure.

The lack of address regularity makes it difficult for conventional prefetching techniques to predict future addresses. The data dependence is a result of pointer indirection. Pointer dereferences are required to generate addresses for successive elements in a linked data structure (LDS). This is commonly called the pointer-chasing problem. This serialization hinders efforts to choose appropriate prefetch distances [1] so that memory latency can be fully overlapped with computation.

We recently proposed a novel data movement model—called push—to overcome the above limitations [2]. The push model performs pointer dereferences at lower levels of the memory hierarchy and pushes data up to the processor.

H. Zima et al. (Eds.): ISHPC 2002, LNCS 2327, pp. 160–174, 2002.

This decouples the pointer dereference from the transfer of the current LDS element up to the processor. Implementations can pipeline these two operations and eliminate the request-response delay required for a conventional pull-based technique where the processor fetches an LDS element before requesting the next element. To realize this push model, we attach a prefetch engine (PFE) to each level of the memory hierarchy. The prefetch engines execute instructions that access LDS elements (LDS traversal kernels), and cache blocks accessed by the prefetch engines are pushed up to the CPU. An important aspect of this model is that it is not required for correct program execution, it is simply a performance enhancement similar to other non-binding prefetching techniques.

The contribution of this paper is that we provide a general architectural solution for this novel data movement model. Our previous work [2] performs a preliminary performance evaluation using a limited implementation of the push model. The initial design can only support linked-list traversals, which simplify the interaction among prefetch engines. This paper presents a flexible implementation of the push model – the push architecture. First, we use a fully programmable processor in each prefetch engine to support a multitude of LDS traversal kernels instead of the specialized hardware used in the initial design. In this way, the push model can also be easily extended to prefetch other data structures, e.g., sparse matrix, or to execute speculative slices [3–5]. Second, to accommodate more sophisticated structures (e.g., trees) we establish a general interaction scheme among three PFEs, particularly how a PFE suspends and resumes execution. Third, we present a throttle mechanism to synchronize the CPU and PFE execution. Fourth, we evaluate two variations of the push architecture which require less hardware resources. One attaches the PFE to the L1 and main memory levels; the other uses only one PFE at the main memory level.

Our simulations show that a simple, single-issue, in-order processor (e.g., ARM-like core) is sufficient to provide significant speedups for most of our benchmarks across a variety of linked data structures. The proposed prefetching scheme reduces the execution time between 13% to 25% for applications that the traditional pull method can not achieve significant speedup. In the specific case of list-based applications, the programmable PFE is able to deliver performance within 10% of the specialized hardware. We also find that the push architecture using the PFE only at the L1 and main memory levels can achieve performance close to using the PFE at each level of the memory hierarchy.

The rest of this paper is organized as follows. Section 2 provides background on the push model. Related work is discussed in Section 3. Section 4 describes the design details of our programmable memory hierarchy. Section 5 presents our experimental methods and results are presented in Section 6. We conclude in Section 7.

2   The Push Model

The conventional data movement model initiates all memory requests (demand fetch or prefetch) by the processor or upper level of the memory hierarchy. We

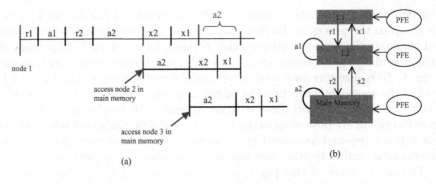

Fig. 1. The Push Model

call it the pull model because cache blocks are moved up the memory hierarchy in response to requests issued from the upper levels. Since pointer dereferences are required to generate addresses for successive prefetch requests (recurrent loads), these techniques serialize the prefetch process. The latency between recurrent prefetches in a pull-based prefetching scheme is the round-trip memory latency.

In contrast, the push model performs pointer dereferences at the lower levels of the memory hierarchy by executing traversal kernels in micro-controllers associated with each memory hierarchy level. The accessed cache blocks are proactively pushed up to the processor. This eliminates the request from the upper levels to the lower level and enables overlapping the data transfer of node i with the RAM access of node i+1 as shown in Figure 1(a). In this way, we are able to request subsequent nodes in the LDS much earlier than a traditional system. In the push model, node i+1 arrives at the CPU a2 cycles (the DRAM access time) after node i. From the CPU standpoint, the latency between node accesses is reduced from the round-trip memory latency to DRAM access time.

To realize the push model we attach a prefetch engine to each level of the memory hierarchy model as shown in Figure 1(b). The prefetch engines (PFE) execute traversal kernels independent of CPU execution. Cache blocks accessed by the prefetch engine in the L2 or main memory level are pushed up to the CPU and stored in a prefetch buffer. The prefetch buffer is a small, fully-associative cache, which can be accessed in parallel with the L1 cache. Our previous work realizes this push model only for list-based applications [2]. Although the existing design could support a few different linked list traversals, it is not flexible enough to accommodate more sophisticated structures (e.g., trees). In this paper, we present a flexible implementation of the push model which is able to support a multitude of LDS traversal kernels.

3   Related Work

Early data prefetching research focuses on array-based applications with regular access patterns [6–9]. Correlation-based prefetching [10, 11] can capture complex access patterns from the address history, but the prediction accuracy relies on the size of the prediction table and stable access patterns.

Several studies [12–14] explore data structure information to insert prefetch instructions at compile time for irregular applications. Chilimbi et al. [15, 16] seek to improve cache performance of pointer-based applications by reorganizing data layouts. Mehrotra et al. [17] extend stride detection schemes to capture both linear and recurrent access patterns. Roth et al. [18, 19] propose a dynamic scheme to capture LDS traversal kernels and a jump-pointer prefetching framework to overcome the pointer-chasing problem. The performance of the jump-pointer approach is limited for applications which change the LDS structure or traversal order frequently. Applications that have short LDS or only traverse the LDS for few times have also limited benefit from jump-pointer prefetching. Karlsson et al. [20] present a prefetch array approach, which aggressively prefetches all possible nodes a few iterations ahead. The downside of this approach is that it could issue many unnecessary prefetches.

Several studies [21–23] also combine processing power and memory in the same chip. The push architecture differs from these studies in that the PFEs do not perform computation except for address calculation and value comparison for control flow. The memory controller in Impulse [24] is also capable of prefetching data. But they only prefetch next cache line and data are not pushed up the memory hierarchy as proposed in this paper. Concurrent with this study, Hughes [25] evaluates memory-side prefetching in multiprocessor systems. His scheme does not provide solutions for two important design issues of the push model: the interaction protocol among prefetch engines at different memory modules and a mechanism to synchronize the CPU and PFE execution.

Recently, several studies [26, 3, 4, 27, 5, 28–30] suggest using pre-execution to improve cache performance for irregular applications. They either employ a seperate processor at the L1 level to execute a speculative slice or simply invoke a helper thread if the CPU is a multithreading processor. This is essentially the pull model that we evaluate. Pre-execution techniques could be used with the push model by executing the speculative slices on the PFEs in the memory hierarchy. Crago et al. [31] propose a hierarchical decoupled architecture that adds a processor between each level of the memory hierarchy. Their framework is still based on the traditional pull model, and they only study array-based applications.

4   The Push Architecture

In this section, we present an architectural design to realize the push model. We adopt a 5 stage pipeline, in-order issue general processor core as the PFE. The PFE implements a root register to store the root address of the LDS being traversed. A write to the root register activates the PFE. The L2/memory PFE contains a TLB for address translation and a local data cache to reduce the performance impact of redundant prefetches, which push up cache blocks that already exist in the upper level of memory hierarchy. The details of the

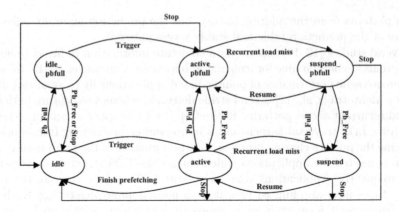

Fig. 2. Interaction Profocol among the PFEs and CPU

PFE microarchitecture can be found in [32]. Below we describe the interaction scheme among 3 level PFEs and a mechanism to synchronize the CPU and PFE execution. We also discuss two variations of the push architecture.

4.1   Interaction Among Prefetch Engines

The main design challenge of the push architecture lies in the interaction among the CPU and three PFEs. In this section, we focus on how three PFEs work together to perform prefetching. The next section addresses how the PFEs synchronize with the CPU execution. The complete interaction protocol is summarized in Figure 2. Below we elaborate on the operations of this protocol.

The L1 PFE is triggered when the root address of the LDS being traversed is written into the root register (i.e. from the idle to active state). The L1 PFE continues execution until a recurrent load (a load that generates the address of the next LDS element) misses in the L1 cache. The idea behind the push model is to perform pointer dereferences at the level where the LDS element resides such that the cache blocks can arrive at the CPU earlier than a pull-based prefetch. Since a recurrent load is responsible for loading the next node address, we use it as a hint for which memory level should perform prefetching. The PFE uses a new flavor of load instruction, ld_recurrent, to distinguish a recurrent load from other loads. When a recurrent load causes a L1 miss, the cache controller signals the L1 PFE to suspend execution (i.e. from the active to suspend state). If this recurrent load is a hit in the L2 cache, the loaded value is written to the root register of the L2 PFE, which triggers the engine. Otherwise, on a L2 miss, the memory PFE begins prefetching.

Consider traversing a binary tree in depth-first order, as shown in Figure 3. The assembly code is the traversal kernel. Assume that the recurrent load x (in Figure 3(a)), between node 1 and 2 in Figure 3(b), misses in both the L1 and L2 cache. The L1 PFE suspends execution and the memory engine starts prefetching. However, the memory PFE has only enough information to prefetch nodes

```
      00400950 addiu $sp[29],$sp[29],-56
      00400958 sw $ra[31],48($sp[29])
      00400960 sw $s8[30],44($sp[29])
      00400968 sw $s0[16],40($sp[29])
      00400970 addu $s8[30],$zero[0],$sp[29]
      00400978 addu $s0[16],$zero[0],$a0[4]
      00400980 beq $s0[16],$zero[0],004009a8
  (x) 00400988 lw $a0[4],4($s0[16])
      00400990 jal 00400950 <K_TreeAdd>
  (y) 00400998 lw $a0[4],8($s0[16])
      004009a0 jal 00400950 <K_TreeAdd>
      004009a8 addu $sp[29],$zero[0],$s8[30]
      004009b0 lw $ra[31],48($sp[29])
      004009b8 lw $s8[30],44($sp[29])
      004009c0 lw $s0[16],40($sp[29])
      004009c8 addiu $sp[29],$sp[29],56
      004009d0 jr $ra[31]
```

(a) Traversal Kernel

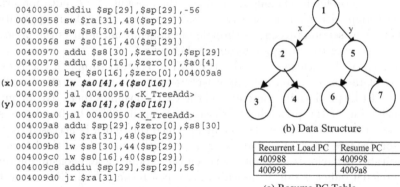

(b) Data Structure

Recurrent Load PC	Resume PC
400988	400998
400998	4009a8

(c) Resume PC Table

Fig. 3. Tree Traversal

2, 3 and 4. The L1 engine should resume execution to prefetch the remaining tree nodes after the memory PFE finishes prefetching.

We assume a dedicated bus, called the PFE Channel, for communication between the PFEs. When the PFE finishes prefetching, it sends a Resume token up the memory hierarchy through the PFE Channel. The first PFE that is in the suspend state keeps the token and resumes execution (i.e. from the suspend to active state). In this example, the L1 PFE resumes execution after receiving Resume token sent by the memory PFE. To correctly prefetch the remaining tree nodes, the L1 PFE needs to set its current program counter (PC) to 0x400998 (i.e. the recurrent load to node 5) and ensure the register values are valid. Because by the time the PFE is signaled to stop execution, it might have progressed to the next tree level and modified some of the registers.

We annotate a PFE recurrent load with its PC (RPC) and stack pointer value (SP) when it is issued. When the cache controller signals the engine to suspend execution, the PFE uses these two pieces of information to find the resume PC and restore the register values. First, the RPC is used to access the Resume PC Table (see Figure 3(c)) to obtain the correct resume PC. This table is easily constructed statically, and is downloaded as part of the traversal kernel. To restore the register values, we can take advantage of the stack maintained by the traversal kernel itself. Recursive programs always save the current register values on the stack before proceeding to the next level. Thus we can restore registers from the stack with the SP information. The implementation details are described in [32].

4.2   Synchronization between the Processor and PFEs

Timing is an important factor for the success of a prefetching scheme. Prefetching too late results in useless prefetches; prefetching too early can cause cache pollution. The prefetch schedule depends on the amount of computation available to overlap with memory accesses, which can vary throughout application execution. This section presents a throttle mechanism that adjusts the prefetch distance according to a program's runtime behavior.

The idea of this scheme is to throttle PFE execution to match the rate of processor data consumption. Our approach is built around the prefetch buffer, where the PFEs produce its contents and the CPU consumes them. We augment each cache line of the prefetch buffer with a free bit [33]. The free bit is set to 0 when a new cache block is brought in. Once this block is accessed by the processor, the free bit is set to 1 and this cache block is moved into the L1 cache. The desirable behavior is a balance between the producing and consuming rate. If the prefetch buffer is full, it indicates there is a mismatch in these two rates so synchronization between the CPU and PFEs needs to occur.

When the prefetch buffer is full, the PFEs could be running too far ahead of the CPU or performing useless prefetching (e.g., mis-predicted or late prefetches), so the active PFE should suspend execution at this point. The L1 cache controller broadcasts a Pb_Full message to the PFEs through the PFE Channel once it detects that prefetch buffer is full. The active engine suspends execution after observing the Pb_Full message (i.e., from the active to active_pbfull state).

If the PFEs are running ahead of the main execution, the CPU will eventually access the prefetch buffer. Once an entry is freed up, the L1 cache controller broadcasts a Pb_Free message, and the active engine resumes execution (i.e., from the active_pbfull to active state). Note that because of the delay for the communication, cache blocks can still arrive at the L1 level even though the prefetch buffer is full. The LRU policy is used in this case to replace blocks in the prefetch buffer. The replaced blocks are stored in the L1 cache for later accesses. If the prefetch buffer is full because of useless prefetches, most of the time, the CPU will never access data in the prefetch buffer, and the PFE should start at a new point. We use a recurrent load miss from the CPU execution as a synchronization point. The implementation details can be found in [32].

4.3   Variations of the Push Architecture

The push architecture described above attaches the PFE to each level of the memory hierarchy (3_PFE). To reduce the hardware cost, we present two design alternatives: 2_PFE and 1_PFE. 2_PFE attaches the PFE to the L1 and main memory levels, while 1_PFE only uses one PFE at the main memory level.

2_PFE performs pull-based prefetching between the L1 and L2 cache until a recurrent load misses in the L2 cache. As a result, data in the L2 cache is *pulled* up to the CPU instead of being *pushed* up as in 3_PFE. For LDS elements that exist in the L2 cache, the push model brings data to the L1 level only $r1 + x1$ cycles earlier than the pull model ($r1$: time to send a request from L1 to L2; $x1$: time to transfer a cache block from L2 to L1). For most computer systems, $r1 + x1$ is only a small portion of the round-trip memory latency. So 2_PFE should perform comparably to 3_PFE. We can further reduce the hardware cost of 2_PFE if the CPU is a multithreading processor. Instead of using a separate processor at the L1 level for prefetching, we can invoke a helper thread to execute the LDS traversal kernel. This approach has been used in several pre-execution studies [4, 5]. In this way, 2_PFE only needs to employ one PFE at the main memory level.

In the 1_PFE architecture, the root address of the LDS being traversed is written into the root register of the memory PFE. All prefetches are issued from the main memory level. 1_PFE greatly simplifies the push architecture design because the interaction issue among the PFEs no longer exists. 1_PFE should work effectively if a large portion of the LDS being traversed exists only in main memory. However, for applications in which the L2 cache is able to capture most of the L1 misses, load instructions are resolved more slowly in the memory PFE than in the CPU (the DRAM access time vs. round-trip L2 latency). So 1_PFE may not be able to achieve any prefetching effect because the memory PFE is very likely to run behind the CPU.

5   Methodology

To evaluate the push model we modified the SimpleScalar simulator [34] to include our PFE implementation. Our base model uses a 4-way superscalar, out-of-order processor. The memory system consists of a 128-entry TLB, a 32KB, 32-byte cache line, 2-way set-associative first level data and instruction caches and a 512KB, 64-byte cache line, 4-way set associative unified second level cache. We implement a hardware managed TLB with a 30 cycle miss penalty. Both the L1 and L2 caches are lock-up free caches that can have up to eight outstanding misses. The L1 cache has 2 read/write ports, and can be accessed in a single cycle. The second level cache is pipelined with an 18 cycle round-trip latency. Latency to main memory after an L2 miss is 66 cycles. We derived the memory system parameters based on the Alpha 21264 design [35], which has a 800MHz CPU and 60ns DRAM access time (i.e., 48 cycles). We also vary the memory system parameters to evaluate the push model for future processors. Simulation results can be found in [32].

To evaluate the performance of the push model, we use an in-order, single-issue processor as the PFE. We simulate a 32-entry fully-associative prefetch buffer with four read/write ports, that can be accessed in parallel with the L1 data cache. Both the L2 and memory PFEs contain a 32-entry fully-associative data cache. We simulate the pull model by using only a single PFE at the L1 cache. This engine executes the same traversal kernels used by the push architecture, but pulls data up to the processor. We use CProf [36] to identify kernels that contribute the most cache misses, and construct only these kernels by hand.

We evaluate the push model using the Olden pointer-intensive benchmark suite [37] and Rayshade [38]. Olden contains ten pointer-based applications written in C. It has been used in the past for studying the memory behavior of pointer-intensive applications [18, 19, 13]. We omit the power application because it has a low (1%) miss rate. Rayshade is a real-world graphics application that implements a raytracing algorithm. We evaluate the push model performance for both a perfect and 128-entry TLB. For this set of benchmark tested, a 128-entry TLB has less than 1% impact on the push model performance except for health. Therefore, we only present simulation results assuming a perfect TLB in this paper.

6   Simualtion Results

In this section, we present simulation results that demonstrate the effectiveness of the proposed prefetching scheme. We first evaluate the performance improvement of the push model on the 3_PFE architecture. We then examine the effect of the PFE data cache and throttle mechanism. We also compare the performance of the push model using a programmable PFE to that of a specialized PFE for list-based applications. We finish by evaluating the performance of the 2_PFE and 1_PFE architecture.

Performance Comparison Between The Push and Pull Model

Figure 4 shows execution time normalized to the base system without prefetching. For each benchmark, the three bars correspond to the base, push and pull models, respectively. Execution time is divided into 2 components, memory stall time and computation time. We obtain the computation time by assuming a perfect memory system. For the set of benchmarks with tight traversal loops (health, mst, treeadd), the push model is able to reduce between 25% and 41% of memory stall time (13% to 25% overall execution time reduction) while the pull model can only reduce the stall time by at most 4%. Perimeter traverses down a quad-tree in depth-first order, but has an unpredictable access pattern once it reaches a leaf. Therefore, we only prefetch the main traversal kernel. Although perimeter performs some computation down the leaves, it has very little computation to overlap with the memory access when traversing the internal nodes. So the pull model is not able to achieve any speedup, but the push model reduces the execution time by 4%.

For applications that have longer computation lengths between node accesses (rayshade and em3d), the pull model is able to achieve significant speedup (25% and 39% execution time reduction), but the push model still outperforms the pull model (31% and 57% execution time reduction). For rayshade, the push model is able to reduce 89% of memory stall time, but the pull model can only reduce 62%. For bh, both the push and pull models reduce up to 100% of memory stall time.

Bisort and tsp dynamically change the data structure while traversing it. The prediction accuracy is low for this type of application because the PFEs will prefetch the wrong data after the CPU has modified the structure. For tsp, we are able to identify some traversal kernels that do not change the structure dynamically. The results presented here prefetch only these traversal kernels. The push model is able to reduce the execution time by 4% and the pull model 1%. For bisort, neither the push or pull model is able to improve performance because

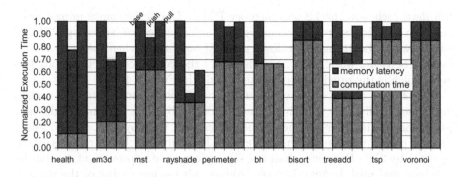

Fig. 4. Performance comparison between the push and pull model

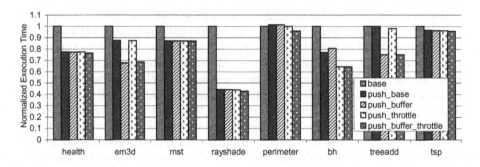

Fig. 5. Effect of the PFE Data Cache and Throttle Mechanism

the prediction accuracy is low (only 20% of prefetched cache blocks are useful). Voronoi uses pointers, but array and scalar loads cause most of the cache misses. Since we did not target these misses, we do not see any performance improvement for either the push or pull model. Since bisort and voronoi do not benefit from the push model, we omit these applications in the following discussion.

From these results we know that the push model is effective even when the applications have very tight loops where the performance of the traditional pull model is limited because of the pointer-chasing problem (e.g., health and mst). For applications with enough computation between node accesses, the push model is able to achieve performance close to a perfect memory system, but the pull model is still not able to deliver comparable performance (e.g., rayshade).

Effect of the PFE data cache and Throttle Mechanism

Recall that the push architecture uses two mechanisms to avoid redundant and early prefetches: the PFE data cache and throttle mechanism. The results presented above show the combined effect of both features. In this section, we evaluate the effect of the PFE data cache and throttle mechanism separately in Figure 5. Push_base is a plain push model with no data caches or throttle mechanism. Push_throttle is push_base with throttling and push_buffer is push_base with data caches in the L2 and memory PFEs. The performance impact from these two techniques are not exclusive. The data caches can speed up PFE execution, while throttling slows down the PEF execution when they run too far ahead. Therefore, push_buffer_throttle shows the combined effect of both features, which is the previously presented results in Figure 4.

From Figure 5 we see that em3d and treeadd benefit most from data caches. Push_base is only able to reduce execution time by 12% for em3d. Adding a data cache (push_buffer) further reduces execution time by 20%. Treeadd does not show performance improvement for push_base but push_buffer reduces execution time by 25%. Perimeter does not see performance improvement comparing push_base and push_buffer. However, adding this feature on top of the throttle mechanism (i.e. push_buffer_throttle and push_throttle) does give performance improvement.

Em3d, perimeter, and treeadd have 30%, 33% and 45% of prefetches that are redundant. The L2/memory PFE data caches are able to capture between 80% to 100% of these redundant prefetches. Treeadd and perimeter are tree traversals in depth-first order, which generates redundant prefetches when the execution moves up the tree. Em3d simulates electro-magnetic fields and processes lists of interacting nodes in a bipartite graph. The intersection of these lists creates locality, which results in redundant prefetches. Note that bh traverses an octree in depth-first order so it also has a significant amount of redundant prefetches (33%). However, bh does not benefit from the PFE data cache as shown in Figure 5 because the PFEs already issue prefetch requests far ahead of the CPU for the push_base configuration (93% of prefetched blocks are replaced before accessed).

Figure 5 shows that the throttle mechanism has the most impact on bh (push_base vs. push_throttle). Bh has long computation between node accesses. So the PFE runs too far ahead of the CPU for push_base. 93% of prefetched cache blocks are replaced before accessed by the CPU. The proposed throttle mechanism successfully prevents early prefetches. Nearly 100% of prefetched cache blocks are accessed by the CPU for push_throttle. Push_base reduces execution time by 23% and push_throttle further reduces it by 13%. It is surprising that push_base is able to reduce execution time by 23% even though most prefetched cache blocks are replaced from the prefetch buffer. Push_base obtains speedup because of better L2 cache performance compared to the base configuration (92% of L2 misses are eliminated). Since the push model deposits data in both the L2 cache and the prefetch buffer, blocks replaced from the prefetch buffer can still be resident in the L2 cache at the time the CPU accesses them.

Programmable vs. Specialized PFE
Our previous work proposes a specialized PFE for traversing linked lists. This specialized hardware traverses the entire linked list until a NULL pointer is encountered. The programmable PFE used in this paper has the flexibility advantage over the specialized hardware, but may not be able to deliver the same performance. Simulations show that for three list-based applications (health, em3d and rayshade), the programmable PFE is able to deliver performance within

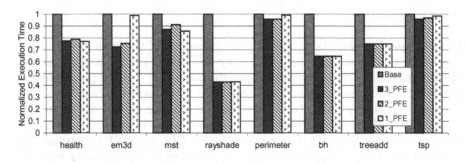

Fig. 6. Variations of the Push Architecture

10% of the specialized hardware. For rayshade, it even slightly outperforms a specialized PFE (1%). Rayshade always traverses the whole linked lists, but the fields of a LDS structure accessed cannot be determined statically. The specialized PFE is not able to handle such dynamic behavior as the programmable one.

Number of PFEs

This section evaluates the performance of various push architectures discussed in Section 4.3: 3_PFE, 2_PFE and 1_PFE. 3_PFE attaches the PFE to each level of the memory hierarchy, which is the push architecture examined so far. 2_PFE attaches the PFE to the L1 and main memory levels and 1_PFE only uses one engine at the main memory level. Figure 6 shows execution time of these three architectures normalized to the base configuration.

From Figure 6 we see that the performance of 2_PFE is comparable to 3_PFE. This indicates that 3_PFE gains most of its performance from pushing data up to the L2 level. 1_PFE achieves similar performance to 3_PFE and 2_PFE for all benchmarks except for em3d. For em3d, 80% of L1 misses are satisfied by the L2 cache. Therefore, most load instructions are resolved more slowly in the memory PFE than in the CPU (the DRAM access time vs. round-trip L2 latency). Thus 1_PFE is not able to produce any prefetching effect. Note that 1_PFE performs better than 2_PFE for mst. Mst implements hash table lookup so the list being traversed is very short. 1_PFE gets some performance advantage over 2_PFE by starting prefetching at the main memory level earlier.

From this experiment we know that 2_PFE achieves comparable performance to 3_PFE for all benchmarks. 1_PFE only needs one prefetch engine but it performs poorly if the L2 cache is able to capture most of the L1 misses, like em3d. As mentioned in Section 4.3, 2_PFE only needs one PFE if the CPU is multi-threaded. Therefore, 2_PFE is the best design choice considering both cost and performance.

172   Chia-Lin Yang and Alvin Lebeck

7 Conclusion

In this paper, we describe a flexible implementation of the push model, which overcomes the pointer-chasing problem by decoupling the pointer dereference from the transfer of the current node up to the processor. We use a programmable processor at each level of the memory hierarchy to cooperatively execute a LDS traversal kernel. The cache blocks accessed by these processors are pushed up to the CPU. We establish a general interaction scheme among three PFEs and propose a throttle mechanism to synchronize the CPU and PFE execution. This architecture could also be used to execute speculative slices or other compiler generated prefetch operations.

Our simulation results show that the push architecture is able to reduce 13% to 25% of the overall execution time for applications with very tight loops, while the traditional pull model is not able to run ahead of the CPU to give significant performance improvement. We have also shown that the proposed throttle mechanism successfully adjusts the prefetch distance to avoid early prefetches. For applications with enough computation between node accesses, the push architecture is able to achieve performance comparable to a perfect memory system. Simulations also show that 2_PFE performs comparably to 3_PFE.

We are currently investigating several extensions to this work. First, is to construct traversal kernels automatically, either by compiler or executable editor. We are also extending the push model to prefetch missing TLB entries and data structures other than LDS. Finally, we must examine issues surrounding context switches.

Acknowledgements

This work supported in part by NSF CAREER Award MIP-97-02547, DARPA Grant DABT63-98-1-0001, NSF Grants CDA-97-2637 and EIA-99-72897, Duke University, and Intel Corporation through a Ph.D. Fellowship and the T4E 2000 program. We thank the anonymous reviewers for comments and suggestions on this work. The views and conclusions contained herein are those of the authors and should not be interpreted as necessarily representing the official policies or endorsements, either expressed or implied, of the U.S. Government.

References

1. Klaiber, A.C., Levy, H.M.: An architecture for software-controlled data prefetching. In: Proceedings of the 18th Annual International Symposium on Computer Architecture. (1991) 43–53
2. Yang, C., Lebeck, A.R.: Push vs. pull: Data movement for linked data structures. In: Proceedings of the ACM International Conference on Supercomputing. (2000) 176–186
3. Collins, J.D., Wang, H., Tullsen, D.M., Christopher, H.J., Lee, Y.F., Lavery, D., Shen, J.P.: Speculative precomputation: Long-range prefetching of delinquent loads. In: Proceedings of the 28th Annual International Symposium on Computer Architecture. (2001) 14–25

4. Roth, A., Sohi, G.: Speculative data-driven multithreading. In: Proceedings of 7th Symposium High-Performance Computer Architecture. (2001) 134–143
5. Zilles, C.B., Sohi, G.: Execution-base prediction using speculative slices. In: Proceedings of the 28th Annual International Symposium on Computer Architecture. (2001) 2–13
6. Jouppi, N.P.: Improving direct-mapped cache performance by the addition of a small fully-associative cache and prefetch buffers. In: Proceedings of the 17th Annual International Symposium on Computer Architecture. (1990) 364–373
7. Baer, J.L., Chen, T.F.: An effective on-chip preloading scheme to reduce data access penalty. In: Proceedings of the 1991 Conference on SuperComputing. (1991) 176–186
8. Callahan, D., Kennedy, K., Porterfield, A.: Software prefetching. In: Proceedings of the Fourth International Conference on Architectural Support for Programming Languages and Operating Systems (ASPLOS IV). (1991) 40–52
9. Mowry, T.C., Lam, M.S., Gupta, A.: Design and evaluation of a compiler algorithm for prefetching. In: Proceedings of the Fifth International Conference on Architectural Support for Programming Languages and Operating System. (1992) 62–73
10. Joseph, D., Grunwald, D.: Prefetching using markov predictors. In: Proceedings of the 24th Annual International Symposium on Computer Architecture. (1997) 252–263
11. Alexander, T., Kedem, G.: Distributed predictive cache design for high performance memory system. In: Proceedings of the 2th International Symposium on High-Performance Computer Architecture. (1996)
12. Lipasti, M.H., Schmidt, W.J., Kunkel, S.R., Roediger, R.R.: Spaid: Software prefeteching in pointer- and call-intensive environments. In: Proceedings of the 28th Annual International Symposium on Microarchitecture. (1995)
13. Luk, C.K., Mowry, T.C.: Compiler based prefetching for recursive data structure. In: Proceedings of the Seventh International Conference on Architectural Support for Programming Languages and Operating Systems (ASPLOS VII). (1996) 222–233
14. Zhang, Z., Torrellas, J.: Speeding up irregular applications in shared-memory multiprocessors: Memory binding and group prefetching. In: Proceedings of the 22nd Annual International Symposium on Computer Architecture. (1995) 188–200
15. Chilimbi, T.M., Hill, M.D., Larus, J.R.: Cache-conscious struture layout. In: Proceedings of the SIGPLAN '99 Conference on Programming Language Design and Implementation. (1999) 1–12
16. Chilimbi, T.M., Davidson, B., Larus, J.R.: Cache-conscious struture definition. In: Proceedings of the SIGPLAN '99 Conference on Programming Language Design and Implementation. (1999) 13–24
17. Mehrotra, S., Harrison, L.: Examination of a memory access classification scheme for pointer-intensive and numeric program. In: Proceedings of the 10th International Conference on Supercomputing. (1996) 133–139
18. Roth, A., Moshovos, A., Sohi, G.: Dependence based prefetching for linked data structures. In: Proceedings of the Eigth International Conference on Architectural Support for Programming Languages and Operating Systems (ASPLOS VIII). (1998) 115–126
19. Roth, A., Sohi, G.: Effective jump-pointer prefetching for linked data structures. In: Proceedings of the 26th Annual International Symposium on Computer Architecture. (1999) 111–121

20. Karlsson, M., Dahlgren, F., Stenstrom, P.: A prefetching technique for irregular accesses to linked data structures. In: Proceedings of Sixth Symposium High-Performance Computer Architecture. (1999) 206–217
21. Patterson, D., Andreson, T., Cardwell, N., Fromm, R., Keaton, K., Kazyrakis, C., Thomas, R., Yellick, K.: A case for intelligent ram. IEEE Micro (1997) 34–44
22. Kang, Y., Huang, W., Yoo, S.M., Keen, D., Ge, Z., Lam, V., Pattnaik, P., Torrellas, J.: Flexram: Toward an advanced intelligent memory system. In: Proceedings of the 1999 International Conference on Computer Design. (1999) 192–201
23. Oskin, M., Chong, F.T., Sherwood, T.: Active pages: a computation model for intelligent memory. In: Proceedings of the 25th Annual International Symposium on Computer Architecture. (1998) 192–203
24. Carter, J., Hsieh, W., Stoller, L., Swanson, M., Zhang, L.: Impulse: Building a smarter memory controller. In: Proceedings of 5th Symposium High-Performance Computer Architecture. (1999) 70–79
25. Hughes, C.J.: Prefetching linked data structures in systems with merged dram-logic, master thesis. Technical Report UIUCDCS-R-2001-2221, Department of Computer Science, University of Illinois at Urbana-Champaign (2000)
26. Annavaram, M.M., Patel, J.M., Davidson, E.S.: Data prefetching by dependence graph precomputation. In: Proceedings of the 28th Annual International Symposium on Computer Architecture. (2001) 52–61
27. Sundaramoorthy, K., Purser, Z., Rotenberg, E.: Slipstream processors: Improving both performance and fault tolerance. (2000) 257–268
28. Luk, C.K.: Tolerating memory latency through software-controlled pre-execution in simultaneous multithreading processors. In: Proceedings of the 28th Annual International Symposium on Computer Architecture. (2001) 40–51
29. Collins, J., Tullsen, D., Wang, H., Shen, J.: Dynamic speculative precomputation. In: Proceedings of the 34st Annual International Symposium on Microarchitecture. (2001)
30. Moshovos, A., Pnevmatikatos, D., Baniasadi, A.: Slice processors: An implementation of operation-based prediction. In: Proceedings of the ACM International Conference on Supercomputing. (2001) 321–334
31. Crago, S.P., Despain, A., Gaudiot, J., Makhija, M., Ro, W., Sricastava, A.: A high-performance, hierarchical decoupled architecture. In: Proceedings of the Memory Access Decoupling for SuperScalar and Multiple Issue Architecture Workship. (2000)
32. Yang, C.L.: The Push Architecture: a Prefetching Framework for Linked-Data Structure. PhD thesis, Department of Computer Science, Duke University (2001)
33. Smith, B.: Architecture and applications of the hep multiprocessor computer system. In: Proceedings of the Int. Soc. for Opt. Engr. (1982) 241–248
34. Burger, D.C., Austin, T.M., Bennett, S.: Evaluating future microprocessors-the simplescalar tool set. Technical Report 1308, Computer Sciences Department, University of Wisconsin–Madison (1996)
35. Kessler, R.E.: The alpha 21264 microprocessor. IEEE Micro (1999) 34–36
36. Lebeck, A., Wood, D.: Cache profiling and the spec benchmarks: A case study. In: IEEE Computer. (1994) 15–26
37. Roger, A., Carlisle, M., Reppy, J., Hendren, L.: Supporting dynamic data structures on distributed memory machines. ACM Transactions on Programming Languages and Sytems **17** (1995)
38. Kolb, C.: The rayshade user's guide. (In: http://graphics.stanford.edu/- cek/-rayshade)

Block Red-Black Ordering Method for Parallel Processing of ICCG Solver

Takeshi Iwashita[1] and Masaaki Shimasaki[2]

[1] Data Processing Center, Kyoto University,
Yoshida-Honmachi Sakyo-ku, Kyoto, 606-8501, Japan
take@kudpc.kyoto-u.ac.jp
[2] Graduate School of Engineering, Kyoto University,
simasaki@kuee.kyoto-u.ac.jp

Abstract. The present paper proposes a new parallel ordering, "block red-black ordering," for a parallelized ICCG solver with fewer synchronization points and a high convergence rate. In the new method, nodes in an analyzed grid are divided into several or many blocks, and red-black ordering is applied to the blocks. Several blocks are assigned to each processor and the substitution is carried out in parallel. Only one synchronization point exists in each parallelized substitution. We performed an analytical investigation using the ordering graph theory and computational tests on a scalar parallel computer. These results show that a convergence rate is improved by an increase in the number of nodes of one block and that a high parallel performance is attained by using an appropriate block size.

1 Introduction

The ICCG (Incomplete Cholesky Conjugate Gradient) method [1] is the most popular solver for symmetric positive-definite linear systems arising in finite element (FE) analyses and finite difference (FD) analyses. Parallel processing of the ICCG method is, however, difficult due to forward-backward substitutions performed for preconditioning. Several strategies have been proposed for the parallelization of the ICCG method [2] [3].

A reordering technique is one of the most effective methods for parallel processing the ICCG solver [2] [3]. This technique, however, entails a trade-off between convergence and parallelism [4]. In order to overcome the trade-off, S. Doi et al. proposed the use of many colors in multi-color ordering [5] [6]. Their technique, which is called the *large-numbered multi-color ordering* method, attains high parallelism with modest increase of CG iterations. The authors have, therefore, intended to apply this technique to electromagnetic field analyses based on FEM [7]. Our research has, however, found that the large-numbered multi-color ordering method has some problems which cause a serious decline in the solver's performance in some computational environments. While the details of these problems are discussed in Sect. 4, the problem that motivates the present work is related to the number of synchronization points. Since the m-color ordering

H. Zima et al. (Eds.): ISHPC 2002, LNCS 2327, pp. 175–189, 2002.

method has m - 1 synchronization points in each parallelized substitution, the use of many colors results in many synchronization points. When the synchronization cost becomes significant, for example in the case of many processors, the parallel speed-up will be saturated. The present paper proposes a new parallel ordering technique in which the number of synchronization points is only one in each substitution. The new method is named the *block red-black ordering* method, which is based on division of an entire grid into blocks and on the application of red-black ordering to the blocks. In this method, increasing the block-size results in an improvement of the convergence. An analytical study using ordering graphs and numerical tests are performed for evaluation of the new method. The numerical test indicates that the new method can attain about the same degree of convergence as the large-numbered multi-color ordering method. It is also shown that the new method achieves a high parallel speed-up by effective utilization of data cache on a scalar parallel computer.

2   ICCG Method

The present paper solves a linear system of equations having a symmetric positive-definite coefficient matrix arising in finite differencing 2-D PDEs. The present discussion is also extended to 3-D problems.

A five-point difference discretization scheme is given by

$$a_{ij,1}u_{ij-1} + a_{ij,2}u_{i-1j} + a_{ij,3}u_{ij} + a_{ij,4}u_{i+1j} + a_{ij,5}u_{ij+1} = f_{ij}, \qquad (1)$$

for each node $(i,j) \in G = \{(i,j)|1 \le i \le nx, 1 \le j \le ny\}$ on a 2-D rectangular grid $nx \times ny$. On an assumption of symmetry $a_{ij,1} = a_{ij-1,5}$, $a_{ij,2} = a_{i-1j,4}$, (1) is transformed into

$$c_{ij-1}u_{ij-1} + b_{i-1j}u_{i-1j} + a_{ij}u_{ij} + b_{ij}u_{i+1j} + c_{ij}u_{ij+1} = f_{ij}, \qquad (2)$$

where $a_{ij} = a_{ij,3}$, $b_{ij} = a_{ij,4}$ and $c_{ij} = a_{ij,5}$. When assembling (2) for all nodes, we obtain a linear system of equations

$$Au = f, \qquad (3)$$

where A is a $n \times n$ symmetric positive-definite matrix ($n = nx \cdot ny$). The arrangement of u_{ij} and f_{ij} in one-dimensional vectors u and f is called the "ordering" (of nodes), on which we focus in the present paper.

The most popular solver for a symmetric positive-definite linear system such as (3) is the ICCG (Incomplete Cholesky Conjugate Gradient) method. The ICCG method consists of incomplete factorization for determining a preconditioner matrix and ICCG iterations to solve the linear system. The ICCG iterations involve forward-backward substitutions (preconditioning) and CG iterations. While the CG iteration kernel is easily parallelized, the parallel processing of incomplete factorization and substitutions is difficult. Thus, the present paper as well as previous other researchers' works [3] [5] discusses mainly the parallelization of these kernels.

In the 5-point finite difference case, the incomplete factorization is defined by

$$A \simeq M = (L_s + D)D^{-1}(L_s + D)^T, \tag{4}$$

where M is the preconditioner matrix and L_s is a strictly lower triangular part of A. The matrix D is a diagonal matrix and is obtained from the forward substitution

$$D = \text{diag}(A) - \text{diag}(L_s D^{-1} L_s^T), \tag{5}$$

where $\text{diag}(X)$ denotes a diagonal part of X.

The preconditioning is given by the solution of linear system:

$$\{(L_s + D)D^{-1}(L_s^T + D)\}z = r \tag{6}$$

where r is a residual vector. Equation (6) is solved by the forward substitution

$$y = D^{-1}(r - L_s y) \tag{7}$$

and backward substitution

$$z = y - D^{-1}L_s^T z. \tag{8}$$

The parallelism of substitutions (5)(7)(8) is determined by the pattern of non-zero entries in the matrix L_s that is ordering. Thus, the preconditioner matrix derived from (4)(5) depends on the ordering. Consequently, ordering affects a preconditioning effect, i.e., the convergence of ICCG method.

3   Parallelized ICCG Method with Multi-color Ordering

3.1   Parallel Ordering and Ordering Graph

Several parallel ordering techniques have been proposed for the parallelization of substitutions (5)(7)(8) [4] [5]. For example, red-black ordering and multi-wavefront ordering are well known. Previous numerical research has, however, found a trade-off problem between parallelism and convergence in parallel ordering techniques [4]. Parallel ordering with higher parallelism results in slower convergence. In order to overcome the trade-off problem, Doi et al. [5] [6] proposed the use of an ordering graph for the estimation of convergence in parallel ordering techniques. Figure 1 shows an example of an ordering graph. In the ordering graph, a directed graph demonstrates the order between two adjust nodes, which is the data-dependency. Parallel ordering methods with identical ordering graphs have identical preconditioning matrices, and their convergence rates are the same.

Doi et al., by their numerical tests, showed that the number of incompatible nodes as shown in Figure 1 greatly affects convergence in the preconditioned iterative method. The higher incompatibility ratio results in the slower convergence. The incompatibility ratio R_{ic} is defined by the ratio of the number of

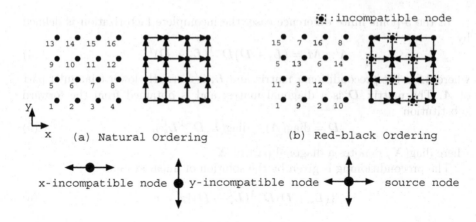

Fig. 1. Examples of ordering graphs

actual incompatible nodes N_{ic} with respect to the number of possible incompatible nodes N. The number N is given by half of total number of nodes, since half of entire nodes (red nodes) are incompatible nodes in the red-black ordering that has the most dense distribution of incompatible nodes. Accordingly, the incompatibility ratio R_{ic} is given by

$$R_{ic} = N_{ic}/N = 2N_{ic}/n. \tag{9}$$

On the other hand, the source node (see Fig. 1), which is a kind of incompatible node, is a potential starting point of parallelized substitutions. Therefore, incompatible nodes are needed for parallelization of the substitutions. The trade-off problem exists here.

3.2   Large-Numbered Multi-color Ordering Method

Utilizing the ordering graph theory, Doi et al. have proposed the use of many colors in the multi-color ordering method in order to obtain a good balance between parallelism and convergence [5]. In multi-color ordering, nodes of the same color, which have no data-relationship to each other, can be independently processed. A brief procedure of the multi-color ordering method is given as follows. The number of colors $m(\geq 2)$ is first set, and the entire nodes G are divided into m subsets $(C(l), l = 1, \cdots, m)$ as

$$C(l) = \{(i,j) \in G | \mod(i + j - 2, m) + 1 = l\}, \tag{10}$$

where $C(l)$ is a set of nodes with color l. In multi-color ordering, nodes are ordered from $C(1)$ to $C(m)$. The nodes with identical colors are, then, independent of each other, and the substitutions (5)(7)(8) are parallelized in each color [8].

Figure 2 shows 4-color ordering, where $C1 \cdots C4$ denote colors and black points represent incompatible nodes. Figure 2 indicates that when the number

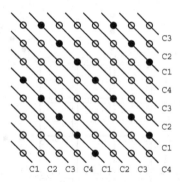

Fig. 2. 4-color ordering

of colors is increased, the incompatibility ratio decreases and the convergence will be improved. In multi-color ordering, the incompatibility ratio is approximately given by $R_{ic} = 2/m$. The parallelism in the ordering is n/m. In order to strike a good balance between the parallelism and the convergence, Doi's research recommended the ranges of $m = 30 \sim 100$ and $R_{ic} = 0.02 \sim 0.13$ [5]. Their technique is called the *large-numbered multi-color ordering method* as distinguished from the conventional multi-color ordering method with $2 \sim 8$ colors.

4 Block Red-Black Ordering Method

4.1 Problems in the Multi-color Ordering Method

The large-numbered multi-color ordering method, which achieves both high parallelism and a slight decline in convergence, is effective in many application fields. The method, however, has some problems that seriously affect the solver performance in some applications or in some computational environments. The problem is given as follows:

A. Number of synchronization points
In the m-color ordering method, m - 1 synchronization points exist in each parallelized substitution. When the synchronization cost or the communication cost occupies a large part of the total computational cost, the parallel speed-up is limited.

B. Difficult selection of optimal number of colors
Although the number of colors affects the convergence, it is difficult to choose the optimal number of colors, because both the convergence rate and the synchronization cost are taken into account. Moreover, when a stride access of m is used in the substitution for vectorization avoiding indirect addressing on a vector computer, the number of colors greatly affects the computational time because of memory bank conflicts. Our numerical tests on a Fujitsu vector parallel computer VPP-800 show that the use of 32 or 64 colors leads to a serious decline in solver performance.

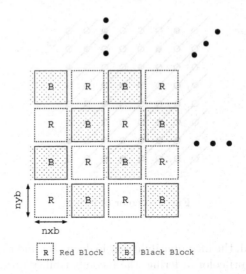

Fig. 3. Colored blocks in block red-black ordering

C. Implementation on scalar parallel computer

The large-numbered multi-color ordering method can obtain very high parallelism; for example, 32-color ordering attains approximately 30000 degrees of parallelism for a 1000×1000 2-D mesh. It is, however, difficult to utilize such high parallelism on scalar (non-vector) computers. On a scalar computer, the effective utilization of data on cache memory is more important than the parallelism.

Considering the above problems, particularly the number of synchronization points, we propose a new parallel ordering that we call *block red-black ordering*. The details of the new method are described in the following subsection.

4.2   Block Red-Black Ordering

The block red-black ordering method has a different strategy for improving the convergence of the red-black ordering method from that in the large-numbered multi-color ordering method. While the latter strategy increases the number of colors, our strategy is based on assembling nodes into a block. Both strategies result in a reduction in the incompatibility ratio and an improvement in convergence.

In the block red-black ordering method, the size of block $nxb \times nyb$ is set first, where nxb and nyb are the number of x-direction and y-direction nodes in one block, respectively. For simplicity, we here assume that

$$\mathrm{mod}(nx, nxb) = 0, \quad \mathrm{mod}(ny, nyb) = 0. \tag{11}$$

The treatment for the case where (11) is not satisfied is described in the Sect. 5.2. Next, the entire grid is divided into several or many blocks of $nxb \times nyb$. Red-black ordering is applied to the generated blocks. Following the order of blocks,

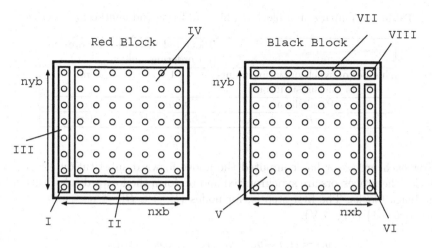

Fig. 4. Sequence of forward substitution in each block
(Partition of nodes in each block)

we order the nodes from red blocks to black blocks, where the natural ordering is used for the internal ordering in the block. Since blocks with identical colors never have a data-relationship, the blocks can be independently processed in each color. Each processor deals with one or several blocks in each color, and forward and backward substitutions are performed in parallel among processors. In the red-black ordering method, the number of synchronization points is only 1 in each substitution. The procedure for parallelized substitutions is given as follows. We here note that the following procedure is sequential and that this sequential computation is simultaneously performed on each processor. For explanation, we here define the local coordinate in the block by

$$(i_b, j_b) = (1 \le i_b \le nxb, \ 1 \le j_b \le nyb). \tag{12}$$

The forward substitution for nodes in red blocks R is given by follows: The first step is, for nodes $\{(i,j) \in R | i_b = 1, \ j_b = 1\}$ (Fig. 4 I),

$$y_{ij} = r_{ij}/d_{ij}, \tag{13}$$

where y_{ij}, r_{ij} and d_{ij} are the entries of y, r and diag(D) corresponding to the node (i,j), respectively. The second step is, for nodes $\{(i,j) \in R | 2 \le i_b \le nxb, \ j_b = 1\}$ (Fig. 4 II),

$$y_{ij} = (r_{ij} - b_{i-1j}y_{i-1j})/d_{ij}. \tag{14}$$

The third step is, for nodes $\{(i,j) \in R | i_b = 1, \ 2 \le j_b \le nyb\}$ (Fig. 4 III),

$$y_{ij} = (r_{ij} - c_{ij-1}y_{ij-1})/d_{ij}. \tag{15}$$

The final step is, for nodes $\{(i,j) \in R | 2 \le i_b \le nxb, \ 2 \le j_b \le nyb\}$ (Fig. 4 IV),

$$y_{ij} = (r_{ij} - b_{i-1j}y_{i-1j} - c_{ij-1}y_{ij-1})/d_{ij}. \tag{16}$$

182   Takeshi Iwashita and Masaaki Shimasaki

Table 1. Comparison of block red-black ordering and multi-color ordering

	Block red-black ordering	m-color ordering
Number of synchronization points in one substitution	1	m
Parallelism	$n/2nb^2$	n/m
R_{ic}	$2/nb - 1/nb^2$	$2/m$

After each processor has computed the above forward substitution for the red blocks, the processors are synchronized and perform the forward substitution for the black blocks. The first step is, for nodes $\{(i,j) \in B | 1 \le i_b \le nxb - 1,\ 1 \le j_b \le nyb - 1\}$ (Fig. 4 V),

$$y_{ij} = (r_{ij} - b_{i-1j}y_{i-1j} - c_{ij-1}y_{ij-1})/d_{ij}. \tag{17}$$

The second step is, for nodes $\{(i,j) \in R | i_b = nxb,\ 1 \le j_b \le nyb - 1\}$ (Fig. 4 VI),

$$y_{ij} = (r_{ij} - b_{i-1j}y_{i-1j} - c_{ij-1}y_{ij-1} - b_{ij}y_{i+1j})/d_{ij}. \tag{18}$$

The third step is, for nodes $\{(i,j) \in R | 1 \le i_b \le nxb - 1,\ j_b = nyb - 1\}$ (Fig. 4 VII),

$$y_{ij} = (r_{ij} - b_{i-1j}y_{i-1j} - c_{ij-1}y_{ij-1} - c_{ij}y_{ij+1})/d_{ij}. \tag{19}$$

The final step is, for nodes $\{(i,j) \in R | i_b = nxb,\ j_b = nyb\}$ (4 VIII),

$$y_{ij} = (r_{ij} - b_{i-1j}y_{i-1j} - c_{ij-1}y_{ij-1} - b_{ij}y_{i+1j} - c_{ij}y_{ij+1})/d_{ij}. \tag{20}$$

After the forward substitution, the backward substitution is performed for black blocks (Figs. 4 VIII → VII → VI → V). After the processors are synchronized, the backward substitution for red blocks is carried out (Figs. 4 IV → III → II → I). The incomplete Cholesky factorization is also parallelized by the same procedure as the forward substitution.

In the block red-black ordering method, the parallelism is given by the number of blocks of one color. On the assumption of $nx = ny$ and $nb = nxb = nyb$, the parallelism is $n/2nb^2$. The incompatible nodes in the block red-black ordering are given by the nodes which belong to the domains I, II and III in Figure 4. Accordingly, the incompatible ration R_{ic} is written by $2/nb - 1/nb^2$.

4.3   Comparison of Block Red-Black Ordering and Multi-color Ordering

Table 1 shows a comparison of the block red-black ordering method (block size: $nb \times nb$) (BRB) and m-color ordering method (MC).

The BRB has an advantage in the number of synchronization points. Next, we examine a comparison of convergence in both methods. When setting $nb = m$, the incompatible ratio of the BRB is less than that of the MC. The BRB, thus, can obtain about the same convergence as the MC, when using an appropriate block size. From the view point of parallelism, the MC is superior to the BRB. The BRB, however, attains sufficient parallelism in actual application problems. For example, the BRB with a block size of 32×32, which has about the same convergence as the 32-color ordering method, attains approximately 1000 degrees of parallelism for a 1400×1400 2-D problem. This problem is not very large and it never requires more than 1000 processors. When considering the problem C in Sect. 4.1, the BRB has another merit in that it can easily set the optimal block size. In the BRB, the larger block size results in better convergence and solver performance, since the number of synchronization points does not depend on the block size. Thus, if the block size is increased until the parallelism is equal to the number of used processors N_p, we then get the best convergence rate for the computational environment. The optimal block size $(nb \times nb)$ is given by

$$nb = \sqrt{\frac{n}{2N_p}}. \tag{21}$$

5   Numerical Experiments

5.1   Test Problem and Computational Environment

The present paper uses the following Poisson equation with the Dirichlet boundary condition as a test problem:

$$\nabla \cdot (k\nabla u(x,y)) = f \ in \ \Omega(0,1) \times (0,1) \tag{22}$$

$$u(x,y) = 0 \ on \ \delta\Omega$$

$$if \ \left(\frac{1}{4} \le x \le \frac{3}{4} \ \& \ \frac{1}{4} \le y \le \frac{3}{4}\right) \ then$$

$$k = 100$$

$$else \quad k = 1.$$

The grid size is 1025×1025 and f is given by 0.5 sin $(i_d + 1)$ where i_d is the number of node (x, y) on natural ordering. When using the five point difference scheme, problem (22) results in a linear system of equations to solve. The elements of the coefficient matrix and the preconditioner a_{ij}, b_{ij}, c_{ij}, d_{ij} are stored in one-dimensional arrays a, b, c, d. The convergence criterion of the ICCG method is given by $||r||_2/||f||_2$. The numerical test is performed on a shared-memory type scalar parallel computer Fujitsu GP-7000F model 900. The program code is written in the FORTRAN and the OPEN MP. A Fujitsu FORTRAN compiler version 5.1 is used with options -KOMP, -O2.

5.2   Implementation of the Block Red-Black Ordering Method

The present analysis uses natural ordering for arrangement of the elements a_{ij} $\cdots$ d_{ij} on the arrays a $\cdots$ d in order to avoid indirect memory access. Washio et al. have proposed a special treatment [8] of the multi-color ordering method in this case, which is used in the present analysis. In this technique, the number of the color m is assumed to satisfy

$$\mod(nx - 1, m) = 0, \tag{23}$$

and m stride memory access is used.

The implementation of the block red-black ordering method (BRB) in the above case is given as follows. In BRB, one block or several blocks in each color are assigned to processors. If the node number in these blocks is known, the processor can perform their part of the computation. Therefore, x- and y- global coordinates of left-lower corner and right-upper corner in each block: (LLx, LLy) and (RUx, RUy) are stored in memory. Then, the nodes belonging to this block are given by

$$\{(i, j) | (LLx \leq i \leq RUx), (LLy \leq j \leq RUy)\}, \tag{24}$$

and the entry number i_{dic} corresponding to the node (i, j) in the array is obtained as

$$i_{dic} = i + nx \cdot j. \tag{25}$$

In this implementation, the program code is written by using the coordinates of original problem, that is, (LLx, LLy) and (RUx, RUy). Accordingly, the blocks need not be of the same size. Then even if (11) is not satisfied, the BRB can be applied by modifying the block size of the right border or upper border of the entire grid. A sample program code of the forward substitution is shown in appendix A.1.

5.3 Numerical Results

Figures 5 and 6 show a comparison of convergence between BRB and MC. Figure 6 shows that the increase of block size improves the convergence without increasing the number of synchronization points in the BRB. These also show that BRB can attain a better convergence than MC when nb is set to be same as m. Table 2 lists the elapsed time, number of iterations and speed-up of both methods. The speed-up is based on a comparison of elapsed time with the implementation of the ICCG method on 1 CPU. Table 2 shows that BRB can attain a higher speed-up than MC. In MC, though the convergence is improved by increasing the number of colors, the speed-up value is limited. This is due to the increase of synchronization costs and the decline in utilization of data cache by stride access of memory. On the other hand, the convergence and the solver performance of the BRB is improved by increasing the block size. Figure 7 depicts the parallel speed-up by the BRB. The mark $\triangle$ represents the result when the optimal block size is used. The optimal number was calculated from (21) by setting n=1024×1024. Figure 7 indicates that the best solver performance is achieved when the optimal block size is used. We also examined the elapsed time in one ICCG iteration in the BRB. It was found that the elapsed time in

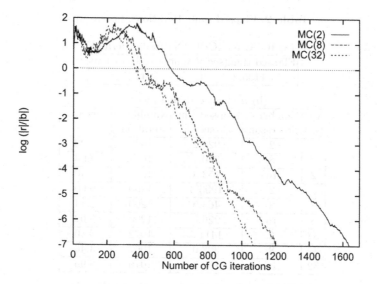

Fig. 5. Convergence behavior in multi-color ordering method

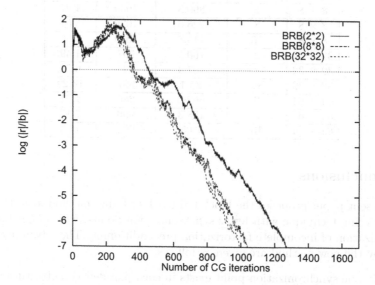

Fig. 6. Convergence behavior in block red-black ordering method

one iteration becomes shorter when the block size is increased. This is due to an improvement of utilization of data cache. The BRB, thus, has an advantage of effective utilization of data cache in addition to an advantage regarding the number of synchronization points. Consequently, the BRB is especially effective on scalar parallel computers.

Table 2. Elapsed time, Iterations, Speed-up

(a) ICCG on 1CPU (Natural ordering)

Elapsed time(sec)	Number of Iterations
1135.2	944

(b) m-color ordering method

MC (m)	Number of processors	Elapsed time (sec)	Number of iterations	Speed-up
2	2	928.2	1638	1.22
2	8	230.1	1638	4.93
2	16	153.3	1638	7.40
8	2	1903.3	1200	0.60
8	8	463.9	1200	2.44
8	16	226.0	1205	5.02
32	2	1411.3	1072	0.80
32	8	328.0	1067	3.46
32	16	163.3	1070	6.95

(c) Block red-black ordering method ($nb \times nb$)

BRB ($nb \times nb$)	Number of processors	Elapsed time (sec)	Number of iterations	Speed-up
8×8	2	890.8	1039	1.27
8×8	8	226.6	1039	5.01
8×8	16	115.8	1039	9.80
32×32	2	760.5	997	1.49
32×32	8	167.4	994	6.78
32×32	16	85.4	997	13.29
64×64	2	662.0	985	1.71
64×64	8	158.3	990	7.17
64×64	16	79.5	992	14.3

6   Conclusions

The present paper proposes the block red-black ordering method as a new parallel ordering technique with fewer synchronization points, which is utilized for parallelization of incomplete factorization preconditioning. The advantages provided by the new method are given as follows:

- Only one synchronization point exists in each parallelized substitution.
- The convergence in the preconditioned iterative method is improved by increasing the block size. The new method attains the same degree of convergence as the large-numbered multi-color ordering method.
- The optimal block size is easily set when the parallelism in the method is equal to the number of processors.

An analytical study using an ordering graph and numerical tests confirmed the above advantages of the new method. Moreover, the numerical tests indicate that a data cache is effectively utilized in the new method.

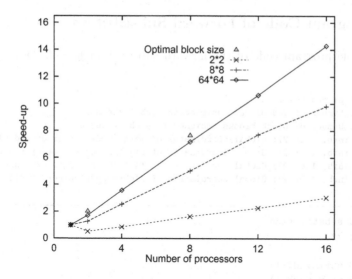

Fig. 7. Speed-up in block red-black ordering method

References

1. J. Meijerink and H. A. van der Vorst, "An Iterative Solution Method for Linear Systems of Which the Coefficient Matrix Is a Symmetric M-matrix," *Mathematics of Computation*, vol. 31, (1977), pp. 148-162.
2. H. A. van der Vorst and T. F. Chan, "Parallel preconditioning for sparse linear equations," ZAMM. Z. angew. Math. Mech., 76, (1996), pp.167-170.
3. I. S. Duff and H. A. van der Vorst, "Developments and trend in the parallel solution of linear systems," Parallel Computing, 25, (1999), pp. 1931-1970.
4. I. S. Duff and G. A. Meurant, "The effect of ordering on preconditioned conjugate gradients, BIT, 29, (1989), pp.635-657.
5. S. Doi and T. Washio, "Ordering strategies and related techniques to overcome the trade-off between parallelism and convergence in incomplete factorization", Parallel Computing, 25, (1999), pp. 1995-2014.
6. S. Doi and A. Lichnewsky, "A graph-theory approach for analyzing the effects of ordering on ILU preconditioning," INRIA report 1452, (1991).
7. T. Iwashita and M. Shimasaki, "Construction and Ordering of Edge Elements for Parallel Computation," *IEEE Trans. Magn.*, vol. 37, (2001), pp. 3498-3502.
8. T. Washio and K. Hayami, "Overlapped multicolor MILU Preconditioning," SIAM Journal of Scientific Computing, 16, (1995), pp. 636-650.
9. T. Osoda, K. Maruyama, T. Washio, S. Doi and S. Yamada, "Vectorization and parallelization technique of block ILU preconditioning for unstructural problems," *IPSJ Trans. HPS*, vol. 41, (2000), pp. 92-99 (in Japanese).
10. R. Barrett, et al., "Templates for the solution of linear systems: building blocks for iterative methods," SIAM, (1994).

A   Program Code of Forward Substitution

The sample program code of the forward substitution $y = L^{-1}r$ is given by follows.

```
myid: Rank of processor
ntr(myid): Number of red blocks assigned to each processor
ntb(myid): Number of black blocks assigned to each processor
rbs(myid, inrb, 1 or 2): Global coordinates of left-lower corner of red blocks
rbe(myid, inrb, 1 or 2): Global coordinates of right-upper corner of red blocks
bbs(myid, inrb, 1 or 2): Global coordinates of left-lower corner of black blocks
bbe(myid, inrb, 1 or 2): Global coordinates of right-upper corner of black blocks
```

```
-----------------------------------------------------------------------
! Forward substitution
! Red blocks

do inrb=0,ntr(myid)-1
  llx=rbs(myid,inrb,1)
  lly=rbs(myid,inrb,2)
  rux=rbe(myid,inrb,1)
  ruy=rbe(myid,inrb,2)

! Domain I
   y(lly*nx+llx)=r(lly*nx+llx)/diag(lly*nx+llx)

! Domain II
   do i=lly*nx+llx+1,lly*nx+rux
     y(i)=(r(i)-b(i-1)*y(i-1))/diag(i)
   enddo

! Domain III
   do i=(lly+1)*nx+llx,ruy*nx+llx,nx
     y(i)=(r(i)-c(i-nx)*y(i-nx))/diag(i)
   enddo

! Domain IV
   do iy=lly+1,ruy
     do ix=llx+1,rux
     i=iy*nx+ix
     y(i)=(r(i)-b(i-1)*y(i-1)-c(i-nx)*y(i-nx))/diag(i)
   enddo
   enddo

enddo

!$OMP BARRIER
! Black blocks
```

```
do inbb=0,ntb(myid)-1
  llx=bbs(myid,inbb,1)
  lly=bbs(myid,inbb,2)
  rux=bbe(myid,inbb,1)
  ruy=bbe(myid,inbb,2)

! Domain V
  do iy=lly,ruy-1
    do ix=llx,rux-1
      i=iy*nx+ix
      y(i)=(r(i)-b(i-1)*y(i-1)-c(i-nx)*y(i-nx))/diag(i)
    enddo
  enddo

! Domain VI
  do i=ruy*nx+llx,ruy*nx+rux-1
    y(i)=(r(i)-b(i-1)*y(i-1)-c(i-nx)*y(i-nx)-c(i)*y(i+nx))/diag(i)
  enddo

! Domain VII
  do i=lly*nx+rux,ruy*nx+rux-nx,nx
    y(i)=(r(i)-b(i-1)*y(i-1)-c(i-nx)*y(i-nx)-b(i)*y(i+1))/diag(i)
  enddo

! Domain VIII
  ii=ruy*nx+rux
  y(ii)=(r(ii)-b(ii-1)*y(ii-1)-c(ii-nx)*y(ii-nx) &
      -b(ii)*y(ii+1)-c(ii)*y(ii+nx))/diag(ii)

enddo
```

Integrating Performance Analysis in the Uintah Software Development Cycle

J. Davison de St. Germain[1], Alan Morris[1], Steven G. Parker[1],
Allen D. Malony[2], and Sameer Shende[2]

[1] School of Computing,
University of Utah
{dav,amorris,sparker}@cs.utah.edu
[2] Department of Computer and Information Science,
University of Oregon
{malony,sameer}@cs.uoregon.edu

Abstract. Technology for empirical performance evaluation of parallel programs is driven by the increasing complexity of high performance computing environments and programming methodologies. This paper describes the integration of the TAU and XPARE tools in the Uintah computational framework. Performance mapping techniques in TAU relate low-level performance data to higher levels of abstraction. XPARE is used for specifying regression testing benchmarks that are evaluated with each periodically scheduled testing trial. This provides a historical panorama of the evolution of application performance. The paper concludes with a scalability study that shows the benefits of integrating performance technology in the development of large-scale parallel applications.

1 Introduction

Modern scientific simulations have become incredibly complex. It is not uncommon for high-performance software systems to have large development teams involving personnel across a broad range of expertise who work simultaneously on different parts of the system. In these programming environments, software developers increasingly turn to industrial tools for managing the complex software process. Tools for revision control, automated testing, and bug tracking are now commonplace. Unfortunately, tools to help achieve the highest performance possible over a broad range of inputs and hardware configurations are not commonly available. As a result, many software development efforts leave performance evaluation and improvement until the end of a long, many-stage development process. Even if performance is studied early in development, tracking the performance of the system as new features are added is often too time-consuming. While the complexity of the software development process may justify these engineering decisions. increased sophistication in high-performance parallel software and platforms rarely reduces performance complexity as development and use of the software proceeds.

Certainly, one very serious problem that arises is when developers of parallel scientific software make design decisions without knowledge or understanding

H. Zima et al. (Eds.): ISHPC 2002, LNCS 2327, pp. 190–206, 2002.

of the performance ramifications. Any code decision, however localized, may have significant impact on performance overall. These performance influences can be difficult to observe and subtle to understand. If a performance engineering methodology is not incorporated in the software design and development process, it will be extremely difficult to achieve the high-performance goals of the project over its lifetime. Moreover, if the methodology is not adequately supported by flexible and robust performance tools, it will be difficult to address all performance problems that arise.

In this paper, we report on our efforts to integrate performance analysis capabilities into one such complex scientific software system: the Uintah Computational Framework. These capabilities support a performance engineering methodology that augments Uintah's current software design process. We describe the Uintah system in sufficient detail to highlight the challenges we have faced in performance measurement and analysis, and in tracking, maintaining, and improving Uintah performance. The TAU and XPARE tools we developed for Uintah performance engineering are then discussed in detail. We demonstrate their benefits to Uintah performance analysis and improvement with several examples. Finally, we outline our plans for future work.

2   Background and Motivation

In 1997, the Center for the Simulation of Accidental Fires and Explosions (C-SAFE) [2] was created at the University of Utah to focus specifically on providing state-of-the-art, science-based tools for the numerical simulation of accidental fires and explosions, especially within the context of handling and storage of highly flammable materials. C-SAFE was created by the Department of Energy's Accelerated Strategic Computing Initiative's (ASCI) Academic Strategic Alliance Program (ASAP) [1].

C-SAFE's objective is to build a problem-solving environment in which fundamental chemistry and engineering physics are coupled fully with non-linear solvers, optimization, computational steering, visualization and experimental data verification. Such a system would allow better evaluation of the risks and safety issues associated with fires and explosions. However, the software needed to model such real-world scientific and engineering problems is very complex, and is further compounded when multiple simulation codes must work together. Likewise, achieving high performance on large-scale computer systems is a necessary, but non-trivial goal.

C-SAFE's Uintah Problem Solving Environment [4] is a massively parallel, compo- nent-based, problem solving environment (PSE) designed to simulate large-scale scientific problems, while allowing the scientist to interactively visualize, steer, and verify simulation results. Uintah is derived from the SCIRun[1] PSE [9,10,11,12], adding support for a more powerful component model on distributed-memory parallel computers. The Uintah PSE is being developed

---

[1] Pronounced "ski-run." SCIRun derives its name from the Scientific Computing and Imaging (SCI) Institute at the University of Utah.

Fig. 1. Visualization of two different simulations from C-SAFE. On the left is a simulation of a heptane fire. On the right is a simulation of stress propagation through a block of granular material. Each of these simulations were performed using the Uintah Computational Framework and were executed on 1000 processors

specifically to study interactions between hydrocarbon fires, structures, and high-energy materials (explosives and propellants), such as those shown in Figure 1.

In designing the Uintah software system, we focused on three guiding properties. First, the complexities of code creation for parallel machines should (as much as possible) be hidden from the scientist. Second, complex simulation components developed by third parties should be tools available for scientists to employ. And third, the scientist should be able to visually monitor and steer his or her simulation while it is running. A software environment that efficiently integrates these properties into a usable system allows scientists to effectively create and use complex simulations in an interactive, exploratory way. The Uintah PSE is such a system. It allows scientists and engineers to focus on algorithm development and data analysis rather than details of the underlying software architecture, without sacrificing the ability to realize the full potential of large parallel computers.

While Uintah is provides a general framework in which a wide variety of large scale, massively parallel simulations can be conducted, the specific problem that has driven its creation is the modeling of the interactions between hydrocarbon fires, structures and high-energy materials (explosives and propellants), as shown in Figure 2. In order to produce realistic simulations of these problems, we must utilize large-scale parallel computers at maximum efficiency. For the largest simulations, we use DOE ASCI computing resources consisting of thousands of processors. A typical simulation consists of billions of degrees of freedom or more.

During simulation software development at C-SAFE, the need for performance analysis became very apparent. In particular, performance measurement and analysis tools were required for three main tasks:

1. Optimization of code kernels for maximum serial performance (micro tuning).
2. Analysis of parallel execution bottlenecks (scalability tuning).
3. Understanding the performance impacts of code modifications over the course of development (performance tracking).

Fig. 2. A Typical C-SAFE Problem

By integrating tools to address these tasks in the Uintah PSE development process, we have created a scalable simulation environment for C-SAFE problems where performance of the overall environment is high and will not diminish unexpectedly due to evolution of the Uintah code.

3 Uintah Architecture

The Uintah PSE provides a component-based environment for developing parallel scientific applications. Uintah is based on the component architecture being developed by the Common Component Architecture (CCA) Forum. The CCA Forum [3] was established to specify a software component architecture that could address the needs of high-performance computing. The CCA architecture aims to provide higher performance, explicit support for multi-dimensional arrays, and support for parallelism. Uintah is a research vehicle for implementing these ideas and for exercising their efficacy on complex scientific applications, such as the C-SAFE simulations.

Solving a typical C-SAFE problem involves running multiple large-scale physically coupled simulations. For example, to investigate the effects of fire on metal structures, a fluid-dynamics-based combustion model might be coupled with a particle-based solid mechanics simulation. The simulation models may involve representations of size 10^9 finite volume cells and 10^8 solid material points. To handle the large number of operations necessary to process such immense datasets, we have designed the *Uintah Computational Framework* (UCF). The UCF is the foundation upon which all C-SAFE simulation components are developed.

The UCF is a set of components and classes that build on the Uintah component model, adding capabilities such as semi-automatic parallelism, automatic checkpoint/restart, load-balancing mechanisms, resource management, and scheduling. The UCF exposes flexibility in dynamic application structure by adopting an execution model based on software or "macro" dataflow. Computations are expressed as directed acyclic graphs of *tasks*, each of which consumes some input and produces some output (input of some future task). These inputs and outputs are specified for each patch in a structured grid. Tasks are organized in a UCF data structure called the *task graph*.

In natural agreement with the functional nature of its pure macro-dataflow execution model, the UCF presents developers with an abstraction of a global single-assignment memory, with automatic data lifetime management and storage reclamation. Storage is abstractly presented to the scientific programmer as a dictionary mapping names to values. The value associated with a name can be written only once, and once written is communicated by UCF to all tasks awaiting that value. Values are typically array-structured. Communication is scheduled by a local scheduling algorithm that approximates the true globally optimal communication schedule. Because of the flexibility of single-assignment semantics, the UCF is free to execute tasks close to data or move data to minimize future communication.

The UCF storage abstraction is sufficiently high-level that it can be mapped efficiently onto both message-passing and share-memory communication mechanisms. Threads sharing a memory can access their input data directly; single-assignment dataflow semantics eliminate the need for complex locking of values. The UCF is free to optimize allocation of physical memory to minimize remote memory accesses. Threads running in disjoint address spaces communicate by message-passing protocol, and the UCF is free to optimize such communication by message aggregation. Tasks need not be aware of the transports used to deliver their inputs and, thus, the UCF has complete flexibility in control and data placement to optimize communication both between address spaces and within the shared ccNUMA memory hierarchy of the Origin 2000 (or other SMP-based distributed memory supercomputers). Solving this optimization problem for C-SAFE simulations is difficult and is a subject of ongoing investigation.

An example UCF taskgraph is shown in Figure 3. Ovals represent tasks, each of which is a simple array algorithm and easily treated by traditional compiler array optimizations. Edges represent named values stored by the UCF. Solid

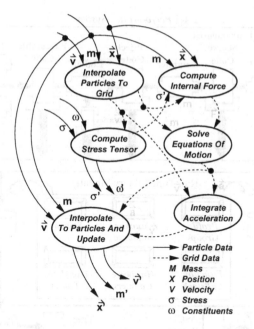

Fig. 3. An Example UCF Task Graph

edges have values defined at each material point (Particle Data) and dashed edges have values defined at each grid vertex (Grid Data). Variables denoted with a prime (') have been updated during the time step. The figure shows a slice of the actual Uintah Material Point Method (MPM) task graph concerned with advancing Newtonian material point motion on a single patch for a single timestep.

4   Performance Technology Integration

The Uintah PSE and the UCF present interesting challenges to performance analysis technology and its integration. The diversity of the Uintah software, including the UCF middleware and simulation code modules, and Uintah's portability objectives requires performance instrumentation and measurement tools that are both cross-language and cross-platform. The performance system must also work at large scales, and be able to analyze performance data captured for the different execution modes (shared-memory, message passing, mixed-mode) that Uintah supports. Perhaps the most important concern is being able to relate multi-level performance data to the high-level task abstractions used within Uintah for simulation programming and during execution by the UCF for task graph scheduling and storage management. Without this capability, it would be extremely difficult to piece apart performance effects across UCF levels and to identify the simulation components responsible for different performance behaviors.

TAU Performance System

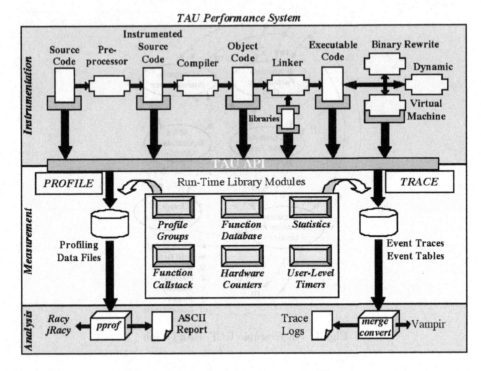

Fig. 4. TAU Performance System Architecture

4.1 TAU Performance System

Performance technology integration in the Uintah PSE is based on the TAU performance system [7]. TAU provides robust technology for performance instrumentation, measurement, and analysis for complex parallel systems. It targets a general computation model consisting of shared-memory computing *nodes* where *contexts* reside, each providing a virtual address space shared by multiple *threads* of execution. The model is general enough to apply to many high-performance scalable parallel systems and programming paradigms. Because TAU enables performance information to be captured at the node/context/thread levels, this information can be mapped to the particular parallel software and system execution platform under consideration.

As shown in Figure 4, TAU supports a flexible instrumentation model that applies at different stages of program compilation and execution. The instrumentation targets multiple code points, provides for mapping of low-level execution events to higher-level performance abstractions, and works with multi-threaded and message passing parallel computation models. Instrumentation code makes calls to the TAU measurement API. The TAU measurement library implements performance profiling and tracing support for performance events occurring at function, method, basic block, and statement levels during execution. Performance experiments can be composed from different measurement modules (e.g.,

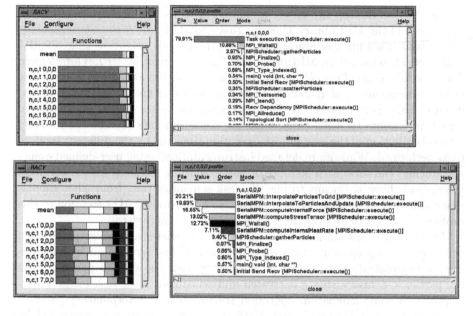

Fig. 5. TAU Performance Profiles Without Mapping (top) and With Mapping (bottom)

hardware performance monitors) and measurements can be collected with respect to user-defined performance groups. The TAU data analysis and presentation utilities offer text-based and graphical tools to visualize the performance data as well as bridges to third-party software, such as Vampir [8] for sophisticated trace analysis and visualization.

4.2 Tau Performance Mapping in Uintah

To evaluate the performance of Uintah applications, we selectively instrument at the source level and the message passing library level. Source-level instrumentation occurs at subroutine and method boundaries, as well as at important code sections using TAU user-defined timers (with *start/stop* semantics) to highlight the time spent in groups of statements. Message passing instrumentation (using a MPI interposition library based on PMPI [6]) shows both execution time spent in message communication and messaging behavior with respect to application level routines. Figure 5 shows two profiles of the execution time of different tasks within the UCF's parallel scheduler for an MPI-only run. The displays were created by TAU's parallel profile visualizer, *Racy*, which can show full profile details across all threads of execution. Here, the right views show the detailed performance profile on "n,c,t (node,context,thread) 0,0,0" (i.e., MPI process with rank 0). The left views show performance for all of the MPI processes in bargraph form.

To generate the top two views, we placed instrumentation in the *MPISched-uler* class and the MPI library. Clearly, *Task execution [MPIScheduler:execute()]* (green bar) takes up a significant chunk of the overall execution time, 79.91% of the total (exclusive) on MPI process 0. The time spent in *MPI_Waitall()* and *MPISchedule::gatherParticles* is also of significance, but the other routines are of less consequence. Unfortunately, these top two views give only a rough break-down of UCF performance. While it is important to see a high percentage of time being spent executing tasks, what the scientist wants to know additionally is the distribution of the overall task execution time among the different types of tasks performed. While more detailed instrumentation (using user-defined events and tracing) can show each instance of task execution, standard instrumentation mechanisms have no means to identify task semantics (i.e., from what simula-tion component the tasks were produced). To understand TAU's solution to this problem, we need to describe how UCF operates in more detail.

During the computation, many individual particles are being partitioned across processing elements (processes or threads) and worked on by the sim-ulation components represented in the task graph. As work is performed on the particles, a *task instance* is created and scheduled. Each task instance cor-responds to some simulation operation (task), such as interpolating particles to the grid in the Material Point Method, and its execution is controlled by its task graph dependencies. We can give each task instance a name (e.g., *Se-rialMPM::interpolateParticlesToGrid*) that identifies its domain-specific charac-ter in the computation (i.e., its specific simulation task relationship). The the number of task types is finite and is typically less than twenty in Uintah ap-plications. In contrast, there are a large number of task instances created and executed during the computation. The association of a task type with a task in-stance occurs at a time different from when the task instance is finally scheduled and executed.

Thus, to provide the desired performance view, we must map the performance of each individual task instance to the task type to which it belongs and then accumulate the performance data at the task level. Using TAU's *Semantic Entity, Association, and Attributes* (SEAA) model of performance mapping [13], we form an association during initialization between a timer for each task (the task semantic entity) and the task name (its semantic attribute). Then, while processing each task instance in the scheduler, a method to query the task name (stored within the task instance object) is invoked and the address of the task name (a static character string) is returned. Using this address, we do an external map lookup (implemented as a hash-table) and retrieve the address of the timer object (i.e., a runtime semantic association). Once the timer is known, it can be started and stopped around the code segment that executes the task instance.

The bottom two views in Figure 5 show the results of this task mapping performance analysis in Uintah. Clearly, there is a significant benefit of the SEAA approach in presenting performance data with respect to high-level semantics of the Uintah application. The performance of all five simulation model components (i.e., tasks) are now clearly distinguished in the profile. With the generation of event traces, the benefits are even more dramatic as this task mapping allows distinct phases of computation to be highlighted based on task semantics. This can be seen in the trace visualization in Figure 7. Although we are looking at individual task instances being executed, the color-coded mapping allows us to view their performance data at a higher level.

4.3   Performance Experiment Reporting and Alerting

With the integration of performance measurement support in the Uintah software system comes the ability to analyze performance throughout Uintah's development lifetime. Typically, performance analysis is done ad hoc, at the convenience of the developer, and only when time permits. When such performance practice is applied across a large, multi-person effort such as C-SAFE, the resulting "performance portfolio" becomes scattered and tends to report performance information only after significant stages of development have been accomplished and software committed. The downside of such a performance methodology is a disengagement of performance knowledge from key software design decisions. The goal of our work is to more tightly couple the reporting of performance experimentation results with timely software testing and alerting to performance problems. We have created the XPARE (eXPeriment Alerting and REporting) system for this purpose.

The Uintah software system was engineered with a regression testing harness to regularly evaluate correctness. At these times, minimal performance benchmarking would be conducted to determine if total execution time was seriously degraded. If so, the tester would notify software developers, but left it up to them to manually run specific instrumented tests to investigate where the performance problems lay. The XPARE system augments the regression tester to conduct a range of performance experiments with fully-instrumented code modules. Multiple experiments can be conducted with different instrumentation layouts to exercise different code regions and behaviors. The TAU performance tools are used for measurement and analysis, allowing execution time and hardware statistics to be used to construct a complete performance portrait.

Once the performance experiments have been conducted, XPARE will automatically interrogate the performance data to determine not only if the overall code has run for longer than expected, but also which tasks and profiled procedures are potential suspects. XPARE accomplishes this by applying alerting "rulesets" (performance difference thresholds) to a historical, multiple experiment performance database. Experiment sets can be selected by the user from the database for evaluation. For each experiment set, specific performance data can be chosen for analysis. Performance regression testing is then done by comparing the current performance with that in the experiment set, using the alerting rulesets constructed by the user to determine performance violations worthy of report.

The XPARE system architecture is shown in Figure 6, with images of the web-based interfaces for experiment selection, performance data selection, and ruleset definition. As also shown, results of regression analysis are automatically

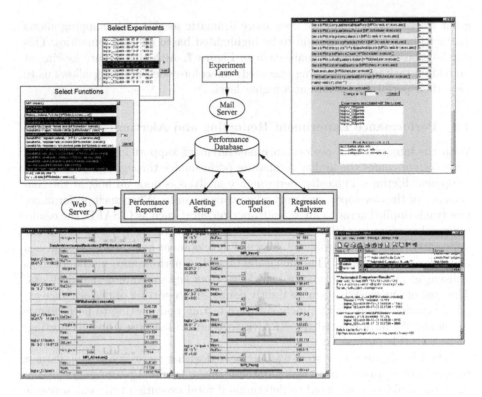

Fig. 6. XPARE System Architecture and Tools

reported to the software developers, who can explore the performance data more fully through the performance reporter, whether or not significant performance shifts have been detected. Because the performance database contains prior performance history, a panoramic view of performance change can be scrutinized based not only on code alteration, but also platform, choice of compiler, different optimizations, and other performance factors.

By scheduling regular performance regression tests, performance knowledge can be closely linked with the Uintah software development cycle. Currently, we use XPARE to run weekly performance tests of small to medium-scale experiments, and monthly evaluations of full-scale experiments. The general construction of XPARE will allow it to easily extend to changes in the Uintah code base and to incorporate new simulation components as they become available.

5 Performance Studies

Contemporary efforts in gathering performance data have focused on function by function analysis. C-SAFE has taken the somewhat novel approach of gathering performance statics on an *algorithmic* basis. This approach provides four major benefits.

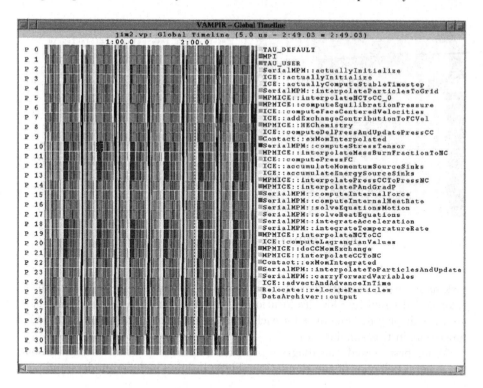

Fig. 7. MPM Simulation Performance (TAU /Vampir)

1. Due to the use of the task abstraction in the UCF, it is straightforward to manually insert the profiling code at one location in the code to capture data on the performance of all tasks.
2. The performance characteristics of each of the algorithmic tasks is clearly displayed in relation to the other simulation tasks.
3. Scientific programmers are allowed to focus on making performance improvements at an algorithmic level.
4. Uintah Computational Framework developers can easily find performance bottlenecks that are not directly associated with application codes (e.g.; MPI communications, task scheduling overhead, and data I/O).

The first step in optimizing Uintah software was to manually instrument the code base with hooks to the TAU system. The event-traces generated were converted to the Vampir trace data format and visualized using Vampir. Figure 7 depicts one of the first visualizations of an early version of the Uintah code running an MPM simulation on 32 processors. The figure shows six time steps with the black lines between the time steps depicting the large MPI communications necessary to transmit boundary data. Listed on the right hand side of the window are each of the specific tasks, delineated by major software component (e.g.; SerialMPM, MPMICE, DataArchiver, Contact, etc.) followed by specific

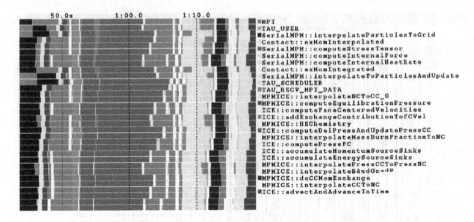

Fig. 8. MPM Simulation (Single Time Step)

task name (e.g.; computeStressTensor, relocateParticles, etc.) Each task can be color coded to easily view its location in the time line. On the left hand side are rows displaying time lines for each process, running in parallel on individual processors in this simulation run.

When first viewed, this diagram provided a number of "Aha!" insights about the general behavior of the simulation. These insights included understanding:

1. the load imbalances we were experiencing with a rudimentary load balancer;
2. that the *computeStressTensor* task constituted a large portion of the execution time; and
3. that there was a significant amount of MPI overhead distributed throughout the computation.

Figure 8 is a zoomed-in view of a single time step in the MPM Simulation. This view provided insight into the parallelization of each of the tasks in a single time step. It also provided us with a visual feedback for how the processors where lining up and how much work each was doing.

Similarly, Figure 9 depicts five time steps of the "Arches" fire simulation within the UCF. This figure portrays explicitly how much time is being spent in the "pressure solving" portion of the simulation. (The pressure solve calculation utilizes a PETSc linear solver.) Figure 10 is a close up view of the *PressureSolver* task within the time step and reveals that a major portion of the solver's time is spent in MPI calls. This visualization has led to focusing performance enhancement resources on determining the best way to use PETSc solvers (including exploring different pre-conditioners).

Once candidate tasks are identified as potential performance bottlenecks, the tasks are inspected from both an algorithmic view and from an implementation view. At this point, it is sometimes necessary to perform additional functional instrumentation of the code. We used this method of performance analysis from late 2000 through the first half of 2001 to investigate performance problems

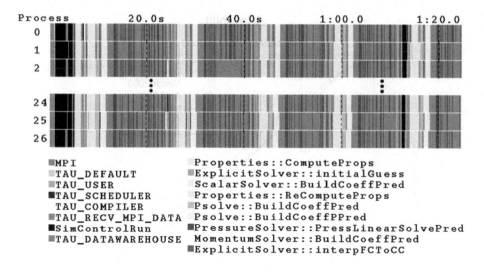

- ■ MPI
- ▨ TAU_DEFAULT
- ▨ TAU_USER
- ■ TAU_SCHEDULER
- ▨ TAU_COMPILER
- ▨ TAU_RECV_MPI_DATA
- ■ SimControlRun
- ▨ TAU_DATAWAREHOUSE

- ▨ Properties::ComputeProps
- ▨ ExplicitSolver::initialGuess
- ▨ ScalarSolver::BuildCoeffPred
- ▨ Properties::ReComputeProps
- ▨ Psolve::BuildCoeffPred
- ▨ Psolve::BuildCoeffPPred
- ■ PressureSolver::PressLinearSolvePred
- ▨ MomentumSolver::BuildCoeffPred
- ■ ExplicitSolver::interpFCToCC

Fig. 9. Arches Task Performance

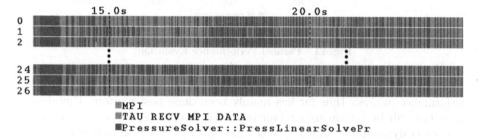

- ■ MPI
- ■ TAU RECV MPI DATA
- ■ PressureSolver::PressLinearSolvePr

Fig. 10. Arches Task Zoomed In

in the Uintah software. This lead to the parallel scaling improvements seen in Figure 11. Successive lines on the graph show the performance improvements after finding and fixing performance bottlenecks.

After directing our efforts at improving the Uintah scalability up to 2000 processors, our focus changed to other aspects of code development. It was at this point that we recognized the need for the XPARE system. Once implemented, it has allowed us to monitor the performance of individual simulation pieces in addition to the overall performance. XPARE has been developed with the goal of keeping the Uintah system efficient as we expand the system and add new features.

6   Lessons Learned and Future Work

The integration of performance measurement in the UCF scheduling component has been extremely useful in exposing bottlenecks and inefficiencies. While the

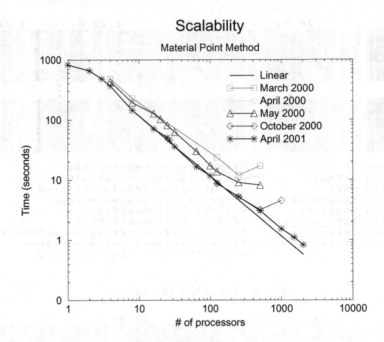

Fig. 11. Parallel Performance Evolution

performance analysis thus far has mainly been done post-mortem, Uintah applications will be increasingly adaptive in the future and will require UCF to implement dynamically adjusting scheduling policies. We plan to develop online performance query and feedback capabilities in TAU that will support adaptive Uintah execution. Also, to enhance online performance analysis, we are developing a runtime infrastructure to visualize dynamic, large-scale performance data using the SCIRun visualization environment.

We will also continue to build on the success of performance mapping in Uintah to attribute execution costs from the simulation component parts. We have recently encountered the need for more flexible performance mapping specification that allows multiple mappings attributions (e.g., for mapping execution costs from component parts to higher-level tasks and patches) to be active simultaneously. The current rudimentary means to support these mappings will be implemented in more robust forms in the near future. Not only is the UCF a target for performance integration, but the individual simulation components can benefit from performance analysis. We will begin to work more closely with the developers of C-SAFE simulation component software to integrate performance measurement, analysis, and regression testing in their codes.

With the completion of a mixed-mode UCF implementation will come the need for performance analysis of integrated multi-threaded and message-based execution. While preliminary tests have demonstrated TAU's ability to observe thread and communication events in mixed-mode Uintah execution, it will be important to develop techniques for cross-mode sharing of instrumentation information so that integrative performance mapping and analysis is possible.

We will greatly enhance the existing prototype XPARE system to play an increasingly important role in Uintah software performance engineering in the future. In particular, we will concentrate on XPARE's performance database which is currently implemented in an ad hoc manner. The TAU project is building a performance database framework (PerfDBF) that will be employed by XPARE for more flexible cross-experiment data query and analyses. PerfDBF will allow for the set of analysis operations to be easily extended by UCF and simulation component developers. XPARE's alerting and reporting tools can then incorporate these expanded analysis options to construct more sophisticate threshold functions and performance data processing for generating performance reports.

Acknowledgments

This work was supported by the DOE ASCI ASAP Program. The work at Oregon was supported by a contract from the DOE 2000 program (Agreement No. DEFC 0398 ER 259 986) and a sub-contract from the University of Utah's DOE C-SAFE ASCI center (Agreement No. B341493). C-SAFE visualization images were provided by Kurt Zimmerman and Wing Yee. Datasets were created by Scott Bardenhagen, Jim Guilkey, and Rajesh Rawat. The DOE ASCI ASAP program also provided computing time for the simulations shown.

References

1. Academic Strategic Alliances Program. http://www.llnl.gov/asci-alliances.
2. Center for the Simulation of Accidental Fires and Explosions. http://www.csafe.utah.edu.
3. Common Component Architecture Forum. http://www.cca-forum.org.
4. Davison de St. Germain, J., McCorquodale, J., Parker, S.G., Johnson, C.R.: Uintah: A Massively Parallel Problem Solving Environment. HPDC'00: Ninth IEEE International Symposium on High Performance and Distributed Computing (2000)
5. Lindlan, K.A., Cuny, J., Malony, A.D., Shende, S., Mohr, B., Rivenburgh, R., Rasmussen, C.: Tool Framework for Static and Dynamic Analysis of Object-Oriented Software with Templates. Proceedings SC'2000, (2000)
6. Message Passing Interface Forum: MPI: A Message Passing Interface Standard. International Journal of Supercomputer Applications (Special Issue on MPI) 8(3/4) (1994)
7. Malony, A., Shende, S.: Performance Technology for Complex Parallel and Distributed Systems. In: Kotsis, G., Kacsuk, P. (eds.): Distributed and Parallel Systems From Instruction Parallelism to Cluster Computing. Proc. 3rd Workshop on Distributed and Parallel Systems, DAPSYS 2000, Kluwer (2000) 37–46
8. Pallas GmbH: VAMPIR: Visualization and Analysis of MPI Resources. http://www.pallas.de/pages/vampir.htm.
9. Parker, S.G., Beazley, D.M., Johnson, C.R.: Computational steering software systems and strategies. IEEE Computational Science and Engineering, 4(4) (1997) 50–59

10. Parker. S.G., *The SCIRun Problem Solving Environment and Computational Steering Software System.* PhD thesis, University of Utah (1999)
11. Parker, S.G., Johnson, C.R.: SCIRun: A scientific programming environment for computational steering. Proc. Supercomputing '95. IEEE Press (1995)
12. Parker, S.G., Weinstein, D.M., Johnson C.R.: The SCIRun computational steering software system. In: Arge, E., Bruaset, A.M., Langtangen, H.P., (eds.): Modern Software Tools in Scientific Computing, Birkhauser Press (1997) 1–44
13. Shende, S.: The Role of Instrumentation and Mapping in Performance Measurement. Ph.D. Dissertation, University of Oregon (2001)
14. Shende, S., Malony, A., Ansell-Bell, R.: Instrumentation and Measurement Strategies for Flexible and Portable Empirical Performance Evaluation. Proc. International Conference on Parallel and Distributed Processing Techniques and Applications, PDPTA '2001, CSREA, (2001) 1150–1156
15. Shende, S., Malony, A., Cuny, J., Lindlan, K., Beckman, P., Karmesin, S.: Portable Profiling and Tracing for Parallel Scientific Applications using C++. Proc. SIGMETRICS Symposium on Parallel and Distributed Tools, SPDT'98, ACM, (1998) 134–145

Performance of Adaptive Mesh Refinement Scheme for Hydrodynamics on Simulations of Expanding Supernova Envelope

Ayato Noro[1], Tomoya Ogawa[1], Takuma Ohta[2], Kazuyuki Yamashita[3],
Shigeki Miyaji[1], and Mitue Den[4]

[1] Graduate School of Science and Technology, Chiba University, Japan
{noro,tomoya,miyaji}@astro.s.chiba-u.ac.jp
[2] Morioka Local Meteorological Observatory, Morioka, Japan
takuma.ohta@met.kishou.go.jp
[3] Institute of Media and Information Technology, Chiba University, Japan
yamasita@imit.chiba-u.ac.jp
[4] Communications Research Laboratory, Japan
den@crl.go.jp

Abstract. We present results of performance measurement of our astrophysical fluid dynamics code with Adaptive Mesh Refinement (AMR) scheme. As an example of its application for astrophysical phenomena, we show the results of 3D simulation of Rayleigh-Taylor instability in expanding supernova envelope and the possibility of reconciliation of the present model with observations. The efficiency of memory and CPU-time savings is discussed and measured with this simulation. Its result shows that our code has the ability to simulate phenomena with wide dynamic ranges even in limited computing resources.

1 Introduction

Non-linear models of gas/fluid dynamics are verified by hydrodynamical numerical simulations. In the case of using Eulerian code, we can simulate models with high spatial accuracy using MUSCL method etc. [1]. However, its resolution is always limited by the available memory size on the computer. This limitation makes it difficult to simulate target phenomenon or gives invalid results for the system in which wide dynamic ranges of scale length are required. One example is the case of cosmological galaxy formation. In this case, a large scale density enhancement accelerates the growth of small clumps. The other example is the case of Rayleigh-Taylor (RT) instability in the expanding supernova envelope. In this case, a perturbation of short wave length grows to form a profile of large scale, i.e., composition mixing by RT instability. In these cases, regions where fine resolution is needed are only determined by the physics included and thus it is hard to predict the locations where meshes should be refined before simulation. Therefore any fixed hierarchical structure scheme as like nested grid scheme is not suitable for these cases. Only Adaptive Mesh Refinement (AMR) scheme,

H. Zima et al. (Eds.): ISHPC 2002, LNCS 2327, pp. 207–218, 2002.

in which high resolution mesh cells are dynamically generated, is the method which can treat such phenomena.

Here we present results of performance evaluation of our AMR scheme hydrodynamic code. As an example of this performance measurement, we focus the simulation of the growth of RT instability in the expanding supernova envelope.

Numerical studies of the growth of RT instability in the expanding supernova envelope which have been performed with 2D and low resolution 3D regimes ([2], [3], [4], and references there in) have been failed to reproduce high expansion velocities observed, i.e., $\sim 3000\,\mathrm{km/s}$ [5]. The explosion velocities simulated by these studies are only about $2000\,\mathrm{km/s}$, i.e., two third of the observed value. One reason of this deficit of the velocity may be due to the lack of spatial resolution. In the 2D simulations, the perturbation along the symmetric direction is ignored, which means the wave length of the perturbation along this direction is assumed to be very large, so that the effective wave length of perturbation is regarded as large as that of low resolution 3D simulations.

The costs of memory and CPU are high for 3D simulations. In the case of RT instability simulation in expanding supernova envelope, this situation is very serious. The ratio of the shock front radius at the initial stage, where it reached He-H discontinuity, to the final stage is about 0.042. If homogeneous mesh is adopted, the maximum mesh size we can prepare is practically 1024^3 at present so that the initial stage of shock front could be modeled by less than 40^3 cells. However, at this stage, outer region is almost vacuum and meaningful calculation is limited only at small region within the shock front.

Recently, Noro et al. [6] carried out 3D simulation with AMR scheme and obtained much faster expanding velocity. Furthermore, they also showed that the difference in the growth of RT instabilities for 2D and 3D simulations is about 20% in the case of simple two layers model in constant gravity.

Therefore, the significance of AMR scheme to simulate this phenomenon is evident. Here we evaluate its performance in detail and show the efficiency of AMR scheme for general subjects.

In section 2, outline of AMR scheme is given. Resource usage of AMR scheme is estimated in section 3. Our simulation of RT instability of expanding supernova envelope is given in section 4. Performance evaluation of AMR code is given in section 5. Section 6 is devoted to discussion.

2   AMR Scheme

The one of the most popular AMR schemes is Fully Threaded Tree (FTT) method (Khokhlov [7]) that gives easy operation of refining and coarsening cells by using list-vectors. The list-vectors point to an "oct", which is a unit of the hierarchical mesh structure and has 8 cells in 3D space (Fig. 1). Each cell has 2^{-l} side length; the index l of the power of 2 is called "level" of the oct. The oct has following information: its level l, physical variables (ρ, v_x, v_y, v_z, P) of the 8 cells, coordinates (x, y, z) of its center, pointers to neighboring octs, pointers to its children octs, the index of its parent cell, and a refinement indicator which

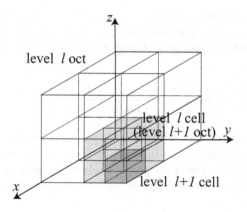

Fig. 1. Hierarchical structure of octs and cells (Fig. 1 of Noro et al. [6]). The biggest box shows an oct of level l. This oct has lower hierarchical (level $l + 1$) octs which are called its "children". The light shaded box shows its child oct of level $l + 1$ which is equivalent to the level l cell. The dark shaded box shows one of level $l + 1$ cells of the child oct.

is used for a criterion whether the oct should be refined or coarsened. We use an FTT code which was implemented by Ogawa et al. [8] and developed by Noro et al. [6]. Our present code differs from Khokhlov's original FTT at following points: using the level-independent time step and distributing the refinement indicator of each oct over surrounding octs by the homogeneous cubic point-spread-function [6]. We adopted Roe-MUSCL scheme [9] for the part of flux evaluation in which the 3rd order accuracy in space is achieved. Time advancing part is done by the operator splitting method [9] and we use 1st order accuracy in time.

Khokhlov [7] designed the time step $\Delta t(l)$, by which the hydrodynamical variables of level l cells are advanced, to be

$$\Delta t(l) = \Delta t(l_{\mathrm{max}})\, 2^{l_{\mathrm{max}} - l} \tag{1}$$

so that the CFL condition should be satisfied minimally at every level l. However, if there exists a hydrodynamical flow at the boundary between cells of different levels, inevitable noise takes place at the boundary and may cause numerical difficulty. In order to avoid occurrence of this unphysical effect, we adopt the level-independent time step [6] determined by the CFL condition for the finest level l_{max} as

$$\Delta t(l) = \Delta t(l_{\mathrm{max}}), \tag{2}$$

instead. Although this treatment may well have disadvantage in the usage of CPU-time, this disadvantage is negligible in some cases (see section 3).

The refinement indicator ξ at every oct is calculated as follows. If there is a shock surface or a contact discontinuity (CD) in the oct, then we set $\xi = 1$, otherwise ξ is given by the maximum gradient function $G(q)$ where q represents hydrodynamical variables. The function $G(q)$ is introduced so that the regions where physical quantities spatially change gradually should be coarsened and is defined as

$$G(q) = \max \left\{ \frac{||q_+| - |q_-||}{\max(|q_+|, |q_-|)} \right\}, \tag{3}$$

where the subscripts "+" and "−" denote a pair of adjoining cells and the "+" cell is the one having larger coordinate value. The maximum function of large parentheses "{}" operates over all pairs of adjoining cells in the oct. Since we use both mass density ρ and pressure P as q in $G(q)$, we can write the formula to calculate ξ as

$$\xi = \begin{cases} 1 & \text{if shock or CD exists in the oct,} \\ \max(G(\rho), G(P)) & \text{otherwise.} \end{cases} \tag{4}$$

Shock and CD are detected by

$$- \text{ shock}: \frac{|P_+ - P_-|}{\min(P_+, P_-)} > \varepsilon_s \text{ and } v_- > v_+,$$

$$- \text{ CD}: \frac{|P_+ - P_-|}{\min(P_+, P_-)} \le \varepsilon_s \text{ and } \frac{|\rho_+ - \rho_-|}{\min(\rho_+, \rho_-)} > \varepsilon_c,$$

where ε_s and ε_c are the parameters to control detection of shock and CD, respectively, and we use $\varepsilon_s = 0.2$ and $\varepsilon_c = 0.1$.

We require that cells having occasional noise should not be refined and that cells should be refined before the arrival of shock etc. According to that, we revise the value by adding ξ of the surrounding 26 octs. By using two thresholds, ξ_{refine} and ξ_{coarsen}, operation of refining or coarsening on cells can be decided. The condition of refining the cells into new octs is $\xi^{\text{rev}} > \xi_{\text{refine}}$, and the condition of coarsening the already refined cells is $\xi^{\text{rev}} < \xi_{\text{coarsen}}$. If $\xi_{\text{coarsen}} \le \xi^{\text{rev}} \le \xi_{\text{refine}}$, then the operation of refining or coarsening is not carried out.

3 Resource Usage

We evaluate the resource usage of 3D AMR code and compare it with that of regular mesh code, here. A normalized volume of "generated" cells at level l is expressed $V_0(l)$. Particularly, we express the volume at level l_{max}, $V_0(l_{\text{max}})$, as α. The volume $V_0(l)$ at level $l_{\text{min}} < l < l_{\text{max}}$ can be rewritten as $V_0(l) = V_0(l+1) + V_{\text{extra}}$. We assume $V_{\text{extra}} \ll V_0(l_{\text{max}})$ here[1]. Then, we get

$$V_0(l) = \begin{cases} 1 & (l = l_{\text{min}}), \\ \alpha & (l_{\text{min}} < l \le l_{\text{max}}), \end{cases} \tag{5}$$

---

[1] This relation holds in the case of the supernova simulation as seen in Table 2.

where the total volume size of simulation space is normalized to be 1. The total number of generated cells is

$$N_{\text{total}} = \sum_{l=l_{\text{min}}}^{l_{\text{max}}} V_0(l) \cdot 8^l = \left(1 - \frac{8}{7}\alpha\right) 8^{l_{\text{min}}} + \frac{8}{7}\alpha 8^{l_{\text{max}}}. \tag{6}$$

On the other hand, the normalized volume of "actually used" cells is given by

$$V_1(l) = \begin{cases} 1 - \alpha & (l = l_{\text{min}}), \\ \alpha & (l = l_{\text{max}}), \\ 0 & (\text{otherwise}), \end{cases} \tag{7}$$

and the number of these cells is

$$N_{\text{used}} = \sum_{l=l_{\text{min}}}^{l_{\text{max}}} V_1(l) \cdot 8^l = (1 - \alpha) 8^{l_{\text{min}}} + \alpha 8^{l_{\text{max}}}. \tag{8}$$

Then, we obtain the ratio of the number of used cells to that of generated cells as

$$\frac{N_{\text{used}}}{N_{\text{total}}} = \frac{7\alpha \left(8^{l_{\text{max}}-l_{\text{min}}} - 1\right) + 7}{8\alpha \left(8^{l_{\text{max}}-l_{\text{min}}} - 1\right) + 7} > \frac{7}{8}. \tag{9}$$

Therefore, we can estimate the wasteful part of the total cells generated in AMR scheme is less than $1/8$.

The total number of "homogeneous mesh" cells with a single resolution level l_{homo} is expressed as $N_{\text{homo}} = 8^{l_{\text{homo}}}$. If we assume the same memory space is used both in homogeneous mesh scheme and AMR scheme, i.e., $N_{\text{homo}} = N_{\text{total}}$, then we obtain

$$8^{l_{\text{homo}}} = \left(1 - \frac{8}{7}\alpha\right) 8^{l_{\text{min}}} + \frac{8}{7}\alpha 8^{l_{\text{max}}}. \tag{10}$$

Here, we define a resolution gain as

$$\beta = l_{\text{max}} - l_{\text{homo}}, \tag{11}$$

and then we obtain

$$\beta = -\log_8\left[\left(1 - \frac{8}{7}\alpha\right) 8^{-(l_{\text{max}}-l_{\text{min}})} + \frac{8}{7}\alpha\right]. \tag{12}$$

This β is approximately equal to $-\log_8(8\alpha/7)$ when we employed enough levels and the practical size of α is large enough, i.e., $8^{-(l_{\text{max}}-l_{\text{min}})} \ll \alpha$. This result implies that if one requires the resolution gain $\beta > \beta_0$, then the volume of the finest cells α should be controlled as

$$\alpha < \frac{7}{8} \cdot 8^{-\beta_0}. \tag{13}$$

Now, we examine the efficiency of CPU-time saved in our FTT code and compare it with that of FTT code in Khokhlov's manner [7]. We define a CPU-time saving factor η as the ratio of the loop count for cells in FTT to that in the single resolution (no-AMR) scheme that has a single level only of $l = l_{\max}$. In our FTT code ($\Delta t(l) = \Delta t(l_{\max})$), it is evaluated as

$$\eta = \frac{N_{\text{used}}}{8^{l_{\max}}} = \alpha + (1 - \alpha) \, 8^{-(l_{\max} - l_{\min})}. \tag{14}$$

As for Khokhlov's FTT code ($\Delta t(l) = \Delta t(l_{\max}) \, 2^{l_{\max} - l}$), the CPU-time saving factor η_{k} is obtained as

$$\eta_{\text{k}} = \frac{\alpha \, 8^{l_{\max}} + (1 - \alpha) \, 8^{l_{\min}} \cdot 2^{-(l_{\max} - l_{\min})}}{8^{l_{\max}}}$$

$$= \alpha + (1 - \alpha) \, 8^{-(l_{\max} - l_{\min})} \cdot 2^{-(l_{\max} - l_{\min})}. \tag{15}$$

If $8^{-(l_{\max} - l_{\min})} \ll \alpha$, then η and η_{k} both are approximately equal to α, that is the volume of level $l_{\max}$. In this case, disadvantage of our FTT code in CPU-time against Khokhlov's code is negligible.

4   Simulation of RT Instability in the Supernova Envelope

When the supernova explosion is triggered deep inside the star, a shock wave is formed at the center and propagates outward. During the propagation of this shock wave through envelope shells, i.e., from inside CO and He shells, and H envelope, RT instability takes place when the shock wave passes these composition discontinuities (see Fig. 2). Because inner heavy shell, inside of the shock wave, has larger expanding velocity, effective gravity towards outward when the shock wave reached these layers. For the cases of inner composition discontinuities, RT instabilities turned stable when rarefaction waves generated at outer discontinuities have arrived [4].

Starting from well established spherical model of SN1987A [10], we traced numerically the spherical propagation of shock wave in the supernova envelope. After 5 seconds from core explosion, the shock wave reached the CO-He discontinuity, and it reached He-H discontinuity after 80 seconds. At each composition discontinuity, RT instability takes place. However, at CO-He discontinuity, rarefaction wave from He-H discontinuity stabilizes RT instability at 200 seconds after core explosion. Therefore, in order to be focused on the growth of RT instability generated at He-H discontinuity, we take the profile of this spherical simulation after 100 seconds from core explosion as an initial structure of 3D simulation.

The simulation box is defined to be a cubic box of $0 \leq x \leq 1$, $0 \leq y \leq 1$, and $0 \leq z \leq 1$ and the explosion occurs at the origin, i.e., $(x, y, z) = (0, 0, 0)$. The boundary condition for the 3 planes of $x = 0$, $y = 0$, and $z = 0$ is reflecting one and that for the rest 3 planes of $x = 1$, $y = 1$, and $z = 1$ is free.

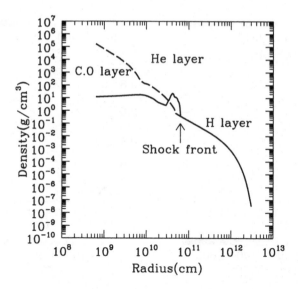

Fig. 2. The radial density profile of the spherical model for 1987A. Dashed line is the density profile at $t = 0$ (before explosion) and solid line is that at $t = 100\,\text{s}$. Arrow indicates the position of shock front at $t = 100\,\text{s}$.

We input 5% perturbation in the radial velocity inside shock front (Fig. 2) as

$$V = v_r(1 + \epsilon \sin(m\theta)\cos(m\phi)), \qquad (16)$$

where v_r is radial velocity at $t = 100\,\text{s}$, ϵ ($= 0.05$) is perturbation amplitude, and m ($= 20$) is wavelength parameter. We employed refining and coarsening thresholds as $\xi_{\text{refine}} = 2.9$ and $\xi_{\text{coarsen}} = 2.1$, respectively.

We started to follow the growth of RT instability with levels 5 – 11 AMR scheme (2048^{-3} resolution) (Fig. 3: simulated region size is $1.5 \times 10^{12}\text{cm}$ and grayscale shows density in the unit of $6 \times 10^{-4}\text{g/cm}^3$). Because of memory limitation, total number of cells are limited to 3.44×10^6. So we reduced resolution down to levels 5 – 10 (1024^{-3} resolution) at 200 seconds; at this time we erase all level 11 cells. Afterward, we gradually reduce the resolution as follows; down to levels 5 – 9 (512^{-3} resolution) at 470 seconds, levels 5 – 8 (256^{-3} resolution) at 1100 seconds, and levels 5 – 7 (128^{-3} resolution) at 2900 seconds.

The final expansion velocity of heavy element, inside of clumps of figure 3 (f), is about 2700 km/s.

5   Results

We divide the whole process in our code into the following parts: a part of creating initial condition, a part of calculating flux including time-advancing, a part of evaluating refinement indicator, refining cells, destroying refined cells, and

214   Ayato Noro et al.

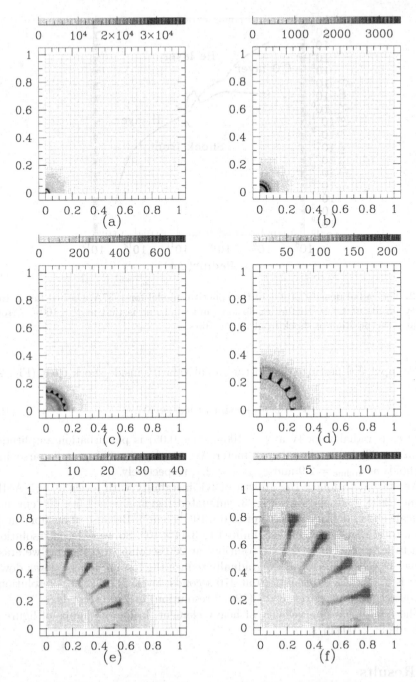

Fig. 3. Six panels show density and mesh profile at the equator of the stages at (a) $t = 100.7\,\mathrm{s}$, (b) $t = 239.5\,\mathrm{s}$, (c) $t = 517.9\,\mathrm{s}$, (d) $t = 1133.8\,\mathrm{s}$, (e) $t = 2937.9\,\mathrm{s}$ and (f) $t = 4608.2\,\mathrm{s}$. Simulation volume is $(1.5 \times 10^{12}\mathrm{cm})^3$. Grayscale shows density in the unit of $6 \times 10^{-4}\mathrm{g/cm}^3$.

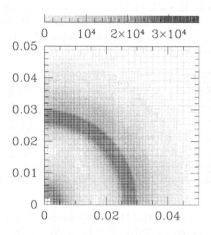

Fig. 4. This panel is magnified central region of Fig. 3 (a) (in the vicinity of the origin). Interface of He/H layers is at $r = 0.04$ (r is the distance from the origin).

re-ordering octs, a part of outputting data, and the rest, those are named as "initialization", "hydrodynamics", "AMR", "output", and "rest", respectively. We clock the CPU-times spent for these procedure groups in a simulation run on the expanding supernova envelope during the first 100 steps and we show the results as a ratio of the CPU-time of each group to the total CPU-time in Table 1, where the simulation is performed on 1 node, 8 processor elements (PEs), of Hitachi SR8000. We measured with following three options; pseudo-vector and parallel (V, P), pseudo-vector and nonparallel (V, N), scalar and nonparallel (S, N) options. The most time-consuming procedure group is the "hydrodynamics" part because the number of floating point operations for this part is much more than that for the other parts. The "AMR" part is the 2nd ranked time-consuming one but is sufficiently small compared with the "hydrodynamics" part. The "output" procedure is called for 21 times during 100 steps and the total data size is only 30 MB since the code outputs only the 2D data, written in ASCII form, sliced at a $z = $ constant surface. If the code outputs the whole 3D data in binary form for these 21 times, the total data size is about 1 GB. There is only a slight difference between CPU-times^{V,N} and CPU-times^{S,N}. This means that pseudo-vector option does not bring benefit to our code. On the other hand, by comparing CPU-times^{V,P} with CPU-times^{V,N}, we found about six times gain at the "hydrodynamics" part.

Next, we show how much memory and CPU-time are reduced in our simulation compared with those for a simulation of regular-mesh having the same resolution as ours. The number of cells prepared in this simulation is 3.44×10^6 but the maximum number of used cells is about 3.2×10^6. These numbers are found to be small enough by comparing them with the required number of cells in a regular mesh simulation, $8^{l_{max}} \sim 8.6 \times 10^9$ for $l_{max} = 11$. The number of cells of each level l, $N(l)$, the total number of cells all levels, N_{total}, and the total

Table 1. CPU-times spent in each procedure group during the first 100 steps and the ratios of those to the total ones. The superscripts V, S, P, and N indicate pseudo-vector option, scalar option, parallel option, and nonparallel option, respectively. The simulation is performed on 1 node (8 PEs) of Hitachi SR8000.

procedure group	CPU-time^{V,P}	ratio^{V,P}	CPU-time^{V,N}	CPU-time^{S,N}
initialization	15 s	2.6%	13 s	13 s
hydrodynamics	517 s	83.3%	3333 s	3390 s
AMR	47 s	7.5%	64 s	90 s
output	37 s	5.9%	37 s	37 s
rest	5 s	0.7%	30 s	32 s
total	621 s	–	3477 s	3562 s

number of cells which are used in the "hydrodynamics" procedure group, N_{used}, at the typical time are shown in Table 2, where N_{total} and N_{used} are obtained by

$$N_{\text{total}} = \sum_{l=l_{\min}}^{l_{\max}} N(l), \qquad (17)$$

and

$$N_{\text{used}} = N_{\text{total}} - \sum_{l=l_{\min}+1}^{l_{\max}} \frac{N(l)}{8}. \qquad (18)$$

The CPU-time acceleration efficiency $(\eta^{-1} = 8^{l_{\max}(t)}/N_{\text{used}})$ is quite high at the early stages when $l_{\max}(t)$ is high. We find that the relation $8^{-(l_{\max}(t)-l_{\min})} \ll \alpha \equiv N(l_{\max}(t))/8^{l_{\max}(t)}$ is satisfied and the approximation of $\eta \sim \alpha$ derived in section 3 is valid at every step.

On the average, we can calculate 580 times faster than with single size meshes. Moreover this saving factor by our method $(\Delta t(l) = \Delta t(l_{\max}))$ is only 1.043 times larger than Khokhlov method $(\Delta t(l) = \Delta t(l_{\max}) \, 2^{l_{\max}-l})$.

6 Discussion

While our simulation is carried out in limited resources, the final expansion velocity of heavy elements is found to be about 2700 km/s. This value is a little short from observed velocity of 3000 km/s, but more than 30% increased from that of previous simulations. Therefore, when we simulate again with greater machine resource in future, we could use more fine resolution and would have more faster expansion velocity. Furthermore, though in this computation we consider only hydrodynamics, if we include thermodynamics[2], the expected expanding velocity is higher than that of hydrodynamic result.

---

[2] At each composition discontinuity, thermonuclear burning takes place and outer components turn into ashes, i.e., inner components. When the shock compression takes place, it enhances thermonuclear reaction rate drastically, releases much nuclear energy, and accelerates the shock.

Table 2. The numbers of cells, $N(l)$, N_{total}, and N_{used}, and the CPU-time acceleration efficiency η^{-1} at certain time steps (see Figs. 3 (a)–(f)). Note $l_{min} = 5$.

step	4	800	1600	3000	6600	9400
t	100.7 s	239.5 s	517.9 s	1133.8 s	2937.9 s	4608.2 s
$N(5)$	32768	32768	32768	32768	32768	32768
$N(6)$	1280	2432	4224	12160	77888	203648
$N(7)$	5824	12544	22592	67264	504704	1475264
$N(8)$	28736	28736	80576	417536	0	0
$N(9)$	43200	94080	524800	0	0	0
$N(10)$	90176	612096	0	0	0	0
$N(11)$	458112	0	0	0	0	0
N_{total}	660096	782656	664960	529728	615360	1711680
N_{used}	581680	688920	585936	467608	542536	1501816
η^{-1}	15000	1600	230	36	3.9	1.4

For the case of supernova simulation, fine structure is developed into the unperturbed region. As this case, if the region where high resolution is required is significantly large and the region where intermediate resolution is rather small, the disadvantage of uniform time step is negligible. Especially when the structure of high resolution region becomes complicated, numerical noise is apt to arise at the surface between cells of different levels. Generation of such a noise can be avoided by applying uniform time step to all levels cells. Therefore when the finest structure plays an important role and its region size has certain amount, our method with uniform time step is suitable.

In this paper, we showed that our AMR scheme reduces memory and CPU-time drastically. So we could argue a good efficiency of AMR scheme for the astrophysical simulation. Further application to other fields with similar situation would also be fruitful.

This computation is carried out by one node with 464 MB Hitachi SR8000. Typical CPU time with pseudo-vector and parallel option is about 100, 000 seconds per one model in the case of 1 node use. This work is supported in part by the Grant of Japan Science and Technology Corporation.

The authors are grateful to anonymous referees for comments.

References

1. van Leer, B.: Towards the ultimate conservative difference scheme. V - A second-order sequel to Godunov's method. J. Comp. Phys. **32** (1979) 101–136
2. Yamada, Y., Yoshida, T., and Den, M.: Two-dimensional Simulation of Rayleigh-Taylor Instability in Supernova Explosions II. Prog. Theor. Phys. **86** (1991) 1211–1225
3. Hachisu, I., Matsuda,T., Nomoto, K., and Shigeyama, T.: Mixing in ejecta of supernovae. I - General properties of two-dimensional Rayleigh-Taylor instabilities and mixing width in ejecta of supernovae. Ap.J. **390** (1992) 230–252

4. Kane, J., Arnett, D., Remington, B. A. Glendinning, S. G., Bazan, G., Müller, E., Fryxell, B. A., and Teyssier, R.: Two-dimensional versus Three-dimensional Supernova Hydrodynamic Instability Growth. Ap.J. **528** (2000) 989–994

5. Kunkel, W., Madore, B., Shelton, I., Duhalde, O., Bateson, F. M., Jones, A., Moreno, B., Walker, S., Garradd, G., Warner, B., Menzies, J.: Supernova 1987A in the Large Magellanic Cloud. IAU Circ. (1987) 4316

6. Noro, A., Ohta, T., Ogawa, T., Yamashita, K., and Miyaji, S.: Implementation of Adaptive Mesh Refinement Scheme into Astrophysical Fluid Dynamics Code – Rayleigh-Taylor Instability in the Supernova Envelope –. IPSJ Symposium Series **2002** (2002) 9–15

7. Khokhlov, A. M.: Fully Threaded Tree Algorithms for Adaptive Refinement Fluid Dynamics Simulations. J. Comp. Phys. **143** (1998) 519–543

8. Ogawa, T., Ohta, T., Matsumoto, R., Yamashita, K., and Den, M.: Hydrodynamical Simulation Using a Fully Threaded Tree. Prog. Theor. Phys. Suppl. **138** (2000) 654–655

9. Hirsch, C.: "Numerical Computation of Internal and External Flows (Vol.2: Computational Methods for Inviscid and Viscous Flows)", A Wiley-Interscience Publication, New York (1992)

10. Shigeyama, T. and Nomoto, K.: Theoretical light curve of SN 1987A and mixing of hydrogen and nickel in the ejecta. Ap.J. **360** (1990) 242–256

An MPI Benchmark Program Library and Its Application to the Earth Simulator

Hitoshi Uehara[1], Masanori Tamura[2], and Mitsuo Yokokawa[1]

[1] Earth Simulator Research and Development Center,
Japan Atomic Energy Research Institute
3173-25, Showa-cho, Kanazawa-ku, Yokohama-shi, Kanagawa 236-0001, Japan
uehara@es.jamstec.go.jp, yokokawa@gaia.jaeri.go.jp
[2] NEC Corporation HPC group,
1-10, Nisshin-cho, Fuchuu-shi, Tokyo 183-8501, Japan
tmr@hpc.bs1.fc.nec.co.jp

Abstract. Parallel programming is essential for large-scale scientific simulations, and MPI is intended to be a de facto standard API for this kind of programming. Since MPI has several functions that exhibit similar behaviors, programmers often have difficulty in choosing the appropriate function for their programs. An MPI benchmark program library named MBL has been developed for gathering performance data for various MPI functions. It measures the performance of MPI-1 and MPI-2 functions under several communication patterns. MBL has been applied to the measurement of MPI performance in the Earth Simulator. It is confirmed that a maximum throughput of 11.7GB/s is obtained in inter-node communications in the Earth Simulator.

1 Introduction

Since the advent of parallel computers, parallelized programs have been developed to make feasible very large-scale scientific simulations that play an important role in solving various kinds of natural phenomena with computers. A parallelization technique that is frequently used in programming is task partitioning. This technique requires data communications between parallel tasks that are coded explicitly using a message-passing library. MPI[1,2], the Message Passing Interface, is one of the major parallel programming interfaces and is intended to be a de facto standard interface world-wide. MPI is available to develop parallel programs on most parallel computers at present. MPI will also be supported on the Earth Simulator, which is currently being built and will be operational in March 2002.

The data communication time in MPI is included in the total run time of parallel programs. Reducing the communication time should have a significant influence on the execution performance of programs. Therefore, optimizing the communication portion of parallel programs is necessary to obtain higher performance. In particular, such optimizations are essential to reduce the run time of parallel programs running on large-scale distributed memory parallel computers such as the Earth Simulator.

H. Zima et al. (Eds.): ISHPC 2002, LNCS 2327, pp. 219–230, 2002.

Actual performance data for MPI implementations is quite useful for optimizing the communication part of the programs and it is important to measure performance of MPI implementations. There are a number of benchmark software packages available for MPI-1 such as the Pallas MPI Benchmarks (PMB)[3] or Beff[4], and a few benchmark packages available for MPI-2. They are limited to measurements of the performance of a few point-to-point communications functions, several collective communication functions, and a few complex communication patterns utilizing several MPI functions. These benchmarks do not measure the performance of all the MPI functions, and they cannot provide sufficient data to compare the performance of similar MPI functions. Moreover, it is difficult for application users to modify the benchmarks when they decide to add their own measurements.

We have developed an MPI benchmark program library (MBL) to measure detailed MPI performance with using various measurement patterns, each of which has several implementations. Also, MBL can be easily modified by users to augment their own measurements. Specifically, MBL has been implemented as a collection of 14 small benchmark modules. Each benchmark module measures the performance of one communication pattern such as a ping-pong communication with different implementations. These benchmark modules of MBL have been implemented using framework technology[5,6].

In this paper, an overview of MBL is presented. The measurement results of MPI performance on the Earth Simulator are also presented as a demonstration of the applicability of MBL to actual parallel computers.

2   MBL: MPI Benchmark Program Library

2.1   Overview of MBL

An MPI benchmark program library called MBL has been developed to measure detailed MPI performance using various measurement patterns, each of which pattern has several implementations. MBL has been developed based on the following policies.

- It can measure not only the performance of MPI-1 functions such as major point-to-point communication functions and major collective communication functions, but also the performance of MPI-2 functions such as the read/write functions of MPI-I/O and all access functions of the one-sided communication i.e. remote memory access (RMA) functions.
- It should be implemented compactly for maintainability and expandability.
- One benchmark module should measure the performance of one communication pattern.
- Each benchmark program is implemented using framework technology.
- It should be coded in FORTRAN for legacy FORTRAN application users.

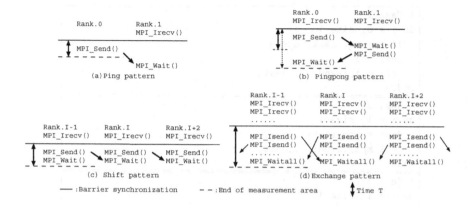

Fig. 1. Measurement patterns in MBL

Measuring MPI-1 Functions. MBL measures the performance of four communication patterns, the ping, ping-pong, shift, and exchange patterns, that were written using MPI-1 functions as shown in Figure 1. Multiple implementations for each pattern are provided with each implementation using a different MPI function, so that the users can find which function is the best to use for their programs. For example, the ping-pong pattern is implemented with both MPI_Send/MPI_Irecv functions and MPI_Isend/MPI_Irecv functions.

When measuring the shift pattern, two implementations are considered. One version is implemented with the non-blocking functions, MPI_Isend and MPI_Irecv. The other version is implemented with the blocking function, MPI_Sendrecv. These implementations for the shift pattern are often found in practical simulation programs. By performing this evaluation, users can determine whether blocking or non-blocking functions should be used for shift communications on their parallel computers.

Three versions of the measurement of the exchange patten are shown in Figure 2. The variable "`size`" in Figure 2 is the number of MPI processes that participate in the communication.

The first version is implemented with the MPI_Isend and MPI_Irecv functions. The message is sent to each process from 0 to the last process in increasing order. The second version is also implemented with the MPI_Isend and MPI_Irecv functions but the message is sent in a different order in each process shown in Figure 2. The first message is sent to the process with the next highest rank until the message has been sent to each process. All MPI_Irecv functions are required to obtain messages sent from other MPI processes and they are invoked before any MPI_Isend functions are issued in the two previous versions. The third version is implemented with the MPI_Sendrecv function. This version determines the efficiency of the implementation of the MPI_Sendrecv function when it is compared with the first and second version.

The time spent executing each communication pattern is measured as shown in Figure 1 with a clock routine such as MPI_Wtime. The execution time for a

```
! Case 1.
do i = 0, size-1
  if (i .eq. rank) cycle
  dest = i
  call mpi_isend(...,dest,...)
enddo
```

```
! Case 2.
do i = 1, size-1
  dest = mod(rank+i,size)
  call mpi_isend(...,dest,...)
enddo
```

```
! Case 3.
do i = 1, size-1
  dest = mod(rank+i,     size)
  src  = mod(rank-i+size,size)
  call mpi_sendrecv(...,dest,...,src,....)
enddo
```

Fig. 2. Three versions of the exchange pattern

message size of 0 bytes is the startup cost. The time T in Figure 1 is called the latency, and the throughput is calculated using the following equation.

$$Throughput = L/T \quad (B/s), \tag{1}$$

where L is the length of the transferred data in each communication pattern. When several messages are processed in one communication such as in the exchange pattern, throughput is calculated using the following equation.

$$Throughput = L \times N/T, \tag{2}$$

where N is the number of processed messages.

Measurements of MPI-2 Functions. MBL also measures the performance of MPI-2 functions such as RMA and MPI-I/O. It measures the performance of the ping pattern, the shift pattern and the exchange patterns for RMA functions, and measures the performance of all file-access functions, file synchronization and file seek for MPI-I/O functions. Processing time includes not only the time to execute the RMA functions but also the execution time of the RMA fence function in the measurement.

For the ping pattern measurement, MBL measures the performance of three RMA functions; MPI_Get, MPI_Put and MPI_Accumulate with the sum and binary-AND functions. For the shift pattern measurement, MBL measures the performance of two implementations; one using MPI_Get and one using MPI_Put.

In the exchange pattern, four versions were considered. These versions can be distinguished by two characteristics. The first characteristic is that the function

```
t1 = MPI_Wtime()
do i = 1, N
  call MPI_Get(...)
enddo
call MPI_Win_fence(...)
t2 = MPI_Wtime()
AverageTime = (t2-t1)/N
```

Fig. 3. Image of measurement style in PMB

```
!(1)
Total = 0
do i=1, N
  t1 = MPI_Wtime()
  call MPI_Get(...)
  t2 = MPI_Wtime()
  call MPI_Win_fence(...)
  Total = Total + t2-t1
enddo
AverageTime = Total/N
```

```
!(2)
Total = 0
do i = 1, N
  t1 = MPI_Wtime()
  call MPI_Get(...)
  call MPI_Win_fence(...)
  t2 = MPI_Wtime()
  Total = Total + t2-t1
enddo
AverageTime = Total/N
```

Fig. 4. Image of measurement style in MBL

used for the implementation is either MPI_Get or MPI_Put. The other characteristic is the sequence of accesses to MPI processes like in versions 1 and 2 shown in Figure 2.

The way execution time is measured in MBL is different from the method used in conventional benchmarks. For example, only a single method is provided in PMB. In the measurement of the overhead of the MPI_Get function in PMB shown in Figure 3, the final average time includes both the execution time of T_{MPI_Get} and that the execution time of $T_{MPI\_Win\_fence}/N$ and therefore it is not an accurate measurement of the latency of the MPI_Get function. MBL provides two methods for measuring the RMA functions as depicted in Figure 4. One method measures the time for only the RMA function calls. The other method measures the execution time for both RMA function calls and the RMA fence call. The actual latency of RMA functions can be obtained by these measurements.

2.2   MBL Implementation

One of the objectives of MBL is to provide an MPI benchmark tool for application users who want to determine MPI performance and improve their programs accordingly. Since the most common programming language is FORTRAN, it is preferable for MBL to be written in FORTRAN as well so that the application programmers can modify the MBL for their needs. Also, MBL should be designed to be portable and maintainable.

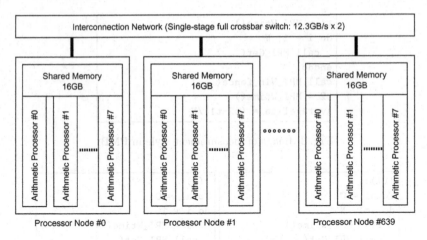

Fig. 5. Overview of the Earth Simulator

Moreover, the concept of detailed performance measurement using various possible implementations is one of the basic policies of MBL. If these complex measurements are implemented without careful consideration, then the MBL codes would not be maintainable and expandable. Object oriented techniques are useful for maintainability and expandability, but they are difficult to use directly in FORTRAN programming.

Therefore, we adopt two basic concepts of framework technology and apply them to FORTRAN programming: hot-spots and frozen-spots. Hot-spots are the parts of the source code that are modified frequently and are also unique to the source code. Frozen-spots are the parts of source code that are commonly shared in all other modules and are rarely changed. Since the hot-spots are modularized, the hot-spots can be easily modified and would not influence other modules.

We assumed that the measurement modules for each communication pattern in MBL are the hot-spots and the modules such as initialization, post-processing for results output and calls to the measurement module are the frozen-spots. Because the MBL implementation has adopted the above concepts, it is likely that MBL could be easily modified and extended.

3   MPI Performance on the Earth Simulator

3.1   MPI Implementation on the Earth Simulator

The Earth Simulator project aims to promote research and development for understanding and prediction of global environment changes. It is necessary to develop ultra high-speed supercomputing suitable to the project.

The Earth Simulator is a distributed memory parallel system that consists of 640 processor nodes (PNs) connected by a 640 × 640 inter-node crossbar switch[7], as shown in Figure 5. Each PN is a shared memory system composed of 8 arithmetic processors (APs), a shared memory system of 16GB, a remote

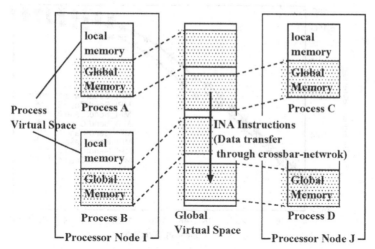

Fig. 6. Global Memory Space

access control unit (RCU), and an I/O processor (IOP). AP is vector processor, and the peak performance of each AP is 8Gflops. Each AP has 32GB/s memory bandwidth, and the total memory bandwidth per node is 256GB/s. The RCU is directly connected to the crossbar switch and controls inter-node data transfers at a maximum rate of 12.3GB/s for each direction of the bi-directional paths with some additional software overhead. An MPI implementation will be supported on the Earth Simulator.

The memory space of a process is divided into two spaces, which are called local memory (LMEM) and global memory (GMEM) as shown in Figure 6. GMEM is addressed globally and can be shared by every MPI processes assigned to different PNs. These areas are used for MPI communications. Buffer areas for message transfer can be allocated to either of them. The behavior of MPI communications differs depending on the memory area the data to be transferred resides in. It is classified into four categories. Here we explain the implementations of the four categories using the example shown in Figure 6.

Category 1 Data that is stored in the LMEM of a process A is transferred to the LMEM of another process B on the same PN. In this case, the GMEM is used as a general shared memory. The data in the LMEM of process A is copied into an area of the GMEM of process A first. Next, the data in the GMEM of process A is copied into the LMEM of process B. Two memory copy operations must be executed.

Category 2 Data that is stored in the LMEM of process A is transferred to the LMEM of process C which is invoked in a different PN. First, the data in the LMEM of process A is copied into the GMEM of process A. Next, the data in GMEM of the process A is copied into the GMEM of process C via the crossbar switch using the INA instruction. Finally, the data in the GMEM of process C is copied into the LMEM of process C. Three memory copy operations must be executed.

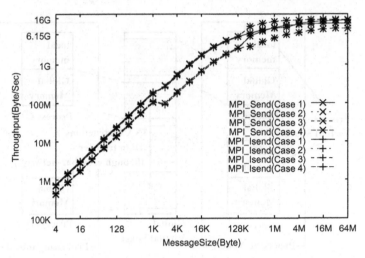

Fig. 7. Throughput of MPI-1 functions on the Earth Simulator

Category 3 Data that is stored in the GMEM of process A is transferred to the GMEM of processor B in the same PN, the data in the GMEM of process A is copied directly into the GMEM of process B. A copy operation is executed.

Category 4 Data that is stored in the GMEM of process A is transferred to the GMEM of processor C which is invoked in a different PN. The data in the GMEM of process A is copied directly into the GMEM of process C via the crossbar switch using the INA instruction. A copy operation is also executed.

3.2   Performance Measurement of MPI-1 Functions

The performance of the ping-pong pattern implemented by either MPI_Send/ MPI_Irecv functions or MPI_Isend/MPI_Irecv functions is evaluated on the Earth Simulator as example of the measurement of MPI-1 functions. The performance measurement is shown in Figure 1(b). Two MPI processes are invoked corresponding to the four categories described in the previous subsection. The performance of the two send functions are measured. Figure 7 shows the measured throughput.

For intra-node communication, the maximum throughput is 14.87GB/s in Category 3; half of the peak is achieved when message length is larger than 256KB, and the startup cost is 5.20microsecond. A vector operation is used to copy the data in Category 1 and vector data load and store operations are issued. Since a load/store pipeline is provided in the AP of the Earth Simulator, half of peak bandwidth is the theoretical maximum which is 16GB/s. The throughput measured by MBL achieves a 92.8% efficiency compared to the peak.

For inter-node communication, the maximum throughput is 11.76GB/s in Category 4; half of the peak is achieved when message length is larger than 512KB, and the startup cost is 8.56 microseconds.

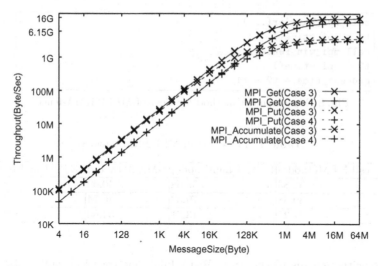

Fig. 8. Throughput of MPI-2 RMA functions on the Earth Simulator

The gap between the message size of 1KB and 128KB in the figure is caused by the change in communication method in the MPI functions, which change according to message length.

3.3   Performance Measurement of MPI-2 Functions

The performance of the ping pattern implemented using three RMA functions (MPI_Get, MPI_Put, and MPI_Accumulate with a sum operation) is measured. Categories 3 and 4 for the allocation of buffer areas to either the LMEM or the GMEM described in section 3.1 are considered in the measurement because RMA functions are only available for the variables in the GMEM. The performance measurement method is shown in Figure 1(a). Two MPI processes are executed within a PN or on two PNs according to the appropriate categories. Figure 8 is the throughput. For intra-node communication, the maximum throughputs are 14.77GB/s (MPI_Get), 14.78GB/s (MPI_Put), and 3.66GB/s (MPI_Accumulate). The startup costs are 6.46 microseconds (MPI_Get), 6.06 microseconds (MPI_Put), and 8.61 microseconds (MPI_Accumulate). Efficiency of intra-node communication using RMA functions, MPI_Get and MPI_put is obtained about 92% of the half memory bandwidth peak. For the internode communication, the maximum throughputs are 11.63GB/s (MPI_Get), 11.63GB/s (MPI_Put), and 3.16GB/s (MPI_Accumulate). The startup costs are 16.52 microseconds (MPI_Get), 15.00 microseconds (MPI_Put), and 14.57 microseconds (MPI_Accumulate). These startup costs are measured when the length of message is 4 bytes and the cost of the RMA fence call does not count towards the startup cost.

The results of the RMA fence cost measurements are shown in Table 1. The costs of MPI_Win_fence reflect the processing time when only MPI_Win_fence

```
call mpi_barrier(...)
t1 = mpi_wtime()
call mpi_win_fence(...)
t2 = mpi_wtime()
process_time = t2 - t1
```

Fig. 9. Measurement method for Cost of MPI-2 RMA-fence

Table 1. The Costs of MPI-2 RMA fence

processes	LMEM(1node)	LMEM(multi-node)	GMEM(1node)	GMEM(multi-node)
2	32.83	53.64	40.09	63.65
4	39.58	79.89	46.74	90.36
8	47.27	113.71	54.50	124.81

is called without calls to the other RMA functions (See Figure 9). The RMA window is allocated to either the LMEM or the GMEM. The measurements are carried out for intra-node and inter-node communications. It was found that the cost of the RMA fence increases as the number of MPI processes increases non-linearly. For a fixed number of MPI process, the fence cost when the RMA window is located in the LMEM and on a node is the smallest of all the implementations. It is assumed that this fact is important for scalability of RMA communications on the Earth Simulator.

3.4 Performance Measurement of Complicated Patterns

In this section, the performance of the exchange communication pattern is measured because it is an example of a performance measurement of a complicated communication pattern which often appears in data transposition in practical simulation programs. One measurement is of the implementation using MPI_Isend/ MPI_Irecv. The other is of the implementation using MPI_Put/ MPI_Win_fence. This is measured as shown in Figure 1(d) with 2, 4, and 8 MPI processes. Each uses the GMEM for the data buffer and RMA-window assignment.

Measurement results are shown in Figures 10 and 11. The execution time of the exchange running on one node is shorter than the execution time for a multi-node configuration because of the significant difference in data copy speed and the difference in cost for the RMA-fence. The cost of the RMA-fence influences the performance if the message length is short as shown in Table 1.

The results for the MPI_Isend/MPI_Irecv implementation are better than those for the MPI_Put/MPI_Win_fence implementation using 2 MPI processes. However, as the number of MPI processes increases, the implementation using MPI_Put/MPI_Win_fence eventually performs better than MPI_Isend/MPI_Irecv. When execution with greater than 8 processes is required, it is appears that the implementation using MPI_Put/MPI_Win_fence is superior for the ex-

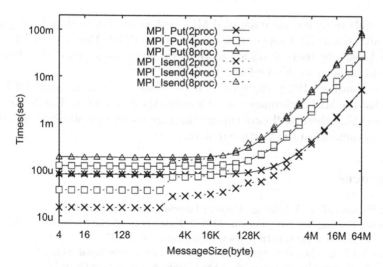

Fig. 10. Performance of exchange pattern on multi nodes

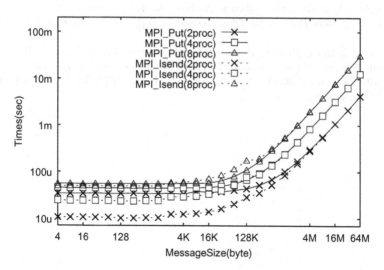

Fig. 11. Performance of exchange pattern within one node

change pattern on the Earth Simulator. This should be examined in more detail in the future.

4   Conclusion

In this paper, we described the MPI benchmark program library (MBL) and its application to the MPI implementation on the Earth Simulator. MBL can measure the detailed performance of MPI implementations using various measurement patterns, each of which has several implementations. 1By using MBL, it is possible to gather more detailed MPI performance data than by measuring with conventional MPI benchmark programs like PMB. Moreover, MBL can be modified easily by users to obtain more useful performance data. We think that MBL will be open to the public in the future.

By evaluating MPI performance on the Earth Simulator using MBL, it is found that the MPI performance is achieved as the design of the Earth Simulator. We believe that these MPI performance data are useful for optimizing simulation programs running on the Earth Simulator.

References

1. MPI Forum. MPI: A Message-Passing Interface Standard, 1995.
2. Message Passing Interface Forum. MPI-2: Extensions to the Message-Passing Interface. Technical report, http://www.mpi-forum.org., 1997.
3. Pallas MPI Benchmarks. http://www.pallas.de/pages/pmbd.html.
4. Effective Bandwidth Benchmark. http://www.hlrs.de/mpi/b_eff/.
5. E. Gamma, R. Helm, R. Johnson, and J. Vlissides. *Design Pattern Elements of Reusable Object-Oriented Software*. Addison Wesley Longman, 1995.
6. W. Pree. *Design Pattern for Object-Oriented Software Development*. ACM Press, 1995.
7. Mitsuo Yokokawa, Shinichi Habata, Shinichi Kawai, Hiroyuki Ito, Keiji Tani, and Hajime Miyoshi. Basic Design of the Earth Simulator. *High Performance Computing*, 1999. Lecture Notes in Computer Science 1625, ISHPC'99, Kyoto, Japan, May 1999 Proceedings.

Parallel Simulation of Seismic Wave Propagation

Takashi Furumura

Earthquake Research Institute, University of Tokyo,
1-1-1 Yayoi, Bunkyo-ku, 113-0032 Japan
furumura@eri.u-tokyo.ac.jp

Abstract. A dense seismic network of strong ground motion instruments in Japan allows a detailed characterization of regional wave propagation during destructive earthquakes. Steadily improving computer power associated with sophisticated parallel algorithms enables a large scale computer simulation of seismic waves in 3D heterogeneous structure. This is illustrated by the 2000 Tottori-ken Seibu earthquake, southwestern Japan. With the aid of assessing potential impact at urban cities expected for future earthquake scenarios (e.g. Nankai earthquake). A close link between strong motion seismology and high-performance computing technology becomes indispensable. The newly developing high-performance parallel computer, Earth Simulator, would provide a key to realistic simulations of strong ground motion.

1 Introduction

Seismic waves from large earthquakes are significantly affected by the presence of 3-D variations in the structure along wave propagation path. The presence of strong lateral heterogeneity in the crust and upper-mantle structure of the Japanese archipelago imposes significant variations in the ground motion pattern as illustrated by observations from the 2000 Tottori-ken Seibu earthquake occurred in southwestern Japan.

Over the last few years very dense arrays of high-resolution strong ground motion networks of K-NET [1] and KiK-net [2] have been installed by NIED, Japan (Fig. 1). Large amount of observations are available to visualize the complex wave propagation in the Japanese archipelago. For example, observed waveforms at 550 strong motion stations realize complex surface wave propagation characteristics during the 2000 Tottori-ken Seibu earthquake, imposed by heterogeneous crustal structure.

Recent parallel computing technologies promising realistic simulation of seismic wave propagation using 3D heterogeneous structure and complex fault rupture model.

In this paper we compare observations and computer simulations for seismic wave propagation in southwestern Japan. We introduce the wave propagation simulation by calculating equations of motion in 3D. We then consider the efficiency of parallel simulation using the hybrid PSM/FDM technique [3], [4]. To show the effectiveness of strong ground motion simulations we conduct

H. Zima et al. (Eds.): ISHPC 2002, LNCS 2327, pp. 231–242, 2002.

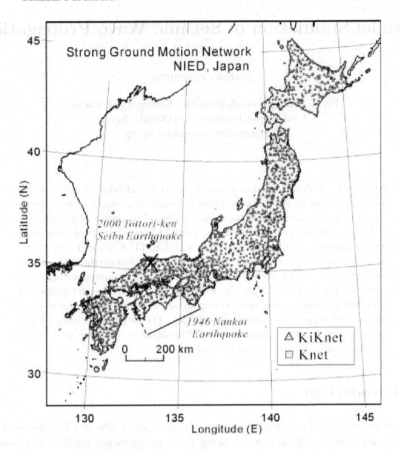

Fig. 1. Distribution of 1550 strong ground motion stations (*K-NET and KiK-net*) of Japan. Locations of the 2000 Tottori-ken Seibu earthquake and the fault rupture area of the 1946 Nankai earthquake (*square*) are shown.

strong motion simulation for the 2000 Tottori-ken Seibu earthquake and expected Nankai megathrust earthquakes (M~8), and compare them with observed strong ground motion characteristics.

2 Numerical Simulation of Seismic Waves by Parallel PSM/FDM Hybrid Method

The seismic wave propagation in an elastic medium is calculated by momentum conservation and by the Hooke's law, relating stresses and displacement derivatives.

In a 3-D Cartesian coordinate system with the z-axis taken position downward, the equations of motion are represented as

$$\rho \ddot{u}_p = \frac{\partial \sigma_{xp}}{\partial x} + \frac{\partial \sigma_{yp}}{\partial y} + \frac{\partial \sigma_{zp}}{\partial z} + f_p, \qquad (p = x, y, z), \tag{1}$$

where σ_{pq}, f_p and ρ stand for a stress, body force and density, respectively. $\ddot{u}_p$ is a particle acceleration, namely the second-order time derivative of a displacement. The stresses in an isotropic medium are given by

$$\sigma_{pq} = \lambda(e_{xx} + e_{yy} + e_{zz})\delta_{pq} + 2\mu e_{pq}, \qquad (p,q = x,y,z), \tag{2}$$

with the Lamé's constants λ, μ and the strains are defined by

$$e_{pq} = \frac{1}{2}\left(\frac{\partial u_p}{\partial q} + \frac{\partial u_q}{\partial p}\right), \qquad (p,q = x,y,z). \tag{3}$$

Equation (1) is numerically integrated with the time step Δt as

$$\dot{u}_p^{n+\frac{1}{2}} = \dot{u}_p^{n-\frac{1}{2}} + \frac{1}{\rho}\left(\frac{\partial\sigma_{px}^n}{\partial x} + \frac{\partial\sigma_{py}^n}{\partial y} + \frac{\partial\sigma_{pz}^n}{\partial z} + f_p^n\right)\Delta t, \qquad (p = x,y,z). \tag{4}$$

where $\dot{u}_p^{n\pm\frac{1}{2}}$ is a particle velocity at the time of $t = (n \pm \frac{1}{2})\Delta t$. The spatial derivatives in Eq. (4) are calculated analytically in the wavenumber domain by use of the fast Fourier transform (FFT). The accurate differentiation using the FFT minimizes the grid dispersion error, so that the demand of computer memory and time consumed for the 3D simulation can be reduced considerably compared with the other schemes such as for FDM or FEM.

The differentiation of Eqs. (2) and (3) with respect to time yields

$$\sigma_{pq}^{n+1} = \sigma_{pq}^n + \left[\lambda\left(\frac{\partial\dot{u}_x^{n+\frac{1}{2}}}{\partial x} + \frac{\partial\dot{u}_y^{n+\frac{1}{2}}}{\partial y} + \frac{\partial\dot{u}_z^{n+\frac{1}{2}}}{\partial z}\right)\delta_{pq}\right. \tag{5}$$

$$\left. + \mu\left(\frac{\partial\dot{u}_p^{n+\frac{1}{2}}}{\partial q} + \frac{\partial\dot{u}_q^{n+\frac{1}{2}}}{\partial p}\right)\right]\Delta t, \quad (p,q = x,y,z),$$

where the spatial derivatives in 6 are again calculated by the FFT.

For marching in time, the stresses at the next time step $t = (n+1)\Delta t$ are calculated by Eq. (6) and the particle velocities result from Eq. (4).

Anelastic attenuation (Q) in the model and absorbing boundaries at the edge of the domain is efficiently incorporated in the time-domain simulation by applying an adequate damper [5].

2.1   Hybrid PSM/FDM Parallel Simulation

In order to model large scale 3D seismic wave propagation a great deal of effort has been expended to implement the Fourier spectral method (PSM) for the high performance parallel computing using Cray X-MP supercomputer [6], Thinking Machine CM-5 parallel computer [7], and IBM POWER parallel 2 [8].

In these studies the whole 3D space is partitioned into subdomains and assigned to processors for concurrently computing. Recall that the spatial differentiation using the FFT requires all the quantities along a coordinate axis of

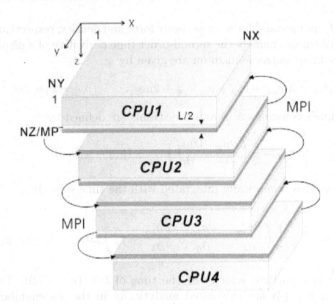

Fig. 2. Schematic illustration of the domain partition for the hybrid PSM/FDM simulation using four CPUs ($MP=4$). The PSM is used to calculate spatial derivatives in the horizontal direction (x,y), and the FDM of order L is used in the vertical (z) direction. The data exchange required for the vertical differentiation in the gray areas is performed by the MPI.

differentiation, so that the communications between processors grow up dramatically with increasing number of processors. Therefore the efficiency of the parallel PSM computing is guaranteed only for slow processors relative to the communication speed.

To overcome this problem we use a 4th-order FDM of staggered-grid formulation in vertical (z) direction and make hybridding with the parallel PSM. Since the localized FDM calculation requires quite less inter-processor communications compared with the conventional parallel PSM, the hybrid PSM/FDM method achieves fairly good speed-up even using a large number of processors. Figure 2 illustrates the domain partition of the hybrid PSM/FDM method, where the field quantities are separated vertically into a number of subregions of equal size with some overlapping between neighboring subdomains. Each processor calculates the wavefield in the assigned domain using the FFT (i.e., PSM) for horizontal (x,y) differentiations and the 4th-order FDM for the z differentiation. At each time step the data in the overlapped area is exchanged with neighbors using the MPI functions of MPI_SEND() and MPI_RECV().

2.2 Parallel Performance of the Hybrid PSM/FDM Simulation

To evaluate the speed-up rate of the hybrid PSM/FDM parallel simulation with respect to the number of processors, we formulate a simple performance prediction model defined by the ratio of CPU speed to communication speed.

We assume that the computation is carried out entirely in single-precision arithmetic (four bytes for each field quantity) for a 3D model of $N_x \times N_y \times N_z$ grid points. The model is divided into MP subdomains with overlapping. For the Mth-order FDM ($L=2,4,8,...$) the field quantities in the overlapped area occupy $N_x \times N_y \times L/2 \times 4$ bytes at the top and at the bottom of each subdomain. We further assume that the communication for each processor pair is performed with the speed of V_c byte/s, and no data collision occurs during communications.

Since z-differentiation appears for 6 variables in equations (3) and (4), the total communication time at each processor is given by

$$T_c = \frac{N_x \times N_y \times L/2 \times 4}{V_c} \times 6 \times 2 . \tag{6}$$

We now measure the total CPU time T_d consumed for the 3D modeling on a single processor, and define the computation (data processing) speed V_d (byte/s) as

$$V_d = \frac{N_x \times N_y \times N_z \times 4}{T_d} . \tag{7}$$

If concurrent computation is carried out by MP processors, the CPU time at each processor is simply expected to be T_d/MP. Thus we can roughly estimate the speed-up rate Ep for parallel computing using MP processors as function of V_d/V_c ratio by

$$Ep = \frac{T_d}{T_d/MP + T_c} = \frac{MP}{1 + MP \times L/N_z \times 6 \times V_d/V_c} . \tag{8}$$

To complement the above theoretical estimate we examined the PSM/FDM hybrid code on a SGI Origin 2000 and HITACHI SR8000/MPP parallel computer and measure the actual speed-up rates by parallel computing using various number of processors. We measured the MPI communication speed for each machine using a simple benchmark program for two processors communications, and obtained V_c =12,700 Kbyte/s and 27,600 Kbyte/s on Origin 2000 and SR8000/MPP, respectively. We also measured the computation speed of V_d =560Kbyte/s and 1320 Kbyte/s for a processor of Origin 2000 and SR8000/MPP, respectively, using a 3D model of N_x=256, N_y=256, N_z=256.

The result in Fig. 3, using 2 to 16 processors of Origin 2000 and 16 to 128 processors of SR8000/MPP, shows good agreement between the actual speed-up rates and theoretically prediction model with V_d/V_c=0.044 and 0.048 for Origin 2000 and SR8000/MPP, respectively.

3   Ground Motion Simulation for the 2000 Tottori-ken Seibu Earthquake

3.1   Observed Ground Motion from Dense Arrays

Large inland earthquake occurred at Tottori on Oct. 6, 2000, is the largest event since the destructive 1995 Kobe earthquake. The ground motions from this earthquake were well recorded by dense arrays of 550 K-NET and KiK-net stations.

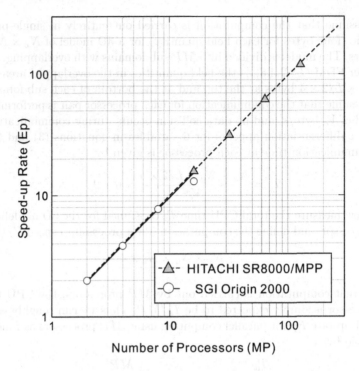

Fig. 3. Speed-up rate of the hybrid PSM/FDM parallel code using 2 to 16 processors of SGI Origin 2000 (*circle*) and 16 to 128 processors of HITACHI SR8000/MPP, relative to single processor computation speed. Theoretical speed-up rate for each system is shown in solid (Origin 2000) and dotted (SR8000/MPP) lines.

Fig. 4 illustrates a snapshot of seismic wave propagation at 60s from the fault rupture initiation, produced by an interpolation of observed 550 waveforms after applying an antialias filter with a cut-off frequency of 0.25 Hz. In the snapshot we can clearly see large amplitude fundamental-mode *Love* wave propagating in southwestern Japan with a relative slower speed of around 2.7 km/s. We find the ground shaking continues over several minutes at urban cities, such as Osaka, Nagoya and Saga, by amplifications and elongations of seismic waves in the sedimentary basins. The maximum ground velocity pattern during the earthquake is shown in Fig. 4b, also indicating that local conditions influence the strong ground motions.

3.2   Numerical Simulation of the 2000 Tottori-ken Seibu Earthquake

The character of observed wavefield can be modeled in 3D simulation of seismic wave using an adequate crust and upper-mantle structural model of southwestern Japan. The 3D model is based on a number of geophysical data such as from reflection and refraction experiments (e.g. [10]), gravity anomaly and travel-time tomography studies (e.g. [11], (Fig. 5). We also use a spatio-temporal source-

(a)   (b)

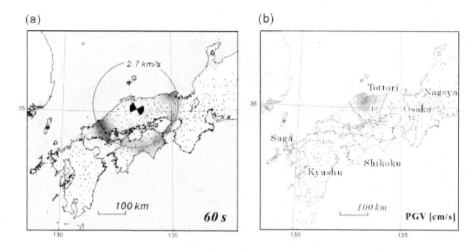

Fig. 4. (a) Snapshot of seismic wave propagation during the 2000 Tottori-ken Seibu earthquake at 60s from the source initiation, derived from the digital records of strong ground motion stations (*dots*). Green circle indicate propagation speed of V=2.7 km/s. (b)Peak ground velocity in cm/s during the earthquake. Strong motions stations are shown by triangles

rupture model [9] for the earthquake, derived from a comprehensive inversion using near-field strong motion records and teleseismic waveforms.

We employed the hybrid PSM/FDM code to calculate the wavefield in this model. The 3D model is 820 km by 410 km by 128 km, is discritized by a uniform interval of $\Delta_x=\Delta_y=1.6$ km in horizontal directions and $\Delta_z=0.64$ km in depth ($N_x=512$, $N_y=256$, $N_z=200$; 26 million grid points, in total). The minimum shear wave velocity in the shallow structure of the sedimentary basin is 1.6 km/s, so that the 3D modeling can accurately treat seismic wave propagation for frequencies below 0.5 Hz. The 3D modeling took computer memory of about 6 GB and wall-clock time of 11 hours using 16 processors of SGI Origin 2000, or 40 minutes using 128 processors of HITACHI SR8000/MPP.

Fig. 6(a) displays a set of snapshots of horizontal ground velocity motion derived by 3D simulation. The simulation results are compared with the observed wavefield (Fig. 6b). In the first frame of snapshots four-lobe pattern of *P*- and *SH*-wave (horizontally polarized *S*-wave) from a shallow strike-slip fault source are clearly seen (20s). As the wave spreading away from the hypocenter, fundamental-mode *Love* wave is gradually built up from multiple *SH*-wave reverberation in the crust. The *Love* wave propagating in the southwestern Japan is a prominent feature at regional wavefield (60s). At the basin the amplitude of the *Love* wave is enhanced further by site amplification effects in the low-velocity materials overlaying rigid bedrock with large impedance contrast. Ground shaking at the basin continues over ten seconds. The simulated seismograms of transverse ground velocity motion at three KiK-net stations are also illustrated in the Fig. 6, and there are good agreements with observed ground motions.

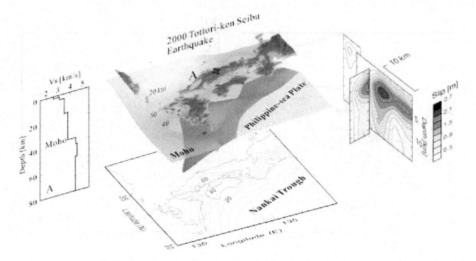

Fig. 5. 3D model for southwestern Japan used in 3-D wavefield calculations. Reference shear wave model at west of Tottori (*marked A in middle panel*), detail of 3-D model showing the configuration of Moho interface, and source slip model for the 2000 Tottori-ken Seibu earthquake [9].

4   Strong Ground Motion Prediction: Nankai Scenario Earthquakes

The results of the 3D simulation for the 2000 Tottori-ken Seibu earthquake using realistic model provide a good representation of ground motion during the earthquake. Therefore the model can be used to investigate the pattern of strong ground motion expected for future scenario earthquakes, such as at Nankai Trough.

The history of the Nankai earthquakes represented as shallow angle thrust faulting over the subducting Philippine-sea plate can be traced back to 684 A.D. [12]. The long documented history of this earthquake indicates that a huge event occurred recurrently at every 100 to 200 years.

Fig. 7a shows a display of the ground motion derived from the wave propagation simulation for the Nankai earthquake. We assumed the fault-slip history of the previous 1946 Nankei event (M8), derived from an analysis of geodetic measurement and Tsunami data [13]. We also assumed unilateral fault rupture propagation from the hypocenter at a constant rupture speed of V_r=2.8 km/s.

The first snapshot of ground motion at 30s after source initiation shows the large amplitude S wavefront propagating at Muroto Promontory on which the effect of fault rupture directivity is clearly seen. The large amplitude S wave energy propagating in west of Shikoku (54s) and producing large ground motion at the east side of Kyushu (around Ohita). However, when the fault rupture initiates at west of the fault plane (near Ashizuri Peninsula; Model 2) and the fault rupture propagates almost unilaterally to the east, the directivity effect

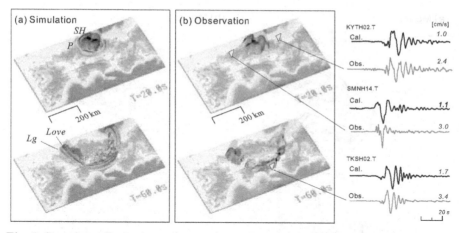

Fig. 6. Snapshots of seismic wave propagation during the 2000 Tottori-ken Seibu earthquake, comparing between (a) simulation results and (b) observed wavefield. Major seismic phases are marked. Three seismograms simulated (*cal.*) and observed (*obs.*) transverse component ground velocity motion at three KiK-net stations are compared at right.

Table 1. Model specifications of current 3D strong ground motion for the Nankai earthquake, using HITACHI SR8000/MPP parallel computer and high-resolution simulation expected for Earth Simulator.

	SR8000/MPP	Earth Simulator
Number of Grid Point	33 million	2.1 billion
Grid Size	1.6/0.8 km	0.4/0.2 km
Frequency Resolution	0.5 Hz	2 Hz
Computer Memory	6 Gbyte	384 Gbyte
Number of CPUs	128	4096
Computation Time	40 min	(60 min)

of fault rupture propagation enhances the ground motions in central Japan, dramatically (see 75s, 120s frames in Fig. 7b). The fault rupture directivity effect imposes significant modulation on the S wave amplitude of propagation in addition to the initial pattern of strong ground motion imposed by the source slip distribution and the 3D subsurface structure.

Fig. 8 shows the potential impact of these earthquakes in terms of the seismic intensities of Japan Meteorological Agency (JMA) scale, and synthetic waveforms of ground velocity motions at three sites. The effect of 3D structure on the wavefield is more apparent as the wave spreads away from the source and imposes significant modulation by the rupture directivity effect. This indicates the importance of 3D modeling including high-resolution structure and complex source-slip model for the realistic simulation of ground motions.

(a) Model 1   (b) Model 2

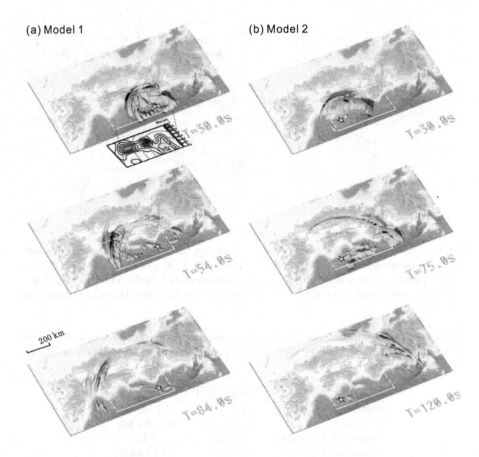

Fig. 7. Snapshot of seismic wave propagation from the Nankai earthquake simulation using the source models of (a) Model 1; for the fault rupture propagation from west to east, and (b) Model 2; fault rupture propagation from west to east. Slip distribution on the fault plane is shown in the top.

5   Conclusions

One of the important issues of strong ground motion seismology is to understand the physics of the seismic waves in heterogeneous structure, and to investigate the ground motion pattern expected for future earthquake scenarios. Towards this goal, a close link between seismological observations and computer simulation technology is indispensable.

It has long been recognized that the numerical simulation of seismic wave propagation in 3D heterogeneous structure is too expensive to conduct due to the limitation of computer power, and most of studies on strong motion prediction has been based on stochastic analysis of seismic waves such as using an attenuation function of observed ground motions. However, the present gener-

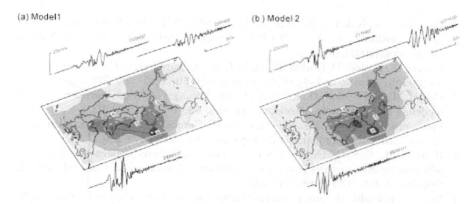

Fig. 8. Expected seismic intensity for the Nankai earthquake derived from the simulation corresponds to two source models (a) for Model 1, (b) for Model 2 (see, Fig. 7). Star mark and arrow indicate hypocenter and source rupture direction, respectively. The seismic intensity scale is shown in the figure with contours. Synthetic seismograms of NS-component ground velocity motion at three sites (*Ohita, Shimane, and Nagoya*) are also displayed.

ation of 3D simulation techniques such as the parallelized hybrid PSM/FDM code allows more reslistic simulation of seismic waves of relatively higher frequencies ($\sim$0.5Hz) at regional wavefield to a few hundred kirometers from the source (Table 1).

Although, it is still difficult to provide adequate resolution of higher-frequencies wavefield which is quite important to the damage of wooden-frame houses and low-rise buildings, it is possible to impliment in next generation high-performance parallel computer, i.e. Earth Simulator.

Acknowledgments

This study is supported by Earth Simulator Project of Ministry of Education, Culture, Sports and Technology. The computation is conducted at Earthquake Information Center, Earthquake Research Institute, University of Tokyo and at Super Computer Center, University of Tokyo. K-NET and KiK-net data are made available by the National Institute for Earth Science and Disaster Research (NIED), Japan. Constructive review by an anonymous reviewer and comments by Dr. Ursula Iturrarán-Viveros have helped to improve the manuscript.

References

1. Kinoshita, S., Kyoshin Net (K-NET) Seism. Res. Lett., **69** (1998) 309–332
2. Aoi, S., Obara, S., Hori, S., Kasahara, K., Okada, Y.: New strong-motion observation network: KiK-net, EOS Trans. Am. Geophys. Union, **81** (2000) F863
3. Furumura, T., Koketsu, K., Takenaka, H.: A hybrid PSM/FDM parallel simulation for 3-D seismic (acoustic) wavefield Butsuri-Tansa (J. SEGJ), **53** (2000) 294–308

4. Furumura, T., Koketsu, K., Wen, K.-L.: Parallel PSM/FDM hybrid simulation of ground motions from the 1999 Chi-Chi, Taiwan, earthquake, Pure and Applied Geophysics, (2002), in press
5. Cerjan, C., Kosloff, D, Kosloff, R., Reshef, M.: A nonreflecting boundary condition for discrete acoustic and elastic wave equations, **50** (1985) 705–708
6. Reshef, M., and Kosloff, D., Edwards, M. and Hsiung, C.: Three-dimensional elastic modeling by the Fourier method, Geophysics, **53** (1988) 1184–1193
7. Furumura, T. Kennett, B.L.N., Takenaka, H.: Parallel 3-D pseudospectral simulation of seismic wave propagation, Geophysics, **63** (1998) 279–288
8. Hung, S.-H., Forsyth, D. W.: Modelling anisotropic wave propagation in oceanic inhomogeneous structures using the parallel multidomain pseudo-spectral method, Geophys. J. Int., **133** (1998) 726–740
9. Yagi, Y., Kikuchi, M.: Source rupture process of the Tottori-ken Seibu earthquake of Oct. 6, 2000, (Mjma 7.3) by usingjoint inversion of far-field and near-field waveform, Abst. Fall Meet. of the Seism. Soc. Japan, **T04** (2001)
10. Yoshii, T., Sasaki, Y., Tada, Y., Okada, H., Shuzo, A., Muramatu, I., Hashizume, I, Moriya, T.: The third Kurayoshi explosion and the crustal structure in western part of Japan, J. Phys. Earth, **22** (1074) 109–121
11. Zhao, D. A., Horiuchi, A., Hasegawa, A.: Seismic velocity structure of the crust beneath the Japan Islands, Tectonophysics, **212**, (1992) 289–301
12. Ando, M.: Source mechanism and tectonic significance of historical earthquakes along the Nankai Trough, Japan, Tectonophysics, **27** (1975) 119–140
13. Tanioka, Y., Satake, K.: Coseismic slip distribution of the 1946 Nankai earthquake and aseismic slips caused by the earthquake, Earth. Planets. Space, **53** (2001) 235–241

Large-Scale Parallel Computing
of Cloud Resolving Storm Simulator

Kazuhisa Tsuboki[1] and Atsushi Sakakibara[2]

[1] Hydrospheric Atmospheric Research Center, Nagoya University
Furo-cho, Chikusa-ku, Nagoya, 464-8601 Japan
phone: +81-52-789-3493, fax: +81-52-789-3436
tsuboki@ihas.nagoya-u.ac.jp
[2] Research Organization for Information Science and Technology
Furo-cho, Chikusa-ku, Nagoya, 464-8601 Japan
phone: +81-52-789-3493, fax: +81-52-789-3436
atsusi@ihas.nagoya-u.ac.jp

Abstract. A sever thunderstorm is composed of strong convective clouds. In order to perform a simulation of this type of storms, a very fine-grid system is necessary to resolve individual convective clouds within a large domain. Since convective clouds are highly complicated systems of the cloud dynamics and microphysics, it is required to formulate detailed cloud physical processes as well as the fluid dynamics. A huge memory and large-scale parallel computing are necessary for the computation. For this type of computations, we have developed a cloud resolving numerical model which was named the Cloud Resolving Storm Simulator (CReSS). In this paper, we will describe the basic formulations and characteristics of CReSS in detail. We also show some results of numerical experiments of storms obtained by a large-scale parallel computation using CReSS.

1   Introduction

A numerical cloud model is indispensable for both understanding cloud and precipitation and their forecasting. Convective clouds and their organized storms are highly complicated systems determined by the fluid dynamics and cloud microphysics. In order to simulate evolution of a convective cloud storm, calculation should be performed in a large domain with a very high resolution grid to resolve individual clouds. It is also required to formulate accurately cloud physical processes as well as the fluid dynamic and thermodynamic processes. A detailed formulation of cloud physics requires many prognostic variables even in a bulk method such as cloud, rain, ice, snow, hail and so on. Consequently, a large-scale parallel computing with a huge memory is necessary for this type of simulation.

Cloud models have been developed and used for study of cloud dynamics and cloud microphysics since 1970's (e.g., Klemp and Wilhelmson, 1978 [1]; Ikawa, 1988 [2]; Ikawa and Saito, 1991 [3]; Xue et al., 1995 [4]; Grell et al., 1994 [5]). These models employed the non-hydrostatic and compressible equations systems

H. Zima et al. (Eds.): ISHPC 2002, LNCS 2327, pp. 243–259, 2002.

with a fine-grid system. Since the computation of cloud models was very large, they have been used for research with a limited domain.

The recent progress in high performance computer, especially a parallel computers is extending the potential of cloud models widely. It enables us to perform a simulation of mesoscale storm using a cloud model. For this four years, we have developed a cloud resolving numerical model which was designed for parallel computers including "the Earth Simulator".

The purposes of this study are to develop the cloud resolving model and its parallel computing to simulate convective clouds and their organized storms. Thunderstorms which are organization of convective clouds produce many types of severe weather: heavy rain, hail storm, downburst, tornado and so on. The simulation of thunderstorms will clarify the characteristics of dynamics and evolution and will contribute to the mesoscale storm prediction.

The cloud resolving model which we are now developing was named "the Cloud Resolving Storm Simulator (CReSS)". In this paper, we will describe the basic formulation and characteristics of CReSS in detail. Some results of numerical experiments using CReSS will be also presented.

2   Description of CReSS

2.1   Basic Equations and Characteristics

The coordinate system of CReSS is the Cartesian coordinates in horizontal x, y and a terrain-following curvilinear coordinate in vertical ζ to include the effect of orography. Using height of the model surface $z_s(x, y)$ and top height z_t, the vertical coordinate $\zeta(x, y, z)$ is defined as,

$$\zeta(x, y, z) = \frac{z_t[z - z_s(x, y)]}{z_t - z_s(x, y)}.$$ (1)

If we use a vertically stretching grid, the effect will be included in (1). Computation of CReSS is performed in the rectangular linear coordinate transformed from the curvilinear coordinate. The transformed velocity vector will be

$$U = u,$$ (2)

$$V = v,$$ (3)

$$W = (uJ_{31} + vJ_{32} + w)\Big/G^{\frac{1}{2}}.$$ (4)

where variable components of the transform matrix are defined as

$$J_{31} = -\frac{\partial z}{\partial x} = \left(\frac{\zeta}{z_t} - 1\right)\frac{\partial z_s(x, y)}{\partial x}$$ (5)

$$J_{32} = -\frac{\partial z}{\partial y} = \left(\frac{\zeta}{z_t} - 1\right)\frac{\partial z_s(x, y)}{\partial y}$$ (6)

$$J_d = \frac{\partial z}{\partial \zeta} = 1 - \frac{z_s(x, y)}{z_t}$$ (7)

and the Jacobian of the transformation is

$$G^{\frac{1}{2}} = |J_d| = J_d \tag{8}$$

In this coordinate, the governing equations of dynamics in CReSS will be formulated as follows. The dependent variables of dynamics are three-dimensional velocity components u, v and w, perturbation pressure p' and perturbation of potential temperature θ'. For convenience, we use the following variables to express the equations.

$$\rho^* = G^{\frac{1}{2}}\bar{\rho}, \ \ u^* = \rho^* u, \ \ v^* = \rho^* v,$$
$$w^* = \rho^* w, \ \ W^* = \rho^* W, \ \ \theta^* = \rho^* \theta'.$$

where $\bar{\rho}$ is the density of the basic field which is in the hydrostatic balance.

Using these variables, the momentum equations are

$$\frac{\partial u^*}{\partial t} = -\underbrace{\left(u^* \frac{\partial u}{\partial x} + v^* \frac{\partial u}{\partial y} + W^* \frac{\partial u}{\partial \zeta} \right)}_{[\mathrm{rm}]}$$
$$-\underbrace{\left[\frac{\partial}{\partial x} \{ J_d (p' - \alpha Div^*) \} + \frac{\partial}{\partial \zeta} \{ J_{31} (p' - \alpha Div^*) \} \right]}_{[\mathrm{am}]}$$
$$+\underbrace{(f_s v^* - f_c w^*)}_{[\mathrm{rm}]} + \underbrace{G^{\frac{1}{2}} \mathrm{Turb}.u}_{[\mathbf{physics}]}, \tag{9}$$

$$\frac{\partial v^*}{\partial t} = -\underbrace{\left(u^* \frac{\partial v}{\partial x} + v^* \frac{\partial v}{\partial y} + W^* \frac{\partial v}{\partial \zeta} \right)}_{[\mathrm{rm}]}$$
$$-\underbrace{\left[\frac{\partial}{\partial y} \{ J_d (p' - \alpha Div^*) \} + \frac{\partial}{\partial \zeta} \{ J_{32} (p' - \alpha Div^*) \} \right]}_{[\mathrm{am}]}$$
$$-\underbrace{f_s u^*}_{[\mathrm{rm}]} + \underbrace{G^{\frac{1}{2}} \mathrm{Turb}.v}_{[\mathbf{physics}]}, \tag{10}$$

$$\frac{\partial w^*}{\partial t} = -\underbrace{\left(u^* \frac{\partial w}{\partial x} + v^* \frac{\partial w}{\partial y} + W^* \frac{\partial w}{\partial \zeta} \right)}_{[\mathrm{rm}]} - \underbrace{\frac{\partial}{\partial \zeta} (p' - \alpha Div^*)}_{[\mathrm{am}]}$$
$$-\rho^* g \left(\underbrace{\frac{\theta'}{\bar{\theta}}}_{[\mathrm{gm}]} - \underbrace{\frac{p'}{\bar{\rho} c_s^2}}_{[\mathrm{am}]} + \underbrace{\frac{q_v'}{\epsilon + \bar{q}_v} - \frac{q_v' + \sum q_x}{1 + \bar{q}_v}}_{[\mathbf{physics}]} \right)$$
$$+ \underbrace{f_c u^*}_{[\mathrm{rm}]} + \underbrace{G^{\frac{1}{2}} \mathrm{Turb}.w}_{[\mathbf{physics}]}, \tag{11}$$

where αDiv^* is an artificial divergence damping term to suppress acoustic waves, f_s and f_c are Coriolis terms, c_s^2 is square of the acoustic wave speed, q_v and q_x is mixing ratios of water vapor and hydrometeors, respectively. The equation of the potential temperature is

$$\frac{\partial \theta^*}{\partial t} = - \underbrace{\left(u^* \frac{\partial \theta'}{\partial x} + v^* \frac{\partial \theta'}{\partial y} + W^* \frac{\partial \theta'}{\partial \zeta} \right)}_{[\mathrm{rm}]}$$

$$\underbrace{- \bar{\rho} w \frac{\partial \bar{\theta}}{\partial \zeta}}_{[\mathrm{gm}]} + \underbrace{G^{\frac{1}{2}} \mathrm{Turb}.\theta}_{[\mathbf{physics}]} + \underbrace{\rho^* \mathrm{Src}.\theta}_{[\mathbf{physics}]}, \qquad (12)$$

and the pressure equation is

$$\frac{\partial G^{\frac{1}{2}} p'}{\partial t} = - \underbrace{\left(J_3 u \frac{\partial p'}{\partial x} + J_3 v \frac{\partial p'}{\partial y} + J_3 W \frac{\partial p'}{\partial \zeta} \right)}_{[\mathrm{rm}]} + \underbrace{G^{\frac{1}{2}} \bar{\rho} g w}_{[\mathrm{am}]}$$

$$\underbrace{- \bar{\rho} c_s^2 \left(\frac{\partial J_3 u}{\partial x} + \frac{\partial J_3 v}{\partial y} + \frac{\partial J_3 W}{\partial \zeta} \right)}_{[\mathrm{am}]} + \underbrace{G^{\frac{1}{2}} \bar{\rho} c_s^2 \left(\frac{1}{\theta} \frac{d\theta}{dt} - \frac{1}{Q} \frac{dQ}{dt} \right)}_{[\mathrm{am}]}, \qquad (13)$$

where Q is diabatic heating, terms of "Turb." is a physical process of the turbulent mixing and term of "Src." is source term of potential temperature.

Since the governing equations have no approximation, they will express all type of waves including the acoustic waves, gravity waves and Rossby waves. These waves have very wide range of phase speed. The fastest wave is the acoustic wave. Although it is unimportant in meteorology, its speed is very large in comparison with other waves and limits the time increment of integration. We, therefore, integrate the terms related to the acoustic waves and other terms with different time increments. In the equations (9)~(13), [rm] is indicates terms which are related to rotational mode (the Rossby wave mode), [gm] the divergence mode (gravity wave mode), and [am] the acoustic wave mode, respectively. Terms of physical processes are indicated by [physics].

2.2   Computational Scheme and Parallel Processing Strategy

In numerical computation, a finite difference method is used for the spatial discretization. The coordinates are rectangular and dependent variables are set on a staggered grid: the Arakawa-C grid in horizontal and the Lorenz grid in vertical (Fig.1). The coordinates x, y and ζ are defined at the faces of the grid boxes. The velocity components u, v and w are defined at the same points of the coordinates x, y and ζ, respectively. The metric tensor J_{31} is evaluated at a half interval below the u point and J_{32} at a half interval below the v point. All scalar variables p', θ', q_v and q_x, the metric tensor J_d and the transform Jacobian are

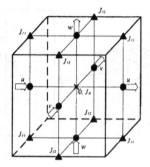

Fig. 1. Structure of the staggered grid and setting of dependent variables.

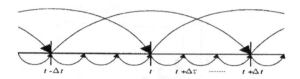

Fig. 2. Schematic representation of the mode-splitting time integration method. The large time step is indicated by the upper large curved arrows with the increment of Δt and the small time step by the lower small curved arrows with the increment of $\Delta\tau$.

defined at the center of the grid boxes. In the computation, an averaging operator is used to evaluate dependent variables at the same points. All output variables are obtained at the scalar points.

As mentioned in the previous sub-section, the governing equation includes all types of waves and the acoustic waves severely limits the time increment. In order to avoid this difficulty, CReSS adopted the mode-splitting technique (Klemp and Wilhelmson, 1978 [1]) for time integration. In this technique, the terms related to the acoustic waves in (9) $\sim$ (13) are integrated with a small time increment $\Delta\tau$ and all other terms are with a large time increment Δt.

Figure 2 shows a schematic representation of the time integration of the mode-splitting technique. CReSS has two options in the small time step integration; one is an explicit time integration both in horizontal and vertical and the other is explicit in horizontal and implicit in vertical. In the latter option, p' and w are solved implicitly by the Crank-Nicolson scheme in vertical. With respect to the large time step integration, the leap-frog scheme with the Asselin time filter is used for time integration. In order to remove grid-scale noise, the second or forth order computational mixing is used.

A large three-dimensional computational domain (order of 100 km) is necessary for the simulation of thunderstorm with a very high resolution (order of less than 1km). For parallel computing of this type of computation, CReSS adopts a two dimensional domain decomposition in horizontal (Fig.3). Parallel processing is performed by the Massage Passing Interface (MPI). Communications between

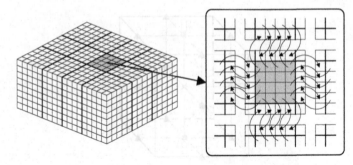

Fig. 3. Schematic representation of the two-dimensional domain decomposition and the communication strategy for the parallel computing.

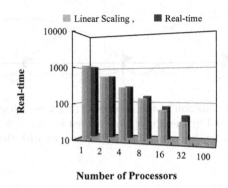

Fig. 4. Computation time of parallel processing of a test experiment. The model used in the test had 67×67×35 grid points and was integrated for 50 steps on HITACHI SR2201.

the individual processing elements (PEs) are performed by data exchange of the outermost two grids.

The performance of parallel processing of CReSS was tested by a simulation whose grid size was 67 × 67 × 35 on HITACHI SR2201. With increase of the number of PEs, the computation time decreased almost linearly (Fig.4). The efficiency was almost 0.9 or more if the number of PEs was less than 32. When the number of PEs was 32, the efficiency decreased significantly. Because the number of grid was too small to use the 32 PEs. The communication between PEs became relatively large. The results of the test showed a sufficiently high performance of the parallel computing of CReSS.

2.3   Initial and Boundary Conditions

Several types of initial and boundary conditions are optional in CReSS. For a numerical experiment, a horizontally uniform initial field provided by a sounding profile will be used with an initial disturbance of a thermal bubble or random noise. Optional boundary conditions are rigid wall, periodic, zero normal-gradient, and wave-radiation type of Orlanski (1976) [6].

CReSS has an option to be nested within a coarse-grid model and performs a prediction experiment. In this option, the initial field is provided by interpolation of grid point values and the boundary condition is provided by the coarse-grid model.

2.4 Physical Processes

Cloud physics is an important physical process. It is formulated by a bulk method of cold rain which is based on Lin et al. (1983) [7], Cotton et al. (1986) [8], Murakami (1990) [9], Ikawa and Saito (1991) [3], and Murakami et al. (1994) [10]. The bulk parameterization of cold rain considers water vapor, rain, cloud, ice, snow, and graupel. Prognostic variables are mixing ratios for water vapor q_v, cloud water q_c, rain water q_r, cloud ice q_i, snow q_s and graupel q_g. The prognostic equations of these variables are

$$\frac{\partial \bar{\rho} q_v}{\partial t} = \text{Adv.}q_v + \text{Turb.}q_v + \bar{\rho}\text{Src.}q_v \tag{14}$$

$$\frac{\partial \bar{\rho} q_x}{\partial t} = \text{Adv.}q_x + \text{Turb.}q_x + \bar{\rho}\text{Src.}q_x + \bar{\rho}\text{Fall.}q_x \tag{15}$$

where q_x is the representative mixing ratio of $q_c, q_r, q_i, q_s and q_g$, and "Adv.", "Turb." and "Fall." represent time changes due to advection, turbulent mixing, and fall out, respectively. All sources and sinks of variables are included in the "Src." term. The microphysical processes implemented in the model are described in Fig.5. Radiation of cloud is not included.

Turbulence is also an important physical process in the cloud model. Parameterizations of the subgrid-scale eddy motions in CReSS are one-order closure of the Smagorinsky (1963) [11] and the 1.5 order closer with turbulent kinetic energy (TKE). In the latter parameterization, the prognostic equation of TKE will be used.

CReSS implemented the surface process formulated by a bulk method. In this process, the surface sensible flux H_S and latent heat flux LE are formulated as

$$H_S = -\rho_a C_p C_h |V_a|(T_a - T_G), \tag{16}$$

$$LE = -\rho_a L C_h |V_a|\beta \left[q_a - q_{vs}^*(T_G)\right], \tag{17}$$

where "a" indicates the lowest layer of the atmosphere and "G" the surface. The coefficient of β is the evapotranspiration efficiency and L is the latent heat of evaporation. The surface temperature of the ground T_G is calculated by the n-layers ground model. The momentum fluxes (τ_x, τ_y) are

$$\tau_x = \rho_a C_m |V_a| u_a, \tag{18}$$

$$\tau_y = \rho_a C_m |V_a| v_a. \tag{19}$$

The bulk coefficients C_h and C_m are formulated by the scheme of Louis et al. (1981) [12].

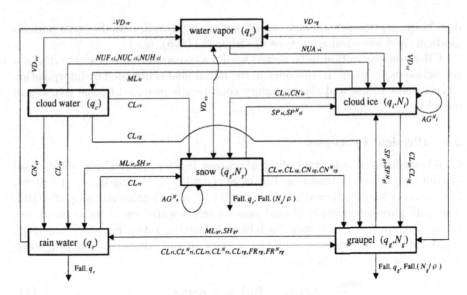

Fig. 5. Diagram describing of water substances and cloud microphysical processes in the bulk model.

3 Dry Model Experiments

In the development of CReSS, we tested it with respect to several types of phenomena. In a dry atmosphere, the mountain waves and the Kelvin-Helmholtz billows were chosen to test CReSS.

The numerical experiment of Kelvin-Helmholtz billows was performed in a two-dimensional geometry with a grid size of 20 m. The profile of the basic flow was the hyperbolic tangent type. Stream lines of u and w components (Fig.6) show a clear cats eye structure of the Kelvin-Helmholtz billows. This result is very similar to that of Klaassen and Peltier (1985) [13]. The model also simulated the overturning of potential temperature associated with the billows (Fig.7). This result shows the model works correctly with a grid size of a few tens meters as far as in the dry experiment.

In the experiment of mountain waves, we used a horizontal grid size of 400 m in a three-dimensional geometry. A bell-shaped mountain with a height of 500 m and with a half-width of 2000 m was placed at the center of the domain. The basic horizontal flow was 10 m s^{-1} and the buoyancy frequency was 0.01 s^{-1}. The result (Fig.8) shows that upward and downwindward propagating mountain waves developed with time. The mountain waves pass through the downwind boundary. This result is closely similar to that obtained by other models as well as that predicted theoretically.

These results of the dry experiments showed that the fluid dynamics part and the turbulence parameterization of the model worked correctly and realistic behavior of flow were simulated.

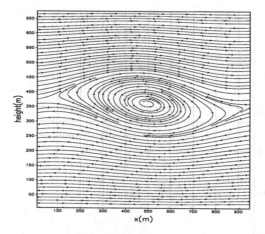

Fig. 6. Stream lines of the Kelvin-Helmholtz billow at 240 seconds from the initial simulated in the two-dimensional geometry.

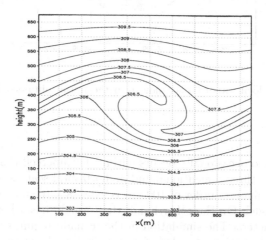

Fig. 7. Same as Fig.6, but for potential temperature.

4 Simulation of Tornado within a Supercell

In a simulation experiment of a moist atmosphere, we chose a tornado-producing supercell observed on 24 September 1999 in the Tokai District of Japan. The simulation was aiming at resolving the vortex of the tornado within the supercell.

Numerical simulation experiments of a supercell thunderstorm which has a horizontal scale of several tens kilometers using a cloud model have been performed during the past 20 years (Wilhelmson and Klemp, 1978 [14]; Weisman and Klemp, 1982, 1984 [15], [16]). Recently, Klemp and Rotunno (1983) [17] attempted to increase horizontal resolution to simulate a fine structure of a meso-cyclone within the supercell. It was still difficult to resolve the tornado.

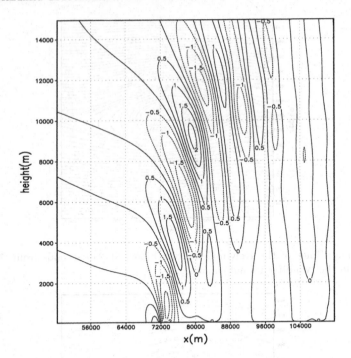

Fig. 8. Vertical velocity at 9000 seconds from the initial obtained by the mountain wave experiment.

An intense tornado occasionally occurs within the supercell thunderstorm. The supercell is highly three-dimensional and its horizontal scale is several tens kilometer. A large domain of order of 100 km is necessary to simulate the supercell using a cloud model. On the other hand, the tornado has a horizontal scale of a few hundred meters. The simulation of the tornado requires a fine resolution of horizontal grid spacing of order of 100 m or less. In order to simulate the supercell and the associated tornado by a cloud model, a huge memory and high speed CPU are indispensable.

To overcome this difficulty, Wicker and Wilhelmson (1995) [18] used an adaptive grid method to simulate tornado-genesis. The grid spacing of the fine mesh was 120 m. They simulated a genesis of tornadic vorticity. Grasso and Cotton (1995) [19] also used a two-way nesting procedure of a cloud model and simulated a tornadic vorticity. These simulations used a two-way nesting technique. Nesting methods include complication of communication between the coarse-grid model and the fine-mesh model through the boundary. On the contrary, the present research do not use any nesting methods. We attempted to simulate both the supercell and the tornado using the uniform grid. In this type of simulation, no complication of the boundary communication. The computational domain of the present simulation was about 50 × 50 km and the grid spacing was 100 m. The integration time was about 2 hours.

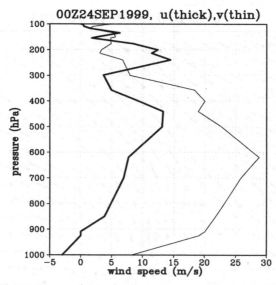

Fig. 9. Vertical profiles of zonal component (thick line) and meridional component (thin line) observed at Shionomisaki at 00 UTC, 24 September 1999.

The basic field was give by a sounding at Shionomisaki, Japan at 00 UTC, 24 September 1999 (Fig.9). The initial perturbation was given by a warm thermal bubble placed near the surface. It caused an initial convective cloud.

After one hour from the initial time, a quasi-stationary super cell was simulated by CReSS (Fig.10). The hook-shaped precipitation area and the bounded weak echo region (BWER) which are characteristic features of the supercell were formed in the simulation. An intense updraft occurred along the surface flanking line. At the central part of BWER or of the updraft, a tornadic vortex was formed at 90 minutes from the initial time.

The close view of the central part of the vorticity (Fig.11) shows closed contours. The diameter of the vortex is about 500 m and the maximum of vorticity is about 0.1 s^{-1}. This is considered to be corresponded to the observed tornado. The pressure perturbation (Fig.12) also shows closed contours which corresponds to those of the vorticity. This indicates that the flow of the vortex is in the cyclostrophic balance. The vertical cross section of the vortex (Fig.13) shows that the axis of the vorticity and the associated pressure perturbation is inclined to the left hand side and extends to a height of 2 km. At the center of the vortex, the downward extension of cloud is simulated.

While this is a preliminary result of the simulation of the supercell and tornado, some characteristic features of the observation were simulated well. The important point of this simulation is that both the supercell and the tornado were simulated in the same grid size. The tornado was produced purely by the physical processes formulated in the model. A detailed analysis of the simulated data will provide an important information of the tornado-genesis within the supercell.

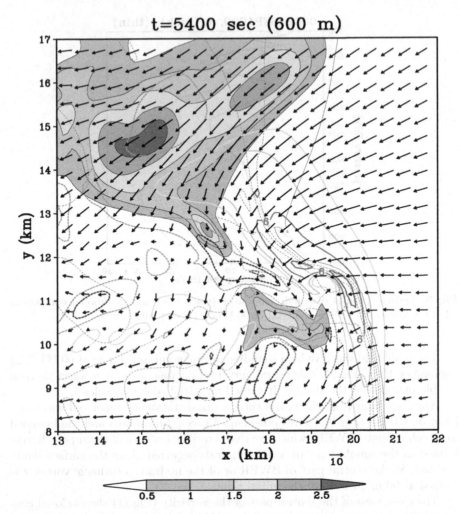

Fig. 10. Horizontal display at 600 m of the simulated supercell at 5400 seconds from the initial. Mixing ratio of rain (gray scales, g kg^{-1}), vertical velocity (thick lines, m s^{-1}), the surface potential temperature at 15 m (thin lines, K) and horizontal velocity vectors.

5   Simulation of Squall Line

A squall line is a significant mesoscale convective system. It is usually composed of an intense convective leading edge and a trailing stratiform region. An intense squall line was observed by three Doppler radars on 16 July 1998 over the China continent during the intensive field observation of GAME / HUBEX (the GEWEX Asian Monsoon Experiment / Huaihe River Basin Experiment). The squall line extended from the northwest to the southeast with a width of a few tens kilometers and moved northeastward at a speed of 11 m s^{-1}. Radar

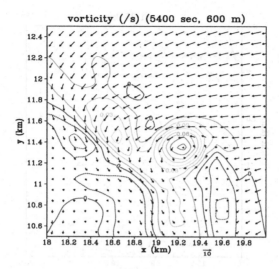

Fig. 11. Close view of the simulated tornado within the supercell. The contour lines are vorticity (s^{-1}) and the arrows are horizontal velocity. The arrow scale is shown at the bottom of the figure.

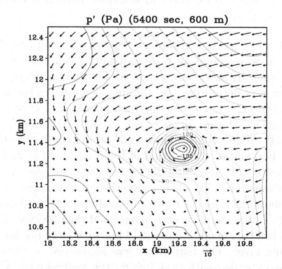

Fig. 12. Same as Fig.11, but for pressure perturbation.

observation showed that the squall line consisted of intense convective cells along the leading edge. Some of cells reached to a height of 17 km. The rear-inflow was present at a height of 4 km which descended to cause the intense lower-level convergence at the leading edge. After the squall line passed over the radar sites, a stratiform precipitation was extending behind the convective leading edge.

The experimental design of the simulation experiment using CReSS is as follows. Both the horizontal and vertical grid sizes were 300 m within a domain

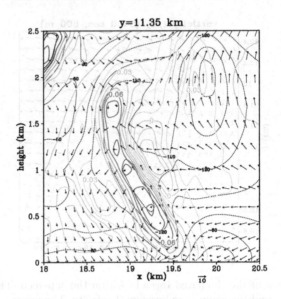

Fig. 13. Vertical cross section of the simulated tornado. Thick lines are vorticity (s^{-1}), dashed lines are pressure perturbation and arrows are horizontal velocity.

of 170 km × 120 km. Cloud microphysics was the cold rain type. The boundary condition was the wave-radiating type. An initial condition was provided by a dual Doppler analysis and sounding data. The inhomogeneous velocity field within the storm was determined by the dual Doppler radar analysis directly while that of outside the storm and thermodynamic field were provided by the sounding observation. Mixing ratios of rain, snow and graupel were estimated from the radar reflectivity while mixing ratios of cloud and ice were set to be zero at the initial. A horizontal cross section of the initial field is shown in Fig.14.

The simulated squall line extending from the northwest to the southeast moved northeastward (Fig.15). The convective reading edge of the simulated squall line was maintained by the replacement of new convective cells and moved to the northeast. This is similar to the behavior of the observed squall line. Convective cells reached to a height of about 14 km with large production of graupel above the melting layer. The rear-inflow was significant as the observation. A stratiform region extended with time behind the leading edge. Cloud extended to the southwest to form a cloud cluster.

The result of the simulation experiment showed that CReSS successfully simulated the development and movement of the squall line.

6 Summary and Future Plans

We are developing the cloud resolving numerical model CReSS for numerical experiments and simulations of clouds and storms. Parallel computing is indispensable for a large-scale simulations. In this paper, we described the basic

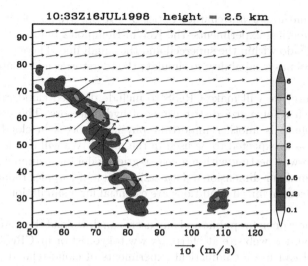

Fig. 14. Horizontal cross section of the initial field at a height of 2.5km at 1033 UTC, 16 July 1998. The color levels mixing ratio of rain (g kg^{-1}). Arrows show the horizontal velocity obtained by the dual Doppler analysis and sounding.

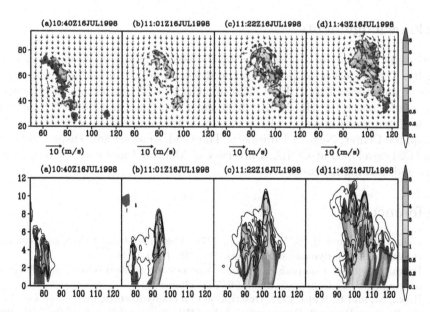

Fig. 15. Time series of horizontal displays (upper row) and vertical cross sections (lower row) of the simulated squall line. Color levels indicate total mixing ratio (g kg^{-1}) of rain, snow and graupel. Contour lines indicate total mixing ratio (0.1, 0.5, 1, 2 g kg^{-1}) of cloud ice and cloud water. Arrows are horizontal velocity.

formulations and important characteristics of CReSS. We also showed some result of the numerical experiments: the Kelvin-Helmholtz billows, the mountain waves, the tornado within the supercell and the squall line. These results showed that the CReSS has a capability to simulate thunderstorms and related phenomena.

In the future, we will make CReSS to include detailed cloud microphysical processes which resolve size distributions of hydrometeors. The parameterization of turbulence is another important physical process in cloud. The large eddy simulation is expected to be used in the model. We will develop CReSS to enable the two-way nesting within a coarse-grid model for a simulation of a real weather system. Four-dimensional data assimilation of Doppler radar is also our next target. Because initial conditions are essential for a simulation of mesoscale storms.

CReSS is now open for public and any users can download the source code and documents from the web site at "http://www.tokyo.rist.or.jp/CReSS_Fujin" (in Japanese) and can use for numerical experiments of cloud-related phenomena. CReSS has been tested on a several computers: HITACHI SR2201, HITACHI SR8000, Fujitsu VPP5000, NEC SX4. We expect that CReSS will be performed on the Earth Simulator and make a large-scale parallel computing to simulate a details of clouds and storms.

Acknowledgements

This study is a part of a project led by Professor Kamiya, Aichi Gakusen University. The project is supported by the Research Organization for Information Science and Technology (RIST). The simulations and calculations of this work were performed using HITACHI S3800 super computer and SR8000 computer at the Computer Center, the University of Tokyo and Fujitsu VPP5000 at the Computer Center, Nagoya University. The Grid Analysis and Display System (GrADS) developed at COLA, University of Maryland was used for displaying data and drawing figures.

References

1. Klemp, J. B., and R. B. Wilhelmson, 1978: The simulation of three-dimensional convective storm dynamics. *J. Atmos. Sci.*, **35**, 1070–1096.
2. Ikawa, M., 1988: Comparison of some schemes for nonhydrostatic models with orography. *J. Meteor. Soc. Japan*, **66**, 753–776.
3. Ikawa, M. and K. Saito, 1991: Description of a nonhydrostatic model developed at the Forecast Research Department of the MRI. *Technical Report of the MRI*, **28**, 238pp.
4. Xue, M., K. K. Droegemeier, V. Wong, A. Shapiro and K. Brewster, 1995: Advanced Regional Prediction System, Version 4.0. *Center for Analysis and Prediction of Storms, University of Oklahoma*, 380pp.
5. Grell, G., J. Dudhia and D. Stauffer, 1994: A description of the fifth-generation of the Penn State / NCAR mesoscale model (MM5). *NCAR Technical Note*, 138pp.

6. Orlanski, I., 1976: A simple boundary condition for unbounded hyperbolic flows. *J. Comput. Phys.*, **21**, 251–269.

7. Lin, Y. L., R. D. Farley and H. D. Orville, 1983: Bulk parameterization of the snow field in a cloud model. *J. Climate Appl. Meteor.*, **22**, 1065–1092.

8. Cotton, W. R., G. J. Tripoli, R. M. Rauber and E. A. Mulvihill, 1986: Numerical simulation of the effects of varying ice crystal nucleation rates and aggregation processes on orographic snowfall. *J. Climate Appl. Meteor.*, **25**, 1658–1680.

9. Murakami, M., 1990: Numerical modeling of dynamical and microphysical evolution of an isolated convective cloud — The 19 July 1981 CCOPE cloud. *J. Meteor. Soc. Japan*, **68**, 107–128.

10. Murakami, M., T. L. Clark and W. D. Hall 1994: Numerical simulations of convective snow clouds over the Sea of Japan; Two-dimensional simulations of mixed layer development and convective snow cloud formation. *J. Meteor. Soc. Japan*, **72**, 43–62.

11. Smagorinsky, J., 1963: General circulation experiments with the primitive equations. I. The basic experiment. *Mon. Wea. Rev.*, **91**, 99–164.

12. Louis, J. F., M. Tiedtke and J. F. Geleyn, 1981: A short history of the operational PBL parameterization at ECMWF. *Workshop on Planetary Boundary Layer Parameterization 25–27 Nov. 1981*, 59–79.

13. Klaassen, G. P. and W. R. Peltier, 1985: The evolution of finite amplitude Kelvin-Helmholtz billows in two spatial dimensions. *J. Atmos. Sci.*, **42**, 1321–1339.

14. Wilhelmson, R. B., and J. B. Klemp, 1978: A numerical study of storm splitting that leads to long-lived storms. *J. Atmos. Sci.*, **35**, 1974–1986.

15. Weisman, M. L., and J. B. Klemp, 1982: The dependence of numerically simulated convective storms on vertical wind shear and buoyancy. *Mon. Wea. Rev.*, **110**, 504–520.

16. Weisman, M. L., and J. B. Klemp, 1984: The structure and classification of numerically simulated convective storms in directionally varying wind shears. *Mon. Wea. Rev.*, **112**, 2478–2498.

17. Klemp, J. B., and R. Rotunno, 1983: A study of the tornadic region within a supercell thunderstorm. *J. Atmos. Sci.*, **40**, 359–377.

18. Wicker, L. J., and R. B. Wilhelmson, 1995: Simulation and analysis of tornado development and decay within a three-dimensional supercell thunderstorm. *J. Atmos. Sci.*, **52**, 2675–2703.

19. Grasso, L. D., and W. R. Cotton, 1995: Numerical simulation of a tornado vortex. *J. Atmos. Sci.*, **52**, 1192–1203.

Routing Mechanism for Static Load Balancing in a Partitioned Computer System with a Fully Connected Network

Hitoshi Oi[1] and Bing-rung Tsai[2]

[1] Dept. of Computer Science & Engineering
Florida Atlantic University
Boca Raton, FL 33431
hitoshi@cse.fau.edu
[2] Kuokoa Networks, Inc.
Santa Clara, CA 95054
btsai@kuokoa.com

Abstract. System partitioning provides the users of high-performance parallel servers with the flexibility in resource allocation and dynamic reconfiguration as well as fault isolation. However, the bandwidth of links that connect different domains can be wasted while links within the same domains are congested. In this paper, we present a routing mechanism that can utilize the bandwidth of otherwise unused links to balance the message traffic and lead to lower message latencies for the latency-sensitive transactions. The performance of the proposed routing mechanism was studied using an analytical model with on-line transaction processing type workload parameters. The results indicated the proposed routing mechanism reduced the congestion on the direct paths significantly and lowered the queuing delay for the links. For example, when a 4-cluster system with a fully connected network with the bandwidth of 3.2GB/s per link is partitioned into two 2-cluster domains, the queuing delay was reduced from 53ns to 37ns and resulted in the improvement of CPI by 2%.

Keywords: System partitioning, distributed shared memory, interconnection network, message routing.

1 Introduction

Cache coherent non-uniform memory architecture (CC-NUMA) has the advantages of both the shared memory space of SMPs and the scalability of loosely-coupled multi-computers. Till mid-90s, CC-NUMA multiprocessors were mainly used for engineering or scientific applications. Nowadays, they are used for running different commercial applications, such as on-line transaction processing (OLTP), decision support systems (DSS), or web-servers, various operating systems. Each system runs different applications time to time, and the required system resources, such as processors, memory, storage devices, and network, also vary.

H. Zima et al. (Eds.): ISHPC 2002, LNCS 2327, pp. 260–270, 2002.

System partitioning enables users to divide a system into several domains. Each domain behaves as a separate system, where users can run different applications with a different operating system. The resource allocation for each domain can be changed without shutting down the system or physical re-configurations (such as reconnecting cables) [1,2,3,4]. In a CC-NUMA multiprocessor, memory modules are physically distributed among nodes and cache coherence is maintained by sending messages between nodes over the interconnect network. When a CC-NUMA multiprocessor is partitioned into domains, message traffic are routed through the links that connect nodes belonging to the same domain. Thus, the bandwidth of the other links (those connecting nodes belonging to different domains) are unused.

In this paper, we present a static routing mechanism that can utilize the bandwidth of such unused links to balance the message traffic and lead to lower message latencies. The proposed routing mechanism classifies message traffic into two categories, latency sensitive and non-latency sensitive transactions. The latter transactions are routed through the indirect paths to balance the message traffic on the direct and indirect paths.

This paper is organized as follows. In Section 2, the proposed routing mechanism is described. In Section 3, the performance advantage of the proposed routing mechanism is evaluated. Section 4 discusses related work and the conclusion is provided in Section 5.

2 Routing Mechanism

In this section, we describe the proposed routing mechanism and its operations for the different transaction categories. A fully interconnected three cluster system is shown in Fig. 1. Each cluster has a CPU node, an I/O node and a router. From a router, there are two links that connect the cluster to others. Thus, in this example, the router is a 4×4 crossbar switch. Small numbers inside the square of the router indicate the ports to which nodes and links are connected. Note that this configuration is simplified in terms of number of nodes and hierarchical level for easier explanation. In Section 3, we will use more realistic configurations.

Consider a case where the system in Fig. 1 is partitioned into two domains: Domain 0 that consists of clusters A and B, and Domain 1 that consists of Cluster C alone. Each domain behaves as an independent CC-NUMA system and has its own address space which is only accessible to the nodes within the domain. Thus, the link between Clusters A and B (the link drawn with a bold line in Fig. 1) is heavily used while other links are not used at all. In the proposed routing mechanism, these unused links are utilized to balance the message traffic in the system.

In the proposed scheme, all the transactions are classified into two categories: either I/O transactions or processor-memory transactions (or non-I/O transactions for short). I/O transactions are less sensitive to the latency. For example, when a file is opened for an application, direct memory access (DMA) trans-

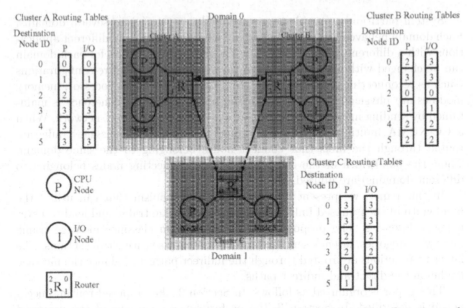

Fig. 1. A Fully Interconnected 3 Cluster System Partitioned into Two Domains and Routing Tables

actions are initiated by the I/O node so that future accesses to the file by the application will hit at the memory buffer. On the other hand, cache fill transactions initiated by cache misses of processors are latency sensitive. The processor could be blocked until the cache fill transaction is completed[1]. By routing I/O transactions through the unused links, it is possible to reduce traffic on the links that connects cluster within the same domain and this leads to a lower latency of non-I/O (i. e. processors-memory) transactions that are latency-critical.

In this mechanism, each cluster has two routing tables, one for I/O and another for non-I/O transactions. In a message packet, there is an I/O bit indicating whether the originator of the message is an I/O node in the cluster or not. This I/O bit is used at the router to choose one of routing tables. Each node in the system (either CPU or I/O node) has a unique Node ID as shown in the table in Fig. 1 and entries in the routing tables are indexed by the Node ID.

Below, operations of the proposed routing mechanism are described with four different transactions: memory access from a processor and its response, and I/O access from a processor and its response.

2.1   Processor-Memory Transactions

A processor in Node 0 accesses a memory module in Node 2 (Processor-Memory Transaction in Fig. 2). First, the processor sends a request message to Node 2:

---

[1] modern microprocessors employ techniques to overlap cache-fill and other operations, such as nonblocking cache, or multi-threading. However, blocking still occurs if the processor reaches to the point where it uses the data in the missed cache block

1. Node 0 sends a request message with $I/O = 0$.
2. Since $I/O = 0$, Non-I/O Routing Table (labeled "P" in the table in Fig. 1) is used at the router in Cluster A. The entry corresponding to Node 2 (destination of the request message) indicates that the message is to be routed to the Port 2 of the router (to the cable directly connected to Cluster B).
3. The message is transmitted over the cable and reaches at the router in Cluster B. Since this message is not originated from an I/O node in the cluster, its $I/O = 0$. Thus, Non-I/O Routing Table is used.
4. Non-I/O table indicates that the message is to be routed to Port 0, which is the destination of the message (Node 2).

After the memory access is completed, Node 2 sends a response message back to the processor in Node 0.

1. Node 2 sends a response message with $I/O = 0$.
2. Since $I/O = 0$, Non-I/O Routing Table is used at the router in Cluster B. The entry corresponding to Node 0 (destination of the response message) indicates that the message is to be routed to the Port 2 of the router (to the cable directly connected to Cluster A).
3. The message is transmitted over the cable and reaches at the router in Cluster A. Since this message is not originated from an I/O node in the cluster, its $I/O = 0$. Thus, Non-I/O Routing Table is used.
4. Non-I/O table indicates that the message is to be routed to Port 0, which is the destination of the response message (Node 2).

2.2 Processor-I/O Transactions

A processor in Node 0 accesses an I/O device in Node 3 (I/O Transaction in Fig. 2). Similar to the previous example, the processor first sends a request message to Node 3:

1. Node 0 sends a request message with $I/O = 0$ since it is a CPU node.
2. Since $I/O = 0$, Non-I/O Routing Table is used at the router in Cluster A. The entry corresponding to Node 3 (destination of the request message) indicates that the message is to be routed to the Port 3 of the router (to the cable connected to Cluster C).
3. The message is transmitted over the cable and reaches at the router in Cluster C. Since this message is not originated from an I/O node in the cluster, its $I/O = 0$. Thus, Non-I/O Routing Table is used.
4. Non-I/O table indicates that the message is to be routed to Port 2, to the cable connected to Cluster C
5. The message is transmitted over the cable and reaches at the router in Cluster B. Since this message is not originated from an I/O node in the cluster, its $I/O = 0$. Thus, Non-I/O Routing Table is used.
6. Non-I/O table indicates that the message is to be routed to Port 1, which is the destination of the request message (Node 3).

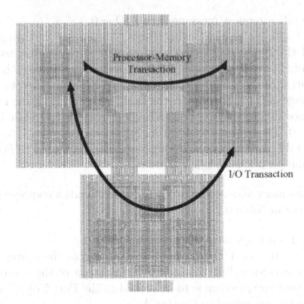

Fig. 2. Message Flow for Processor-Memory and I/O Transactions

After the I/O access is completed, Node 3 sends a response message back to the processor in Node 0.

1. Node 3 sends a response message with $I/O = 1$ (since Node 3 is an I/O node).
2. Since $I/O = 1$, I/O Routing Table (labeled "I/O" in the table in Fig. 1) is used at the router in Cluster B. The entry corresponding to Node 0 (destination of the response message) indicates that the message is to be routed to the Port 3 of the router (to the cable connected to Cluster C).
3. The message is transmitted over the cable and reaches at the router in Cluster C. Since this message is not originated from an I/O node in the cluster, its $I/O = 0$. Thus, Non-I/O Routing Table is used.
4. Non-I/O table indicates that the message is routed to Port 3, to the cable connected to Cluster A.
5. The message is transmitted over the cable and reaches at the router in Cluster A. Since this message is not originated from an I/O node in the cluster, its $I/O = 0$. Thus, Non-I/O Routing Table is used.
6. Non-I/O table indicates that the message is routed to Port 0, which is the destination of the response message (Node 0).

3   Performance Evaluation

In this section, we study the effectiveness of the proposed routing mechanism using an analytical model. First, we describe the target system configurations. Then the methodology of the evaluation and the results follow.

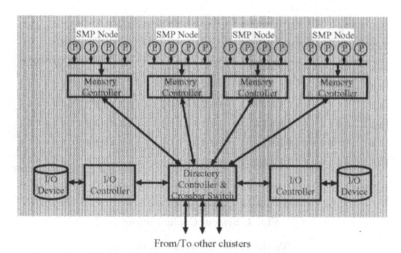

From/To other clusters

Fig. 3. Cluster Architecture

3.1   Architecture Model

We assume hierarchical CC-NUMA system configurations that can be partitioned into multiple domains. The system consists of four of clusters and clusters are fully connected by the point-to-point links. A cluster consists of four SMP nodes, two I/O nodes, a directory controller, a crossbar switch that connects the nodes within the cluster as well as the inter-cluster network. An SMP node has four processors, a memory controller and memory modules (Fig. 3).

We evaluate the performance of the proposed routing mechanism in two configurations in Fig. 4: there are four clusters in the system and they are partitioned into two domains. In Fig. 4 (a), there are two domains: one consisting of clusters 0 and 1 and another of clusters 2 and 3 Links between clusters 0 and 1 and clusters 2 and 3 (drawn with bold lines in the figure) are used for processor-memory transactions and other links are used for I/O transaction. Similarly, in Fig. 4 (b), clusters 0 to 2 form a domain while cluster 3 is used as a single cluster domain. Table 1 lists key system parameters, which are chosen in consideration of the current implementation technologies. Memory access latencies are for read accesses to unmodified memory blocks. Using an analytical model, which is described in the next section, the utilization and queuing delay of the links connecting clusters and clock per instruction (CPI) are derived and used as performance indices.

3.2   Performance Model

We developed an analytical model based on the one used in our previous project at HAL Computer Systems [5].

The primary index of the performance is clock per instruction, CPI which is obtained as follows:

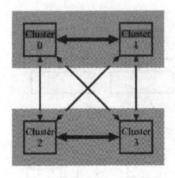 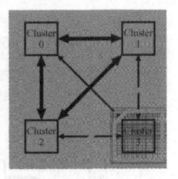

(a) 4 Clusters, 2 Domains (2 + 2)   (b) 4 Clusters, 2 Domains (3 + 1)

Fig. 4. System Configurations

Table 1. Key System Parameters

Parameter	Value
Proc/SMP Node	4
Proc Clock	1.1GHz
Bus Clock	200MHz
Local Memory	250ns
Remote Memory (same cluster)	385ns
Remote Memory (other cluster)	550ns
Link Bandwidth	3.2, 3.6, 4GB/s

$$CPI = CPI_{perfectL2} + f_b \times mpi \times cpm \tag{1}$$

where,

$CPI_{perfectL2}$ is the processor CPI with perfect (no miss) L2 cache,
f_b is a *blocking factor*, which reflects the effectiveness of latency hiding techniques in both hardware and software,
mpi is the average L2 miss rate (miss per committed instruction),
cpm is the average miss penalty in clock cycles.

To determine the value of cpm, L2 miss is classified into various transaction scenarios by the access mode (read/write), state of the main memory, state of other L2 caches, locations of the accessed block's home and the sharer or owner L2 cache. Each transaction provides the usage of resources in the system, such as memory controller, directory controller, shared bus in the SMP node, links between clusters.

Next, we derive cpm in (1), which is given by

$$cpm = cpm_{NoContention} + QueingDelay \tag{2}$$

where $cpm_{NoContention}$ is the L2 cache miss penalty when there is no contention on the system resources, while $QueingDelay$ is the delay caused by the contention on the system resources. Note that $QueingDelay$ is a function of CPI:

Table 2. Configuration (a) Result

	Link Bandwidth					
	3.2GB/S		3.6GB/S		4.0GB/S	
	DP	Prop	DP	Prop	DP	Prop
Link Utilization	82.8%	77.7%	76.2%	71.0%	68.7%	63.7%
Queuing Delay	53.0ns	36.8ns	31.8ns	23.3ns	19.4ns	14.8ns
CPI	4.155	4.083	4.061	4.023	4.006	3.986

the lower CPI, the more frequently resources are used and higher contention. Thus

$$CPI = CPI_{perfectL2} + f_b \times mpi \times cpm(CPI). \qquad (3)$$

The queuing delay of each resource is approximated by M/D/1 formula. The probability of each transaction scenario was chosen from our experiences in the past projects together with an OLTP-type workload [6,7]. Also, the following assumption are used:

- Memory modules and I/O devices are uniformly distributed.
- The ratio of read and write access is $2 : 1$.
- DMA I/O traffic is assumed $100MB/(20000Transaction/min)$

With these workload assumptions, the derivation of (3) is iterated until it is converged.

3.3   Results

Configuration (a) in Fig. 4 is a 4-cluster system partitioned into two 2-cluster domains . Table 2 shows the link utilization and queuing delay on the direct paths (drawn with bold lines in Fig. 4) and CPI, derived from the analytical model. In the table, DP columns are the values when only the links on the direct paths are used while $Prop$ are for the proposed scheme. The proposed routing scheme reduced the link utilization significantly and resulted in the lower queuing delay. At the link bandwidth of 3.2GB/s, the queuing delay was reduced by 30% and the CPI was improved by 2%. At the higher bandwidth values, 3.6GB/s and 4.0GB/s, the queuing delay was reduced by 25% and 24%, and CPI was improved by 1% and 0.5%, respectively.

Configuration (b) in Fig. 4 is a 4-cluster system partitioned into two (3 and 1-cluster) domains. Table 3 shows the link utilization, queuing delay, and CPI values of the 3-cluster domain derived from the analytical model. The link bandwidth of 3.2GB/S was not sufficient for the high transaction rate assumed in the workload. The link was saturated (link utilization exceeded 100% in the analytical model) in both DP and $Prop$. In these cases, the system behavior is non-linear and the performance cannot be predicted by the analytical mode. At the bandwidth of 3.6GB/s, however, DP and $Prop$ behaved differently. with DP the links are still saturated while with $Prop$ the link utilization is far below saturation. At the bandwidth of 4.0GB/s, the queuing delay was reduced by 30% and CPI was improved by 1.5%.

Table 3. Configuration (b) Result

	Link Bandwidth					
	3.2GB/S		3.6GB/S		4.0GB/S	
	DP	Prop	DP	Prop	DP	Prop
Link Utilization	sat	sat	sat	84.1%	81.9%	76.9%
Queuing Delay	N/A	N/A	N/A	45.3ns	35.8ns	25.4ns
CPI	N/A	N/A	N/A	4.760	4.696	4.626

4   Related Work

Modern high-performance NUMA systems with system partitioning feature include HP Superdome [1], NEC AZUSA [2], Compaq AlphaServer [3], Unisys Enterprise Server ES 7000 [4]. In particular, HP Superdome has a similar structure as the architectural mode we assumed in this paper: a number of processors form a cell, and cells are fully connected by the network of crossbar switches. However, as to the best of our knowledge, none of the existing systems including above can utilize the inter-domain links to alleviate the traffic on the intra-domain links.

In addition to the normal interconnection network for the inter-processor communications, NUMA-Q of IBM [8] has a dedicated network of fiber channel switches for connecting I/O devices and processor nodes, called Multipath I/O. In principle, Multipath I/O and the proposed routing mechanism are the same approach in the sense that both try to reduce the traffic on the inter-processor network by routing I/O traffic to alternative paths. The difference is in the cost-effectiveness: while NUMA-Q invested in the extra hardware (fiber channel switches) for the I/O traffic, the proposed scheme utilizes the bandwidth of unused links resulted from the system partitioning.

In a parallel computer system with the message-passing communication model, which has a much higher inter-processor communication latency than the shared memory model, there are many sophisticated dynamic (adaptive) routing algorithms [9]. They are, however, rarely used for the interconnection network of NUMA systems. Sophisticated adaptive routing algorithms inherently increase the routing latencies. This conflicts with one of the objectives in the interconnection network of NUMA systems, which is to reduce the remote access latency.

In this paper, we assumed the indirect paths were to be used for I/O transactions, which are normally implemented as uncached DMA operations [10]. If the proposed routing scheme is used for cache coherent memory transactions, the race conditions caused by the multiple paths between nodes must be solve [11].

5   Conclusion

Most of current commercial parallel servers have cluster structures and they can be partitioned into several domains to better utilize the system resources to fit users need as well as fault isolation. In this paper, we have presented a static routing scheme that balances the message traffic in a partitioned system. To minimize the latency of remote memory accesses by processors, it is preferable to route messages over the links that directly connect clusters in the same domain. However, this routing scheme leads to an unbalance of the message traffic: the links connecting clusters in the same domain are heavily congested while other links are not used. As an initial study, we have evaluated the performance of the proposed scheme with an analytical model using workload parameters that modeled an OLTP-type application.

In addition to balancing the message traffic, the proposed scheme has several advantages. The hardware resource required for its implementation is small and complexity is fairly low. Thus, it is expected that the effect on the latency for the transaction on the direct paths is minimum. Another advantage is, since the proposed scheme is static, it does not require to keep track of strayed transactions, which is a common problem in more sophisticated dynamically adaptive routing mechanisms.

The proposed routing scheme can be applied to a system in which each cluster (in the terminology used in this paper) is fully connected to every other cluster with a dedicated link. For example, Superdome of Hewlett Packet [1] is met with this structural requirement and the proposed scheme can be applied to Superdome.

The topics of further research include studying dynamic network behavior using execution driven simulations or other methods, use of more various workload such as web-servers or decision support systems (DSS). In this paper, I/O transactions represented latency-insensitive messages. The proposed scheme can be used for other types of transactions, such as prefetching memory blocks into processors' caches, writing back replaced cache blocks to the main memory, and the effectiveness of the proposed scheme for these transaction should also be studied.

Acknowledgment

The proposed routing mechanism was developed in the course of the ASURA project which was conducted at HAL Computer Systems Inc., Campbell, California. The authors would like to appreciate the encouragement, discussion, and valuable comments from the colleagues in the project.

References

1. "HP Virtual Partitions", *a white paper from Hewlett-Packard Company*, September 2000
2. Fumio Aono and Masayuki Kimura, "The Azusa 16-way Itanium Server", *IEEE Micro*, Vol. 20, Issue. 5, 54–60, Sep-Oct 2000.
3. "AlphaServer GS80, GS160, and GS320 Systems Technical Summary", Compaq.
4. "Cellular Multiprocessing Shared Memory", White Paper, Unisys, September 2000.
5. "Distributed Shared Memory System Performance", HAL internal document, February 1998.

6. Russell M. Clapp, "STinG Revisited: Performance of Commercial Database Benchmarks on a CC-NUMA Computer System", in *Proceedings of Fourth Workshop on Computer Architecture Evaluation using Commercial Workloads*, Monterrey, Mexico, January 2001.
7. Don DeSota, "Characterization of I/O for TPC-C and TPC-H workloads", in *Proceedings of Fourth Workshop on Computer Architecture Evaluation using Commercial Workloads*, Monterrey, Mexico, January 2001.
8. "The IBM NUMA-Q enterprise server architecture", IBM, January 2000.
9. M. Boari, A. Corradi, C. Stefanelli and L. Leonardi, "Adaptive Routing for Dynamic Applications in Massively Parallel Architectures", *IEEE Parallel & Distributed Technology: Systems & Applications*, Vol. 3 Issue 1, Spring 1995, pp61–74.
10. D. Lenoski and W.-D. Weber, "Scalable Shared-Memory Multiprocessing", Morgan Kaufmann Publishers, 1995.
11. D. Dai and D. Panda, "Exploiting the benefits multiple-path network in DSM systems: architectural alternatives and performance evaluation", *IEEE Trans. on Computers*, Vol. 48 Issue 2 , Feb. 1999 pp236–244.

Studying New Ways for Improving Adaptive History Length Branch Predictors

Ayose Falcón[1], Oliverio J. Santana[1], Pedro Medina[2], Enrique Fernández[2],
Alex Ramírez[1], and Mateo Valero[1]

[1] Dpto. de Arquitectura de Computadores, Universidad Politécnica de Cataluña
{afalcon,osantana,aramirez,mateo}@ac.upc.es
[2] Dpto. de Informática y Sistemas, Universidad de Las Palmas de Gran Canaria
{pmedina,efernandez}@dis.ulpgc.es

Abstract. Pipeline stalls due to branches limit processor performance significantly. This paper provides an in depth evaluation of Dynamic History Length Fitting, a technique that changes the history length of a two-level branch predictor during the execution, trying to adapt to its different phases. We analyse the behaviour of DHLF compared with fixed history length gshare predictors, and contribute showing two factors that explain DHLF behaviour: *Opportunity Cost* and *Warm-up Cost*. Additionally, we evaluate the use of profiling for detecting future improvements. Using this information, we show that new heuristics that minimise both opportunity cost and warm-up cost could outperform significantly current variable history length techniques. Especially at program start-up, where the algorithm tries to learn the behaviour of the program to better predict future branches, the use of profiling reduces considerably the cost produced by continuous history length changes.

Keywords: branch prediction, dynamic history length, warm-up, opportunity cost.

1 Introduction

Presence of control hazards in the processor pipeline can significantly reduce ILP in programs execution because the outcome of a branch is not known until several cycles after it has been fetched. As modern processors increase the number of pipeline stages and the number of instructions on the fly, it becomes more critical to overcome the problem caused by a control dependency, since the number of stall cycles increases.

In this paper we focus on two-level branch predictors [12,5,8,4,6] as being one of the most efficient solutions for solving this problem.

Juan et al. [3] argued that the use of a global fixed history length during the execution of a program can hinder the accuracy of two-level branch predictors, and proposed an algorithm (DHLF) for dynamically fitting the history length according to different phases of program execution.

The goals of this paper are: (i) to analyse the behaviour of the dynamic history length fitting algorithm, and (ii) to define new variations in the method

H. Zima et al. (Eds.): ISHPC 2002, LNCS 2327, pp. 271–280, 2002.
© Springer-Verlag Berlin Heidelberg 2002

that could lead to more accurate predictions, either by adopting new factors that determine when it is necessary a history length change or by using profile feedback. Our purpose is to provide new data about the behaviour of this kind of algorithms, analysing their advantages and drawbacks. Although we will deal with DHLF, our study can be extended to oth er variable history length methods. Along the paper, we will study not only the misprediction rates that different aproximations can achieve, but also their effect on IPC (Instructions Per Cycle).

The rest of this paper is organised as follows: In section 2 we explain other related work; section 3 motivates our work, showing the potential improvement of using variable history length; section 4 analyses DHLF, studying the basis of its way of operating; in section 5 we study how DHLF could be enhanced using profiling and present the results obtained. Finally, in section 6 we conclude this paper and present gu idelines for future work.

2   Related Work

Most of the predictors used in current processors are based on the two-level adaptive branch predictor proposed by Yeh & Patt [12,11]. In particular, in this paper we will use the *gshare* predictor by McFarling [5]. The low-order branch instruction address bits and global history bits (*Branch History Register*) are xor-ed together to form an index to the *Pattern History Table*, thus reducing interferences that appear when using a global second level table.

Two main characteristics affect the accuracy of a *gshare* predictor: the second level table size and the number of bits taken to xor the branch address. Two level table sizes of 2^{12}–2^{16} are integrated in current microprocessors and designers usually select history lengths of maximum size.

Chang et al. [1] questioned the assumption that correlation among branches is always benefitial. They showed that the accuracy of two-level branch predictors tends to decrease when increasing the first level register length for branches that are mostly taken or not-taken. Tarlescu et al. [10] also proposed variations in the gshare predictor, assigning different fixed history lengths to different static branches using profiling information.

Juan et al. [3] propose an algorithm (DHLF) for dynamically adjusting the history length to the portion of code considered. During an interval of a number of *step* branches the history length remains unchanged, and after this interval the algorithm can change to a better history length. When the history length changes, the algorithm adds a warm-up interval to adapt to the new change.

Figure 1.a shows prediction rates when using a gshare branch predictor with history lengths ranging from 0 to 16 and a PHT with 2^{16} entries, for each SPECint95. The last point of each curve corresponds to the results obtained with the DHLF algorithm. DHLF monitors branch mispredictions produced during the execution of the program (in intervals of *step* branches) and changes the history length to the best one according to the past behaviour of the program. This is based on the observation that history lengths that can lead to high prediction rates in one benchmark, can lead to low prediction rates in others. For example,

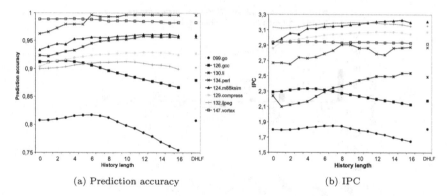

Fig. 1. Prediction accuracy and IPC of gshares with fixed history length vs. DHLF

the use of the appropriate history length compared to the worst one can lead to speedups of 20.82% for *li*, 12.36% for *go* or 10.40% for *m88ksim* (see figure 1.b).

Stark et al. [9] apply the same DHLF idea to path-based branch predictors using profiling to determine which is the best path length for each static branch.

For more information, an extended version of this paper can be found in [2].

3   Motivation

Figure 2 shows the speedup obtained by the DHLF mechanism over several fixed history length gshares, and over an ideal variable history length gshare. The first bar corresponds to the speedup obtained with a DHLF over a gshare with the longest history length (16); the second bar, over a gshare with the shortest history length (0, like a bimodal predictor [7]); the third bar represents the speedup of DHLF over the fixed history length gshare that achieves the best IPC (accordin g to the results of figure 1.b). In some cases the use of the longest history does guide to the best results, and in other cases the best performance is obtained with the shortest history length. In all cases, compared with the worst gshare, DHLF can improve the performance significantly, as in *li*, where we obtain a 18.5% speedup. But compared with the fourth bar (against the best fixed gshare) there is a negative speedup in all benchmarks, up to 6.6% in the case of *gcc*. That confirms that the election of the optimum history length when using a fixed history length gshare is a critical factor and that the choice is different for each benchmark.

The last bar of figure 2 shows the speedup obtained by DHLF over an ideal variable history length gshare. This gshare works as an oracle DHLF which knows the best history length in each interval and uses an independent PHT for each history length. This involves that no aliasing will appear due to history changes. The data of this fourth bar shows us which is the maximum potential that can be achieved. DHLF obtains a −4.2% speedup in average, and a peak of −15% in *gcc*. Compa red with the best gshare, the ideal predictor achieves a 2.6% speedup in average, with peaks of 11% in *gcc* and 8% in *perl*.

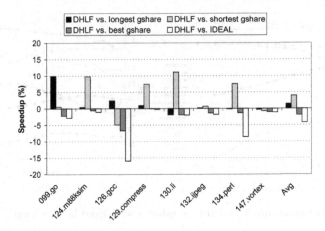

Fig. 2. Speedup of DHLF compared with gshare with the longest, shortest and best history length, and with an ideal variable history length gshare

Table 1. Baseline configuration

Processor core		Memory Hierarchy	
Inst. Fetch Queue	8 insts.	L1 data cache	64 KB, 2-way (LRU), 1-cycle lat.
Fetch width	8 insts./cycle	L1 instruction cache	64 KB, 2-way (LRU), 1-cycle lat.
Issue width	8 insts./cycle	L2 cache	Unified, 2MB, 4-way (LRU),
Decode width	8 insts./cycle		4-cycle lat.
Commit width	8 insts./cycle	Memory	75-cycles lat.
Register Update Unit	128 entries	L1 cache ports	2
Load/Store queue	64 entries	Branch prediction	
ALU	8 int + 8 fp	BTB	2048-entry, 4-way
Multiplier/Divider	2 int + 2 fp	Return Address Stack	8 entries

3.1 Simulation Environment

For the purpose of our study, we have selected SPECint95 benchmarks, compiled on a DEC Alpha AXP-21264 using Compaq's C compiler. All benchmarks were run until completion using a reduced input. The timing simulator used is derived from the SimpleScalar 3.0a Toolkit.

The processor configuration used in our simulations is shown in table 1. In all cases, the BHR length determines the PHT size: for N bits of history, 2^N PHT entries are allocated. As in [3], we will use a DHLF step of 16000 branches.

4   Analysis of History Length Changes Effects

4.1   Warm-Up and Opportunity Cost: When Is Good to Change the History Length?

Each time the history length changes many mispredictions are introduced because the mapping of branches in PHT entries change radically, and a branch that was previously predicted by a particular 2 bit counter will be now predicted by another counter. This warm-up period involves a large payment and causes the number of mispredictions in the next interval to grow considerably.

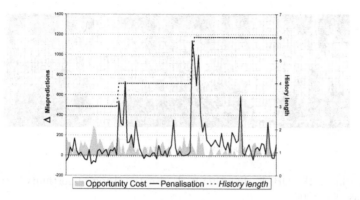

Fig. 3. Effect of history length changes in *penalisation* and *opportunity cost* in **go**

To alleviate this problem, DHLF stops counting mispredictions during an interval to allow the predictor to adapt to the new history length. During this warm-up interval, the misprediction counter remains unchanged and after this period begins the count of mispredictions of the real interval.

The control algorithm will be the responsible for deciding which will be the history length to use in the next interval. For evaluating the cost of these decisions, we define the following terms:

- *Opportunity Cost* (OC): Number of additional mispredictions payed due to being in the history length B instead of being in the best history length C.
- *Penalisation* (P): Number of mispredictions payed due to change from history length A to history length B [$\equiv$ *warm-up cost*].
- *Penalised Opportunity Cost* (POC): Number of mispredictions obtained with the current history length, due to *opportunity cost* and *penalisation* (POC = P + OC).

Using these terms, a fixed history length gshare will have a null penalisation and its penalised opportunity cost will be due to opportunity cost. At the contrary, if we change the history length continuously to the best one, the opportunity cost will be null but the penalisation paid on each history change will be the responsible of the penalised opportunity cost. The best solution will be an intermediate proposal that minimises penalised opportunity cost.

Figure 3 shows intervals 2100 to 2200 during the execution of *go*. The dotted line represents the history length used. Note that after a history length change the opportunity cost tends to decrease because the algorithm is reaching the optimum state. The penalisation (due to warm-up) increases significantly after the history length change but tends to decrease in the next intervals because two-bit counters begin to adapt.

The effect of the warm-up reaches in average 20 intervals, which means that after 320,000 branches a PHT of 2^{16} should have learned the new situation. With a smaller PHT, the effect reaches less intervals. However, the penalisation

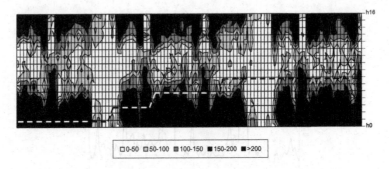

Fig. 4. Plot of the DHLF algorithm path over the **ijpeg** opportunity cost 2D graph (intervals 900–1000)

effect is larger if the history length increment or decrement is bigger than 1 unit, because the number of PHT entries affected after a change is larger and the warm-up requires more time.

Note that the best results of opportunity cost are obtained with history length 6, the best for *go* according to figure 1.

4.2   DHLF Path Analysis

In this subsection we will study the behaviour of DHLF, comparing the decisions taken by the algorithm about which history length must be used in each interval, and the errors that will obtain gshare predictors with fixed history length in the same intervals.

The methodology used is the following: first of all, we execute all the program using the DHLF algorithm, and take a record of which history length has been used in each interval. Later we execute 17 gshare predictors, each one with a fixed history length from 0 to 16, and we record the number of mispredictions obtained in each interval. With this information, we plot the 2D opportunity cost values (figures 4 and 5) and plot over them the evolution of the history length used b y the DHLF algorithm. The X axis represents intervals between the number displayed below the graphs (100 intervals ≡ 1,600,000 branches). The Y axis represents the history lengths used by each gshare, ranging from 0 to 16. The level of grey of the zones represents the opportunity cost and the dotted line represents the time evolution of the DHLF algorithm, i.e., which history length is being used in each interval.

Looking at figures 4 and 5 we can evaluate how well is performing the control algorithm of DHLF. Further details of this analysis can be found in [2].

With these graphics we have tried to show the behaviour of DHLF during the execution. From this data, we can conclude that:

- During the execution of the code there are zones very difficult-to-predict in which choosing the history length that fits better becomes critical.
- In some cases DHLF behaves well, adapting the history length to the changes in the code and trying to avoid the black and dark grey zones we show above.

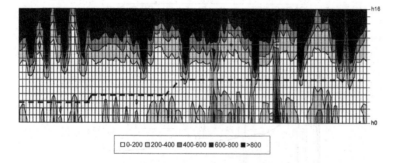

Fig. 5. Plot of the DHLF algorithm path over the **go** opportunity cost 2D graph (intervals 2100–2200)

- In other cases DHLF does not behave well, taking a lot of intervals to reach an optimum zone, or not realising that it is crossing a bad zone.
- The heuristics used by DHLF, based on mispredictions obtained in the previous parts of the code, could be improved using more information.

5   Algorithm Feedback

In previous sections we have shown why DHLF works, and the reasons of its good performance. However, we consider that results obtained in some programs under some situations can still be improved. One of the reasons of the low prediction rate in some intervals during the execution of the code is due to the DHLF mechanism, which tries to learn the past to predict the future. But sometimes, the future does not behave in the same way, especially in programs with very different phases, and it causes the DHLF al gorithm to obtain too many mispredictions due to a high opportunity cost. Other times, the causes of mispredictions come from continuous changes in the history length, specially in programs with very different and very difficult-to-predict phases, because the warm-up time until the 2 bit counters become ready to predict is too long. In these cases, penalisation cost paid overcomes opportunity cost saved.

For solving this, we have used feedback data extracted from previous complete executions of the program. Our first test has been feeding the DHLF algorithm with a record of the history lengths that had the best prediction rate in each interval when considering executions of the individual gshares. From the results of figure 2 ('*DHLF vs. IDEAL*' bar) we know that an ideal variable gshare can achieve great IPC values, and now we study the gain obtained when applying this history le ngths record in a DHLF predictor with an only PHT.

Figure 6.a shows the results of the normal DHLF algorithm in terms of opportunity cost and penalisation. We use an accumulative value (both opportunity cost and penalisation are divided by the number of branches taken until the current interval) for a better view. The first intervals, that corresponds to the first phase of the execution, show a penalised opportunity cost that reaches the

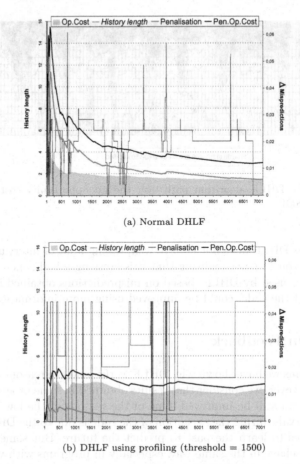

(a) Normal DHLF

(b) DHLF using profiling (threshold = 1500)

Fig. 6. Evolution of Penalisation, Opportunity Cost and Penalised Opportunity Cost (*accumulated form*) in *go*

6%. In that zone the cost due to penalisation is much larger than the opportunity cost. Looking at this gr aphic it is clear that improving the performance of this zone we could improve the performance of the overall execution. Taking the total execution, the penalised opportunity cost has a value near to 1.2%, to which contribute almost equally the opportunity cost and the penalisation.

Figure 6.b shows the results of employing the heuristic described below with distance 1500. This algorithm pays few due to the history length change, but pays a lot due to not being in the best history length each time —low penalisation, but high opportunity cost—. When using distance 1000 the opportunity cost decreases, but arises the penalisation due to the larger number of history length changes. Finally, using distance 400, the effect is the contrary: the payment due to opportunity cos t is smaller but there is more penalisation due to the history length changes. Note that using these heuristics we are outperforming normal DHLF in the first intervals, those related with the algorithm start-up (< 3%

Table 2. Use of profiling with different error distances (Prediction accuracy and IPC)

		d0	d200	d400	d600	d800	d1000	d1250	d1500	DHLF
099.go	PrAcc	0.79	0.80	0.80	0.80	0.81	0.81	0.80	0.81	0.81
	IPC	1.72	1.75	1.78	1.79	1.80	1.80	1.81	**1.81**	1.81
124.m88ksim	PrAcc	0.96	0.96	0.96	0.96	0.94	0.94	0.94	0.94	0.96
	IPC	3.18	3.14	**3.22**	**3.22**	2.92	2.92	2.92	2.92	3.21
126.gcc	PrAcc	0.89	0.90	0.91	0.91	0.91	0.91	0.91	0.91	0.88
	IPC	2.19	2.26	2.27	2.28	2.28	2.28	2.28	**2.29**	2.17
129.compress	PrAcc	0.93	0.93	0.93	0.93	0.92	0.90	0.90	0.90	0.93
	IPC	3.05	3.07	**3.08**	3.06	3.02	2.86	2.86	2.86	3.07
130.li	PrAcc	0.95	0.96	0.95	0.95	0.95	0.95	0.96	0.92	0.96
	IPC	2.41	2.50	2.34	2.29	2.29	2.29	**2.50**	2.24	2.48
132.ijpeg	PrAcc	0.90	0.91	0.92	0.90	0.90	0.90	0.90	0.90	0.90
	IPC	3.14	3.18	**3.20**	3.15	3.13	3.13	3.13	3.13	3.15
134.perl	PrAcc	0.99	0.99	0.99	0.96	0.96	0.96	0.96	0.96	0.99
	IPC	2.84	**2.88**	2.81	2.67	2.67	2.67	2.67	2.67	2.87
147.vortex	PrAcc	0.99	0.99	0.99	0.99	0.99	0.99	0.99	0.99	0.99
	IPC	2.91	**2.94**	**2.94**	**2.94**	**2.94**	**2.94**	**2.94**	**2.94**	2.92

in the worst case). The reason is that the cold start makes DHLF to traverse all the history length to begin PHT warm-up, and both opportunity cost and penalisation are too high.

Table 2 shows the results obtained when feeding DHLF with profiling information. For each SPEC, the first row corresponds to the prediction rate accuracy and the second row represents the IPC obtained. The last column shows the results of the normal DHLF for each SPEC. Of course, as we have mentioned during this paper, the effects of the history length change limit the overall performance. In such cases, we will try to process the history lengths record to get flater curves, chang ing the history length only when the difference in number of mispredictions between staying in the same history length and changing to the best in the same interval is over a determined threshold. We have considered 9 different thresholds: 0, 100, 200, 400, 600, 800, 1000, 1250 and 1500 mispredictions. Logically, greater threshold values suppose smoother curves and the algorithm will repeat many times the history length during several intervals, instead of changing. Our purpose is to minimize penalisation d ue to warm-up mispredictions without improving significantly mispredictions due to opportunity cost. The average speedup obtained is 1.2% and yields 5.53% in *gcc*.

6 Conclusions

This paper has examined the behaviour of DHLF technique, analysing the impact of changing or not the history length in the overall performance. Using the terms *opportunity cost* and *penalisation* we have analised the effect of the decisions taken by the DHLF algorithm. We have shown that the high penalisation cost limits dynamic fitting capability, so it is desirable to employ new heuristics that improve performance by avoiding quick history length variations.

Another contribution of this study is the use of profiling information for feeding back DHLF. We found that profile information can improve the bad decisions taken by DHLF in some zones of the execution, especially at the startup. This is a particular application that shows up that our approach is useful to improve dynamic history length mechanisms.

We have analised DHLF, but our study can be applied to other variable history length mechanisms. Future work includes new schemes that reduce the effect of warm-up, maintaining the adaptativity of changing the history length dynamically. In addition, we will study the influence of changing history length in other two-level schemes based on the factors analised in this paper, as well as the benefits of combining heuristics decisions and profiling information to help these methods to select the next history l ength to use.

Acknowledgments

This work has been supported by the Ministry of Education of Spain under contract TIC–2001–0995–C02–01, CEPBA and an Intel fellowship. A. Falcón is also supported by the Ministry of Education of Spain grant AP2000–3923. O.J. Santana is also supported by the Generalitat de Catalunya grant 2001FI–00724–APTIND. E. Fernández is supported by the Gobierno de Canarias.

References

1. P.-Y. Chang, E. Hao, T.-Y. Yeh and Y. N. Patt. Branch classification: a new mechanism for improving branch predictor performance. *Proceedings of the 27th Intl. Symp. on Microarchitecture*, pp 22–31, 1994.
2. A.Falcón, O. J. Santana, P. Medina, E. Fernández, A. Ramirez and M. Valero. Analysis of dynamic history length changes effect in two-level branch predictors. Technical Report UPC-DAC-2002, Universitat Politècnica de Catalunya, 2002.
3. T. Juan, S. Sanjeevan and J. J. Navarro. Dynamic history-length fitting: A third level of adaptivity for branch prediction. *Proceedings of the 25th Intl. Symp. on Computer Architecture*, pp 155–166, June 1998.
4. Ch.-Ch. Lee, I-Ch. K. Chen and T.r N. Mudge. The bi-mode branch predictor. *Proceedings of the 30th Intl. Symp. on Microarchitecture*, pp 4–13, December 1997.
5. S. McFarling. Combining branch predictors. TN-36, Compaq WRL, June 1993.
6. P. Michaud, A. Seznec and R. Uhlig. Trading conflict and capacity aliasing in conditional branch predictors. *Proceedings of the 24th Intl. Symp.on Computer Architecture*, pp 292–303, 1997.
7. J. E. Smith. A study of branch prediction strategies. *Proceedings of the 8th Intl. Symp. on Computer Architecture*, pp 135–148, 1981.
8. E. Sprangle, R. S. Chappell, M. Alsup and Y. N. Patt. The agree predictor: A mechanism for reducing negative branch history interference. *Proceedings of the 24th Intl. Symp. on Computer Architecture*, pp 284–291, 1997.
9. J. Stark, M. Evers and Y. N. Patt. Variable length path branch prediction. In *Proceedings of the 8th Intl. Conference on Architectural Support for Programming Languages and Operating Systems*, pp 170–179, San José, October 3–7, 1998.
10. M.-D. Tarlescu, K. B. Theobald and G. R. Gao. Elastic history buffer: A low-cost method to improve branch prediction accuracy. In *International Conference on Computer Design: VLSI in Computers and Processors (ICCD '97)*, pp 82–87, Washington - Brussels - Tokyo, October 1997.
11. T.-Y. Yeh and Y. N. Patt. A comparison of dynamic branch predictors that use two levels of branch history. *Proceedings of the 20th Intl. Symp. on Computer Architecture*, pp 257–266, 1993.
12. T.-Y. Yeh and Y. N. Patt. Two-level adaptive branch prediction. *Proceedings of the 24th Intl. Symp. on Microarchitecture*, pp 51–61, 1991.

Speculative Clustered Caches
for Clustered Processors

Dana S. Henry[1], Gabriel H. Loh[2], and Rahul Sami[2]

[1] Yale University, Department of Electrical Engineering
[2] Yale University, Department of Computer Science
New Haven, CT, 06520, USA

Abstract. Clustering is a technique for partitioning superscalar processor's execution resources to simultaneously allow for more in-flight instructions, wider issue width, and more aggressive clock speeds. As either the size of individual clusters or the total number of clusters increases, the distance to the first level data cache increases as well. Although clustering may expose more parallelism by allowing a greater number of instructions to be simultaneously analyzed and issued, the gains may be obliterated if the latencies to memory grow too large. We propose to augment each cluster with a small, fast, simple *Level Zero* (L0) data cache that is accessed in parallel with a traditional L1 data cache. The difference between our solution and other proposed caching techniques for clustered processors is that we do not support versioning or coherence. This may occasionally result in a load instruction that reads a stale value from the L0 cache, but the common case is a low latency hit in the L0 cache. Our simulation studies show that 4KB, 2-way set associative L0 caches provide a 6.5-12.3% IPC improvement over a wide range of processor configurations.

1 Introduction

The trend in modern superscalar uniprocessors is toward microarchitectures that extract more instruction level parallelism (ILP) at faster clock rates. To increase ILP, the processor execution cores use multiple functional units, buffers, and logic for dependency analysis to support a large number of instructions in various stages of execution. Such a large window of execution requires very large and complex circuits in traditional superscalar designs. Between the increasing logic complexity and wire delays, traditional processor microarchitectures consisting of unified instruction buffers and schedulers can not extract enough parallelism to overcome the increase in latency associated with these large structures. Techniques such as pipelining and clustering can help maintain aggressive clock speeds, but fail to address a crucial component of the performance equation: cache latency. As the processor core increases in size, and the clock cycle time decreases, the number of cycles required to load a value from the cache continues to grow. In this paper, we present an effective, yet simple technique to address the cache latency problem in clustered superscalar processors.

H. Zima et al. (Eds.): ISHPC 2002, LNCS 2327, pp. 281–290, 2002.

The increasing size of the processor core forces the L1 data cache to be placed further away, resulting in longer cache access latencies. Larger on-chip caches further exacerbate the problem by requiring more area (longer wire delays) and decode and selection logic. Modern processors are already implementing clustered microarchitectures where the execution resources are partitioned to maintain high clock speeds. Two-cluster processors have been commercially implemented [5], [10], and designs with larger numbers of clusters have also been studied [1], [15]. We propose to augment each cluster with a small *Level Zero* (L0) data cache. The primary design goal is to maintain hardware simplicity to avoid impacting the processor cycle time, while servicing some fraction of the memory load requests with low latency. To avoid the complexity of maintaining coherence or versioning between the clusters' L0 caches, a load from a L0 cache may return erroneous values. The mechanisms that already exist in superscalar processors to detect memory-ordering violations of speculatively issued load and store instructions can be used for recovery when the L0 data cache provides an incorrect value.

The paper is organized as follows. In Section 2, we briefly review related research in clustering processors and caching in processors with distributed execution resources. Section 3 details the base processor configuration used in our simulation studies, and also explains the simulation methodology. Section 4 describes our speculative L0 cache organization and the behavior of the caching protocol. Section 5 presents our performance results. Finally, Section 6 concludes with a short summary.

2   Related Work

Clustering breaks up a large superscalar into several smaller components [3], [9], [17], [18]. To the degree that most register results travel only locally within their cluster, the average register communication delay is reduced. At the same time, smaller hardware structures associated with each cluster run faster.

Palacharla et al. studied the critical latencies of circuits in superscalar processors and showed that the circuits do not scale well [14]. They suggested dividing the processor core into two clusters to address the complexity of more traditional organizations. The Alpha 21264 implemented a two-cluster microarchitecture [5], [10]. For highly-clustered processors, the manner in which instructions are assigned to clusters may play an important role in determining overall performance. Baniasadi and Moshovos explored different instruction distribution heuristics for a quad-clustered superscalar processor with unit inter-cluster register bypassing delays [1].

In many of these studies, the focus is on the communication of register values between clusters and how this additional delay affects overall performance. Therefore to isolate these effects, the assumptions about the cache hierarchy are a little relaxed. Although the cache configurations (size and associativity) used in these studies are reasonable (32KB to 64KB, 2- or 4-way set associative), the cache access latencies of one cycle [14], [15] and two cycles [1] are unrealistically fast. The aggressive clock speeds of modern processors force the computer architect to choose either smaller and faster caches (for example the 2 cycle, 8KB, 4-way L1 cache on the Pentium 4 [8]) or larger and slower caches (for example the 3 cycle, 64KB, 2-way L1 cache on the AMD Athlon [13]). In either case, the average number of clock cycles needed to service a load instruction is likely to increase due to increased miss rates or longer latencies.

For clustered superscalars with in-order instruction distribution, Gopal et al. [6] propose the Speculative Versioning Cache to handle outstanding stores. In their approach, small per-cluster caches, which we will refer to as *level zero* or *L0* caches, and the L1 run a modified write-back coherence protocol. The modifications allow different caches to cache different data for the same address. A chain of pointers link different versions of a memory address. A cluster that issues a load from an uncached address initiates a read request along a snooping bus. The responses from all clusters' L0's and the global L1 are combined by a global logic, the Version Control Logic, and the latest version is returned. Similarly a cluster's first store to a given address travels across the snooping bus and invokes the global logic that inserts the store into the chain of pointers and invalidates any mispredicted loads between that store and its successor in the chain. Both operations require that data travel back and forth across all the clusters and to L1. Hammond et al. [7] proposed a similar solution for the Hydra single-chip multiprocessor.

3   Simulation Methodology

We start by briefly describing our processor parameters and our simulation environment. We loosely model a single integer cluster of our processor on the Alpha 21264 [5], [10] integer core. The processor executes the Alpha AXP instruction set and uses resources summarized in Table 1. Table 1 also describes the parameters of our initial instruction and data memory hierarchies, without any clustered L0 data caches. The data caches are sized for future generation processors, and therefore the caches are somewhat larger than what is typical in current superscalar processors. Table 1 only describes a base configuration; in Section 5, we explore many other design points to demonstrate the effectiveness of the clustered L0 data caches.

The simulated processor aggressively speculates on loads, issuing them as soon as their arguments are ready, even if there are earlier unresolved stores. We use the *selective reissue* method of recovering from misspeculated loads [17]. One cycle after a load instruction is informed that it has mispredicted, it rebroadcasts its result causing any dependent instructions to iteratively reissue. The mispredicted load's immediate children must first request to be scheduled before reissuing. If the dependencies are across clusters, we charge additional cycles corresponding to the interconnect delay.

We base our studies of ILP on the cycle-level out-of-order simulator from the SimpleScalar 3.0 Alpha AXP toolset [2]. We simulated the SPEC2000 integer benchmarks [19]. The benchmarks were compiled on a 21264 with full optimiza-

Table 1. Default processor parameters.

Cluster Window Size	32 instructions		
Cluster Issue Width	4 instructions	Instruction TLB	64 entries 4-way set associative 30 cycle miss latency
Cluster Functional Units	4 Integer ALUs 1 Integer Multiplier 2 Memory Ports	Unified L2 Cache	2MB 4-way set associative 12 cycle latency
Cluster Interconnect	Unidirectional Ring 1 cycle per cluster	Trace Cache [4], [16]	512 traces 20 instruction per trace
Instruction Distribution	Sequential ("First Fit")	Main Memory	76 cycle latency
L1 D-Cache	16 banks, 128KB total 4-way set associative 5 cycle latency	Branch Prediction	6KB McFarling [12] (Bi-Mode[11]/local)
		Branch Misprediction Penalty	6 cycles
Data TLB	128 entries 4-way set associative 30 cycle miss latency	Decode/Rename Bandwidth	10 instructions per cycle
Load Store Unit	16 entries per L1 bank	Instruction Fetch Queue	32 instructions
		Load Misspeculation Recovery	Selective Re-execution
L1 I-Cache	16KB 2-way set associative		

tions. We used the "test" data set for all simulations. We skipped the first 100 million instructions of each benchmark, and then simulated the next 100 million instructions.

4   The Speculative L0 Cache

We address the problem of increasing memory load latencies by including a small *Level Zero* (L0) data cache with each cluster. To keep the cache implementation as simple as possible, mechanisms such as versioning and coherence are avoided. Our proposed design allows for the possibility of erroneous results from the L0 cache, and we use existing load speculation recovery mechanisms to ensure correct program execution in these instances. In this section, we describe the L0 data cache organization, and the rules governing how loads are serviced and how the data in the L0 caches are updated.

4.1   Protocol

In designing the cache model, we stress the importance of hardware simplicity, requiring minimal changes to existing structures, so as not to impact the cycle time. Each cluster of the processor has its own private L0 data cache, in addition to a shared global L1 cache. The L0 caches contain only values generated by retired instructions; this eliminates the extra hardware required to support multiple versions of a value. We also require that values loaded from the L0 caches are always treated as *speculative* values, which must be verified later. This greatly simplifies the design, because it obviates the need for a coherence mechanism between the multiple L0 caches and the L1 cache. Thus, the L0 caches are truly local structures, which do not require any new global structures for correct execution.

The size of a L0 cache line is the same as the largest unit of memory that load and store instructions can operate on (i.e. the width of a register). For our simulated processor which uses the Alpha AXP instruction set, the data held in a L0 cache line is 64 bits, or 8 bytes. This simplifies the implementation by not having to retrieve additional values from L1 to fill a larger cache line, but prevents the L0 cache from leveraging any spatial locality.

As stores execute, the address and data value are broadcasted to other clusters where the information about the store instruction may be buffered. If the buffer in a particular cluster is full, then the incoming store is simply dropped and no attempt is made to rebroadcast it. For our simulations in the next section, we use 32 entry incoming store buffers. The data from the store instructions are not written into the L0 caches until the store has retired. The cluster does not use any mechanisms for searching these buffered stores to forward data to load instructions. The tradeoff is that occasionally, a load instruction will speculatively execute and load an incorrect value despite the fact that the correct value is sitting in this uncommitted store buffer. The results of stores are also written into the L1 data cache upon retirement, and so there is no need for a writeback from a L0 cache to the L1 cache. This further simplifies the implementation of the L0 caches.

In addition to L0 data caches, the memory hierarchy includes a global load store unit (LSU), and the L1 data cache. Both are banked sixteen ways. The global LSU and L1 data cache are collectively referred to as *Level 1*. When a memory instruction is sent to the LSU, the instruction's dynamic sequence number is included so the LSU can identify the correct order of the instructions.

All loads and stores behave according to the following protocol. The configuration without L0 caches can be treated as a special case in which the L0 caches have zero size.

Load Issue: When a load issues, the L0 data cache in the load's cluster is accessed. If the load hits in the L0 cache, the value is returned in a single cycle. Whether or not the L0 cache hits, the load simultaneously issues to Level 1. The load arrives at the LSU sometime later. The LSU is then scanned for an earlier store to the same address. If such a store is found, the value is sent back to the load's cluster. Otherwise, the data is retrieved from the L1 data cache or higher levels of the memory hierarchy, and sent back to the cluster. If the load hits in the L0 cache, then when the load's data arrives from Level 1, the value is compared against the previously used L0 value. If they differ, a load misspeculation is flagged.

Store Issue: When a store issues, the address and data are sent to the LSU. Upon arrival, the address is compared against newer loads that have already reached the LSU. The search is truncated if a newer store to the same address is encountered. If a conflicting load is found, a load misspeculation is flagged and the store's value is forwarded to the dependent load's cluster.

An issuing store is also broadcasted to other clusters. Each cluster maintains a buffer of these stores. This buffer is used *only* to keep stores until they can be written into the L0 cache on retirement; we do not add hardware to search for and forward values from this buffer.

Store Retirement: When a store retires, its value is written into the L1 data cache and removed from the LSU. The store's value is also written into the L0 data caches.

Load Retirement: When a load retires, the correct value is written into the local L0 cache. If the load misspeculated due to a stale value in the L0 cache, this write updates the cache with the correct value.

Load Misspeculation: When a load misspeculation is detected, its value is updated with the value sent from Level 1, and all dependent instructions are eventually reissued.

5   Results

Our L0 caching solution can be applied to a wide range of processor configurations. In this section, we examine the performance impact for different degrees of clustering, cluster sizes, instruction distribution policies, and intercluster register bypassing.

For each processor configuration, we simulated the processor without any L0 caches, and with 2KB, 4KB and 8KB L0 caches. We also varied the L0 cache associativity: 1-way, 2-way or 4-way. Figure 1 shows the mean IPCs achieved without L0 caches, and with L0 caches of different sizes and associativities. For the 2KB caches, only a small improvement is seen; however, the 4KB and 8KB caches yield a significant improvement of 6.7-14.4%.

We have shown that our L0 caches provide increases in ILP for processors with varying numbers of clusters. The size of the clusters, the issue width of the clusters, and the instruction-to-cluster distribution rules were all held constant to observe the benefit provided by the L0 caches for those design points. We now explore a larger design space to demonstrate that our L0 caching solution is a general solution that provides performance improvements across a wide variety of processor configurations. The 4KB 2-way set associative L0 cache appears to be a reasonable tradeoff between capacity and associativity, and we use this for all remaining configurations.

The 4-issue cluster configurations used thus far may not be the only interesting design point. The cluster configurations in future highly clustered superscalar processor may have smaller issue queues and fewer functional units to achieve higher clock rates. On the other hand, the trend in the organization of the processor clusters may go the other direction towards larger and more complex cores.

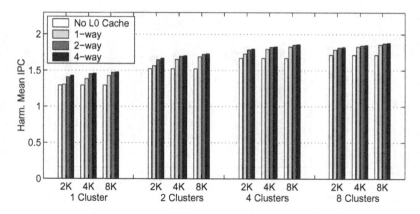

Fig. 1. Impact of L0 caches of different sizes and associativities.

Table 2. The processor parameters for our smaller and larger clusters.

	Small Cluster	Medium Cluster	Large Cluster
Cluster Window Size	16	32	64
Issue Width	2	4	6
Integer ALUs	2	4	6
Integer Multiplier	1	1	2
Memory Ports	2	2	2
All other parameters are the same as in Table 1.			

We simulated processor configurations for both of these design possibilities. We group these into *small cluster* and *large cluster* configurations, as listed in Table 2. The original configuration from Table 1 is also included for reference, and is called the *medium cluster* configuration.

The IPC performance results for the different sized clusters are plotted in Figure 2(a). The key observation is that our L0 caches provide a relatively consistent performance improvement across all configurations, regardless of the number or size of clusters. Furthermore, increasing the cluster sizes to the large configuration does not provide substantial gains over the medium configurations. For the 8-cluster configurations, the additional execution resources of the large clusters uncover so little additional parallelism that it is better to simply use the medium cluster configuration augmented with the L0 caches.

The processor configurations that we have analyzed so far dispatch instructions to the clusters in program order using a First Fit distribution rule. There are other possible ways to assign instructions to clusters. By attempting to group dependent instructions into the same clusters, inter-cluster register communication can be decreased. On the other hand, distributing instructions across multiple clusters allows better utilization of the execution resources and issue slots of the other clusters.

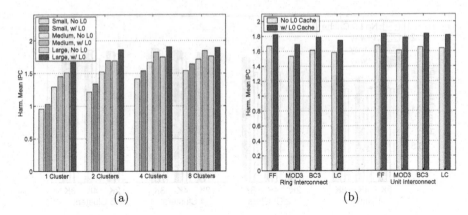

(a)   (b)

Fig. 2. (a) Varied cluster sizes. (b) Different instruction distribution policies and interconnects.

Baniasadi and Moshovos investigated a variety of instruction distribution heuristics for a quad-clustered superscalar processor [1]. In their study, the intercluster communication mechanism was assumed to have a single cycle delay between any two clusters regardless of how near or far the clusters are physically located from each other. We call this the *Unit Interconnect*. Depending on the implementation details, this may not be feasible for processors with four or more clusters, which is why we have used the ring network earlier in this paper.

We implemented a few of the instruction distribution rules from [1] to test the sensitivity of the L0 cache performance to instruction distribution. In particular, we used the MOD_n, BC, and LC distribution rules. The MOD_n rule assigns the first n instructions to one cluster, and then the next n instructions to the next cluster, and so on. The BC (Branch Cut) rule assigns all instructions to the same cluster until a branch instruction is reached. All subsequent instructions are directed to the next cluster until a branch is reached again, and so on. From our simulations, we found that switching clusters at every third branch (BC3) is more effective. The LC (Load Cut) rule is similar to the BC rule, except that a cluster switch occurs when a load instruction is encountered. One difference is that back-to-back loads are assigned to the same cluster.

Figure 2(b) shows the IPC performance for configurations using different instruction-to-cluster distribution rules, as well as different inter-cluster register bypass networks. The results show that among the instruction distribution techniques and inter-cluster register bypass networks simulated, the L0 data caches do provide consistent performance improvements. Because the cluster sizes and organizations differ from [1], the relative performance of the distribution rules also differ from the original study.

In all of the simulations presented in this section, we have not imposed a limit on the number of stores broadcasted per cycle. Our intuition is that this should not be a problem because we found that the average number of stores per cycle is well under one. For example, we observed 0.32 stores per cycle on average across all benchmarks with a four cluster, 4KB 2-way L0 cache configuration. On the other hand, broadcasting stores can still become a bottleneck if the broadcasts occur in a bursty fashion. The hardware to support a large number of stores per cycle would be complex to implement and unlikely to scale to a large number of clusters. Therefore, we need to evaluate the situation where the number of broadcasted stores per cycle is limited.

To quantify the effects of reducing the available store broadcast bandwidth, we simulated a processor configuration with a single store broadcast bus; that is, every cycle at most one store may broadcast to the other clusters. In situations where multiple stores request to broadcast, the oldest (in program order) store receives the broadcast bus. For the four and eight cluster, 4KB 2-way L0 cache configurations, this bus limitation decreases the mean IPC by only 0.34% and 0.35%, respectively. These results show that limiting the processor to only a single broadcast bus has minimal impact on the benefit of the L0 caches.

6 Summary

As superscalar processors are designed to handle a larger number of in-flight instructions, and as the processor clock cycle continues to decrease, the cache access latency will continue to grow. Longer delays to service load instructions result in degraded performance. We address this problem in the context of clustered superscalar processors by augmenting each execution cluster with a small, speculative Level Zero (L0) data cache. The hardware is very simple to implement because we allow the cache to occasionally return erroneous values, thus obviating the need for coherence or versioning mechanisms.

We have shown that a small 4KB 2-way set associative L0 data cache attached to each cluster can yield an increase in the mean instruction level parallelism by 11.3% for a 2-cluster processor, and 9.1% for a 4-cluster processor. By varying many important processor parameters, we have demonstrated that the L0 caches can be gainfully employed in a large variety of clustered superscalar architectures. As the processor-memory speed gap increases, techniques such as our L0 caches that address the cache access latency will become increasingly important.

Acknowledgments

This research was supported by NSF Career Grant MIP-9702281. We are also grateful to Bradley Kuszmaul and Vinod Viswanath for helpful discussions, and to the referees for their suggestions.

References

1. Amirali Baniasadi and Andreas Moshovos. Instruction Distribution Heuristics for Quad-Cluster, Dynamically-Scheduled, Superscalar Processors. In *Proceedings of the 33rd International Symposium on Microarchitecture*, 2000.

2. Doug Burger and Todd M. Austin. The SimpleScalar Tool Set, Version 2.0. Technical Report 1342, University of Wisconsin, June 1997.
3. Keith I. Farkas, Paul Chow, Norman P. Jouppi, and Zvonko Vranesic. The Multi-cluster Architecture: Reducing Cycle Time Through Partitioning. In *Proceedings of the 30th International Symposium on Microarchitecture*, 1997.
4. Daniel H. Friendly, Sanjay J. Patel, and Yale N. Patt. Alternative Fetch and Issue Techniques From the Trace Cache Mechanism. In *Proceedings of the 30th International Symposium on Microarchitecture*, 1997.
5. Bruce A. Gieseke, Randy L. Allmon, Daniel W. Bailey, Bradley J. Benschneider, and Sharon M. Britton. A 600MHz Superscalar RISC Microprocessor with Out-Of-Order Execution. In *Proceedings of the International Solid-State Circuits Conference*, 1997.
6. Sridhar Gopal, T. N. Vijaykumar, James E. Smith, and Gurindar S. Sohi. Speculative Versioning Cache. In *Proceedings of the 4th International Symposium on High Performance Computer Architecture*, 1998.
7. Lance Hammond, Ben Hubbert, Michael Siu, Manohar Prabhu, Mike Chen, and Kunle Olukotun. The Stanford Hydra CMP. *IEEE Micro Magazine*, March–April 2000.
8. Glenn Hinton, Dave Sager, Mike Upton, Darrell Boggs, Doug Karmean, Alan Kyler, and Patrice Roussel. The Microarchitecture of the Pentium 4 Processor. *Intel Technology Journal*, Q1 2001.
9. G. A. Kemp and Manoj Franklin. PEWs: A Decentralized Dynamic Scheduler for ILP Processing. In *Proceedings of the Proceedings of the International Conference on Parallel Processing*, 1996.
10. R. E. Kessler. The Alpha 21264 Microprocessor. *IEEE Micro Magazine*, March–April 1999.
11. Chih-Chieh Lee, I-Cheng K. Chen, and Trevor N. Mudge. The Bi-Mode Branch Predictor. In *Proceedings of the 30th International Symposium on Microarchitecture*, pages 4–13, December 1997.
12. Scott McFarling. Combining Branch Predictors. TN 36, Compaq Computer Corporation Western Research Laboratory, June 1993.
13. Dirk Meyer. AMD-K7 Technology Presentation. *Microprocessor Forum*, October 1998.
14. Subbarao Palacharla, Norman P. Jouppi, and James E. Smith. Complexity-Effective Superscalar Processors. In *Proceedings of the 24th International Symposium on Computer Architecture*, June 1997.
15. Narayan Ranganathan and Manoj Franklin. An Empirical Study of Decentralized ILP Execution Models. In *Proceedings of the Symposium on Architectural Support for Programming Languages and Operating Systems*, October 1998.
16. E. Rotenberg, S. Bennett, and J. E. Smith. Trace Cache: A Low Latency Approach to High Bandwidth Instruction Fetching. In *Proceedings of the 29th International Symposium on Microarchitecture*, December 1996.
17. Eric Rotenberg, Quinn Jacobson, Yiannakis Sazeides, and Jim Smith. Trace Processors. In *Proceedings of the 30th International Symposium on Microarchitecture*, December 1997.
18. Gurindar S. Sohi, Scott E. Breach, and T. N. Vijaykumar. Multiscalar Processors. In *Proceedings of the 22nd International Symposium on Computer Architecture*, pages 414–425, 1995.
19. The Standard Performance Evaluation Corporation. WWW Site. http://www.spec.org.

The Effects of Timing Dependence and Recursion on Parallel Program Schemata

Yasuo Matsubara and Takahiro Shakushi

Bunkyo University, Faculty of Information and Communications,
Namegaya 1100, Chigasaki-shi, 253-8550 Japan
matubara@shonan.bunkyo.ac.jp

Abstract. We are interested in the effects of timing dependence and recursion in the class of parallel program schemata. We compare the expression power of some classes of dataflow schemata: UDF, DF_π, RDF etc. DF_π includes a π-gate which introduces timing dependence, and RDF has a facility for recursion and UDF has both devices. In conclusion, we show some inclusion relations between classes, including the existence of a class standing between EF and EF^d. These relations reveal the role of timing dependence and recursion.

1 Introduction

Highly expressive programming languages are necessary to extract the power of high performance computers. Currently, to determine the expression power of a programming language, a comparison is made using the functional classes which can be realized by the program schemata classes [1].

It is well known that in many cases recursion enhances expressiveness. However, in parallel programming languages, timing dependence may also affect the expression power. We are interested in the effects of recursion and timing dependence on parallel program schemata.

We take the class of dataflow schematas as our basic class. We must discriminate between cases where each arc of the dataflow schema can hold at most one token and also those that can hold any number of tokens. The former case is indicated by attaching '(1)' after the class name and the latter case is indicated by '(∞)'. The class EF of effective functionals [2] is known to have the greatest expression power. Jaffe [3] investigated the expressiveness of the class $DF(\infty)$ of dataflow schemata for the first time, and showed that $DF(\infty)$ is equivalent to EF in total interpretations. However, Matsubara and Noguchi [4] showed that $DF(\infty)$ is equivalent to EF^d in partial interpretations. This means that $DF(\infty)$ has an expression power which can be realized by a class of sequential program schemata. Next, Matsubara and Noguchi [5] proposed the class $ADF(1)$ which is enhanced with arbiters and recursion, and showed that $ADF(1)$ is equivalent to EF. Previously, we have shown some equivalent relations and inclusion relations [8]. The equivalences are as follows: $ADF(1) \equiv ADF(\infty) \equiv EF$, $DF(\infty) \equiv RDF(\infty) \equiv RDF(1) \equiv EF^d$ and $DF(1) \equiv P$, where P is the class

H. Zima et al. (Eds.): ISHPC 2002, LNCS 2327, pp. 291–300, 2002.

of simple program schemata including only simple variables. The inclusions are as follows: $ADF(1) > DF(\infty) > DF(1)$ and $ADF(1) > DF_{arb}(1) > DF(1)$.

However we could not show the correct position where the class $DF_{arb}(\infty)$ should be inserted in the inclusions.

We have been using the arbiter as a device to introduce timing dependence. In this paper, we change the device from an arbiter to the π-gate, a new device to introduce timing dependence.

In conclusion, we show that the class $DF_\pi(\infty)$ should be inserted between EF and EF^d and that $DF_\pi(1)$ is properly included in $DF_\pi(\infty)$.

For definitions of EF and EF^d, the reader is referred to [8].

2   Dataflow Schemata

The dataflow schema is a kind of program schema, but differs from ordinary program schemas in the way that parallel execution proceeds in accordance with the flow of the data.

In the class of dataflow schemata, it is necessary to distinguish the case where an arc can hold at most one token from the case where an arc can hold an arbitrary numbers of tokens. The two classes are equal from the standpoint of their syntax. When an arc holds an arbitrary number of tokens, the token order is maintained, i.e. the arc plays the role of a FIFO queue.

Definition 2.1. *Only the types of nodes used in schemata belonging to the class udf are depicted in Fig.1. Besides nodes used in ordinary dataflow schemata, we permit the node to call a procedure(Fig.1.(o)). "Procedure name" in the graph is defined as a graph of the class udf.*

White arcs are control arcs, black ones are data arcs, and mixed ones are either control or data arcs.

In Fig. 1(o), x', y', etc. are the arcs at declaration time. The input arcs and the output arc of the schema, are clearly depicted as Fig.1.(p),(q).

In the description below, let the figure and statement be used equivalently.

To define the semantics of the graph, it is necessary to designate $\mu = 1$ or ∞, where $\mu = 1$ means that each arc can hold at most one token, and $\mu = \infty$ means that each arc can hold an arbitrary number of tokens. We attach (μ) after the class name.

Definition 2.2. *Semantics of udf(1): the schema of udf(1) is driven into action as shown below when interpretation I is given and input data is provided to the individual input variables.*

Let it be assumed that the execution is done in discrete time with t = 0, 1, 2, ... On the assumption that u is a predicate symbol or function symbol and on the supposition that $d_1, . . . , d_{Ru} \in D$, $\tau(u, d_1, . . . , d_{Ru})$ gives the evaluation time of $u(d_1, . . . , d_{Ru})$. A time function τ returns an infinite evaluation time when the value of $u(d_1, . . . , d_{Ru})$ becomes undefined and that returns a positive integer evaluation time value when the value is defined in the interpretation I; it is called a time function consistent with the interpretation I.

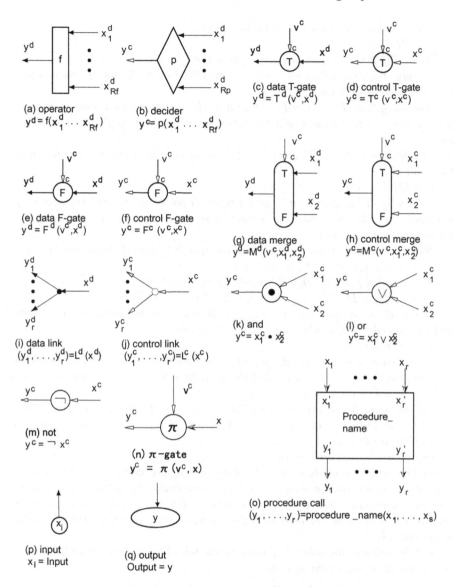

Fig. 1. Dataflow diagram and it's statement

In the following description, the behavior of the individual statements is explained.

$y=u(x_1^d,\ldots,x_{Ru}^d)$, where u is a function symbol or predicate symbol:

When tokens $d_1,\ldots,d_{Ru}$ are placed on the individual input arcs and concurrently no tokens are placed on y at the time $t = \tau_0$, then at $t=\tau_0+\tau(u,d_1,\ldots,d_{Ru})$, the tokens on the individual input arcs are removed and the tokens of the evaluation result are placed on the output arc y.

Other nodes take only unit time to execute.

$y = T(v^c, x)$:

Assuming that 'T' is placed on v^c and a token is placed on x and concurrently y is empty, then the token on x will be transfered on to y and 'T' on v^c will be removed after a unit time.

Assuming that 'F' is placed on v^c and a token is placed on x and concurrently y is empty, the token on x and the token on v^c are removed after one unit of time.

The behavior of $y = F(v^c, x)$ is obtained by reversing the truth value of the token on v^c.

$y = M(v^c, x_1, x_2)$:

Assuming 'T' is placed on v^c and a token is placed on x_1, and concurrently y is empty, then the token on v^c is removed and the token on x_1 will be transferred on to y after a unit time.

When 'F' is placed on v^c, the token on x_2 will be transfered instead of x_1.

$y^c = x_1^c \cdot x_2^c$, $y = x_1^c \vee x_2^c$, $y^c = \neg \, x^c$:

These nodes perform logical operations for input tokens.

$(y_1^d, \ldots, y_r^d) = L^d(x^d)$, $(y_1^c, \ldots, y_r^c) = L^c(x^c)$:

When all the output arcs are empty, the token on the input arc is removed and the copied token is placed on each output arc.

$y^c = \pi(v^c, x)$:

In one case, the token on the top of v^c is removed and 'F' is added to the tail of y^c at time $t = \tau_0 + 1$, provided that there exists a token on the input arc v^c and there is no token on x at the time τ_0.

On the other hand, the token on the top of v^c and x are removed and 'T' is added to the tail of y^c at time $t = \tau_0 + 1$, provided that there exists a token on each of the input arcs at $t = \tau_0$.

$(y_1, \ldots, y_r) = <ProcedureName>(x_1, \ldots, x_s)$:

In lieu of this procedure call, the body part of the declared procedure is expanded. Here, let the arc name be changed systematically so that no arc having the same arc name will come into existence. Let the input/output arc at the time of declaration be replaced with the corresponding input/output arcs of the procedure call.

Let the value of the token be placed on the top of the output arc in the schema be the output result of the schema.

As to the semantics for $udf(\infty)$, it is the same as that of $udf(1)$ but the behavior disregards whether there is a token or not on the output arc(s).

Definition 2.3. *For $\mu = 1$, ∞, $UDF(\mu)$ is the class comprised of schemata satisfying the two conditions shown below in the schemata of the class $udf(\mu)$.*

(1) When interpretation I and input values are given, the same result is provided for an arbitrary time function τ consistent with I.

(2) In procedures declared recursively, there is no statement that can be driven into action in an initial state. (That is to say, no action is started until a token arrives from another part. Thanks to this condition, the procedure can be expanded when necessary.)

Some related classes of dataflow schemata are defined here.

Definition 2.4. *For $\mu = 1, \infty$, the class $RDF(\mu)$ is comprised of the schemata which do not use the π-gate in the class $UDF(\mu)$.*

Definition 2.5. *For $\mu = 1, \infty$, the class $DF(\mu)$ is comprised of the schemata which do not use the π-gate, and the call and declaration of procedures in the $UDF(\mu)$.*

Definition 2.6. *For $\mu = 1, \infty$, the class $DF_\pi(\mu)$ is comprised of the schemata excluding procedure call or procedure declaration, in the $UDF(\mu)$.*

Definition 2.7. *A Boolean graph is an acyclic graph comprised exclusively of logical operators and control links from Fig.1, and has the following property: If and only if a token is inputed into each input arc, a token is outputted from each output arc.*

For any arbitrary logical function, an equivalent Boolean graph realizing the function can easily be composed. As is shown in [4,5], any kind of finite state machine can be composed by connecting the output arcs to the input arcs of a Boolean graph realizing an appropriate logical function.

Because an arbiter can be easily simulated using π-gate, it is obvious that the following proposition(For the definitions of arbiter and the classes $ADF(\mu)$, see [8]).

Proposition 2.1. *$UDF(\mu) \equiv ADF(\mu)$ for $\mu = 1, \infty$.*

3 $DF_\pi(1) > P$

Since $DF_\pi(1)$ can use the π-gate as a device (and it cannot be used by $DF(1)$) it is evident that $DF_\pi(1) \geq DF(1)$. Therefore, it is easy to see that $DF_\pi(1) \geq P$ by the relation $P \equiv DF(1)$. In this section, we consider whether the inclusion is proper or not.

Next, we introduce an example which can be used for testing our hypothesis. Example A is depicted in Fig.2. When both the predicates $P_1(x)$ and $P_2(x)$ have the value T, the value of the output token is x regardless of which predicate outputs the result the fastest. Therefore, the example is an element of the class $DF_\pi(1)$.

Lemma 3.1. *Example A cannot be simulated by a schema of EF^d.*

Proof. Here, we assume that a schema S equivalent to example A belongs to the class EF^d. S should have the same sets of input variables, predicate symbols and function symbols, and the same output variable as example A.

We introduce three interpretations I_1, I_2 and I_3 which have the same data domain $D = \{a\}$.

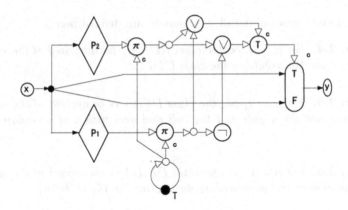

Fig. 2. Example A

On I_1, $P_1(a) = T$ and $P_2(a) =$ undefined.

On I_2, $P_1(a) =$ undefined and $P_2(a) = T$.

On I_3, $P_1(a) = P_2(a) =$ undefined.

On I_1, the first computation in S which has a value is assumed to be the i_1-th computation. On I_2, the i_2-th.

First, we assume that i_1 equals i_2. This computation should not include $P_1(x)$ or $P_2(x)$, because if $P_1(x)$ is included, then the computation is undefined on I_2. If $P_2(x)$ is included, then the computation is undefined on I_1.

If these predicates are not included, the computation should have a value on I_3. This contradicts the assumption that S is equivalent to example A. Therefore i_1 should not be equal to i_2.

Next, we assume that $i_1 < i_2$. On I_2, the i_1-th computation should not have a value. Therefore, it should include $\neg P_2(x)$. This means that i_1-th computation becomes undefined on I_1. When $i_2 < i_1$ is assumed, a similar contradiction is deduced. □

Therefore, we can prove the following theorem.

Theorem 3.1. $(DF_\pi(1) > P)$: *The class $DF_\pi(1)$ properly includes the class P.*

Proof. It is evident that example A cannot be simulated by a schema of P by Lemma 3.1 and the relation $EF^d \geq P$. Therefore we can conclude the result by $DF_\pi(1) \geq P$. □

4   The Power of Class $DF_\pi(\infty)$

In this section, we investigate the ability of the class $DF_\pi(\infty)$. It is self-evident that the relation $DF_\pi(\infty) \geq DF(\infty)$ is satisfied. However, we are interested in whether the inclusion is proper or not. To answer the question, we present the relation $DF_\pi(\infty) \geq DF_\pi(1)$.

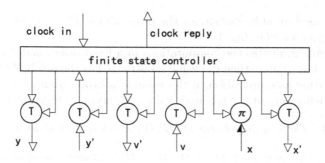

Fig. 3. Simulation of the statement $y = \pi(v, x)$ of $DF_\pi(1)$ in the schema of $DF_\pi(\infty)$

Theorem 4.1. $(DF_\pi(\infty) \geq DF_\pi(1))$: *For an arbitrary schema S of the class $DF_\pi(1)$, an equivalent schema S' of the class $DF_\pi(\infty)$ can be constructed.*

Proof. To imitate the behavior of a node of S', it is important to assure that the output arc is empty before execution. For that purpose, we use a pair of arcs a and a' for each arc a of S. On one hand, when a is empty a' holds 'T'. On the other hand, when a holds a token, a' is empty.

Each statement is simulated by the statement simulator which is composed of the corresponding statement and a finite state controller. When a is an output of a statement, a' becomes an input for the finite state controller of the simulator of the statement. Each finite state controller provides the input tokens for the corresponding statements only when the output arc's emptiness is assured.

To make sure that all the statement simulator takes the next action after all the previous action finish, we have prepared a central clock unit. The unit provides a control token for each statement simulator as a clock and takes in the reply tokens from each simulator. After all the tokens are received, the next clock tokens are provided. The simulator for the π-gate is depicted in Fig.3. Simulators of other statementes can be constructed easily. Thus, S' simulates the behavior of S precisely.   □

To reveal whether the inclusion is a proper one or not, we introduce Example B.

Example B: This is a schema of EF^d with x as an input variable; L and R are the function symbols and P is the predicate symbol. The computations enumerated by the schema are as follows:

1st:$< P(x), x >$
2nd:$< PL(x), x >$
3rd:$< PR(x), x >$
4th:$< PLL(x), x >$
5th:$< PLR(x), x >$
6th:$< PRL(x), x >$
$\vdots$

Here, we used an abbreviation for the propositions. For example, $PRL(x)$ is the abbreviation of $P(R(L(x)))$.

This schema evaluates the computations in a fixed order, and if it finds the computation has a value, then the value is outputted as the result of the schema. While evaluating a computation, if the value of a proposition or an expression becomes undefined, then the schema's value also becomes undefined.

Lemma 4.1. *There is no schema of $DF_\pi(1)$ which is equivalent to Example B.*

Proof. We assume S is a schema of the class $DF_\pi(1)$ which is equivalent to Example B. We can imagine an interpretation I which evaluates to 'T' for propositions of a given length and evaluates to 'F' for other propositions. If the length is greater than the number of arcs included, S cannot hold all of the intermediate values. This is a contradiction. □

Now, we can conclude some proper inclusion relations.

Corollary 4.1. *($DF_\pi(\infty) > DF_\pi(1)$): The class $DF_\pi(\infty)$ properly includes the class $DF_\pi(1)$.*

Proof. If we assume $DF_\pi(\infty) \equiv DF_\pi(1)$, $DF_\pi(1) \geq DF(\infty) \equiv EF^d$ should be satisfied from $DF_\pi(\infty) \geq DF(\infty)$. This contradicts Lemma 4.1. □

Corollary 4.2. *($DF_\pi(\infty) > DF(\infty)$): The class $DF_\pi(\infty)$ properly includes the class $DF(\infty)$*

Proof. If we assume $DF_\pi(\infty) \equiv DF(\infty)$, $EF^d \equiv DF(\infty) \geq DF_\pi(1)$ should be satisfied by Theorem 4.1. This contradicts the fact that Example A cannot be simulated by a schema of EF^d. □

In the next step, we investigate whether the class $DF_\pi(\infty)$ is strong enough to be equivalent to EF. To answer this question we introduce the third example which is commonly known as the Leaf Test.

Example C: This is a schema of EF with x as an input variable, L and R are the function symbols and P is the predicate symbol. The schema gives a value of x, if and only if there is at least a proposiotion composed of L, R, P and x which have the value T.

Lemma 4.2. *($DF_\pi(\infty) \leq EF$): For any schema S of the class $DF_\pi(\infty)$, there is an equivalent schema S' of EF.*

Proof. The behavior of S can be simulated by a nondeterministic Turing machine $\mathcal{N}$ in a symbolic manner. The behaviors of $\mathcal{N}$ which stop can be enumerated and the corresponding computation can be constructed. It is possible to construct a deterministic Turing machine $\mathcal{T}$ which takes a natural number i as input and outputs the i-th computation. S' can be constructed using $\mathcal{T}$. □

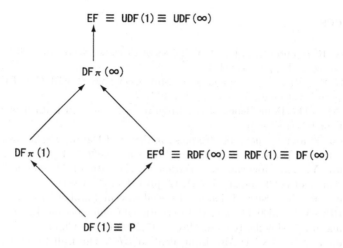

Fig. 4. Inclusion relations between classes

Theorem 4.2. *$(DF_\pi(\infty) < EF)$: The inclusion of $DF_\pi(\infty)$ by EF is the proper one.*

Proof. A schema of $DF_\pi(\infty)$ includes only a finite number of predicate symbols and function symbols. However, Example C includes an infinite number of expressions and propositions. Therefore Example C can not be simulated by a schema of $DF_\pi(\infty)$. Therefore, we conclude that the inclusion is proper.   □

5   Conclusions

We determined the structure of inclusion relations between some dataflow schemata classes. After adding some relations from [4,5,6,7,8], the whole structure is illustrated in Fig. 4. Here, $\alpha \to \beta$ means that α is properly included by β.

We list several important points which can be concluded from the figure:

(1) Both timing dependence and recursion are necessary for the class of dataflow schemata to become equivalent to the class EF of effective functionals.
(2) The timing dependence and recursion take on different roles in dataflow schemata. Recursion is strong enough to make $RDF(1)$ stronger than $DF(1)$, but it can not make $RDF(\infty)$ stronger than $DF(\infty)$. Timing dependence makes $DF_\pi(1)$ stronger than $DF(1)$ and makes $DF_\pi(\infty)$ stronger than $DF(\infty)$.
(3) The value of μ takes rather a minor role. It has an effect only when recursion is not included.
(4) It is shown that the class $DF_\pi(\infty)$ stands between the class EF and the class EF^d.

We have investigated the expressiveness of dataflow languages by the classes of realizable functionals. The conclusions seem very suggestive for the language design of any kind of parallel processing.

References

1. Constable, R. L. and Gries, D., "On Classes of Program Schemata", SIAM Journal on computing, 1, 1, pp.66-118(1972).
2. Strong, H. R., "High Level Languages of Maximum Power", PROC. IEEE Conf. on Switching and Automata Theory, pp.1-4(1971).
3. Jaffe, J. M., "The Equivalence of r.e. Program Schemes and Data flow Schemes", JCSS, 21, pp.92-109(1980).
4. Matsubara, Y. and Noguchi, S., "Expressive Power of Dataflow Schemes on Partial Interpretations", Trans. of IECE Japan, J67-D, 4, pp.496-503 (1984).
5. Matsubara, Y. and Noguchi, S., "Dataflow Schemata of Maximum Expressive power", Trans. of IECE Japan, J67-D, 12, pp.1411-1418(1984).
6. Matsubara, Y. "Necessity of Timing Dependency in Parallel Programming Language", HPC-ASIA 2000,The Fourth International Conference on High Performance Computing in Asia-Pacific Region. May 14-17,2000 Bejing,China.
7. Matsubara,Y.,"$ADF(\infty)$ is Also Equivalent to EF", The Bulletin of The Faculty of Information and Communication, Bunkyo University. (1995).
8. Matsubara,Y. and Miyagawa,H. "Ability of Classes of Dataflow Schemata with Timing Dependency", Third International Symposium, ISHPC2000, Tokyo, Japan, October 2000 Proceedings, "HighPerformance Computing", Lecture Notes in Computer Science 1940 Springer.

Cache Line Impact on 3D PDE Solvers

Masaaki Kondo, Mitsugu Iwamoto*, and Hiroshi Nakamura

Research Center for Advanced Science and Technology, The University of Tokyo
4-6-1 Komaba, Meguro-ku, Tokyo, 153-8904, Japan
{kondo,mitsugu,nakamura}@hal.rcast.u-tokyo.ac.jp

Abstract. Because performance disparity between processor and main
memory is serious, it is necessary to reduce off-chip memory accesses
by exploiting temporal locality. Loop tiling is a well-known optimization
which enhances data locality. In this paper, we show a new cost model
to select the best tile size in 3D partial differential equations. Our cost
model carefully takes account of the effect of cache line. We present
performance evaluation of our cost models. The evaluation results reveal
the superiority of our cost model to other cost models proposed so far.

1   Introduction

Because performance disparity between processor and main memory is serious, it
is necessary to reduce off-chip memory accesses by exploiting temporal locality.
Cache blocking or tiling [1] is a well-known optimization to increase temporal
locality. Making good use of caches by exploiting temporal locality is a promising
way to improve performance.

In scientific/engineering computing, solving partial differential equation
(PDE) using finite differencing techniques is one of the most important computations. Historically, PDE solvers have targeted 2D domain and cache optimization techniques in 2D code have been proposed so far. Recently, solving
PDEs in 3D domain has become popular as computation power increases. Because performance degradation caused by poor memory behavior in 3D code
is more serious than 2D code, tiling optimization is the key to achieve high
performance especially in 3D PDE solvers.

Rivera and Tseng showed tiling and padding transformation techniques for
3D PDE codes [2]. Their techniques are based on the *cost function* which gives the
goodness of a given tile size and tile shape. Using their cost function, programmer
or compiler can find appropriate tile size and shape straightforwardly. However,
they did not consider the effect of cache line.

In this paper, we propose a new cost model for tile size selection in 3D PDE
codes. We carefully take account of the effect of cache line. We also present
performance evaluation of our cost model compared with Revera's cost model.

---

* Presently with The Department of Mathematical Informatics, Graduate School of
Information Science and Technology, The University of Tokyo

H. Zima et al. (Eds.): ISHPC 2002, LNCS 2327, pp. 301–309, 2002.

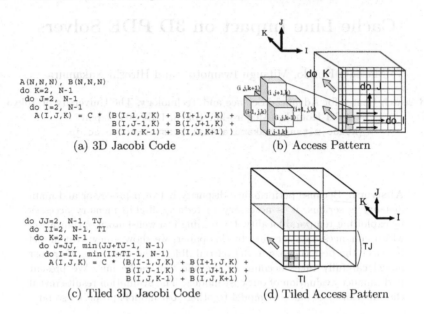

```
A(N,N,N), B(N,N,N)
do K=2, N-1
  do J=2, N-1
    do I=2, N-1
      A(I,J,K) = C * (B(I-1,J,K) + B(I+1,J,K) +
                      B(I,J-1,K) + B(I,J+1,K) +
                      B(I,J,K-1) + B(I,J,K+1) )
```

(a) 3D Jacobi Code   (b) Access Pattern

```
do JJ=2, N-1, TJ
  do II=2, N-1, TI
    do K=2, N-1
      do J=JJ, min(JJ+TJ-1, N-1)
        do I=II, min(II+TI-1, N-1)
          A(I,J,K) = C * (B(I-1,J,K) + B(I+1,J,K) +
                          B(I,J-1,K) + B(I,J+1,K) +
                          B(I,J,K-1) + B(I,J,K+1) )
```

(c) Tiled 3D Jacobi Code   (d) Tiled Access Pattern

Fig. 1. 3D Jacobi Kernel

2 Previous Tiling Optimization for 3D PDE Solvers

In this section, we explain overview of PDE solvers and tiling optimization for 3D stencil codes presented in [2].

2.1 Access Pattern of 3D PDEs

Fig. 1-(a) shows the code of 3D Jacobi kernel which is a simple PDE solver and Fig. 1-(b) shows the data access pattern of it. In 3D Jacobi kernel, neighboring six elements of $B(I\pm1, J\pm1, K\pm1)$ are required to obtain a value of $A(I, J, K)$. These set of data accessed for computing one value is called *stencil* pattern.

Considering data access pattern of 3D Jacobi Kernel, we can fully exploit reusability of array B if three adjacent planes (that is, $(K-1)$-th, K-th and $(K+1)$-th I-J plains) of the array can fit into cache[1].

2.2 Tiling Optimization for 3D Stencil Codes

In 3D stencil codes, neighboring several plains are desired to reside in cache for making full use of reusability. However, the current caches are too small to hold several plains in many realistic applications even if we have a few MBytes of cache. Thus, to extract reusability of 3D stencil codes, it is necessary to reduce the size of the plain by dividing I-J plain into sub-plains so that the divided sub-plain can fit into cache. This is accomplished by tiling I and J loops of Fig. 1-(a)

---

[1] Exactly speaking, $(2N^2 + 1)$ is the minimum required cache size.

[2]. Fig. 1-(c) shows the tiled version of 3D Jacobi Kernel and Fig. 1-(d) shows the schematic view of the access pattern.

2.3   Cost Function

To apply tiling transformation, we must decide tile size and tile shape, that is TI and TJ in Fig. 1-(c). These factors greatly affect performance because off-chip traffic heavily depends on them. Though the reusability of inner elements of tiled plains can be exploited, boundary elements cannot be reused. Therefore, ratio of boundary elements to inner elements should be kept as small as possible. Rivera and Tseng presented the formula (*cost function*) which reflects the above intuition [2]. Their cost function is based on the number of distinct cache lines fetched into cache. The followings are the simple explanation of their cost function.

In the tiled Jacobi Kernel, $TI \times TJ \times (N - 2)$ elements are calculated in a K loop iteration assuming N^3 problem size. For calculating each block in K loops, we need to access approximately $(TI + 2)(TJ + 2)(N - 2)$ elements of array B. The total number of blocks is $N^2/(TI \times TJ)$. We can obtain the total number of elements brought into cache by multiplying the number of blocks and the number of elements in each block: $N^3(TI + 2)(TJ + 2)/(TI \times TJ)$. The number of cache lines brought into cache is estimated from dividing this formula by the number of words in cache line L. Here, N^3/L is invariant under different tile sizes. To divide out this constant, we can obtain the following cost function:

$$Cost_{Riv} = \frac{(TI + 2)(TJ + 2)}{TI \times TJ} \tag{1}$$

When $TI \times TJ$ is constant, this cost function gets minimum when TI and TJ are the same. This means if TI and TJ are the same value, we can minimize the number of lines fetched into cache from lower memory hierarchy (Hereafter, we call it *cache traffic*). Thus, their cost function favors square tiles.

2.4   Avoiding Conflicts

In the previous section, we described the tiling optimization for 3D stencil codes. Higher performance is expected by the tiling optimizations. However, many works [1,3,4] discovered that performance is greatly degraded due to line conflicts even if tiling optimization is applied.

For avoiding line conflicts, there have been proposed many techniques, for example, copy optimization technique [5], using small fraction of the cache [6], non-conflicting tile size selection algorithm [1,3], and padding technique [4].

As for 3D stencil codes, three algorithms, Euc3D, GcdPad, and Pad, were proposed for avoiding line conflict [2]. The first algorithm "Euc3D" is based on the non-conflicting tile size selection algorithm. Using this algorithm, several candidates of non-conflicting tile dimensions are derived. The second algorithm "GcdPad" applies padding transformation after calculating the tile size. Tile is

Table 1. Non Conflicting Tile Sizes by Rivera's Method [2] (Problem Size: $200 \times 200 \times M$)

TK	1	1	1	1	2	2	2	2	3	3	3	4	4	4	...
TJ	1	10	41	256	1	4	5	15	5	11	15	4	15	56	...
TI	2048	200	48	8	960	200	160	40	72	40	24	72	16	8	...

selected to be a factor of cache size. The third algorithm "Pad" is a mixture of the first and the second algorithms. In this paper, the details of these algorithms are omitted. See [2] for the detail description of them.

After applying these algorithms, cost function is used to select the best tile size among candidates. Table 1 illustrates several non-conflicting tiles obtained by Euc3D algorithm in the case of $200 \times 200 \times M$ array (M can be any value) assuming 16KB cache (2048 elements of double precision). Because the cost function $Cost_{Riv}$ gets minimum in tile size of $(TI, TJ) = (24, 15)$, this tile size is selected as the best tile dimension.

3 Proposed Cost Model

The cost function $Cost_{Riv}$ described in the previous section is very simple. However, it is too simple to select the best tile dimension because $Cost_{Riv}$ does not reflect the effect of cache line. It selects nearly or completely square tile as the best one no matter what cache line is. In this section, we propose a new cost function which carefully takes account of cache line size.

3.1 Additional Cache Traffics

In the previous section, we described $N^3(TI+2)(TJ+2)/(TI \times TJ)$ is the total number of elements brought into cache. If we do not divide the whole data set into sub blocks and cache size were infinite, only N^3 elements would be needed to bring into cache. Therefore, the term of $(TI+2)(TJ+2)/(TI \times TJ)$ (cost function) represents additional cache traffic due to division of the planes.

We discuss the cost function from the viewpoint of additional cache traffic. The cost function of $Cost_{Riv}$ implies that additional cache traffic caused at each tile boundary is only one element. This is true when I-J plain is divided for J-direction. However, if I-direction (assuming I-direction is consecutive on memory) is divided, the number of elements transferred into cache at each boundary is not one but at most the number of words in cache line L.

Fig. 2 illustrates tiled I-J plain. *boundary 1-2* represents data which is in tile 2 and required for calculating tile 1. For example, when we calculate tile 1 (as illustrated in Fig. 2-(a)), we require not only elements within the tile 1 but also elements located in *boundary 1-2* and *boundary 1-3*. Because L elements are transferred when cache miss occurs, circled elements are transferred in addition to elements located in the tile 1. After a while, the computation reaches tile 2 (illustrated in Fig. 2-(b)). At this time, elements used in calculation of tile

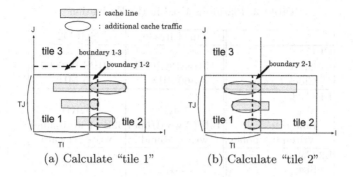

(a) Calculate "tile 1"   (b) Calculate "tile 2"

Fig. 2. I-J plain

1 are not expected to remain in the cache since K loop iterates across all the tiled I-J plain. Thus, cache lines including *boundary 2-1* are transferred into cache again, which causes additional cache traffics (circled elements in Fig. 2-(b)). Consequently, the total additional cache traffic in each boundary is usually L. Note that if a boundary of the tile is the same as that of cache line, $2L$ extra elements are additionally transferred.

3.2   New Cost Function

Let m be the number of divided blocks for I-direction and n be the number of divided blocks for J-direction. Assuming N^3 problem size, m and n are defined as $m = N/TI$ and $n = N/TJ$ (when it cannot be divided, it is rounded up). Dividing I-direction into m makes additional $L \times (m-1)$ cache traffic and dividing J-direction into n makes additional $2 \times (n-1)$ cache traffic. Therefore, we need to transfer approximately $N^2(L(m-1)+2(n-1))$ elements of array B in addition to total problem size. We can estimate total cache traffic of array B at $N^3 + N^2((m-1)L+2(n-1))$. Here, N^3 and N^2 are invariant under different values of m and n. Omitting these constants, we can obtain the following new cost function:

$$Cost_{New}* = L(m-1) + 2(n-1)$$
$$(m = \lceil \tfrac{N}{TI} \rceil, n = \lceil \tfrac{N}{TJ} \rceil) \qquad (2)$$

When $TI \times TJ$ is constant, this cost function $Cost_{New}*$ gets minimum when $TI : TJ = L : 2$.

Next, we consider array A in Fig 1-(c). If we assume a write around cache, array A is not brought into cache. However, if cache is a write allocate, elements of array A are brought into cache in addition to array B. Because array A has no stencil access pattern, dividing J-direction does not yield additional cache traffics for array A. However, dividing I-direction can cause additional cache traffics of approximately $N^2 L(m-1)$ elements[2].

---

[2] If TI is multiple of cache line size and the array is aligned, additional cache traffic does not occur.

Table 2. Platforms Used in the Evaluation

	Sun Ultra30	SGI O2
CPU		
- Chip	UltraSPARC-II	MIPS R10000
- Clock Cycle	360MHz	175 MHz
L1 Cache		
- Size	16KB	32KB
- Line Size	16B (2 word/line)	32B (4 word/line)
- write-miss policy	write around	write allocate
L2 Cache		
- Size	4MB	1MB
- Line Size	64B (8 word/line)	64B (8 word/line)
- write-miss policy	write allocate	write allocate

Moreover, if a kernel has additional array, additional cache traffic of $N^2 L(m - 1)$ can occur. To include these issues in cost function, we generalize our cost function $Cost_{New}*$ as follows:

$$Cost_{New} = P \times L(m-1) + Q \times 2(n-1) \tag{3}$$

Here, P is the number of arrays brought into cache. In the Jacobi Kernel, array A and array B are accessed. In write around cache, because only array B is brought into cache, P is set to 1. If cache is write allocate, P is set to 2. Q is the number of arrays accessed with stencil pattern. For example of Jacobi Kernel, because only array B is accessed with stencil pattern, Q is set to 1. In the case of 32B line (4 double words) and write allocate cache, (in this case, $L = 4, P = 2, Q = 1$), we select tile size of $(TI, TJ) = (40, 11)$ as the best tile dimension from Table 1 by applying $Cost_{New}$.

4   Performance Evaluation

4.1   Evaluation Methodology

Rivera and Tseng presented their performance evaluation using three kernels, 3D Jacobi, Red-Black SOR, and RESID subroutine. The detailed description and tiling transformation of these kernels are described in [2]. We compare our cost function $Cost_{New}$ with their $Cost_{Riv}$ using the same three kernels. Performance is measured on two real workstation, Sun Ultra30 and SGI O2. Specifications of both machines are shown in Table 2.

Because previous research [7] has revealed that tiling only targeted the L1 cache also improves L2 cache performance, we applied tiling transformation with targeting the L1 cache.

$N \times N \times 30$ problem size was used for the three kernels. N was varied from 400 to 600 for Sun Ultra30 and from 200 to 400 for SGI O2. When N is small, L2 cache can hold a few $N \times N$ plain. However, when N gets larger, L2 cache cannot hold such plains. Since the size of the third dimension (K dimension)

Table 3. Value of P and Q

Platform	Program	for L1		for L2	
		P	Q	P	Q
	Jacobi	1	1	2	1
Sun	Red-Black	1	1	1	1
	RESID	2	1	3	1
	Jacobi	2	1	2	1
SGI	Red-Black	1	1	1	1
	RESID	3	1	3	1

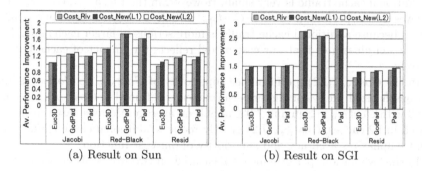

(a) Result on Sun   (b) Result on SGI

Fig. 3. Evaluation Result

does not affect comparison between two cost models, it was fixed at 30 to reduce experiment time. We also evaluated the performance of original codes (without tiling transformation) for comparison.

As for our cost model, the values of P and Q in $Cost_{New}$ are derived as mentioned in Section 3.2. Table 3 shows these values used in each kernel for both machines.

In the evaluation, the tile size was decided as follows. First, the candidates of non-conflicting tile size were obtained using three algorithms, Euc3D, GcdPad, and Pad, which are roughly mentioned in Section 2.4. Next, the best tile dimension was selected from the candidates by using each cost function, $Cost_{Riv}$ or $Cost_{New}$.

4.2   Evaluation Results

Fig. 3 presents average performance improvement of $Cost_{Riv}$ and $Cost_{New}$ compared with original codes for each program. The average number is the arithmetic means over the range of problem sizes. In the figure, Cost_Riv represents the result when the cost function of $Cost_{Riv}$ is used for selecting the tile size. In Cost_New(L1), we use cost function $Cost_{New}$ in which L1 cache line size is used as L of $Cost_{New}$, whereas L2 cache line size is used as L in C_New(L2).

Firstly, we discuss performance improvement brought by tiling. Higher performance is obtained by tiling in all the cases except for "Euc3D" algorithm using $Cost_{Riv}$ on Sun. Even in that case, performance improves in most of the problem sizes. However, average performance is degraded due to terrible performance degradation in a few problem sizes caused by pathological tile size. In the rest cases, results show good performance improvements (1.04 – 1.9 times higher performance). This indicates the effectiveness of tiling transformation in 3D PDE solvers.

Secondly, we compare $Cost_{Riv}$ with our $Cost_{New}$. As seen from the Fig. 3-(a), there is little performance difference between $Cost_{Riv}$ and $Cost_{New}$(L1). Because cache line size of Sun UltraSPARC 2 is 16B (2 double words), impact of additional cache traffic is very small. However, in the Fig. 3-(b), $Cost_{New}$(L1) achieves better performance than $Cost_{Riv}$ in SGI O2 whose cache line is 32B. This is because additional cache traffic is more serious in larger cache line. This comparison illustrates that our cost model reflects the impact of cache line accurately.

In the $Cost_{New}$(L2), performance superiority of $Cost_{New}$ to $Cost_{Riv}$ is further emphasized. 11.5% of higher performance is obtained in $Cost_{New}$(L2) compared with $Cost_{Riv}$ in Sun and 7.0% in SGI on average. Because large cache line is used in L2 caches, $Cost_{New}$(L2) greatly reduces additional cache traffic. This result emphasizes accuracy and superiority of our cost function.

In this evaluation, performance is measured on two workstations which have 64B of L2 cache line. When L2 cache line is larger than 64B, for example, SGI Origin 2000 has 128B of L2 cache line, the impact of cache line gets more significant. Recently, L2 cache line size tends to become larger for tolerating large memory latency. Thus, it is important to consider the effect of cache line when tiling is applied. Therefore, our cost function is very useful for tiling transformation of 3D PDE solvers.

5   Concluding Remarks

In this paper, we proposed a new cost model for tile size selection of 3D PDE solvers. Our cost model carefully takes account of the effect of cache line and reduces cache traffics. Though our cost function is slightly complex compared with conventional cost function, it is still very simple. Thus, our cost function is easy enough to be applied to compiler tiling optimization.

We presented performance evaluation of our cost function. The results revealed that our cost function successfully achieves better performance than conventional cost function. Our cost model greatly contributes to reduction of cache traffics especially in large cache line. Therefore, it is concluded that our proposed cost function is very effective for optimization of 3D PDE solvers.

Acknowledgement

This work is partially supported by the Research for the Future Program of JSPS in the "Computational Science and Engineering" Field (Project No. JSPS-RFTF 97P01102).

References

1. M. Lam, E. Rothberg and M. Wolf, "The cache performance and optimizations of Blocked Algorithms", Proc. ASPLOS-IV, pp.63–74, 1991
2. G. Rivera and C.-W. Tseng, "Tiling Optimizations for 3D Scientific Computations", In Proceedings of Supercomputing 2000, November, 2000.
3. S. Coleman and K. S. McKinley, "Tile size selection using cache organization and data layout", Proc. of PLDI, June 1995.
4. G. Rivera and C.-W. Tseng, "Data Transformations for Eliminating Conflict Misses", In Proceedings of SIGPLAN'98 Conference on Programming Language Design and Implementation, June, 1998.
5. O. Temam, E. Granston, and W.Jalby, "To Copy or Not to Copy: A Compile-time technique for assessing when data copying should be used to eliminate cache conflicts", In Proceedings of Supercomputing '93, November 1993.
6. M.J. Wolfe, D. Maydan, and D.-K Chen, "Combining Loop Transformations Considering Caches and Scheduling", In Proceedings of the 29th IEE/ACM International Symposium on Microarchitecture, December 1996.
7. G. Rivera and C.-W. Tseng, "Locality Optimizatioins for Multi-Level caches", In Proceedings of Supercomputing '99, November 1999.

An EPIC Processor
with Pending Functional Units

Lori Carter, Weihaw Chuang, and Brad Calder

Department of Computer Science and Engineering
University of California, San Diego
{lcarter,wchuang,calder}@cs.ucsd.edu

Abstract. The Itanium processor, an implementation of an Explicitly
Parallel Instruction Computing (EPIC) architecture, is an in-order processor that fetches, executes, and forwards results to functional units
in-order. The architecture relies heavily on the compiler to expose Instruction Level Parallelism (ILP) to avoid stalls created by in-order processing.

The goal of this paper is to examine, in small steps, changing the in-order
Itanium processor model to allow execution to be performed out-of-order.
The purpose is to overcome memory and functional unit latencies. To
accomplish this, we consider an architecture with *Pending Functional
Units* (PFU). The PFU architecture assigns/schedules instructions to
functional units in-order. Instructions sit at the pending functional units
until their operands become ready and then execute out-of-order. While
an instruction is pending at a functional unit, no other instruction can
be scheduled to that functional unit. We examine several PFU architecture designs. The minimal design does not perform renaming, and
only supports bypassing of non-speculative result values. We then examine making PFU more aggressive by supporting speculative register
state, and then finally by adding in register renaming. We show that the
minimal PFU architecture provides on average an 18% speedup over an
in-order EPIC processor and produces up to half of the speedup that
would be gained using a full out-of-order architecture.

1   Introduction

The Itanium processor, the first implementation of the IA64 Instruction Set Architecture, relies on the compiler to control the scheduling of instructions through
the processor, which reduces the complexity and increases the scalability of the
processor. In-order processing has severe IPC performance limits due to the in-
ability to allow execution to continue past an instruction with an outstanding
register use, where the register is being produced by a long latency instruction
currently executing. In this situation the whole front-end of the processor stalls
and it cannot issue any more instructions until the oldest instruction in the issue window has both of its operands ready. Out-of-order processors have the
capability to allow execution to continue, but at the cost of increased hardware
complexity.

H. Zima et al. (Eds.): ISHPC 2002, LNCS 2327, pp. 310–320, 2002.
© Springer-Verlag Berlin Heidelberg 2002

This paper describes our work in examining how to adapt the Itanium processor model to overcome the limitations caused by unpredictable load latencies as well as long latency instructions that the compiler could not hide. Our goal is to examine moving towards an out-of-order processor without adding a large amount of additional complexity. Rau [8] suggested the idea of small-scale reordering on VLIW processors to support object code compatibility across a family of processors. We describe a method that makes use of features that currently exist in the Itanium model.

In the Itanium, each cycle can have up to 6 instructions scheduled to proceed down through the pipeline and begin executing together. If there is an outstanding dependency for a member of the scheduled group, then all of the instructions in that group stall at the functional units and wait there until all instructions in that scheduled group can start to execute at the same time. To facilitate this, the Itanium already has functional units that have bypassing logic to allow the values being produced to be directly consumed in the next cycle by another functional unit.

In this paper, we examine *Pending Functional Units* (PFU) that expose a small window of instructions (those that have been allocated function units) to be executed out-of-order. The PFU model implements a simple in-order instruction scheduler. Instructions in this new architecture are issued exactly the same as in a traditional in-order VLIW architecture (instructions cannot issue if there are WAW dependencies). In this paper we examine three PFU models. Each model increases the complexity of the processor, but also increases the amount of ILP exposed. All of these form their scheduled group of instructions in-order, and the only instruction window that is exposed for out-of-order execution are the instructions pending at the functional units.

2   Pending Functional Units

In this section we describe the baseline in-order processor modeled, along with the different variations of the PFU architecture.

2.1   Baseline In-order EPIC Processor

We model the Itanium processor as described in [9]. The IA64 ISA instructions are *bundled* into groups of three by the compiler. A template included with the bundle is used to describe to the hardware the combination of functional units required to execute the operations in the bundle. The template will also provide information on the location of the stop bits in the bundle. Like Itanium, our *baseline in-order EPIC processor* has up to two bundles directed or *dispersed* [2] to the functional units in the EXP stage, subject to independence and resource constraints. All instructions dispersed are guaranteed to be independent of each other. We use the bundles and the stop bits to determine this independence. We call this in-order group of instructions sent to the functional units together a *scheduled group* of instructions. If any of the instructions in the scheduled group must stall because a data dependence on an instruction outside of this group is not yet satisfied, the rest of the instructions in the scheduled group will stall in the functional units as well.

2.2 Limited Out-of-order Execution Provided by Pending Functional Units

The goal of this research is to examine, in increasing complexity, the changes needed to allow the Itanium processor to execute out-of-order, for the purpose of overcoming memory and long latency functional unit delays. We provide this capability by turning the functional units of the Itanium processor into *Pending Functional Units*. These are similar to reservation stations [12], but are simpler in that no scheduling needs to be performed when the operands are ready, because the instruction already owns the functional unit it will use to execute. We limit our architecture to 9 functional units in this paper, since we model the PFU architecture after the Itanium processor.

In providing the PFU limited out-of-order model, we need to address how we are going to deal with the following three areas:

1. **Register Mapping.** The Itanium processor provides register renaming for rotating registers and the stack engine, but not for the remaining registers. We examine two models for dealing with registers for the PFU architecture. The first approach does not rename any registers, and instead the scheduler makes sure there are no WAW dependencies before an instruction can be issued in-order to a functional unit. Note, that WAR dependencies do not stall the issuing of instructions, since operands are read in-order from the register file and there can be only one outstanding write of a given register at a time. The second approach we examine allows multiple outstanding writes to begin executing at the same time by using traditional out-of-order hardware renaming.

2. **Scheduling.** When forming a schedule for the baseline Itanium processor, the dispersal stage does not need to take into consideration what has been scheduled before it in terms of resource constraints. In this model, the whole scheduled group goes to the functional units and they either all stall together, or all start executing together. For the PFU architecture, the formation of the scheduled group of instructions needs to now take into consideration functional unit resource constraints, since the functional unit may have a pending instruction at it. The scheduling of instructions (assigning them to functional units) is still performed in-order. To model the additional complexity, we add an additional scheduling pipeline stage (described in more detail below) to the PFU architecture. It is used to form the scheduled groups taking into consideration these constraints. Flush replay is used when the schedule is incorrect.

3. **Speculative State.** The Itanium processor can have instructions complete execution out-of-order. To deal with speculative state, the Itanium processor does not forward/bypass its result values until they are non-speculative. For our PFU architecture we examine two techniques for dealing with speculative

state. The first approach is similar to the Itanium processor, where we do not forward the values until they are known to be non-speculative. The second approach examines using speculative register state (as in a traditional out-of-order processor), and updating the non-speculative state at commit. This allows bypassing speculative state at the earliest possible moment – as soon as an instruction completes execution.

2.3   Instruction Scheduling

As stated above, for the Itanium processor, the formation of the scheduled group of instructions occurred without caring about the state of the functional units. Therefore, all of the functional units are either stalled or available for execution in the next cycle. This simplifies the formation of execution groups in the dispersal pipeline stage of Itanium.

In our PFU architecture we add an additional pipeline stage to model the scheduling complexity of dealing with pending functional units. The scheduler we use includes instructions *in-order* into a scheduled group and *stops* when any of the three conditions occur:

- An instruction with an unresolved WAW dependency is encountered.
- There will be no free functional unit N cycles into the future (N is 2 for the architecture we model) for the next instruction to be included into the group.
- The scheduled group size already includes 6 instructions, which is the maximum size of a scheduled group.

The second rule above is accomplished by keeping track of when each functional unit will be free for execution. This is the same as keeping track, for each instruction pending at a functional unit, when the operands for that instruction will be ready to execute. All functional units are pipelined, so the only time a functional unit will not be free is if the instruction currently pending at the functional unit has an unresolved dependency. The only dependencies that are non-deterministic are load's, which result in cache misses. The schedule is formed assuming that each load will hit in the cache.

When the schedule is wrong a replay needs to occur to form a legal scheduled group of instructions. The Alpha 21264 processor uses *Flush Replay* [5] (also referred to as squash replay) for schedule recovery. If one of the instructions in the scheduled group of instructions cannot be assigned because a functional unit is occupied, up to X groups of instructions scheduled directly after this scheduled group are flushed, and a new schedule is issued. The X cycle penalty is determined by the number of cycles between formation of the schedule and the execution stage. In this paper we assume that the penalty is two cycles.

2.4   Managing Speculative State

The out-of-order completion of instructions creates result values that are speculative and they may or may not be used depending upon the outcome of a prior unresolved branch.

The baseline EPIC processor we model does not forward its result values until they are non-speculative. Therefore, its pipeline up through the start of execution does not have to deal with speculative result values.

In the PFU architecture we examine two methods of dealing with speculative result values:

- Non-Speculative – The first approach is similar to the Itanium processor, where we do not forward the values until they are known to be non-speculative.
- Speculative – The second approach forwards the values in writeback right after the instruction finishes execution. This means that the register state dealt with in the earlier part of the pipeline is considered speculative register state (as in a traditional out-of-order processor), and the non-speculative register state is updated at commit. Note, multiple definitions are not an issue with speculative bypassing, since WAW dependencies must be resolved prior to an instruction entering a functional unit.

2.5   Summary of PFU Configurations Examined

The above variations are examined in 3 PFU configurations, which we summarize below:

- PFU – The base PFU configuration adds one stage to the pipeline for in-order scheduling as described earlier, and flush replay occurs when a functional unit predicted to be idle in the schedule has a pending instruction at it. It uses PFUs with no renaming, following the example of Thorton's simplification [11] of Tomasulo's early out-of-order execution design [12]. This configuration only forwards results and updates register state when they are known to be non-speculative.
- PFU Spec – Builds on PFU allowing bypassing of values as soon as they finish execution. The processor therefore needs to support speculative and non-speculative register state.
- PFU Spec Rename – Builds on PFU Spec adding traditional hardware renaming for predicates and computation registers. This eliminates having to stop forming scheduled group of instructions because of a WAW dependency.

3   Methodology

We created an EPIC simulator by modifying SimpleScalar [4]. It performs its simulation using IA64 traces. We extended the baseline SimpleScalar to model the Itanium stages and to simulate the IA64 ISA. Our extension focused on 5 areas. First, we added the ability to execute in-order and include the detection and enforcement of false WAW dependencies. Second, the simulator had to be extended to support predicated execution. One of the most significant changes in this area was the need to include the possibility of multiple definitions for

Table 1. Presents description of benchmarks simulated including instruction count in millions where the initialization phase of the execution ended and where the trace began recording.

benchmark	suite	end init	begin trace	description
go	95 C	6600	9900	Game playing; artificial intelligence
ijpeg	95 C	4900	7200	Imaging
li	95 C	1200	1800	Language interpreter
perl	2k C	4900	7200	Shell interpreter
bzip2	2k C	400	600	Compression
mcf	2k C	1300	1900	Combinatorial Optimization

the use of a register due to the fact that the same register could be defined along multiple paths that are joined into one through if-conversion. The IA64 supports software pipelining, so we implemented the functionality of rotating registers along with specialized instructions that implicitly re-define predicate registers. Another important feature of the IA64 ISA is its support for control and data speculation. To support this feature, we modeled the implementation of the ALAT, with its related operations and penalties. Finally, we added the ability to detect bundles and stop bits and appropriately issue instructions using this information.

Our traces are generated on IA64 machines running Linux through the ptrace system interface [6]. This allows a parent program to single-step a spawned child. Each trace is built using the *ref* data set, and includes 300 million instructions collected well after the initialization phase of the execution has finished. Table 1 shows the end of the initialization phase and the beginning of the trace in millions of instructions. The end of the initialization phase was determined using the Basic Block Distribution Analysis tool [10].

For each instruction in the trace range, we record the information necessary to simulate the machine state. For all instructions, we record in the trace file the instruction pointer, the current frame marker [1] and the predicate registers. In addition, we record the effective-address for memory operations and the previous function state for return. IA64SimpleScalar then reads in the trace file to simulate the program's execution.

IA64SimpleScalar decodes the traces using an IA64 opcode library containing a record for each IA64 instruction, and a library that interprets each instruction including templates and stop bits. This was built from the GNU opcode library that contains opcode masks to match the instruction, operand descriptions, mnemonic, and type classification. We enhanced this by adding a unique instruction identifier, quantity of register writers, and target Itanium functional units.

In this paper we used a number of SpecInt2000 and SpecInt95 benchmarks, four of which were compiled using the Intel IA64 electron compiler (go, ijpeg, mcf, bzip2), and two were compiled using the SGI IA64 compiler (li,perl). We used the "-O2" compilation option with static profile information. This level of optimization enables inline expansion of library functions, but confines op-

Table 2. Baseline Simulation Model.

L1 I Cache	16k 4-way set-associative, 32 byte blocks, 2 pipeline cycles
L1 D Cache	16k 4-way set-associative, 32 byte blocks, 2 cycle latency
Unified L2 Cache	96k 6-way set-associative, 64 byte blocks, 6 cycle latency
Unified L3 Cache	2Meg direct mapped, 64 byte blocks, 21 cycle latency
Memory Disambiguation	load/store queue, loads may execute when all prior store addresses are known
Functional Units	2-integer ALU, 2-load/store units, 2-FP units, 3-branch
DTLB, ITLB	4K byte pages, fully associative, 64 entries, 15 cycle latency
Branch Predictor	meta-chooser predictor that chooses between bimodal and 2-level gshare, each table has 4096 entries
BTB	4096 entries, 4-way set-associative

timizations to the procedural level. It produces a set of instructions in which an average of 24% found in the trace were predicated, with an average of 4.3% predicated for reasons other than branch implementation and software pipelining. This can arise due to if-conversion, or predicate definitions for use in floating point approximation. Table 2 shows the parameters we used for the simulated microarchitecture, which were modeled after the Itanium.

4   Results

This section presents results for adding PFUs to an EPIC in-order processor, and then compares the performance of this architecture to a traditional out-of-order processor.

4.1   Adding PFUs to an EPIC In-order Processor

Figure 1 shows the performance benefit of allowing the in-order EPIC processor to execute out-of-order using Pending Functional Units. It shows IPC results for the baseline EPIC processor, the various PFU configurations, and a full out-of-order processor with issue buffer sizes of 9 and 64.

The improvement for using the minimal PFU architecture over the baseline EPIC processor ranged from 1% (go) to 27% (bzip2). The average improvement of the PFU model over the EPIC model was 18%. The low IPC and limited improvement seen for go can be attributed to poor cache and branch prediction performance as shown in Table 3. This table describes, for each benchmark, information about cache and branch predictor activity for the baseline EPIC configuration. In particular, the second column describes the number of load misses that occurred in the L1 data cache per 1000 executed instructions. The third column is the number of instructions that missed in the L1 instruction cache per 1000 executed instructions. The next two columns refer to the L2 and

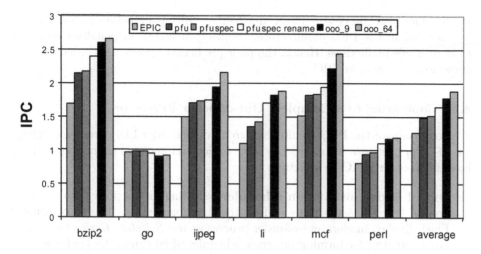

Fig. 1. Comparing the IPCs of the various PFU models and the out-of-order models with issue buffer sizes of 9 and 64 to the EPIC model.

Table 3. Presents misses per 1K instructions in the various caches. Levels 2 and 3 are unified caches. The last column shows the branch prediction accuracy.

benchmark	dl1 miss/1K	il1 miss/1K	ul2 miss/1K	ul3 miss/1K	Br Pred Hit Rate
bzip2	.279	.005	.138	.131	91%
go	11.242	47.930	13.592	1.771	69%
ijpeg	2.629	21.193	1.058	.493	87%
li	6.336	3.451	3.066	.071	83%
perl	11.780	41.791	13.327	2.357	83%
mcf	30.949	.002	32.017	15.893	84%

L3 caches, which are unified. The last column provides the branch prediction hit rate produced by the number of hits divided by the number of updates to the predictor. The predictor is updated for every non-speculative branch executed. As can be seen from this table, the branch prediction accuracy for go was significantly lower than that of the other benchmarks at 69%. In addition, go also had a high L1 instruction cache miss rate, where there were 48 misses for every 1000 executed instructions.

The second configuration considered is PFU Spec. This approach adds an additional 2% improvement, bringing the average improvement in IPC over the EPIC model up to 20%. The largest additional improvement achieved using speculative state is seen in the li benchmark, at 6%.

For the final PFU configuration, we added full register renaming to the model just described. On average, renaming improved the performance of the previous model by 10% as compared to the baseline EPIC, bringing the total average improvement to 30%. Li saw the biggest improvement over EPIC with a gain in IPC of 56%. In this case, the IPC produced by go was actually lower than

the baseline EPIC processor. This is due to the additional pipeline stages, which
we modeled in the PFU architecture, resulting in increased penalties related
to branch mis-predictions. Hence, the poor prediction accuracy recorded by go
translated into reduced IPC.

4.2   Comparing to a Complete Out-of-order Processor

We now compare the PFU architecture to an out-of-order IA64 processor. The
out-of-order processor we model has three major architectural changes over the
baseline in-order EPIC architecture:

1. **Out of Order instruction scheduler**, selecting instructions from a window of 9 or 64 instructions to feed the functional units. Instead of using
   Flush Replay, modern out-of-order processors use *Selective Replay* to avoid
   large penalties for forming incorrect schedules often caused by load misses.
   The scheduler in the Pentium 4 [7] uses a replay mechanism to selectively
   re-issue only those instructions that are dependent on the mis-scheduled instruction. This adds more complexity in comparison to using flush replay as
   in the PFU architecture. We modeled flush replay for out-of-order execution,
   but the IPC results were the same or worse on average than the PFU architecture. Therefore, we only present out-of-order results in this paper for
   selective reply.
2. **Register renaming** to eliminate WAW dependencies
3. **Out-of-order completion** of execution, so speculative register state needs
   to be maintained (a reorder buffer or register pool).

An out-of-order pipeline architecture tries to make use of the available runtime information and schedules instructions dynamically. In an out-of-order environment, instructions effectively wait to execute in either an issue buffer, active list of instructions, or reservation stations. We use the notion of an issue
buffer that is dynamically scheduled to supply instructions to all functional units.
In this section, we examine configurations of an out-of-order architecture issue
buffer sizes of 9 and 64. We chose 9 in order to make a comparison to the PFU
model with 9 pending functional units.

The main difference between PFU spec rename and the OOO architecture is
that PFU spec rename forms its schedule in-order constrained by predicting if
there is going to be a free functional unit for the next instruction to be included
into the schedule. In comparison, the OOO architecture forms its schedule pick-
ing any instruction from the 9 or 64 oldest un-issued instructions that it predicts
will have their operands ready by the time they reach the functional units.

Our out-of-order version of the EPIC pipeline varies from the traditional
out-of-order architecture only as was required to support EPIC features such as
predication and speculation. For example, in the renaming process, each definition of a register is given a unique name, and then that name is propagated to
each use. However, predication allows for multiple possible reaching definitions.
Consequently, this process must be extended to include knowledge of guarding

predicates in conjunction with definitions. We implement out-of-order renaming as described in [13]. As in that study, we add 4 stages to the baseline EPIC pipeline, 2 for renaming, and 2 for scheduling, to model the additional complexity for out-of-order execution.

Figure 1, provides IPC results for the 2 different configurations that were evaluated for the out-of-order model. As mentioned, these varied by the size of the schedulable window of instructions. In the out-of-order architecture, up to six instructions can be scheduled/issued to potentially 9 functional units. The results show that an issue window of 64 instructions produces an additional speedup of 31% over the base PFU configuration. For bzip2, the baseline PFU architecture experienced half of the improvement seen by the full out-of-order model. Go is again an interesting case. The out-of-order models perform worse than all other configurations. This is attributed to the high penalties due to a deeper pipeline and poor branch prediction.

For several of the programs, the out-of-order model with an issue window of 9 performs only slightly worse than the out-of-order model with 64. For the benchmarks bzip2, go, li, and perl, the difference in IPC was under 3%. The average difference in IPC between the two out-of-order configurations was only 5%. For bzip2, ijpeg and perl, the baseline PFU architecture achieves half the performance improvement that is seen by the aggressive OOO model with a window of 9 instructions.

5   Conclusions

In-order processors are advantageous because of their simple and scalable processor pipeline design. Performance of the program is heavily dependent upon the compiler for scheduling, load speculation, and predication to expose ILP. Even with an aggressive compiler, the amount of ILP will still be constrained for many applications.

To address this problem we explored adding the ability to start executing instructions out-of-order to an in-order scheduled processor. We examined the Pending Functional Unit (PFU) architecture, where instructions are allocated to the functional units in order (as in the Itanium), but they can execute out-of-order as their operands become ready. Augmenting the EPIC in-order processor with PFUs provides an average 18% speedup over the EPIC in-order processor. The additional cost in complexity to achieve this speedup is the dynamic in-order scheduling of the functional units and the implementation of flush replay for when the schedule is incorrect.

When comparing the base PFU architecture to a full out-of-order processor, our results show that the PFU architecture yields 44% of the speedup, on average, of an out-of-order processor with an issue window of 9 instructions and renaming. The out-of-order processor with a window of 9 instructions achieved results within 5% of having a window of 64 instructions.

Executing instructions in a small out-of-order window is ideal for hiding L1 cache misses, and long latency functional unit delays. Another way to attack

these stalls would be to increase the amount of prefetching performed by the
processor, and reduce the latency of the L1 cache and functional units. The
plans for the next generation IA64 processor, McKinley, includes reducing latencies, increasing bandwidth, and larger caches [3]. While latency reduction via
prefetching and memory optimizations will provide benefits, a small out-of-order
will most likely still be advantageous, since not all L1 misses will be able to be
eliminated. Performing this comparison is part of future work.

Acknowledgements

We would like to thank the anonymous reviewers for providing useful comments
on this paper. This work was funded in part by NSF grant No. 0073551, a grant
from Intel Corporation, and an equipment grant from Hewlett Packard and Intel
Corporation. We would like to thank Carole Dulong and Harpreet Chadha at
Intel for their assistance bringing up the Intel IA64 Electron compiler.

References

1. IA-64 Application Instruction Set Architecture Guide, Revision 1.0, 1999.
2. Itanium Processor Microarchitecture Reference:for Software Optimization, 2000.
   http: //www.developer.intel.com/design/ia64/itanium.htm.
3. 2001 - a processor odyssey: the first ever McKinley processor is demonstrated by
   hp and Intel at Intel's developer forum, February 2001.
   http: //www.hp.com/products1/itanium/news_events/archives/NIL0014KJ.html.
4. D.C. Burger and T.M. Austin. The Simplescalar Tool Set, version 2.0. Technical
   Report CS-TR-97-1342, University of Wisconsin, Madison, Jun 1997.
5. COMPAQ Computer Corp. Alpha 21264 microprocessor hardware reference manual, July 1999.
6. S. Eranian and D. Mosberger. The Linux/IA64 Project: Kernel Design and Status
   Update. Technical Report HPL-2000-85, HP Labs, June 2000.
7. G. Hinton, D. Sager, M. Upton, D. Boggs, D. Carmean, A. Kyker, and P. Roussel.
   The microarchitecture of the pentium 4 processor. *Intel Technology Journal Q1*,
   2001.
8. B. R. Rau. Dynamically scheduled vliw processors. In *Proceedings of the 26th
   Annual Intl. Symp. on Microarchitecture*, pages 80–92, December 1993.
9. H. Sharangpani and K. Arora. Itanium processor microarchitecture. In *IEEE
   MICRO*, pages 24–43, 2000.
10. T. Sherwood, E. Perelman, and B. Calder. Basic block distribution analysis to find
    periodic behavior and simulation points in applications. In *Proceedings of the 2001
    International Conference on Parallel Architectures and Compilation Techniques*,
    September 2001.
11. J. E. Thornton. Design of a Computer, the Control Data 6600, 1970. Scott,
    Foresman, Glenview, Ill.
12. R. M. Tomasulo. An Efficient Algorithm for Exploiting Multiple Arithmetic Units.
    IBM Journal of Research and Development, 11(1):25–33, 1967.
13. P. H. Wang, H. Wang, R. M. Kling, K. Ramakrishnan, and J. P. Shen. Register
    renaming for dynamic execution of predicated code. In *Proceedings of the 7th
    International Symposium on High Performance Computer Architecture*, February
    2001.

Software Energy Optimization
of Real Time Preemptive Tasks
by Minimizing Cache-Related Preemption Costs

Rakesh Kumar[1], Tusar Kanti Patra[2], and Anupam Basu[2]

[1] Department of Computer Science and Engineering,
University of California, San Diego
La Jolla, CA 92093-0114, USA
rakumar@cs.ucsd.edu
http://www.cse.ucsd.edu

[2] Department of Computer Science and Engineering,
Indian Institute of Technology, Kharagpur
WB-721302, India
{Tpatra,Anupam}@cse.iitkgp.ernet.in

Abstract. An attempt has been made to optimize the software energy of real time preemptive tasks by minimizing the cache related preemption costs which are primarily incurred due to the inter-task interference in the cache. We have presented an algorithm that outputs an "optimum" task layout which reduces the overall inter-task interference in cache and thus reduces the preemption costs of each task of a given task set. We have compared the result of our estimated layout with that of the random layout generated for benchmark examples for demonstrating the performance of our algorithm.

1   Introduction

Software energy optimization is an important issue in real time embedded system design. Different techniques have been developed by several researchers for energy optimization of embedded software. In our previous work [7], we have proposed a technique of cache sizing, based on instruction level energy model, for optimizing software energy of embedded system in non-preemptive environment. In this work we attempt to reduce the software energy of the preemptive tasks by reducing the energy losses incurred in the preemption process.

Task preemption introduces an indirect cost, due to caching, which is known as cache-related preemption cost. When a task is preempted, a large number of memory blocks belonging to that task are displaced from the cache between the time the task is preempted and the time the task resumes execution. When the preempted task resumes its execution, it again reloads the cache with the useful blocks (at that point of execution) which might have been previously displaced. Such cache reloading greatly increases the task execution time and also incurs energy losses.

H. Zima et al. (Eds.): ISHPC 2002, LNCS 2327, pp. 321–328, 2002.

This problem has been addressed in some recent studies by incorporating the cache related preemption costs into schedulability analysis of real time periodic tasks [1, 2, 3, 4]. In both [1] and [2], it is assumed that all the memory blocks (of a preempted task), which have been displaced from the cache by a preempting task, are to be reloaded in cache when the preempted task resumes its execution. This leads to a loose estimation of cache related preemption cost since the replaced block may not be useful any more to the preempted task at that point of execution. In [3], the concept of useful cache block is introduced. A useful cache block at an execution point is defined as a cache block that contains a memory block that may be referenced before it is replaced by another cache block. The number of useful cache blocks at an execution point is taken as an upper bound on the cache related preemption cost that may be incurred if the task is preempted at that point. [4] presents a task layout generation technique which minimizes the cache related preemption cost by minimizing inter task interference in the cache. It presents an integer linear programming (ILP) formulation based on response time equation. The solution of the ILP formulation generates the desired layout of the tasks in primary memory and gives estimated WCRT of the task whose response time equation is the objective function in the formulation. In this approach the WCRT as well as the cache related preemption delay (CRPD) of one task is minimized while satisfying the deadlines of the others. This is a very useful in the context of schedulability analysis, but does not necessarily minimize the CRPD of each of the tasks.

We focus on minimization of inter-task interference of each task in the cache, and thus on overall energy saving by all the tasks of a given task-set. The idea is that if a cache block is used by one task and not used by any other task in the task-set, then that cache block will not contribute to the cache related preemption loss of that task. We have used a Backtrack algorithm for generating an appropriate task layout in cache with the minimum overlapping among the tasks. The reverse mapping from cache to memory is trivial. In our experiments, we have worked with Instruction Cache only and used the instruction set architecture of 486DX2 processor for simulation

This paper is organized as follows. In section 2 we describe the energy formulation for preemptive tasks. Section 3 describes the cache interference matrix.In section 4 we present the problem description. The Backtrack Algorithm, which generates the appropriate layout, is presented in section 5. In section 6 we present the results of the experiments and finally, in section 7, we conclude this paper.

2   Energy Formulation for Preemptive Tasks

Vivek Tiwary et al in [5] first introduced the energy consumed by a given program as follows:

$$E_p = \sum_i (B_i \times N_i) + \sum_{i,j} (O_{i,j} \times N_{i,j}) + \sum_k E_k \qquad (1)$$

The base cost, B_i, of each instruction, i, multiplied by the number of cycles it takes to be executed, N_i, is added up to give the base cost of the program, P. $O_{i,j}$

is the switching overhead occurred when instruction j executed after instruction i and $N_{i,j}$ is the number of times j executed after i. E_k is the energy contribution of other instruction effects, k, (stalls and cache misses) that will occur during the execution of the program. The model presented above did not deal with in details about the energy consumed by the memory accesses and cache misses.[7] and [8] extended this model to take these parameters into account.

However, the energy figure obtained in [7] and [8] is the energy consumed by a program or task while executing in a non-preemptive environment. The energy consumed by a task τ_i of a task set consisting n periodic tasks $\tau_1, \ldots \tau_n$, executing in a preemptive environment, thus, can be formulated:

$$(E_i)_p = (E_i)_{wp} + \sum_{j \in hp(i)} \lceil R_i/T_j \rceil * CRPL_{i,j} \qquad (2)$$

where $(E_i)_{wp}$ is the energy consumed by τ_i when executed without preemption. R_i denotes the response time of task τ_i. T_j is the time period of $\tau_j(j \leq i)$, which belongs to the set of tasks that have higher priority than τ_i. $CRPL_{i,j}$ is the energy loss incurred when τ_i is preempted by a set of higher priority task $\{\tau_j\}$ and is termed as cache related preemption loss(CRPL). Hence $CRPL_{i,j}$ can be defined as follows:

Let U_m be the set of cache blocks that are used by task τ_m and $|U_m|$ denote the number of elements in the set U_m. Then $CRPL_{i,j}$ is given by:

$$CRPL_{i,j} = CRT * \left| U_i \bigcap \left(U_{i-1} \bigcup U_{i-2} \bigcup \ldots \bigcup U_j \right) \right| * I_c \qquad (3)$$

where CRT and I_c are the cache refill time and cache miss current penalty respectively.

The middle term at RHS in the above equation gives the upper bound on the number of cache blocks of τ_i that interfered with the set of cache blocks occupied by higher priority tasks which preempt the task τ_i this time and to be reloaded when τ_i resumes its execution after the preemption. This number is multiplied by cache refill time and cache miss current penalty to give the estimate on energy loss incurred in this preemption. The value of $\lceil R_i/T_j \rceil$ gives the number of times task τ_i is preempted by the higher priority tasks in $\{\tau_j\}$. The value of R_i is computed using response time equation presented in [4] as:

$$R_i = C_i + \sum_{j \in hp(i)} \lceil R_i/T_j \rceil (C_j + \gamma_{ij}) \qquad (4)$$

where R_i denotes the WCRT for τ_i, C_i the WCET for τ_i without preemption, T_j the time period of τ_j and γ_{ij} is the cache related preemption delay (CRPD) that task τ_j imposes on lower priority tasks ((j+1, ..., (i) and defined as

$$\gamma_{ij} = CRT * \left| U_j \bigcap \left(U_{j+1} \bigcup \ldots \bigcup U_i \right) \right|. \qquad (5)$$

Now, we define a runtime as the period of time between the start and stop of a system on which a set of periodic tasks is executed in preemptive environment.

For periodic tasks, many instances of a task may be executed at a particular runtime. Since energy be the figure of merit, we have to consider all the instances of tasks executed in that runtime. If $(E_i)_{wp}^k$ and $(E_i)_p^k$ be the energy consumed by τ_i in k-th instance while executing in non-preemptive and preemptive environment respectively and R_i^k be the response time of τ_i in in k-th instance, then the equation (2) can be rewritten as,

$$(E_i)_p^k = (E_i)_{wp}^k + \sum_{j \in hp(i)} \lceil R_i^k/T_j \rceil * CRPL_{i,j} \tag{6}$$

Therefore, the total energy consumed by each task τ_I, in that runtime, is given by summing the energy consumed by the task in each instance of execution as follows,

$$(E_i)_{total} = \sum_k (E_i)_p^k \tag{7}$$

and the total energy consumed by all the tasks executed in that run time is given below.

$$E_{TOTAL} = \sum_i (E_i)_{total} \tag{8}$$

3   Cache Interference Matrix

Before describing the problem in detail, we define a Cache Interference Matrix, $M_{N \times C}$, to capture the inter-task interference in the cache, where:

$M_{ij} = 1$, if τ_i uses the j_{th} cache block.
$M_{ij} = 0$, otherwise.

Here,

N = No. of tasks
C = No. of cache blocks(or cache lines)

Let us consider a real-time system consisting of three tasks $\tau_i (0 \leq i \leq 2)$ whose layout in main memory and cache mapping are shown in figure 1.

We assume that the size of each task is 2-instruction memory block. The smaller i denotes higher priority task. With this layout, task τ_0 and τ_2 conflict with each other while τ_1 doesn't conflict with any other tasks in the task-set. Then the $CRPL_{1,0}$ is given by:

$$CRPL_{1,0} = CRT * |U_0 \bigcup U_1| * I_c = CRT * |\{c2, c3\} \bigcup \{c0, c1\}| * I_c = 0.$$

where $CRPL_{1,0}$ is the cache related preemption loss of τ_1 when it is preempted by τ_0.

4   Problem Formulation

To formulate the problem of minimizing cache-related preemption loss, we use the Cache Interference matrix, $M_{N \times C}$, which is a binary matrix where $M_{ij} = 1$, if τ_i uses the j-th cache block.

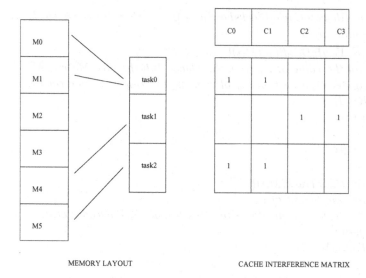

MEMORY LAYOUT   CACHE INTERFERENCE MATRIX

Fig. 1. An example of task layout and the corresponding cache mapping

Let us consider the parameter P_i, where

$$P_i = \sum_{j=0}^{C-1} min \left(1, \sum_{k \in hp(i)} M_{ij} M_{kj} \right)$$

P_i denotes the number of cache blocks that are used by τ_i and atleast one of the higher priority tasks. Therefore, it gives an upper bound on the number of cache blocks that need to be reloaded once τ_i is preempted. Hence, we focus on the problem that is to Minimize: P_i.

5   The Backtrack Algorithm

Here, we use a backtrack algorithm to minimize the cache related preemption loss of each tasks by reducing the overall task interference in the cache. In this approach the cache blocks are allocated starting from the task with the highest priority. At every stage, we choose the allocation that minimizes P_i and verify whether the task is schedulable or not by comparing the response time with deadline. If not, a backtracking strategy is used.

```
* Procedure BacktrackAlgo( ) {
    M_{N×C} = [0]_{N×C}
    for i = 1 to N do {
    AllocateCacheBlocksForTask(M, i)
    if(VerifyDeadline(M, i) = = false)
    BackTrack(M, i)
    }
}
```

* procedure *AllocateCacheBlocksForTask(M, i)*: Allocates cache blocks to τ_i such that P_i is minimized.
* Procedure *VerifyDeadline(M,i)*{
    - Obtain the value of τ_{ij} from the Cache Interference Matrix, M
    - Compute response time R_i of τ_i using the response time equation
    - if($R_i \leq T_i$)
    return true
    else
    return false
    }
* Procedure *BackTrack(M,i)*{
    - change the layout of τ_{i-1}
    - try to allocate cache blocks to τ_i such that VerifyDeadline(M, i) = = true.
    - On failure,
    BackTrack(M,i-1)
    }

It may be noted that while the worst-case complexity of this algorithm is exponential in the number of cache lines, it will run much faster if the layout problem does not suffer from the local minima problem.

6   Incorporating Usefulness in the Formulation

To determine the usefulness of cache blocks in the formulation we follow the per-task analysis technique presented in [3]. The per-task analysis will produce a Usefulness Matrix $U_{N \times O}$, where N is the number of tasks and O is the number of blocks in the task of maximum size. $U_{N \times O}$ is a binary matrix such that:

* $U_{ij} = 1$ iff the jth memory block of τ_i is a useful block.
* $U_{ij} = 0$ iff the jth memory block of τ_i is not a useful block.

Now, let us consider the parameter P_i, where

$$P_i = \sum_{j=0}^{C-1} min \left(1, \sum_{k \in hp(i)} U_{ij} M_{kj} \right)$$

P_i denotes the number of cache blocks that are useful to τ_i and used at least one of the higher priority tasks. The algorithm in this case remains the same as presented above.

7   Experimental Results and Discussion

We selected a set of popular DSP applications in our example task-set. The details of these task-set alongwith the cache memory details are provided as input to an estimator, which uses the Backtrack Algorithm to output the optimum task-layout. Then this layout is used in a simulator,developed by us, to observe the estimated energy consumption by each task and cache related pre-emption losses and also the total energy consumption by the entire task-set over a given runtime (i.e., simulation period). The simulator provides a simulation environment for real time shared-memory architecture and a scheduler. In our experiment, we have worked with I-Cache only and used the ISA of 486DX2 processor for simulation. We have used the data for instruction current cost, switching overhead and cache miss current penalty from [6]. To show the reduction in the cache related preemption loss, we have compared the energy estimate observed by using the optimum task-layout generated by the estimator with the energy observed by task layouts generated randomly.

The two task sets along with their specifications are given in the table below.

Task-Set	Task	Task-size(mem. blocks)	Period(cycles)	$E_{wp}(1_{st}$ instance)	E_{wp}(other instances)
Γ_1	τ_1:compress.s	76	4000	600.8	438.9
Γ_1	τ_2:Laplace.s	114	8000	691.0	447.0
Γ_2	τ_1:Scott.s	20	1000	117.4	76.3
Γ_2	τ_2:Array.s	19	2000	119.8	80.9
Γ_2	τ_3:Fibcall.s	49	4000	331.5	297.0

Γ_1 runs for 8000 cycles and Γ_2 runs for 4000 cycles. We obtained the energy consumption of each task in non-preemptive case by executing them on the simulator individually. The results are shown below:

Layouts	layout 1	layout 2	layout 3	layout 4	layout 5	layout 6	layout 7	layout 8	Estimators
CRPL of τ_2	67.0	108.1	114.6	97.3	71.4	43.4	32.5	114.6	4.5
$E_{T}OTAL$	1929	2001	1977	1925	1873	1839	1860	1994	1802

Corresponding results for the preemptive case are.

Layouts	layout 1	layout 2	layout 3	layout 4	layout 5	layout 6	layout 7	layout 8	Estimators
CRPL of τ_2	45.3	57.9	77.7	82.1	73.4	60.4	48.2	48.2	38.9
CRPL of τ_3	71.2	51.9	28.1	36.7	73.3	62.4	54.0	47.4	45.0
$E_{T}OTAL$	1090	1089	1079	1101	1142	1090	1044	1042	1018

Results show that the "optimum" layout, which is generated by minimizing the inter-task interference in cache, gives the minimum cache related preemption loss and earns a maximum saving in the total energy consumption. In the above results, we see that on an average at most 10.4% of saving can be achieved in the total energy consumption, by the given task sets, by using the optimum task-layout generated by our algorithm. By analyzing the results of task-set Γ_2, we observe that CRPL of τ_2 is minimum at the Estimated layout and CRPL of τ_3 is minimum at the layout 3, whereas the overall CRPL of τ_2 and τ_3 is minimum at the Estimated layout only. Therefore the "optimum" layout may not necessarily the layout for which the CRPL of each task of the task-set will be minimum, but this is the layout for which the overall CRPL of a given task-set will be minimum which is more important in the case of energy consideration.

8   Conclusions

In this work we have done a formulation for estimating the software energy of preemptive real-time tasks and focused on software energy minimization by minimizing the cache related preemption loss that is again achieved by minimizing

the inter-task interference by generating optimum task layout in cache while
multiple tasks are executed in real time preemptive environment.

References

1. Basumallick,S.,Nilsen,K.: Cache Issues in Real-Time Systems. Proceedings of the
   1st ACM SIGPLAN Workshop on Language, Compiler, and Tool Support for Real-
   time Systems, June 1994.
2. Busquets-Mataix,J.V., Serrano-Martin,J.J.,Ors,R.,Gil,P.,Wellings,A.:Adding Instruction cache Effect to Schedulability Analysis of Preemptive Real-Time Systems".Proceeding of the 2nd Real-Time Technology and Applications Symposium,
   June 1996.
3. Lee,C.G.,Hahn,J.,Seo,Y.M.,Min,S.L.,Ha,R., Hong,S.,Park,C.Y.,Lee,M., Kim,C.S.:
   Analysis of Cache Related Preemption Delay in fixed-priority Preemptive schedul-
   ing.IEEE Transactions on Computers, 47(6), June 1998.
4. Datta,A.,Choudhury,S.,Basu,A.,Tomiyama,H.,Dutt,N.: Task Layout Generation
   To Minimize Cache Miss Penalty For Preemptive Real Time tasks: An ILP Approach. Proceedings of the ninth Workshop on Synthesis And System Integration
   of Mixed Technologies (SASIMI), pp.202-208, April 2000
5. Tiwary,V.,Malik,S.,Wolfe,A. Power Analysis of Embedded Software: A First Step
   Towards Software Power Minimization. IEEE Transactions on VLSI systems, December 1994.
6. Lee,M.T.,Tiwary,V.,Malik,S.,Fujita,M.:Power analysis and Minimization Techniques for Embedded DSP Software. IEEE Transactions on VLSI Systems, December 1996.
7. Patra,T.,Kumar,R.,Basu,A.: Cache Size Optimization For Minimizing Software
   Energy in Embedded Systems. Proceedings of International Conference in Communications, Computers and Devices(ICCCD), pp262-266, December 2000.
8. Shuie,W.T.,Chakraborty,C: Memory Exploration for Low Power Embedded Systems. Proceeding of 36th ACM/IEEE Design Automation Conference, June 1999.
9. Shin,Y.,Choi,K.:Power Conscious Fixed Priority Sheduling for Hard Real-Time
   systems. Proceeding of 36th ACM/IEEE Design Automation Conference, June
   1999.

Distributed Genetic Algorithm
with Multiple Populations Using Multi-agent

Jung-Sook Kim

Dept. of Computer Science, Kimpo College
San 14-1, Ponae-ri, Wolgot-myun, Kimpo-si, Kyounggi-do, 415-870, Korea
kimjs@kimpo.ac.kr

Abstract. This paper designs a distributed genetic algorithm in order to reduce the execution time and obtain more near optimal using master-slave multi-agent model for the TSP. Distributed genetic algorithms with multiple populations are difficult to configure because they are controlled by many parameters that affect their efficiency and accuracy. Among other things, one must decide the number and the size of the populations (demes), the rate of migration, the frequency of migrations, and the destination of the migrants. In this paper, I develop two dynamic migration window methods, increasing dynamic window and random dynamic window, that control the size and the frequency of migrations. In addition to this, I design new genetic migration policy that selects the destination of the migrants among the slave agents.

1 Introduction

In the traveling salesman problem, a set of N cities is given and the problem is to find the shortest route connecting them all, with no city visited twice and return to the city at which it started. In the symmetric TSP, $distance(c_i, c_j) = distance(c_j, c_i)$ holds for any two cities c_i and c_j, while in the asymmetric TSP this condition is not satisfied.

Since TSP is a well-known combinatorial optimization problem and belongs to the class of NP-complete problems, various techniques are required for finding optimum or near optimum solution to the TSP. The dynamic programming and the branch-and-bound algorithm could be employed to find an optimal solution for the TSP. In addition to the classical heuristics developed especially for the TSP, there are several problem-independent search algorithms which have been applied to the TSP for finding near optimum solution, such as genetic algorithm[1, 5, 8], ant colonies[2], neural networks[2] and simulated annealing[2, 6]. And parallel and distributed systems have emerged as a key enabling technology in modern computing. During the last few years, many distributed algorithms are designed in order to reduce the execution time for the TSP with exponential time complexity in the worst case[7].

Software agents can be classified in terms of a space defined by three dimensions of intelligence, agency and mobility. Currently agent systems generally do not exploit all the capabilities classified by three dimensions. For example, multi-agent systems of distributed artificial intelligence try to execute a given task using a large number of possibly distributed but static agents that collaborate and cooperate in an intelligent manner with each other[4].

Under the right circumstances, I propose a new distributed genetic algorithm with multiple populations using master-slave multi-agent model for the TSP. Distributed

H. Zima et al. (Eds.): ISHPC 2002, LNCS 2327, pp. 329–334, 2002.

genetic algorithm is based on the distributed population structure that has the potential of providing better near optimal value and is suited for parallel implementation. And distributed genetic algorithm executes a conventional genetic algorithm on each of the distributed populations. But the distributed genetic algorithm with multiple populations is difficult to configure because they are controlled by many parameters that affect their efficiency and accuracy. Among other things, one must decide the number and the size of the populations, the rate of migration, the frequency of migrations, and the destination of the migrants[1, 12, 14]. In this paper, I develop two dynamic migration window methods that control the size and the frequency of migrations. In addition to this, I design new genetic migration policy that selects the destination of the migrants among the slave agents.

The rest of this paper is organized as follows. Section 2 presents the related works and section 3 explains two dynamic migration window methods and the frequency of migrations, and section 4 describes the distributed genetic algorithm using multi-agent. Section 5 summarizes the methods of the experiments and the results. The remaining section presents the conclusion and outlines areas for future research.

2   Related Works

2.1   Genetic Algorithms

Distributed genetic algorithms have received considerable attention because of their potential to reduce the execution time in complex application. One common method to parallelize genetic algorithm is to use multiple demes (populations) that occasionally exchange some individuals in a process called migration.

The studies on genetic algorithms for the TSP provide rich experiences and a sound basis for combinatorial optimization problems. Major efforts have been made to achieve the following. 1. Give a proper representation to encode a tour. 2. Devise applicable genetic operators to keep building blocks and avoid illegality. 3. Prevent premature convergence. Permutation representation has an appeal not only for the TSP but also for other combinatorial optimization problems. This representation may be the most natural representation of a TSP tour, where cities are listed in the order in which they are visited. The search space for this representation is the set of permutations of the cities. For example, a tour of a 17-city   TSP $1 - 2 - 3 - 4 - 5 - 6 - 7 - 8 - 9 - 10 - 11 - 12 - 13 - 14 - 15 - 16 - 17$ is simply represented as follows: <1 2 3 4 5 6 7 8 9 10 11 12 13 14 15 16 17>. The strength of genetic algorithms arises from the structured information exchange of crossover combination of highly fit individuals. Until recently, three crossovers were defined for the path representation, PMX(Partially-Mapped), OX(order), and CX(cycle) crossovers. OX proposed by Davis builds the offspring by choosing a subsequence of a tour one parent and preserving the relative order of cities from the other parent. During the past decade, several mutation operators have been proposed for permutation representation, such as inversion, insertion, displacement, and reciprocal exchange mutation. Reciprocal exchange mutation selects two positions at random and swaps the cities on these positions.

The reproduction simply copies the selected parents of all genes without changing the chromosome into the next generation. Selection provides the driving force in a genetic algorithm, and the selection pressure is critical in it. In rank selection, the individuals in the population are ordered by fitness and copies assigned in such a way that the best individual receives a predetermined multiple of the number of copies than the worst one. TSP has an extremely natural evaluation function: for any potential solution (a permutation of cities), we can refer to the table with distances between all cities and we get the total length of the tour after n-1 addition operations.

2.2 Multi-agent

The various agent technologies existing today can be classified as being either single-agent or multi-agent systems. In single-agent systems, an agent performs may communicate with the user as well as with local or remote system resources, but it will never communicate with other agents. In contrast, the agents in a multi-agent system not only communicate with the user and system resources, but they also extensively communicate and work with each other, solving problems that are beyond their individual capabilities.

Distributed problem solving is applied to a multi-agent system in order to solve the coordination and cooperation problems central to the society of agents. Moreover, negotiation, which plays a fundamentals role in human cooperative activities, is required to resolve any conflicts that may arise. Such conflicts may arise while agents are trying to complete against each other or trying to collaborate with each other to build a shared plan, thus giving rise to two different types of agents participating in a multi-agent systems, competitive agents and collaborative agents. Competitive agents: Each agent's goal to maximize its own interests, while attempting to reach agreement with other agents.

Collaborative agents: In contrast to the above, collaborative agents share their knowledge and beliefs to try maximize the benefit of the community as a whole.

3   Two Dynamic Migration Window Methods and the Frequency of Migrations

3.1   Increasing Dynamic Migration Window

The starting dynamic window size for the migration is 1. Whenever the migration occurs, the migration window size increases with a fixed size until the size is equal to 20% of population size.

3.2   Random Dynamic Migration Window

Whenever the migration occurs, dynamic window size varies at random. The variation is from 1 to θ. The θ value is generated at random within 20% of the population size each migration. For example, first window size is 3 and second is 5, etc.

3.3   Frequency of Migrations

There is interesting question being raised: when is the right time to migrate? If migration occurs too early during the run, the number of correct building blocks in the migrants may be too low to influence the search on the right direction and we would be wasting expensive communication resources.

In the distributed genetic algorithm, migration occurs at predetermined constant interval and occurs after the demes converge completely, but I experiment with a "delayed" migration scheme. As the number of generation increase, migration interval is short in order to alleviate the problem of converging prematurely to a solution of low quality.

4   Distributed Genetic Algorithm Using Multi-agent

A new distributed genetic algorithm with multiple populations using master-slave multi-agent model for the TSP is illustrated in Figure 1.

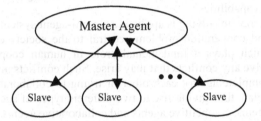

Fig. 1. Distributed genetic algorithm structure

In the master-slave multi-agent model, master agent maintains the lists of partial results that they are sent from slaves. At first, master agent generates the population and divides them into subpopulations. It sends them to slave agents. The slaves execute a conventional genetic algorithm on their subpopulation: fitness evaluation, selection, crossover, mutation and periodically return their best partial results to the master agent. The master stores the partial results in lists. Then, master agent searches the lists and selects two slaves that have bad partial results and sends the migration window size to selected slaves. The selected slaves exchange the subpopulations according to the migration window methods. After the first migration occurs, slave agents execute the conventional genetic algorithm on their fraction of the population and then send a result to master again. To select the destination of migrants again, master agent compares the lists and picks up λ slave agents that have both bad partial result and less difference than threshold. The i denotes the migration number.

I compute the difference as follows:

$$\text{For each slave, } | (i-1)_{th} \text{ partial result} - i_{th} \text{ partial result} | < \text{threshold} \qquad (1)$$

Threshold is variable. As the number of generation increase, threshold is decreasing. The master agent generates neighbors according to all possible permutations of selected slave agents and evaluates all neighbors and then selects the best one as destination of the migrants to avoid the premature convergence. Master sends the migration window size to selected slaves. The selected slaves again exchange the subpopulations according to the migration window methods. Continues these works to find near optimal for the TSP. The migration events occur per the given migration frequency rather than every generation to reduce the communication time.

5   Experiments and Results

The hardware used is a collection of the PCs connected with a 100Mbit/sec Ethernet. The parameters of the genetic algorithm described in this experiment are shown in Table 1.

Table 1. Parameters

	Size and rates
Population size	Variable size
The number of generation	Variable size (50–100)
Crossover rate	0.6
Mutation rate	0.1
Inversion rate	0.1

TSP instances are taken from the TSPLIB[11]. In the experiment, dynamic migration window size starts with 1, after the first migration window size increase 1. So, window size become 2, continue until window size is equal to 20% of population. The second method, whenever migration occurs, the dynamic migration window size varies at random from 1 to θ. The θ value is generated at random within 20% of the population size. The λ size is 3. And the migration occurs per 30, 20 or 10 generations in order to reduce communication time rather than each generation. Table 2 shows the results using variable migration rates.

Table 2. The results using variable migration rates

Migration rates	0.1	0.15	0.2
The rear optimal rates	1.01	1.03	1.08

To find the takeover times of the different migration policies Erick Cantú-Paz [14] simply iterate the difference equations and count the number of iterations until P_t reaches or exceeds 1. We see from [14] that when the migration rate is 0.2, takeover time is 2.5 and the migration rate is 0.1 then takeover time is 5. The choice of migrants is a greater factor in the convergence speed than the choice of replacements.

Also, we obtain slightly better results from random dynamic migration window method than increasing dynamic migration window method.

6   Conclusion and Future Works

TSP is a well-known combinatorial optimization problem and belongs to the class of NP-complete problems, various techniques are required for finding optimum or near optimum solution to the TSP.

This paper designs a distributed genetic algorithm in order to reduce the execution time and obtain more near optimal using master-slave multi-agent model for the TSP. Distributed genetic algorithms with multiple populations are difficult to configure because they are controlled by many parameters that affect their efficiency and accuracy. Among other things, one must decide the number and the size of the populations, the rate of migration, and the destination of the migrants. In this paper, I develop two dynamic migration window methods that control the size and the frequency of migration. Also I propose new migration policy that selects the destination of the migrants among the slave agents.

In the future work, I will find more optimal dynamic migration window size, θ, and more efficient migration policy.

References

1. Erick Cantú-Paz, David E. Goldberg, A Summary of Research on Parallel Genetic Algorithms, *IlliGAL Report No. 95007*, (1995).
2. A.Colorni, M. Dorigo, F. Maffioli, V. Maniezzo, G. Righini, M. Trubian, Heuristics from Nature for Hard Combinatorial Optimization Problems, *Int. Trans. in Operational Research, 3, 1*, 1-21.
3. C. Cotta, J.F. Aldana, A.J. Nebro, J.M. Troya, Hybridizing Genetic Algorithms with Branch and Bound Techniques for the Resolution of the TSP, Springer-Verlag, *Artificial Neural Nets and Genetic Algorithms*, (1997) 277-279.
4. http://www.cs.tcd.ie/research_groups/aig/iag/areab.html, "Distributed Problem Solving in Multi-Agent Systems.
5. David E. Goldberg, Genetic Algorithms: in Search and Optimization, *Addison-Wesley*, (1998) 1-125.
6. Terry Jones, Stephanie Forrest, Genetic Algorithms and Heuristic Search, *International Joint Conference on Artificial Intelligence*, (1995).
7. J. Kim and Y. Hong, "A Distributed Hybrid Algorithm for the Traveling Salesman Problem", *Journal of KISS, Vol.25 No. 2*, (1998) 136-144.
8. Z. Michalewicz, Genetic Algorithms + Data Structures = Evolution Programs, *Springer-verlag*, (1995) 209-237.
9. Gerhard Reinelt, The Traveling Salesman Computational Solutions for TSP Applications, *SpringerVerlag*, (1994.)
10. T. C. Sapountzis, The Traveling Salesman Problem, *IEEE Computing Futures*, (1991) 60-64.
11. TSPLIB, http://www.iwr.uni-heidelberg.de/iwr/comopt/soft/TSPLIB95/ATSP.html.
12. Erick Cantú-Paz, "Designing Efficient Master-Slave Parallel Genetic Algorithms, *IlliGAL Technical Report No. 97004*, (1997).
13. Erick Cantú-Paz, "A Survey of Parallel Genetic Algorithm", *IlliGAL Technical Report 97003*, (1997).
14. Erick Cantú-Paz, "Migration policies, Selection Pressure, and Parallel Evolutionary Algorithms", *IlliGAL Technical Report 99015, (1999)*
15. Marco Tomassini, "Parallel and Distributed Evolutionary Algorithms: A Review", http://www-iis.unil.ch
16. D.Kinny, M.P. Georgeff, "Modeling and Design of Multi-Agent Systems", *Proc. ATAL'97.* (1997) 1-20.

Numerical Weather Prediction
on the Supercomputer Toolkit

Pinhas Alpert[1], Alexander Goikhman[2,*],
Jacob Katzenelson[2], and Marina Tsidulko[1]

[1] Dept. of Geophysics and Planetary Sciences, Tel-Aviv University,
Tel Aviv 69978, Israel
[2] Dept. of Electrical Engineering, Technion – Israel Inst. of Technology,
Haifa 32000, Israel

Abstract. The Supercomputer Toolkit constructs parallel computation
networks by connecting processor modules. These connections are set
by the user prior to a run and are static during the run. The Technion's
Toolkit prototype was used to run a simplified version of the PSU/NCAR
MM5 mesoscale model [9]. Each processor is assigned columns of the
grid points of a square in the (x,y) space. When $n \times n$ columns are
assigned to each processor its computation time is proportional to n^2
and its communication time to n. Since the Toolkit's network computes
in parallel and communicates in parallel, then, for a given n, *the total
time is independent of the size of the two dimensional array or the area
over which the weather prediction takes place.* A mesoscale forecast over
the eastern Mediterranean was run and measured; it suggests that were
the Toolkit constructed from ALPHA processors, 10 processors would
do a 36 h prediction in only about 13 minutes. A 36 hours prediction
with full physics for the whole earth will require 2 hours for 80 ALPHA
processors.

1 Introduction

This is a joint work of two groups: a group that developed the prototype of the
Supercomputer Toolkit and a group that studied numerical weather prediction.
We started from the question "How many processors are required to predict the
weather over the east Mediterranean?" We ended with an estimate for the number of processors to predict the weather over the whole earth. Our cooperation
began once we realized that there is a match between the computational structure of a domain decomposition numerical weather prediction scheme and the
architecture of a Supercomputer Toolkit computational network. It is proper,
therefore, that we start the introduction with this match. Moreover, our main
results, in particular, time and performance estimations, follow readily from this
match and from the performance of the Toolkit processor.

The Supercomputer Toolkit [2] constructs special purpose computers by in-
terconnecting standard modules. The main module is a processor (including

---

* *Current address:* IBM Research Lab, Haifa, Israel

H. Zima et al. (Eds.): ISHPC 2002, LNCS 2327, pp. 335–345, 2002.
© Springer-Verlag Berlin Heidelberg 2002

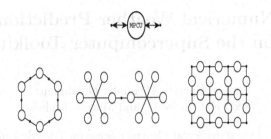

Fig. 1. The abstraction of the Toolkit processor and its interconnection graphs. Each Toolkit processor is a memory-CPU (MCPU) unit with two bidirectional I/O ports. Any graph can be formed as long as the cardinality of each node is less than 8. A ring, a communication cluster and the covering of 2D space are illustrated.

CPU and memory) that has two I/O ports. Processors modules are interconnected to networks by physical connection of ports. Figure 1 depicts a graphical abstraction of such a processor and some networks generated by interconnection of ports. Processors whose ports are connected together are called *neighbors*; neighbors can exchange information between them through the connected ports.

Processors of such a computational network can compute in parallel. Moreover, a processor can read/write through its left- and right-port simultaneously. Thus, considering the ring structure of Fig. 1, a processor can send information to its left neighbor while at the same time read information sent from its right neighbor; each of its neighbors can do the dual action with its own neighbors. In other words, the network's processors can communicate in parallel as well as compute in parallel.

Consider the numerical weather prediction problem. As is well known, weather prediction is a $3D$ partial differential equation problem. The height z corresponds to the atmospheric height and is often taken as 15-20km. The sizes of the other two dimensions, call them x and y, depend on the area to be covered and it can range from 50-5000 km, for mesoscale modeling. A grid is defined over the $3D$ space. Typically, the height is sampled (unevenly) 30-50 times; the x and y dimensions are sampled evenly each 10 to 60 km, depending on the area and the detail to be covered.

The partial derivatives with respect to x, y and z are replaced by differences resulting in ordinary differential equations in time with the state variables being the values of the pressure, velocity, etc., at the grid points. These equations are integrated numerically in what can be considered as a mixed method: The grid points along the z direction at each given x_i, y_j form a *column* that is considered a 'cell'. A time increment Δt is chosen to satisfy the CFL criterion[1].

---

[1] Courant-Friedrichs-Levy stability criterion; see, for example, [10] p.442, with respect to the speed of sound (3 sec per km) and with respect to the x,y distance between grid points.

Each cell calculates the value of the state at each of its vertical sample points at time $t + \Delta t$ using an *implicit* integration formula taking into account the states of the neighboring cells at time t as constants. Finally, the result obtained at $t + \Delta t$ is declared as the state for $t + \Delta t$ and the process repeats for the next time increment.

Since the speed of sound, which together with the grid step determines the integration time step, does not change significantly with the weather, the CFL criterion implies a constant bound on the time step size and (when the grid is fixed in space) the above step can be performed by any cell independently of time or place[2].

Thus, the $3D$ weather prediction problem has the following computational structure: it is a *two* dimensional array of cells in which each cell communicates with its neighbors only and calculates the next state from its previous one plus the states of all its neighbors[3].

If we assign each cell a Toolkit processor and connect the processors to a grid that covers the $2D$ region (see the right most part of Fig. 1), then the computation can be performed by each processor doing the computation of a cell and, once values are computed, the processor exchanges the results with its neighbors (only!) and starts working on the computation for the next time step.

It is easy to see that if we partition the $2D$ space to regions, say squares, each containing $n \times n$ cells, the property that a cell communicates with its neighbors only is preserved or, inherited by the region. I.e., the processor with the cells of a region has to communicate only with its neighbors for the state of the cells that are on the region's boundary. Even on this level of abstraction one can deduce two major properties of the above Toolkit's computation network, i.e., a network for which the $2D$ space is partitioned to squares and each processor contains $n \times n$ cells: (a) Since the computation is done in parallel and the computing time of each processor is a function of the number of cells 'residing' in it, *the total network computation-time does not increase with the size of the network.* (b) Since the communication is done in parallel and the communication-time of each processor is proportional to n, *the total network communication-time does not increase with the size of the network.*

Section 2 describes the Toolkit in somewhat more detail; it explains how the implementation supports (a) and (b) above. Section 3 describes the MM5 model that we ported to the Toolkit and whose performance we measured. The actual porting of the MM5 program to the Toolkit appears in section 4. The experimental results relevant to prediction over the East Mediterranean and over the complete earth is described in section 5. The last section compares the Toolkit with numerical weather prediction programs running on other computing systems and summarizes our conclusions.

---

[2] Actually, to reduce the truncation error the Δt derived from the CFL criterion and the x, y grid step is divided by some integer n, typically 4, and $\frac{\Delta t}{n}$ is used as the integration time step. In our program results were accumulated each Δt only.

[3] One might call MM5 a $2\frac{1}{2}D$ problem.

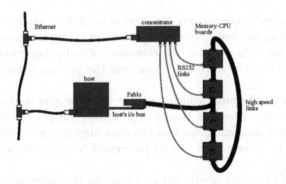

Fig. 2. A general description of the Technion's Toolkit system and its modules. The host is a conventional UNIX workstation. Four memory-CPU modules are shown; they are connected in a ring via wide-band links. The Pablo is an standard extension card connecting one of the memory-CPU modules (called the master) with a standard connector of the host. Each memory-CPU module has a relatively slow connection with the host through a RS232 link that connects to a concentrator and the Ethernet.

2   The Supercomputer Toolkit

Introduction: The Supercomputer Toolkit is a family of hardware modules (processors, memory, interconnect, and input-output devices) and a collection of software modules (compilers, simulators, scientific libraries, and high-level front ends) from which high-performance special-purpose computers can be easily configured and programmed. Although there are many examples of special-purpose computers, see [5], the Toolkit approach is distinguished by constructing these machines from standard, reusable parts. These parts are combined by the user; especially, the connections are set by the user prior to a run and remain static throughout the run. Thus, the Toolkit is a general method for building special-purpose computers for heavy scientific/engineering computing. The following is a brief description of the Toolkit system. It presents the minimum information for an application programmer. The Technion's Toolkit description appears in [2].

Description: Figure 2 is a general description of the Technion's Toolkit system. The system consists of a host, which is a conventional workstation, Toolkit processor boards (memory-CPU boards; four in the figure) and communication components. Fast communication links connect the Toolkit boards. These links are user connected ribbon cables. The fast links are used for communication at computing time. The user uses the host for program development, for loading the network of Toolkit boards, for starting the system and for collecting the results. The memory-CPU board contains an Intel i860 processor (60 advertised Mflops double precision arithmetics), memory, a special communication controller and two ports. The Toolkit processor (board) is considered as a device with two ports. Processors can be connected by connecting their ports together. Figure 1 illustrates such connections. Thus, arbitrary graphs can be formed where processors are the *branches* and the *nodes* are the actual connections.

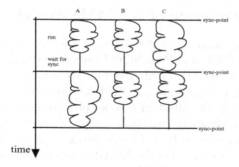

Fig. 3. Programming model – the user's image of the Toolkit processes is illustrated with three processors A, B and C. The synchronization implies a programming model that features *asynchronous* programs, between *synchronization barriers* in which the communication among processors takes place, i.e. *sync-point* in the graph.

Communication: The connections, or the fast links of Figure 2, use a communication method modified from the MIT Toolkit [1][2]. For communication, processors synchronize their instructions (using a synchronization line, see [1][2]) and deliver/accept blocks whose sizes are known *a priori*. This yields a fast transfer rate between neighboring processors – 0.5 giga bit per second per port. The maximum information rate that this method can deliver is half the memory speed per port as there are two ports that operate simultaneously sharing the same bus and the same memory. The synchronization is needed even when running the same programs on all processors since the i860 arithmetic unit timing is data dependent.

The above synchronization implies a simple yet powerful programming model that features *asynchronous* programs, written in, say, C or FORTRAN, between *synchronization barriers* in which the communication among processors takes place. This model of computation is illustrated in Figure 3. According to this model, each processor program looks like a 'normal' program in a high level language that has calls to synchronization routines embedded within it. The calls in different processors have to correspond. For example, to implement the first synchronization barrier (sync-point) each processor program has to have a procedure-like call at an appropriate place. If, for example, processor A has to send information to its neighbor, processor B, at a certain point, then processor B must have a call accepting the information at the same point. This scheme should be repeated for all sync-points and all processor programs.

Referring to standard computer architecture terminology the Toolkit network forms a distributed memory computer; each processor accesses its own memory only; the Toolkit is certainly a MIMD (multiple instruction multiple data) machine; it is not inherently a VLIW (very large instruction word) machine even if it can run as such. Its processor is conventional while its communication method is unique and attains extreme speed for communication between neighbors due to synchronization and the use of properly terminated ribbon cables.

3   A Brief Description of the MM5 Model

The MM5 mesoscale model was developed at the National Center for Atmospheric Research (NCAR) and Penn State University (PSU). The model is based on the integration of primitive hydrodynamic equations [7] and was originally developed for CRAY computers as a FORTRAN program. The MM5 is the fifth-generation model; it is the latest in a series of mesoscale models developed in NCAR starting from the early 70's [7]. A complete model system consists of the preprocessing programs as well as the calculation routines. The MM5 model is the first non-hydrostatic model in the NCAR series, but has also the hydrostatic option [9]. Second-order centered differences are used for spatial finite differencing for most equations. A fourth order scheme is applied for the diffusion terms. A second-order leapfrog time-step scheme is used for temporal finite differencing. The fast terms in the momentum equation that are responsible for sound waves have to be calculated on a shorter time step. The horizontal grid is an Arakawa–Lamb B grid, where the velocity variables are staggered with respect to the scalar variables [8]. A large number of physical parameterization schemes is used in the MM5 model. However, for simplicity, in the runs presented here, only the dry model without the boundary layer physics is employed.

4   Porting MM5 to the Toolkit

General: The code of the MM5 model was modified in order to adapt it to the Toolkit. This modification was done in a semi-automatic way: macros defined the partition of the grid of the domain into symbolic sub-domains. Loops along the domain were identified by hand and their ranges changed according to the sub-domains. The resulting program and a sub-domain index were submitted to a macro processor that yielded the program to be run on the corresponding Toolkit processor. During a time step of integration each processor calculates over its own sub-domain, taking into account at most two auxiliary rows (the number of rows depends on the variable; only one of them requires two auxiliary rows; all the rest require one row only) of cells attached to each sub-domain along its boundary whose values are taken from its neighboring sub-domains. Although for most finite difference algorithms of the model only one neighboring row of cells (in each direction) is required, here, two neighboring rows were necessary because of the fourth-order diffusion scheme.

After the state at time $t + \Delta t$ is calculated, the local communication phase takes place; i.e., the communication among the processors is performed. Processors that were assigned neighboring sub-domains are directly connected to each other (consider again Fig. 1 for covering a 2D space by a network of Toolkit processors). Since the MM5 system with full physics is a very large system, in order to try it out, we removed some options such as the handling of humidity in the air (including convection and micro-physics) and the planetary boundary layer. We had some code modified because the MM5 was written to run on the CRAY with a large amount of memory, and with a dialect of Fortran that we did not have.

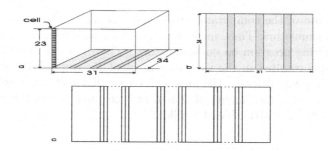

Fig. 4. Sub-domains and data communication in the Toolkit simulation of the Antalya cyclogenesis, (a) The model structure where the cell represents a column of grid-points in meteorological nomenclature. (b) The horizontal model domain and its four sub-domains; The dark shading area consists of four rows of cells. The left two rows belongs to the left domain but their state is used by the right domain; the right two rows belong to the right domain and their state is used by the left domain. The total number of cells is 31x34. (c) Parallelization: two rows of cells, shadow cells, are added to each side of sub-domain to hold the state of the two external rows of the adjacent sub-domain.

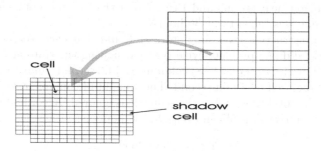

Fig. 5. The covering of a square area by square sub-domains and the cells and shadow cells of a sub-domain.

The case of the Antalya shallow lee cyclogenesis described by [6], was chosen for the Toolkit run. Figure 4 illustrates the overlapping of the sub-domains and the data communication in this particular run. Figure 5 illustrates the covering of a $2D$ space and the shadow cells.

Simulation: The simplified MM5 program was flow analyzed by hand. We found a fair amount of dead code, and a generous amount of arrays used as storage for intermediate results. The dead code was eliminated but for lack of time we did nothing about the intermediate arrays. This extra storage was a problem: the example for which we prepared the initial data had a grid of $31 \times 34 \times 23$ (the last number is the number of vertical levels) partitioned between four processors ($8 \times 34 \times 23$, $8 \times 34 \times 23$, $8 \times 34 \times 23$, $7 \times 34 \times 23$, these numbers are without the overlapping), see Fig. 5.When the code was ready we found that it required about 25% more memory than we had on a Toolkit processor. Since re-writing the code to reduce its volume significantly is a considerable amount of work, we decided to take advantage of the computational structure of the MM5 and get an estimate of the computation time and communication time by running the code on simulators. Thus, instead of re-writing the code, it was decided to run the resulting program on our simulators[4]. The run of the i860 instruction level simulator resulted in the time measurement for running the problem on the current Toolkit. Running the process level simulator (Fortran source) gave a good indication of the amount of time that is required to run the problem on an ALPHA 200 (166 MHz) based Toolkit.

5   MM5 over the East Mediterranean

To get insight into the ability of the Toolkit to run the MM5 we return to the following question: How many processors are required to predict the weather over the east Mediterranean?

The model domain for the 'East Mediterranean' is taken to be an area of 3480km by 2700km, i.e. 59×46 grid-points with 60 km grid interval. The vertical domain is about 16 km, i.e. top-pressure of 100 hPa, with highly-nonuniform spacing (50 meter near ground and 1 to 2 km at the top; total of 23 levels). The forecasting time was taken to be 36 hours.

Using the instruction level simulator, we found that each Toolkit processor (33MHz clock) did one problem-cycle (i.e., calculating the state at $t + \Delta t$ from the state at t), with 272 (i.e., 34×8) columns, in 14.3 second; running one column for one cycle required 52.6 mili-second. The amount of communication time was about 2% of the total time and in the sequel it is ignored. According to the CFL criterion, the number of problem-cycles for 36 hours is:

$$\frac{T}{\Delta t} \approx \frac{T * c}{\Delta x} = \frac{36 * 3600}{60 * 3} = 720$$

where T and Δx are the time interval and the horizontal grid distance, respectively and c is the speed of sound. The time to predict the weather for 36 hours is: $720 * 14.3 = 2h51$ minutes. If one can use processors with 272 columns per processor to cover the $59 \times 46 = 2714$ grid points needed for this problem, one needs approximately 10 Toolkit processors.

The process-level simulator was then run with the same structure as described before. Since this was an ALPHA 200 processor running native code generated

---

[4] We use two simulators, [3]: The first simulator simulates Toolkit system processes by running corresponding host processes. For example, to simulate a Fortran program running on, say, three Toolkit processors, the program is translated using the host's Fortran translator and run, by the simulator, as three host processes that communicate via send/receive, etc. This simulator enables finding bugs in the programs and, most importantly, in the communication between the processes. The second simulator is an i860 instruction-level simulator extended to include the communication system as well as to run a Toolkit system, i.e., several processors and their communication. The simulator includes the timing of instructions and the timing of the communication. It is the kind of simulator used to debug kernel code and we judge the accuracy of its timing to be pretty good.

from Fortran, the measured performance can be considered as the amount of time the problem would have taken if the Toolkit were constructed from ALPHA processors. The time for one cycle, with 272 columns, was found to be 1.06 seconds. That means that with an ALPHA 200 based system, 10 processors using 272 column each will do the prediction in $720 * 1.06 = 763$ seconds or 12 minutes and 43 seconds. We believe that the above are conservative estimates and that the program could be speeded-up considerably. We attribute the fifteen fold speed up of the ALPHA to a faster processor (166 MHz clock versus 33 MHz clock) and to a better compiler.

6   Summary and Conclusions

The Toolkit's computation network most significant qualitative property relevant to this application is that the Toolkit's computation network computes in parallel and communicates in parallel; with equal number of cells per processor computation time, communication time and the *total computation time are independent of the size of the two dimensional array or the area over which the weather prediction takes place.*

The dependency of the number of computers on the distance of the XY grid is another qualitative property. Assume that the domain area is given, and the grid distance is divided by two. Clearly, the number of columns is increased four fold. From the CFL condition follows that the integration step, Δt, has to be halved. Thus, if the number of cells per processor is kept and the number of processors is increased fourfold, the total time is doubled. If it is assumed that the time per step is linear with the number of cells per processor (which is not quite so, see below), to achieve the same run time the number of processors has to be increased by eight. This is a property of the problem expressed in terms of processor numbers.

Some detailed qualitative properties are the comparison of the Toolkit system with other parallel processing systems suitable for running the MM5 [12] [14] [13] and the role of the microprocessor cache in such computation [4]. This discussion in not included here for lack of space.

Conservative estimates performed for a typical mesoscale 36 hour forecast over the eastern Mediterranean suggest that were the Toolkit constructed out of ALPHA processors, 10 processors would do the prediction in only about 13 minutes.

This estimate can be extended to a global General Circulation Model (GCM) run covering the whole earth that is about 40 times larger then our eastern Mediterranean model region. Hence, a Toolkit system with 80 ALPHA processors will do a global mesoscale (i.e. 60 km grid interval and 23 vertical levels) 36 hours prediction within about 1 hour only. Another factor of 2 in the computing time is required in order to account for the full physics that was not incorporated in our MM5 Toolkit experiments. The factor of 2 was determined by independent simulations with and without the full physics on other computers (CRAY and SP2) using the same structure as on the Toolkit. In other words, 80 ALPHA processors will do a prediction over the whole globe in about 2 hours with a resolution comparable to the finest grid achievable today with GCMs in the largest operational weather prediction centers. The Toolkit application performed here uses the MM5 explicit schemes in the horizontal plane.

In summary, the unique property of the Toolkit that computes and communicates in parallel with communication time being negligible, indicates thst the long sought goal of numerical weather prediction with high resolution depends only on the number of available processors with a proper communication scheme and that the number of processors is not that large.

Acknowledgments

The authors thank Professors H. Abelson and G. J. Sussman, MIT, for supportive listening, encouraging and criticism. Professor A. Chorin, UC Berkeley, for the inspiration to tackle partial differential equation. Dr. Gerald N. Shapiro, Fire*star Consulting Co., for a hand with the hardware work and many helpful discussions. Dr Reuven Granot, Israel Department of Defense, for bringing the weather prediction problems to our (JK) attention.

This work was supported by Maf'at, Israel Ministry of Defense, and by the Israel Ministry of Science and Technology. Seed money for this work was given by Intel Corp. We thank them all.

We thank the National Center for Atmospheric Research (NCAR) the MESO-USER staff, and particularly Drs. Sue Chen, Wei Wang and Bill Kuo for the help in adopting the PSU/NCAR meso-scale model and its new versions at Tel-Aviv University. Thanks are due also to the Israel Met. Service for the initial and boundary conditions. The study was partly supported by the US-Israel Binational Science Foundation grant No. 97-00448.

References

1. Abelson, H., et al., "The Supercomputer Toolkit: A general framework for special-purpose computing," International Journal of High Speed Electronics, **3**, Nos. 3 and 4, 337–361, 1992.
2. J. Katzenelson, et al., "The Supercomputer Toolkit and Its Applications," Int. Journal of High Speed Electronics and Systems, Vol. 9, No. 3, 807–846, 1999.
3. Goikhman, A., "Software for the Supercomputer Toolkit," M.Sc. thesis, Dept. of Electrical Eng., Technion – Israel Inst. of Technology, Haifa, Israel, 1997.
4. Goikhman, A. and J Katzenelson, "Initial Experimentation with the Supercomputer Toolkit," EE memo 996, Dept. of Electrical Engineering, Technion – Israel Institute of Technology, Haifa, Israel, 1995.
5. Adler, B.J., "*Special Purpose Computers*," Academic Press, Inc. 1988.
6. Alpert, P., "Implicit filtering in conjunction with explicit filtering". *J. Comp.Phys.*, **44**, 212-219, 1981.
7. Anthes, R. A., and T. T. Warner, "Development of hydrodynamic models suitable for air pollution and other meso-meteorological studies," *Mon. Wea. Rev.*, **106**, 1045-1078, 1978.

8. Arakawa, A., and V. R. Lamb, "Computational design of the basic dynamical process of the UCLA general circulation model," *Methods in Computational Physics*, **17**, 173-265, 1977.

9. Grell, G. A., et al., "A description of the Fifth-generation Penn State/NCAR Mesoscale Model (MM5)," Nationl Center for Atmospheric Research Tech. Note 398+STR, Dept. of Meteorology, The Pennsylvania State Univ., 1994.

10. Holton, J. R., "An Introduction to Dynamic Meteorology," *Academic Press*, NY, pp. 511, 1992.

11. Pielke, R. A., "Mesoscale Meteorological Modeling," *Academic Press*, NY, 1984.

12. Russell, R. M., "The CRAY-1 Computer System," Comm. of the ACM, vol 21, no 1, January 1978, pp 63-72.

13. Sterling, T.L, et al., "How to build a Beowulf," MIT Press, 1999.

14. IBM staff, Special issue on SP2, IBM Systems Journal, Vol 34, No2, 1995.

OpenTella: A Peer-to-Peer Protocol for the Load Balancing in a System Formed by a Cluster from Clusters

Rodrigo F. de Mello[1], Maria Stela V. Paiva[2], Luís Carlos Trevelin[3], and Adilson Gonzaga[2]

[1] RadiumSystems.com
mello@radiumsystems.com.br
[2] University of São Paulo
Engineering School of São Carlos
{mstela,agonzaga}@sel.eesc.sc.usp.br
[3] Federal University of São Carlos
Department of Computer
trevelin@dc.ufscar.br

Abstract. The search for the high performance in the traditional computing systems has been related to the development of specific applications executed over parallel machines. However, the cost of this kind of system, which is high due to the hardware involved in these projects, has limited the continuing development in these areas, as just a small part of the community has access to those systems. With the aim of using a low cost hardware to achieve a high performance, this paper presents the OpenTella, a protocol based on the peer-to-peer models to update the information related to the occupation of resources and an analysis of this occupation for a post migration of processes among computers of a cluster.

Keywords. Peer-to-peer, cluster computing, load balancing, high-performance, process migration.

1 Introduction

The industries and the academic environment are always searching for the increase of the application performance in the several areas of knowledge. In order to increase the application performance in a significant way, in general, they have been written in a way specific for the execution of parallel machines. However, the high cost has limited the continuing development in this area and, besides that, just a small part of the community has access to those systems.

In order to make the cost decrease, several solutions have been found based on off-the-shelf computers, such as the Beowulf project [3]. One of the most important issues in this kind of system is the occupation of the available computing systems. This occupation should be balanced among the computers and organized in the most productive way.

The computing resources management area is distinguished for analyzing more optimized usage aspects in the system. Through these techniques, the system

H. Zima et al. (Eds.): ISHPC 2002, LNCS 2327, pp. 346–353, 2002.

development can be amplified and, besides that, the increase of a cluster can be foreseen and defined in relation to the capacity the application has to use it.

In view to achieve an efficient load balancing, this paper aims to present a model to create virtual sub-nets called OpenTella. Each sub-net is a cluster, i.e. the system is composed by clusters from clusters. The creation of sub-nets is defined through a peer-to-peer protocol that interchanges load information in the system, executes calculation over such information and, after this stage, accomplishes the migration of processes, if necessary and possible.

This paper is composed by section 2 that presents clusters; section 3 that describes peer-to-peer protocols; section 4 that presents the OpenTella model proposed in this paper; section 5 with the conclusions; and section 6 with the bibliography.

2   Cluster

Frequently the applications need higher computing power. A way of increasing the system performance is the new techniques that can improve the existing processors and devices. However, there are a number of laws that restrict the performance increase of the systems such as light speed, thermodynamic laws and the manufacturing costs [8, 9].

Besides the improvement of processors and devices to achieve a higher performance, there are other techniques that use the existing resources to obtain the same results.

In order to meet the requirements of the high performance systems, it has been created a taxonomy of computing systems based on how the processors, memory and other devices are interconnected. The most common systems are:

1) MPP – Massively Parallel Processors.
2) SMP - Symmetric Multiprocessors.
3) CC-NUMA – Cache-Coherent Non-uniform Memory Access.
4) Distributed Systems.
5) Clusters.

The massive parallel processors are, generally, systems composed by several processing items. These processing items are interconnected through a high-speed network and each one of these items can be composed by a series of devices, but, in general, they have a main memory and one or more processors. Each one of these processing items executes a copy of the operating system. It is an architecture that does not share devices.

The symmetric multiprocessors make a kind of architecture that share all the resources and in these systems all processors have access to the global resources. A single copy of the operating system is executed in this kind of architecture.

The systems based on the CC-NUMA architecture are multiprocessed and present non-uniform memory accesses. All processors that make the system have access to the whole memory, which is global. However, there are local memories in the processors that present a lower latency to the access, and that's the reason for the name "non-uniform".

The distributed systems can be considered as a network made of independent computers. There are multiple images of the system, and each computer executes its own operating system. These computers, which are connected to the network, may be symmetric multiprocessors, massive parallel processors or computers for general purposes.

Clusters are systems composed by a collection of workstations or personal computers interconnected by any network technology. These systems work as a collection of integrated resources for the solution of a same computing problem. They offer, in general, a single image of the system.

During the '80s, it was believed that the only way to obtain higher performance systems was through the development of more efficient processors. However, this idea has been changed since the advent of the parallel systems that could use several computers to solve the same problem. During the '90s, there was a great change regarding the construction of low cost computing systems that could use computer nets, by using the parallelism techniques.

Several reasons have directed the studies in the high performance computer net area, such as the fact that the processing capacity was being increased in the computers, increase in the bandwidth of the nets, developing of new communication protocols, computers being easily aggregated to the nets, high cost owner parallel computing systems, computing nets that can be easily increased by adding new computers. Studies in this are lead to the development of clusters.

The clusters may be composed by mono or multiprocessed computers, referring to two or more integrated computers, showing the vision of a single computing system.

The main components of a cluster based system are [1]:

1) Several high performance computers (personal computers, workstations or symmetric multiprocessors).
2) Operating systems based on layers or microkernel.
3) High performance interconnection networks.
4) Network communication interfaces.
5) High performance communication protocols and services (such as Fast and Active Messages).
6) Middlewares that allow a single image of the system and availability infrastructure.
7) Environments and tools for the development of applications (such as MPI and PVM).
8) Sequential, parallel or distributed applications.

The main advantage of the clusters are the computing performance, the system extension, the high availability, the redundancy of hardware and software, the access to the multiple resources that can be paralleled and the large scale. The clusters still can provide:

1) Larger configuration flexibility when compared to the dedicated systems executed over multiprocessed computers.
2) Usage of the existing hardware, characterized by the COTS concept (computers-off-the-shelf).
3) Minimize the existing hardware loss.
4) Lower cost when compared to the multiprocessed computers.
5) Higher scale.
6) High availability is easier to be achieved as there are several computers interconnected and each of them can provide redundancy in the system.
7) High performance.

For the construction of a cluster based system, there are a lot of relevant items, such as: the applied netware technologies, operating systems, single system image, middleware, system administration, parallel I/O, high availability, libraries and applications.

3   Peer-to-Peer Protocols

The peer-to-peer protocols show some advantages, as they do not overload a single server through requests, as it happens with the client-server model. Both protocols showed in [2] add advantages to an efficient communication model, targeting the load balancing.

Napster has as its greatest advantage the larger scale capacity, since it distributes the number of messages in the netware, in a linear way, even with an increasing number of clients. However, Napster does not have an efficient support to keep the system always available. Gnutella, on the other hand, shows a significant increase in the number of messages in the network in relation to the number of connected clients and it has an efficient support to keep the system available.

Besides this two protocols, there is the mixed model proposed in [2] that creates a network composed by sub-nets, where each sub-net is managed by the Napster model and communicates to one another by using the Gnutella model.

Based on these presented peer-to-peer models and on the sending and receiving information sets described on [7], this paper shows a fourth model, which is an alteration of the Gnutella protocol. In this modified Gnutella model, called OpenTella, each computer communicates to a group of computers, as well as in the Gnutella protocol, by using the techniques for creation of sending and receiving sets defined in [7].

Each computer that communicates to its group makes a calculation about the occupied resources. This creation of groups for the interchange of information decreases the number of messages in the system. When a process is submitted to execution, it is forwarded to the network root, this root searches for the most adequate branch to execute certain process, in accordance with the available resources. Each computer controls its sub-net that is the set of computers to which it is connected, decreasing the time spent to define the migration of processes.

When a process is received for execution, the root of the OpenTella model analyzes two issues: the occupation of its own resources and the relative occupation of its sub-nets, through the information of resources load previously obtained. If the root is available, it executes the process, but if it is occupied it sends a request for the migration of process to an available sub-net. The most available sub-net repeats the steps.

When the most available computer is found out, it happens a migration of the process, which is sent, from the root computer, to its destination.

Figure 1 presents the OpenTella model. In this model computers A1 and A2 update their own resources occupation table, and after this updating they send a certain figure to the previous computer, i.e. A, that identify the calculation of its full resources. The A computer creates a table, that contains the whole calculation for the occupation of its own resources and a general calculation about the occupation of all resources of its sub-net. This continues happening, respectively, up to the model root.

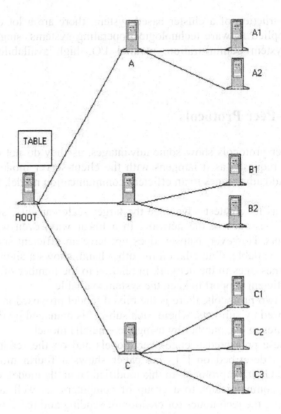

Fig. 1. OpenTella communication model

In order to define with whose computers a certain computer of the system will communicate, the technique of sending and receiving set describe on [7] is used during the connection phase.

When a process is submitted to the system root, the load on the root computer and its sub-nets is analyzed. If the root computer is the most available one, it executes the process by itself. On the contrary, it sends a message for the most available computer of its sub-net requesting for a load balancing analysis. This computer repeats the steps until the computer with the less occupied level of resources is found out. When this stage is completed, the migration process between the root and the less occupied computer is accomplished.

It can be observed that the migration of processes described above works based on the peer-to-peer model, while the most available sub-nets and computer works based on the weight of the sub-nets and computers involved.

4   OpenTella

This project came about because of the motivation to have more efficient techniques regarding the resources occupation of a cluster. From the analyses of the existing systems and methods, it was observed that the increase of clusters has been an important study issue to define the scale of the systems.

The clusters scale is limited by the information updating related to the system occupation and by the calculation of the best occupation of the existing resources. With this objective, this paper presents a model that uses a peer-to-peer protocol to interchange information on the system load and migration of processes, searching for the maximum occupation of the available computing resources.

The peer-to-peer protocol provides the creation of virtual sub-nets that represent each cluster, i.e. the system is composed by clusters from clusters. Each sub-net is managed by a master computer that keeps the information about the system load. Each master computer may manage other computers, such as a sub-net or other sub-nets.

When the process starts up, the root computer of the system analyzes the resources allocation table of the sub-nets and sends a request to a master computer from the most adequate sub-net. When this request is received, the master computer of a sub-net analyses its resources allocation table and sends a request to another sub-net or computer. When a computer receives a request, it analyzes its load compared to the other sub-nets and, if it is the most available one, it starts up the execution of the process. The structure created by the peer-to-peer protocol, called OpenTella, is presented on figures 1 and 2.

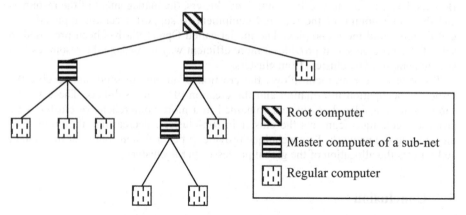

Fig. 2. Tree communication model of OpenTella

Each master computer of a sub-net, when receives a request sent by a computer capable to execute certain process, analyzes the resources allocation table of the virtual sub-net and pass on the request to this computer. The computer that receives the request may pass it on to a sub-net or start up the execution of the process.

The proposed model contributes to minimize the network load, through the creation of groups that send and receive information about the allocation of the available computing resources.

Ni and others [4] proposed a schedule to minimize the number of messages that contain information about the load of the systems, by optimizing the load balancing operations without affecting the system. In this method, the processors are analyzed based on three load levels: available, normal and overloaded. The computers would

352   Rodrigo F. de Mello et al.

interchange the minimum number of messages and would send information describing the status of the processors based on these three levels of load.

Ryou and Wong proposed in [5] a schedule that segments a network of processors in a set to send and in another set to receive load messages. In this model, each processor sends information about its load to N/2, where N is the number of processors in the network. This method makes the migration of tasks whenever it is possible.

Later on, Suen and Wong made some improvements in [6] by using plans of finite projection for the creation of groups that send and receive load information in the system. The size of these sets is $k \approx \sqrt{N}$, where N is the number of processors in the network, and

$k - 1$ is the potency of the prime number, and $N = k (k - 1) + 1$. In this method, the migration of processes is assured whenever is possible.

Continuing the works of Ryou, Wong and Suen, Matthew and Wong proposed in [7] a technique to create groups of sending and receiving messages about the load of the system, based on the processing capacity and using the coterie concept, that was previously used on the mutual distributed exclusion areas and selection of rights on resources.

By observing the presented techniques it was noted that the number of messages interchanged to balance the load is a relevant matter that may affect the whole system. Besides that, with the increasing number of clusters, the management of the resources and the dimensioning of the required equipment to support a certain application is getting more and more complex. The model called OpenTella has been proposed to solve these problems as it provides a more efficient way to manage the resources in a system composed by clusters from clusters.

The peer-to-peer protocol allows the creation of groups to send and receive the resources occupation information and the creation of virtual sub-nets. These virtual sub-nets are considered as clusters contained in a more comprehensive cluster. This proposed technique segments the cluster in sub-clusters, decreasing the number of messages in the network. This technique provides a better dimensioning of the system and reflects the allocation of the global processes in the cluster.

5   Conclusion

This paper has presented a peer-to-peer protocol for the management of information interchange on the load of computers of a system composed by clusters from clusters. This model minimizes the network overload, updates the information about the occupation of the computing resources and provides support to a more optimized occupation of these resources, through the migration of processes.

References

1. Sterling, T. *An Introduction to PC Clusters for High Performance Computing.* Contribution to the Cluster Computing White Paper, version 2.0, 28 December 2000. Obtained from the URL: http://www.csm.port.ac.uk/~mab/tfcc/WhitePaper/ Consulted on: 10 October 2001.

2. Howe, A.J. *Napster and Gnutella: a Comparison of two Popular Peer-to-Peer Protocols.* Universidade de Victoria, 11 December 2000.
   Obtained from the URL: http://www.csc.uvic.ca/~ahowe/research.html
   Consulted on: 05 November 2001
3. Ridge, D., Becker, D., Merkey, P., Sterling, T. *Beowulf: Harnessing the Power of Parallelism in a Pile-of-PCs.*
   Obtained from the URL: http://www.bewoulf.org
   Consulted on: 10 November 2001
4. Ni, L.M., Xu, C., Gendreau, T.B. *A distributed draft algorithm for load balancing.* IEEE Transactions on Software Engineering, 1985, pp. 1153-1161.
5. Ryou, J., Wong, J.S.K. *A task migration algorithm for load balancing in a distributed system.* Proceedings do XXII Annual Hawaii International Conference on System Sciences, Janeiro de 1989, volume II, pp. 1041-1048.
6. Suen, T.T.Y, Wong, J.S.K. *Efficient task migration algorithm for distributed systems.* IEEE Transactions on Parallel and Distributed Systems, 1992, pp. 488-499.
7. Tiemeyer, M., Wong, J.S.K. *A task migration algorithm for heterogeneous distributed computing systems.* The Journal of Systems and Software, Elsevier, 1998, pp. 175-188.
8. Buyya, R. *High Performance Cluster Computing – Architectures and Systems*. Volume 1, 1999, Prentice Hall – PTR.
9. Ng, J. Rogers, G. *Clusters and Node Architecture.*
   Obtained from the URL: http://www.csse.monash.edu.au/~rajkumar/csc433/
   Consulted on: 12 November 2001

Power Estimation of a C Algorithm Based on the Functional-Level Power Analysis of a Digital Signal Processor

Nathalie Julien, Johann Laurent, Eric Senn, and Eric Martin

LESTER, University of South Brittany - Lorient, France
Nathalie.Julien@univ-ubs.fr

Abstract. A complete methodology to estimate power consumption at the C-level for on-the-shelf processors is introduced. It relies on the Functional-Level Power Analysis, which results in a power model of the processor that describes the consumption variations relatively to algorithmic and configuration parameters. Some parameters can be predicted directly from the C-algorithm with simple assumptions on the compilation. Maximum and minimum bounds for power consumption are obtained, together with a very accurate estimation; for the TI C6x, a maximum error of 6% against measurements is obtained for classical digital signal processing algorithms. Estimation results are summarized on a *consumption map*; the designer can compare the algorithm consumption, and its variations, with the application constraints.

1 Introduction

Power consumption has become a critical design constraint in many embedded applications. Amount all the causes of dissipation, the software can have a substantial impact on the global consumption [1]. As two codes can have the same performances but different energy consumption [2], only a reliable power consumption estimation at the algorithmic level can efficiently guide the designer. Therefore, a C-level estimation can be worthwhile if it proposes a fast and easy estimation of the consumption characteristics of the algorithm. As there is no need of compiling, direct comparisons on different processors can be done before choosing the target and then without purchasing the development tools. It also allows to compare the consumption of different versions of the same algorithm to check if the application constraints are respected. More precisely, it allows to locate the 'hot points' of the program on which the writing efforts have to be focused.

When the applications are running on commercial processors, details on the architecture implementation are unavailable, and methods based on a cycle-level simulation [3][4] are then inappropriate. In this case, the classical approach to evaluate the power consumption of an algorithm is the instruction-level power analysis (ILPA) at the assembly-level [5]. This method relies on the measurements of the power consumption for each instruction and inter-instruction (a pair of successive instructions); but for complex processor architectures, the number of required measures becomes unrealistic. Many studies are currently focusing on a functional

H. Zima et al. (Eds.): ISHPC 2002, LNCS 2327, pp. 354–360, 2002.

analysis of the power consumption in the processor [6][7]. But all of them are also made at the assembly-level. Only one attempt of algorithmic estimation has already been made, concluding that a satisfying model could not be provided [8].

This paper presents an accurate estimation of the power consumption of an algorithm directly at the C-level. A simple power model for a complex processor is proposed, including all the important phenomena like pipeline stalls, cache misses, parallelism possibilities... This model expresses the software power dissipation from parameters, determined through a functional analysis of the processor power consumption. It has already been validated with an estimation method at the assembly-level. In this case, the parameters are computed from the code by profiling [9]. To estimate at the C-level, some parameters are predicted assuming different ways of compiling the code. For classical digital signal processing algorithms, C-level estimates are obtained with an average error of 4% against measurements, which is an accuracy similar to the other methods at the assembly-level. We also provide the user with a *consumption map*, representing the consumption behavior of the C algorithm.

The estimation methodology is sketched in Section 2 with the general framework, and the case study of the functional analysis on the TMS320C6201, resulting in the power model of the processor. Section 3 presents the assumptions to predict the parameters required as inputs of the power model of the processor. The application results for digital signal processing algorithms are discussed in Section 4. The conclusion summarizes the conditions and limits of this algorithmic power estimation method and indicates the future works.

2   Functional-Level Power Analysis and Model Definition

2.1   The Estimation Framework

The complete estimation methodology, represented in Figure 1, is composed of two steps: the *Model definition* and the *Estimation process*.

The *Model definition* is done once and before any estimation to begin. It is first based on a Functional Level Power Analysis (FLPA) of the processor, that allows discerning which parameter has a significant impact on the global power consumption. This step conducts to define the power model of the processor: this is a set of consumption rules describing the evolution of the processor core power related to algorithmic and configuration parameters. These consumption rules are computed from a reduced set of measurements performed on the processor for various elementary programs.

The *Estimation Process* is done every time the power consumption of an algorithm has to be evaluated. This step allows to determine the parameter values to apply in input of the power model of the processor. Some algorithmic parameters are estimated through simple assumptions about compilation: the prediction models.

2.2   The Model Definition for the TMS320C6201

As a case study, the power model of the C6x from Texas Instruments has been developed. This processor has a complex architecture with a deep pipeline (up to 11 stages), VLIW instructions set, the possibility of up to 8 parallel instructions, an internal program memory and an External Memory Interface (EMIF), dedicated to load data and program from the external memory. Four memory modes are available: mapped, bypass, cache and freeze [10]. On this processor, there is no significant difference in power consumption between an addition or a multiplication, or a read or a write instruction in the internal memory. Moreover, the data correlation have no more effect than 2% on the global energy consumption. It seems that the architecture complexity of the C6x hides many power variations.

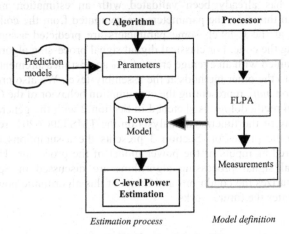

Fig. 1. The Estimation Method Flow

The first part in the *Model Definition* is the FLPA: it consists in a functional analysis of the processor architecture to determine which parameter is relevant from a consumption viewpoint. The complete analysis has already been presented in previous works [9]. The result of this step for the C6x is summarized in Figure 2, with two class of inputs: the algorithmic parameters and the configuration parameters. The configuration parameters, known with the application, are the clock frequency F and the memory mode MM. The algorithmic parameters represent activity rates between functional blocks: the parallelism rate α, the processing rate β, the program cache miss rate γ, and the Pipeline Stall Rate PSR.

The second step in the *Model Definition* is to establish the consumption rules, expressing the core power consumption P_{CORE} from the input parameters. Measurements have been performed for different values of these parameters by elementary assembly programs. Final consumption rules have been obtained by curve fitting. The present processor model does not include external memory yet; to add it will be part of future works.

Here is given by Equation 1 the consumption rule for the mapped mode:

$$P_{CORE} = V_{DD} * ([a\beta(1-PSR) + b_m] F + \alpha(1-PSR) [a_m F + c_m] + d_m) \qquad (1)$$

where V_{DD} is the supply voltage and $a=0.64$, $a_m=5.21$, $b_m=4.19$, $c_m=42.401$, and $d_m=7.6$. Details on the other cases can be found in [9]. The expressions obtained are more complex than those derived from a linear regression analysis; that explains why, in [8], the model conducts to very important errors.

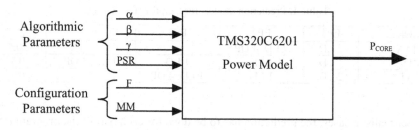

Fig. 2. Power Model for the C6x

3   Estimation Process

To compute the power consumption from the power model, the algorithmic parameters must be determined. Both α and β can be estimated directly from the C code by a prediction model that anticipates the compilation. In some particular cases, γ and/or PSR can be defined, like in the mapped memory mode where $\gamma = 0$. If not, these parameters will be considered as variables.

In the C6x, 8 instructions, forming a *fetch packet* (FP), are fetched at the same time. In this fetch packet, operations are gathered in *execution packets* (EP) depending on the available resources and the parallelism capabilities [10]. The parallelism rate α and the processing rate β are obtained from Equation 2:

$$\alpha = \frac{NFP}{NEP} \leq 1 \; ; \beta = \frac{1}{8}\frac{NPU}{NEP} \leq 1 \qquad (2)$$

NFP and *NEP* stand for the average number of FP and EP. *NPU* is the average number of processing units (every instruction except the NOP). To estimate α and β, it is then necessary to predict *NFP*, *NPU* and *NEP*. Four prediction models were developed according to the processor architecture and requiring a slight knowledge of the compiler:

- SEQ model: all the operations are executed sequentially.
- MAX model: all the architecture possibilities are fully exploited.
- MIN model: the load/store instructions can never be executed in parallel.
- DATA model: the load/store instructions are executed in parallel only if they involve different data.

For (i = 0; i < 512; i++)
$Y = X[i] * (H[i] + H[i+1] + H[i-1]) + Y[i]$;

Fig. 3. Program example for the prediction models

As illustration, a trivial example is presented in Figure 3. The operations out of the loop body are neglected. In the loop nest, 4 loads (LD), and 4 other operations (OP) are needed. Then, we predict 8 instructions, gathered in one single FP, so *NFP* = 1. Because no NOP operation is involved, *NPU* = 8 and $\alpha = \beta$. In the *SEQ* model, all instructions are assumed to be executed sequentially; then *NEP* = 8, and $\alpha = \beta = 0.125$. Results for the other models are summarized in Table 1.

Table 1. Prediction models for the program example

MODEL	EP1	EP2	EP3	EP4	α, β
MAX	2 LD	2 LD, 4 OP	-	-	0.5
MIN	1 LD	1 LD	1 LD	1 LD, 4 OP	0.25
DATA	2 LD	1 LD	1 LD, 4 OP	-	0.33

For realistic cases, the prediction has to be done for each part of the program (loop, subroutine…) to obtain local values. The global parameter values, for the complete C source, are computed by averaging all the local values. Such an approach also permits to spot 'hot points' in the program.

4   Applications

4.1   Estimation Validation

The power estimation method is applied on classical digital signal processing algorithms: a FIR filter, a LMS filter, a Discrete Wavelet Transform with two image sizes: 64*64 (DWT1) or 512*512 (DWT2) and an Enhanced Full Rate vocoder for GSM (EFR v.). The results for these applications are presented in Table 2 for different memory modes (MM) : mapped (M), bypass (B) and cache (C) and different data placement (DP): in internal or external memory (Int/Ext). The nominal clock frequency F is 200MHz. The estimates at the C-level for the different prediction models are presented and compared against measurements. The estimates at the assembly level (Asm) are also provided to confirm the previous validation of the power model of the processor.

Table 2. Comparison between measurements and estimations

Applications			Measurements			Power estimation (W)						
Algo	MM	DP	T_{EXE}	P(W)	Energy	Asm	%	Seq	Max	Min	Data	%
FIR	M	Int	6.885µs	4.5	30.98µJ	4.6	2.3	2.745	4.725	3.015	4.725	5
FFT	M	Int	1.389ms	2.65	3.68mJ	2.585	2.5	2.36	2.97	2.57	2.58	-2.6
LMS	B	Int	1.847s	4.97	9.18J	5.145	3.5	5.02	5.12	5.07	5.12	3
LMS	C	Int	165.75ms	5.665	939mJ	5.56	-1.8	2.55	6	4.76	6	5.9
DWT1	M	Int	2.32ms	3.755	8.71mJ	3.83	1.9	2.82	4.24	3.27	3.53	-6
DWT1	M	Ext	9.19ms	2.55	23.46mJ	2.548	-0.2	2.295	2.63	2.4	2.46	-3.5
DWT2	M	Ext	577.77ms	2.55	1.473J	2.53	-1	2.27	2.61	2.37	2.45	-3.9
EFR v.	M	Int	39µs	5.078	198µJ	4.935	-2.8	2.54	5.636	3.86	5.13	1
error							2%	25%	7%	13%		3.9%

The aim is to provide the designer with accurate estimates about all the possible consumption variations, including the particular point representing the real case. Then we have set γ = 0 and the global power consumption is computed with the PSR obtained after compilation. Validations of the power model at the assembly level for various values of the cache miss rate can be found in [9].

The results of the SEQ model prove that it is not possible to provide estimation without any knowledge about the targeted architecture. Except in one particular case, the MAX and the MIN models always overestimates and underestimates respectively the power consumption of the application. For the LMS in bypass mode, all the models overestimate the power consumption with close results; in this marginal memory mode, pipeline stalls are so dominant that all the instructions become sequential. The DATA model is shown as a very accurate estimation with a maximum error of 6% and an average error of 4%. Although this fine grain model implies to also consider the data placement, it remains very easy to determine.

4.2   Algorithm Power Consumption Exploration

If the *cache miss rate* γ and/or the *pipeline stall rate* (PSR) are unknown, it is possible to give to the programmer a *'consumption map'*. This map represents the power consumption variations of the algorithm. Moreover, in many applications, the designer can evaluate the realistic domain of variation for PSR and γ. It is thus possible to locate, on the *consumption map*, the more probable power consumption limits. In particular, the major part of the embedded applications have a program size (after compilation) that can easily be contained in the internal memory of the C6x (64 kbytes).

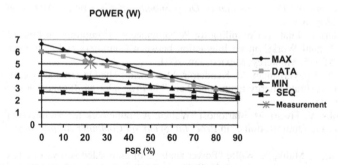

Fig. 4. Power Consumption Exploration for the DWT1 in mapped mode.

Let us reconsider the application of the EFR vocoder, in the mapped mode (γ = 0). In Figure 4, its *consumption map* is represented together with the measurement. Once again, the DATA prediction model is close to the reality.

The power consumption exploration tells the programmer if the algorithm respects the application consumption constraints (energy and/or power). Since at the C level the execution time is unknown, the energy could be evaluated from the execution time constraint (given by the programmer). If the algorithm consumption estimation is always under the constraints then the C code is suitable. On the contrary, the programmer can focus on the more dissipating parts of the algorithm (selected with local values of α and β parameters), being aware on pipeline stalls and cache misses. At last, several versions of the same algorithm could be efficiently compared through their *consumption maps*.

5   Conclusion

In this paper, an accurate power consumption estimation of a C-algorithm is proposed for DSP applications. The accuracy of this estimation directly depends on the considered information about the processor: (i) it is not possible to determine precisely the power consumption without any knowledge about the targeted processor. (ii) an estimation with a coarse grain model taking into account only the architecture possibilities provides the maximum and minimum bounds of the power consumption for the algorithm. (iii) the fine grain model including both elementary information on the architecture and data placement offers a very accurate estimation with a maximum error of 6% against measurements.

Current works include first the development of an on-line power consumption estimation tool. FLPA will also be applied on other processors in order to provide other estimation examples at both C-level and assembly-level. Finally, the development of a power model for the external memory will be an important part of future works.

References

1. K. Roy, M. C. Johnson "Software Design for Low Power," in *NATO Advanced Study Institute on Low Power Design in Deep Submicron Electronics,* Aug. 1996, NATO ASI Series, chap. 6.3.
2. M. Valluri, L. John "Is Compiling for Performance == Compiling for Power?," presented at the 5th Annual Workshop on Interaction between Compilers and Computer Architectures INTERACT-5, Monterey, Mexico, Jan. 2001.
3. W. Ye, N. Vijaykrishnan, M. Kandemir, M.J. Irwin "The Design and Use of SimplePower: A Cycle Accurate Energy Estimation Tool," in *Proc. Design Automation Conf.,* June 2000, pp. 340-345.
4. D. Brooks, V. Tiwari, M. Martonosi "Wattch: A Framework for Architectural-Level Power Analysis and Optimizations," in *Proc. Int. Symp. on Computer Architecture,* June 2000, pp. 83-94.
5. V. Tiwari, S. Malik, A. Wolfe "Power analysis of embedded software: a first step towards software power minimization," *IEEE Trans. VLSI Systems,* vol.2, n°4, Dec. 1994, pp. 437-445.
6. L. Benini, D. Bruni, M. Chinosi, C. Silvano, V. Zaccaria, R. Zafalon "A Power Modeling and Estimation Framework for VLIW-based Embedded Systems," in *Proc. Int. Workshop on Power And Timing Modeling, Optimization and Simulation PATMOS,* Sept. 2001, pp. 2.3.1-2.3.10.
7. G. Qu, N. Kawabe, K. Usami, M. Potkonjak "Function-Level Power Estimation Methodology for Microprocessors," in *Proc. Design Automation Conf,* June 2000, pp. 810-813.
8. C. H. Gebotys, R. J. Gebotys "An Empirical Comparison of Algorithmic, Instruction, and Architectural Power Prediction Models for High Performance Embedded DSP Processors," in *Proc. ACM Int. Symp. on Low Power Electronics Design,* Aug. 1998, pp. 121-123.
9. J. Laurent, E. Senn, N. Julien, E. Martin "High Level Energy Estimation for DSP Systems," in *Proc. Int. Workshop on Power And Timing Modeling, Optimization and Simulation PATMOS,* Sept. 2001, pp. 311-316.
10. TMS320C6x User's Guide, Texas Instruments Inc., 1999.

Irregular Assignment Computations on cc-NUMA Multiprocessors

Manuel Arenaz, Juan Touriño, and Ramón Doallo

Computer Architecture Group
Department of Electronics and Systems, University of A Coruña
Campus de Elviña, s/n, 15071 A Coruña, Spain
{arenaz,juan,doallo}@udc.es

Abstract. This paper addresses the parallelization of loops with irregular assignment computations on cc-NUMA multiprocessors. This loop pattern is distinguished by the existence of loop-carried output data dependences that can only be detected at run-time. A parallelization technique based on the inspector-executor model is proposed in this paper. In the inspector, loop iterations are reordered so that they can be executed in a conflict-free manner during the executor stage. The design of the inspector ensures load-balancing and uniprocessor data write locality exploitation. Experimental results show the scalability of this technique, which is presented as a clear alternative to other existing methods.

1 Introduction

The development of irregular parallel applications is a difficult task. The problem can be significantly simplified if the compiler is able to translate a sequential program into its parallel equivalent automatically. A general parallelization technique based on speculative parallel execution of irregular loops was proposed by Rauchwerger and Padua [9]. General methods can be applied to any loop with irregular computations. However, real codes usually contain costly irregular kernels that demand higher efficiencies. Lin and Padua [6] describe some loop patterns abstracted from a collection of irregular programs, and propose efficient parallelization techniques for some of those patterns. The parallelization of irregular reductions [3, 4, 8, 13] has been an active research area in recent years. From the point of view of this paper, special mention is deserved by the work by Knobe and Sarkar [5]. They describe a program representation that enables the parallel execution of irregular assignment computations. An *irregular assignment* (Fig. 1(a)) consists of a loop with f_{size} iterations, f_{size} being the size of the subscript array f. At each iteration h, the array element $A(f(h))$ is assigned the value $rhs(h)$, A being an array of size A_{size}. The right-hand side expression $rhs(h)$ does not contain occurrences of A, thus the code is free of loop-carried true data dependences. However, as the subscript expression $f(h)$ is loop-variant, loop-carried output data dependences may be present at run-time (unless f is a permutation array). This loop pattern can be found, for instance, in computer

H. Zima et al. (Eds.): ISHPC 2002, LNCS 2327, pp. 361–369, 2002.

362 Manuel Arenaz, Juan Touriño, and Ramón Doallo

graphics algorithms [2], finite element applications [11], and routines for sparse matrix computations [10].

In this work, we use the inspector-executor model to parallelize the irregular assignment loop pattern on cc-NUMA multiprocessors. Our technique is designed to take advantage of uniprocessor data write locality, and to preserve load-balancing. This run-time technique is embedded in our compiler framework [1] for the automatic detection and parallelization of irregular codes.

The rest of the paper is organized as follows. The technique proposed by Knobe and Sarkar [5] and our parallelization method are presented in Section 2. Experimental results conducted on an SGI Origin 2000 are shown in Section 3. Finally, conclusions are discussed in Section 4.

2 Irregular Assignment Parallelization Techniques

2.1 Approach Based on Array Expansion

A program representation that provides compiler support for the automatic parallelization of irregular assignments is described in [5]. Parallel code for $\#PE$ processors is constructed as follows. Each array definition A in the sequential code, is replaced by two new array variables, $A_1(1 : A_{size})$ and $@A_1(1 : A_{size})$. The value of the array elements defined in the sentence are stored in A_1. The @-array stores the last loop iteration at which the elements of array A_1 were modified. Each processor p is assigned a set of iterations of the sequential loop, and the arrays are expanded [12] as $A_1(1 : A_{size}, 1 : \#PE)$ and $@A_1(1 : A_{size}, 1 : \#PE)$. As a result, distinct processors, p_{k_1} and p_{k_2}, can write into different memory locations concurrently, $A_1(f(h_1), p_{k_1})$ and $A_1(f(h_2), p_{k_1})$, while computing different iterations h_1 and h_2, respectively. The entries of array A are computed by a reduction operation. This method replaces the sequential loop by the three-phase parallel code presented in Fig. 1(b). In the first stage, each processor p initializes its region of the expanded arrays, $A_1(1 : A_{size}, p)$ and $@A_1(1 : A_{size}, p)$. Next, processor p executes the set of loop iterations that it was assigned, $iterations(p)$, preserving the same relative order as in the sequential loop. At the end of the second stage, partial results computed by different processors are stored in separate memory locations. In the last stage, the value of each array element $A(j)$ ($j = 1, ..., A_{size}$) is determined by means of a reduction operation that obtains the value of the element of $A_1(j, 1 : \#PE)$ with the largest iteration number in $@A_1(j, 1 : \#PE)$. Each processor computes the final value of a subset of array elements $A(j)$. Note that processors must be synchronized at the end of the execution stage to ensure that the computation of arrays A_1 and $@A_1$ has finished before performing the reduction at the finalization stage.

The parallelization technique described above can be optimized to perform element-level dead code elimination in those loops that only contain output data dependences. In such codes, the same array element may be computed several times during the execution phase, though only the last value is used after the loop ends. In order to remove the computation of intermediate values, the parallel code shown in Fig. 1(b) is modified as follows. The computation of

```
A(...) = ...
DO h = 1, f_size
    ...
    A(f(h)) = rhs(h)
    ...
END DO
... = ...A(...)...
```

(a) Loop pattern.

```
DOALL p = 1, #PE
    ! —— Initialization stage ——
    A_1(1 : A_size, p) = 0
    @A_1(1 : A_size, p) = 0

    ! ——— Execution stage ———
    DO h ∈ iterations(p)
        A_1(f(h), p) = Exp
        @A_1(f(h), p) = h
    END DO
END DOALL

! —— Finalization stage ——
DOALL p = 1, #PE
    DO j ∈ array_elements(p)
        p_max = MAXLOC(@A_1(j, 1 : #PE))
        IF (@A_1(j, p_max) > 0) THEN
            A(j) = A_1(j, p_max)
        END IF
    END DO
END DOALL
```

(b) Array expansion approach.

Fig. 1. Irregular assignment.

array A_1 is removed, and the $MAXLOC$ operation at the finalization phase is replaced with a MAX operation that determines the last iteration h_{max} in the sequential loop at which the array element $A(j)$ was modified. Furthermore, the condition of the *if* statement $(@A_1(j, p_{max}) > 0)$ is rewritten as $(h_{max} > 0)$, and the sentence $A(j) = A_1(j, p_{max})$ as $A(j) = rhs(h_{max})$. Classical dead code elimination typically removes assignment statements from the source code. This technique eliminates unnecessary array element definitions at run-time.

The performance and applicability of the array expansion technique is quite limited for several reasons. Memory overhead ($\mathcal{O}(A_{size} \times P)$) grows linearly with the number of processors due to the expansion of the arrays, A_1 and $@A_1$. The scalability is poor due to the computational overhead introduced by the computation of both the @-arrays and the MAX-like reduction operation. Furthermore, the method may present load-balancing problems when the element-level dead code elimination is applied. The reason is that, in the finalization stage, some processors may be idle because they may be assigned a subset of array elements, $A(j)$, that do not involve computations. In Section 3, we present experimental results that confirm the limitations stated above.

2.2 Approach Based on the Inspector-Executor Model

In this section, we propose a run-time technique to parallelize irregular assignments. The basic idea lies in reordering loop iterations so that write locality is exploited on each processor. The method consists of two stages that follow the inspector-executor model. In the *inspector*, array A is divided into subarrays of consecutive locations, A_p ($p = 1, ..., \#PE$), and the computations as-

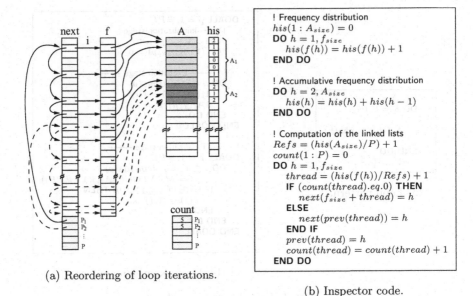

(a) Reordering of loop iterations.

! Frequency distribution
$his(1 : A_{size}) = 0$
DO $h = 1, f_{size}$
    $his(f(h)) = his(f(h)) + 1$
END DO

! Accumulative frequency distribution
DO $h = 2, A_{size}$
    $his(h) = his(h) + his(h - 1)$
END DO

! Computation of the linked lists
$Refs = (his(A_{size})/P) + 1$
$count(1 : P) = 0$
DO $h = 1, f_{size}$
    $thread = (his(f(h))/Refs) + 1$
    IF $(count(thread).eq.0)$ **THEN**
        $next(f_{size} + thread) = h$
    ELSE
        $next(prev(thread)) = h$
    END IF
    $prev(thread) = h$
    $count(thread) = count(thread) + 1$
END DO

(b) Inspector code.

Fig. 2. Inspector-executor approach.

sociated with each block are assigned to different processors. Thus, the loop iteration space $(1, \ldots, f_{size})$ is partitioned into sets f_p that perform write operations on different blocks A_p. In the *executor*, each processor p executes the conflict-free computations associated with the loop iterations contained in a set f_p. Figure **2**(a) shows a graphical description of the method. Sets f_p are implemented as linked lists of iteration numbers using two arrays, $count(1 : \#PE)$ and $next(1 : f_{size} + \#PE)$. Each processor p has an entry in both arrays, $count(p)$ and $next(f_{size} + p)$. The entry $next(f_{size} + p)$ stores the first iteration number h_1^p assigned to the processor p. The next iteration number associated to the processor is stored in the location $next(h_1^p)$. This process is repeated $count(p)$ times. A related technique is described in [3] for the parallelization of irregular reductions.

Load-balancing is a critical factor in the design of parallel applications. The performance of the technique described in this work is limited by the irregular access pattern defined by the subscript expression $f(h)$. In irregular applications, array f usually depends on input data and, thus, cannot be determined until run-time. In our method, load-balancing is achieved by building subarrays A_p of different sizes in the inspector. The inspector (see Fig. **2**(b)) begins with the computation of the frequency distribution, $his(1 : A_{size})$, i.e., the number of write references to each element of array A. The linked lists f_p for reordering the computations are built using the accumulative frequency distribution to determine the size of blocks A_p. In Fig. **2**(a), the shaded regions of array A represent two blocks, A_1 and A_2, that have been associated with processors P_1 and P_2, respectively. Note that although A_1 and A_2 have different sizes (7

```
! Frequency distribution
iter(1 : A_size) = 0
his(1 : A_size) = 0
DO h = 1, f_size
    iter(f(h)) = h
    his(f(h)) = 1
END DO

! Accumulative frequency distribution
DO h = 2, A_size
    his(h) = his(h) + his(h − 1)
END DO

! Computation of the linked lists
Refs = (his(A_size)/P) + 1
count(1 : P) = 0
DO h = 1, A_size
    IF (iter(h).gt.0) THEN
        thread = (his(h)/Refs) + 1
        IF (count(thread).eq.0) THEN
            next(f_size + thread) = iter(h)
        ELSE
            next(prev(thread)) = iter(h)
        END IF
        prev(thread) = iter(h)
        count(thread) = count(thread) + 1
    END IF
END DO
```

```
A(...) = ...
DOALL p = 1, #PE
    h = next(f_size + p)
    DO k = 1, count(p)
        A(f(h)) = ...
        h = next(h)
    END DO
END DOALL
... = ...A(...)...
```

(a) Executor code.

(b) Inspector with dead code elimination.

Fig. 3. Inspector-executor approach with dead code elimination.

and 3, respectively), processors P_1 and P_2 are both assigned 5 iterations of the sequential loop. Linked lists of processors P_1 and P_2 are depicted using solid and dashed arrows, respectively. Figure 3(a) shows the code of the executor that traverses the linked lists built by the inspector.

Element-level dead code elimination can be implemented in the inspector-executor model, too. In this case, the linked-lists only contain the last iteration at which array elements, $A(j)$, are modified. The code of the optimized inspector (the executor does not change) is shown in Fig. 3(b). The frequency distribution array, $his(1 : A_{size})$, contains the value 1 in those array elements that are modified during the execution of the irregular assignment; otherwise, it contains 0. Furthermore, array $iter$ stores the iteration numbers corresponding to the last update of the array elements $A(j)$. Finally, the phase that computes the linked lists is rewritten accordingly.

This technique based on the inspector-executor model overcomes the limitations of the array expansion approach. Memory overhead ($\mathcal{O}(f_{size} + P)$) is associated with the implementation of the linked lists. In practice, it is constant with regard to the number of processors as $f_{size} \gg P$. On the other hand, as the executor performs conflict-free computations, only the inspector introduces computational overhead. Note that this overhead can be amortized in dynamic irregular applications, where the inspector is reused.

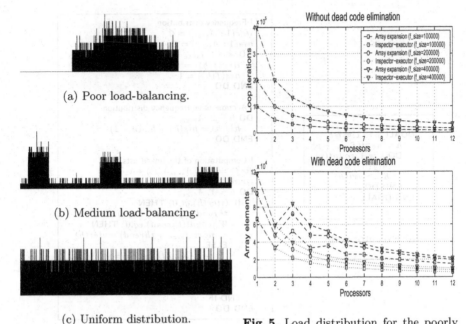

(a) Poor load-balancing.

(b) Medium load-balancing.

(c) Uniform distribution.

Fig. 4. Frequency distributions.

Fig. 5. Load distribution for the poorly balanced pattern.

3   Experimental Results

In this section we present experimental results to compare the performances of our technique and the array expansion method. The target machine was an SGI Origin2000 cc-NUMA multiprocessor. OpenMP [7] shared memory directives have been used in the parallel implementation.

Load-balancing has been taken into account in the design of the inspector-executor approach (see Section 2.2). Figure 4 shows three frequency distributions with different load-balancing properties: poor balancing, medium balancing and uniform distribution. These histograms were selected because they represent irregular access patterns typically found in a well-known rasterization algorithm from computer graphics: the generic convex polygon scan conversion [2]. This algorithm presents output dependences that arise from the depiction of a set of polygons, which represent an image/scene, on a display buffer, A. A typical size for the display buffer is $A_{size} = 512 \times 512 = 262144$ pixels. We have considered $5,000$, $10,000$ and $20,000$ polygons to cover a reasonable range typically found in rasterization. Assuming a fixed mean number of 20 pixels per polygon, the total number of references, f_{size}, to the array A is $100,000$, $200,000$ and $400,000$, respectively.

When dead code elimination is not considered, both the inspector-executor technique and the array expansion method preserve load-balancing by assigning approximately the same number of loop iterations to each processor (see Fig. 5). Experimental results for static irregular codes are presented in Fig. 6(a). Similar results are obtained with the three access patterns depicted in Fig. **4**. Execution times show that our technique outperforms the array expansion approach independent of the frequency distribution and the problem size. In fact, for any frequency distribution, the reduction in execution time is 54%, 40% and 24% on 12 processors for problem sizes of $100,000$, $200,000$ and $400,000$, respectively. Note that the difference in performance rises when the problem size decreases because the executor performs overhead-free computations, and the overhead of the array expansion approach, which depends on the display buffer size and the number of processors, is greater for small problem sizes. The speed-up of the inspector-executor approach grows linearly with the number of processors, while that of the array expansion method grows slower due to the computational overhead introduced in the finalization stage. For totally dynamic codes, where the irregular access pattern is not reused, the sum of the inspector and the executor running times is still lower than that of the array expansion approach. The corresponding reduction in execution time are 40%, 26% and 9% on 12 processors.

When dead code elimination is applied, the performance of the array expansion approach decreases because, as the number of loop iterations actually executed is decreased (the maximum would be one iteration per array element, A_{size}), the computational overhead of the finalization stage dominates the total execution time. Furthermore, the array expansion method with dead code elimination presents load-balancing problems when the irregular access pattern is unbalanced (see Fig. **5**). Note that load-balance is measured in terms of the number of array elements that are computed by each processor. It is clear that some processors may be assigned most of the computations, while others may remain idle most of the time. Consequently, the difference in performance between both techniques increases (see Figs. **6**(b), **6**(c) and **6**(d)). The reduction in execution time increases to 69%, 61% and 54% on 12 processors for the problem sizes $100,000$, $200,000$ and $400,000$, respectively. The speed-up of the inspector-executor technique again grows linearly with regard to the number of processors. In totally dynamic irregular codes, percentages are 54%, 44% and 34% on 12 processors, as opposed to 40%, 26% and 9%, respectively, when dead code is not considered. Better results are obtained because the computational overhead introduced by the inspector is negligible and does not depend on the number of processors nor on the frequency distribution. Furthermore, this overhead is lower than the overhead corresponding to the finalization stage of the array expansion technique.

4 Conclusions

In this paper we have compared two methods for parallelizing loops that contain irregular assignment computations. The approach based on array expansion is easy to implement. However, its scope of application is limited by its high memory and computational overhead, and by load-balancing problems when dead code elimination is considered. In contrast, we present a technique based on the inspector-executor model that is scalable and preserves load-balancing.

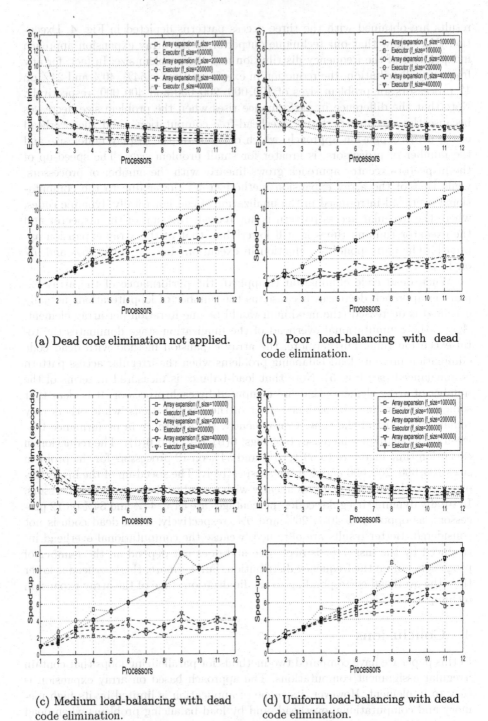

(a) Dead code elimination not applied.

(b) Poor load-balancing with dead code elimination.

(c) Medium load-balancing with dead code elimination.

(d) Uniform load-balancing with dead code elimination.

Fig. 6. Execution times and speed-ups for static irregular codes.

Experimental results show that the inspector-executor model is appropriate for parallelizing irregular assignments because, on the one hand, the overhead is negligible when compared with that of the array expansion method. On the other hand, all the overhead is associated with the inspector (the executor speeds-up linearly) and, thus, it can be amortized over several iterations in dynamic irregular applications.

Acknowledgements

We gratefully thank CSC (Complutense Supercomputing Center, Madrid) for providing access to the SGI Origin 2000 multiprocessor. This work was supported by the Ministry of Science and Technology of Spain and FEDER funds of the European Union (Project TIC2001-3694-C02-02).

References

1. Arenaz, M., Touriño, J., Doallo, R.: A Compiler Framework to Detect Parallelism in Irregular Codes. In Proceedings of 14th International Workshop on Languages and Compilers for Parallel Computing, LCPC'2001, Cumberland Falls, KY (2001)
2. Glassner, A.: Graphics Gems. Academic Press (1993)
3. Gutiérrez, E., Plata, O., Zapata, E.L.: Balanced, Locality-Based Parallel Irregular Reductions. In Proceedings of 14th International Workshop on Languages and Compilers for Parallel Computing, LCPC'2001, Cumberland Falls, KY (2001)
4. Han, H., Tseng, C.-W.: Efficient Compiler and Run-Time Support for Parallel Irregular Reductions. Parallel Computing 26(13-14) (2000) 1861–1887
5. Knobe, K., Sarkar, V.: Array SSA Form and Its Use in Parallelization. In Proceedings of the 25th ACM SIGACT-SIGPLAN Symposium on the Principles of Programming Languages (1998) 107–120
6. Lin, Y., Padua, D.A.: On the Automatic Parallelization of Sparse and Irregular Fortran Programs. In: O'Hallaron, D. (ed.): Languages, Compilers, and Run-Time Systems for Scalable Computers. Lecture Notes in Computer Science, Vol. 1511, Springer-Verlag (1998) 41–56
7. OpenMP Architecture Review Board: OpenMP: A proposed industry standard API for shared memory programming (1997)
8. Ponnusamy, R., Saltz, J., Choudhary, A., Hwang, Y.-S., Fox, G.: Runtime Support and Compilation Methods for User-Specified Irregular Data Distributions. IEEE Transactions on Parallel and Distributed Systems 6(8) (1995) 815–831
9. Rauchwerger, L., Padua, D.A.: The LRPD Test: Speculative Run-Time Parallelization of Loops with Privatization and Reduction Parallelization. IEEE Transactions on Parallel and Distributed Systems 10(2) (1999) 160–180
10. Saad, Y.: SPARSKIT: A Basic Tool Kit for Sparse Matrix Computations. http://www.cs.umn.edu/Research/darpa/SPARSKIT/sparskit.html (1994)
11. Turek, S., Becker, Chr.: Featflow: Finite Element Software for the Incompressible Navier-Stokes Equations. User Manual. http://www.featflow.de (1998)
12. Wolfe, M.J.: Optimizing Supercompilers for Supercomputers. Pitman, London and The MIT Press, Cambridge, Massachussets (1989) In the series, Research Monographs in Parallel and Distributed Computing.
13. Yu, H., Rauchwerger, L.: Adaptive Reduction Parallelization Techniques. In Proceedings of the 14th ACM International Conference on Supercomputing, Santa Fe, NM (2000) 66–77

Large System Performance
of SPEC OMP2001 Benchmarks

Hideki Saito[1], Greg Gaertner[2], Wesley Jones[3], Rudolf Eigenmann[4],
Hidetoshi Iwashita[5], Ron Lieberman[6], Matthijs van Waveren[7], and Brian Whitney[8]
SPEC High-Performance Group

[1] Intel Corporation, SC12-301, 3601 Juliette Lane, Santa Clara, CA 95052 USA
Phone: +1-408-653-9471, Fax: +1-408-653-5374
hideki.saito@intel.com,
[2] Compaq Computer Corporation
[3] Silicon Graphics Incorporated
[4] Purdue University
[5] Fujitsu Limited
[6] Hewlett-Packard Corporation
[7] Fujitsu European Centre for IT Limited
[8] Sun Microsystems, Incorporated

Abstract. Performance characteristics of application programs on large-scale
systems are often significantly different from those on smaller systems. SPEC
OMP2001 is a benchmark suite intended for measuring performance of modern
shared memory parallel systems. The first component of the suite, SPEC
OMPM2001, is developed for medium-scale (4- to 16-way) systems. We
present our experiences on benchmark development in achieving good
scalability using the OpenMP API. This paper then analyzes the published
results of SPEC OMPM2001 on large systems (32-way and larger), based on
application program behavior and systems' architectural features. The ongoing
development of the SPEC OMP2001 benchmark suites is also discussed. Its
main feature is the increased data set for large-scale systems. We refer to this
suite as SPEC OMPL2001, in contrast to the current SPEC OMPM2001
(medium data set) suite.

Keywords. SPEC OMP2001, Benchmarks, High-Performance Computing,
Performance Evaluation, OpenMP

1 Introduction

SPEC (The Standard Performance Evaluation Corporation) is an organization for
creating industry-standard benchmarks to measure various aspects of modern
computer system performance. SPEC's High-Performance Group (SPEC HPG) is a
workgroup aiming at benchmarking high-performance computer systems. In June of
2001, SPEC HPG released the first of the SPEC OMP2001 benchmark suites, SPEC
OMPM2001. This suite consists of a set of OpenMP-based application programs.
The data sets of the SPEC OMPM2001 suite (also referred to as the medium suite) are

H. Zima et al. (Eds.): ISHPC 2002, LNCS 2327, pp. 370–379, 2002.
© Springer-Verlag Berlin Heidelberg 2002

derived from state-of-the-art computation on modern medium-scale (4- to 16-way) shared memory parallel systems. Aslot et al. [1] have presented the benchmark suite. Aslot et al. [2] and Iwashita et al. [3] have described performance characteristics of the benchmark suite.

As of this writing, a large suite (SPEC OMPL2001), focusing on 32-way and larger systems, is under development. SPEC OMPL2001 shares most of the application code base with SPEC OMPM2001. However, the code and the data sets are modified to achieve better scaling and also to reflect the class of computation regularly performed on such large systems.

Performance characteristics of application programs on large-scale systems are often significantly different from those on smaller systems. In this paper, we characterize the performance behavior of large-scale systems (32-way and larger) using the SPEC OMPM2001 benchmark suite. In Section 2, we present our experiences on benchmark development in achieving good scalability using the OpenMP API. Section 3 analyzes the published results of SPEC OMPM2001 on large systems, based on application program behavior and systems' architectural features. The development of SPEC OMPL2001 is discussed in Section 4, and Section 5 concludes the paper.

2   Experiences on SPEC OMPM2001 Benchmark Development

2.1   Overview of the SPEC OMP2001 Benchmark

The SPEC OMPM2001 benchmark suite consists of 11 large application programs, which represent the type of software used in scientific technical computing. The applications include modeling and simulation programs from the fields of chemistry, mechanical engineering, climate modeling, and physics. Of the 11 application programs, 8 are written in Fortran, and 3 are written in C. The benchmarks require a virtual address space of about 1.5 GB in a 1-processor execution. The rationales for this size were to provide data sets significantly larger than those of the SPEC CPU2000 benchmarks, while still fitting them in a 32-bit address space.

The computational fluid dynamics applications are APPLU, APSI, GALGEL, MGRID, and SWIM. APPLU solves 5 coupled non-linear PDEs on a 3-dimensional logically structured grid, using the Symmetric Successive Over-Relaxation implicit time-marching scheme [4]. Its Fortran source code is 4000 lines long. APSI is a lake environmental model, which predicts the concentration of pollutants. It solves the model for the mesoscale and synoptic variations of potential temperature, wind components, and for the mesoscale vertical velocity, pressure, and distribution of pollutants. Its Fortran source code is 7500 lines long. GALGEL performs a numerical analysis of oscillating instability of convection in low-Prandtl-number fluids [5]. Its Fortran source code is 15300 lines long. MGRID is a simple multigrid solver, which computes a 3-dimensional potential field. Its Fortran source code is 500 lines long. SWIM is a weather prediction model, which solves the shallow water equations using a finite difference method. Its Fortran source code is 400 lines long.

AMMP (Another Molecular Modeling Program) is a molecular mechanics, dynamics, and modeling program. The benchmark performs a molecular dynamics simulation of a protein-inhibitor complex, which is embedded in water. Its C source code is 13500 lines long.

FMA3D is a crash simulation program. It simulates the inelastic, transient dynamic response of 3-dimensional solids and structures subjected to impulsively or suddenly applied loads. It uses an explicit finite element method [6]. Its Fortran source code is 60000 lines long.

ART (Adaptive Resonance Theory) is a neural network, which is used to recognize objects in a thermal image [7]. The objects in the benchmark are a helicopter and an airplane. Its C source code is 1300 lines long.

GAFORT computes the global maximum fitness using a genetic algorithm. It starts with an initial population and then generates children who go through crossover, jump mutation, and creep mutation with certain probabilities. Its Fortran source code is 1500 lines long.

EQUAKE is an earthquake-modeling program. It simulates the propagation of elastic seismic waves in large, heterogeneous valleys in order to recover the time history of the ground motion everywhere in the valley due to a specific seismic event. It uses a finite element method on an unstructured mesh [8]. Its C source code is 1500 lines long.

WUPWISE (Wuppertal Wilson Fermion Solver) is a program in the field of lattice gauge theory. Lattice gauge theory is a discretization of quantum chromodynamics. Quark propagators are computed within a chromodynamic background field. The inhomogeneous lattice-Dirac equation is solved. Its Fortran source code is 2200 lines long.

2.2 Parallelization of the Application Programs

Most of the application programs in the SPEC OMPM2001 benchmark suite are taken from the SPEC CPU2000 suites. Therefore, parallelization of the original sequential program was one of the major efforts in the benchmark development. The techniques we have applied in parallelizing the programs are described in [1]. We give a brief overview of these techniques and then focus on an issue of particular interest, which is performance tuning of the benchmarks on non-uniform memory access (NUMA) machines.

The task of parallelizing the original, serial benchmark programs was relatively straightforward. The majority of code sections were transformed into parallel form by searching for loops with fully independent iterations and then annotating these loops with OMP PARALLEL DO directives. In doing so, we had to identify scalar and array data structures that could be declared as private to the loops or that are involved in reduction operations. Scalar and array privatization is adequately supported by OpenMP. When part of an array had to be privatized, we created local arrays for that purpose. Parallel array reduction is not supported by the OpenMP Fortran 1.1 specification [9], and thus these array operations have been manually transformed. As we studied performance and scalability of the parallelized code, we realized that LAPACK routines and array initialization should also be parallel. Note that parallel array reduction is supported by the OpenMP Fortran 2.0 specification [10].

In several situations we exploited knowledge of the application programs and the underlying problems in order to gain adequate performance. For example, linked lists were converted to vector lists (change of data structures). In GAFORT, a parallel shuffle algorithm we decided to use is different from the original sequential one (change of algorithms). In EQUAKE, the global sum was transformed to a sparse global sum by introducing a mask. Classical wisdom of parallelizing at a coarser grain also played a role. We have fused small loops, added NOWAIT at the end of applicable work-sharing loops, and even restructured big loops so that multiple instances of the loop can run in parallel. When memory could be traded for a good speedup, we allowed a substantial increase in memory usage.

Our parallelization effort resulted in good theoretical and actual scaling for our target platforms [1]. In all but one code, over 99% of the serial program execution is enclosed by parallel constructs. In GALGEL the parallel coverage is 95%. Figure 1 shows the speedups computed from Amdahl's formula.

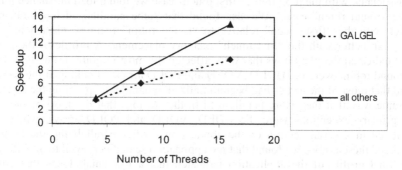

Fig. 1. Amdahl's Speedup for 4, 8, and 16 threads.

2.3   Parallelization of the SWIM Application Program with Initialization for NUMA Architectures

SWIM represents one of the more simple benchmarks in the SPEC OMPM2001 suite and makes an easy example of the techniques used to parallelize the applications in the SPEC OMPM2001 suite using the OpenMP Standard. It also exemplifies the use of the parallelization of the initialization of arrays and memory to get good, scalable performance on a number of cache-coherent non-uniform memory access (NUMA) architectures.

Swim has 3 subroutines that contain more than 93% of the work as measured at runtime using a performance analysis tool. Each of these subroutines contained loops, one of which contained most of the work of that subroutine. These are two-dimensional loops, indexed as U(I,J). So, in the initial parallelization of SWIM, these three loops were parallelized over the J-index to reduce false sharing of cache lines and provide enough work in the contiguous I-direction. The application scaled well to 8 CPUs.

Beyond eight CPUs, a number of problems occurred. We will discuss the 32 CPU case. One of the remaining, serial routines took about 6% of the time on one CPU. By Amdahl's law, this routine would consume over 60% of the execution time on 32 CPUs. A performance analysis tool running on all 32 CPUs with the application indicated that most of this time was spent in one loop. We were able to parallelize this loop over the J index with the help of an OpenMP reduction clause.

At this point the application ran reasonably well on 32 CPUs. But, considering that nearly all of the work had been parallelized, a comparison of the 8-CPU time with the 32-CPU time indicated that the application was not scaling as well as expected. To check how sensitive the application was to NUMA issues, we ran 2 experiments on the SGI Origin3000. In the first, we used the default "first touch" memory placement policy where virtual memory is placed on the physical memory associated with the process that touched the memory first, usually via a load or store operation. In the second experiment we used a placement policy where the memory is spread around the system in a round robin fashion on those physical memories with processes that are associated with the job. In the "first touch" case we found load imbalance relative to the "round robin" case. We also found that some parallelized loops ran much slower with the "first touch" relative to "round robin" memory placement, while others ran better with the "first touch" memory placement. A profile of the parallel loops indicated that the arrays that were affected the most in going from "first touch" to "round robin" were the UOLD, VOLD and POLD arrays. Looking at the code, we found that most of the arrays had been initialized in parallel loops along the J index in the same way that they were parallelized in the sections that did most of the work. The primary exceptions were the UOLD, VOLD and POLD arrays, which were initialized in a serial section of the application. After parallelizing the loop that initialized these arrays, we found that the application scaled very well to 32 CPUs.

A final profile of the application indicated that a few small loops that enforce periodic conditions in the spatial variation of the arrays may also be parallelized for an additional improvement in performance on 32 CPUs. However this modification is not part of the current OMPM2001 suite.

3   Large System Performance of SPEC OMPM2001 Benchmarks

Performance characteristics of application programs on large-scale systems are often significantly different from those on smaller systems. Figure 2 shows a scaling of Amdahl's speedup for 32 to 128 threads, normalized by the Amdahl's speedup of 16 threads. The graph now looks notably different from Figure 1.

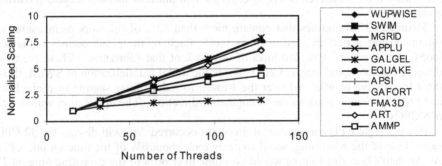

Fig. 2. Normalized scaling of Amdahl's Speedup for 16 to 128 threads.

Amdahl's speedup assumes perfect scaling of the parallel portion of the program. Actual programs and actual hardware have additional sources of overhead, which degrade the performance obtained on a real system relative to the upper bound given by Amdahl's law. Figures 3-6 show the scaling data for published benchmark results of SPEC OMPM2001.

3.1   Benchmarks with Good Scalability

The numbers listed in the following figures have been obtained from the results published by SPEC as of December 23, 2001. For the latest results published by SPEC, see http://www.spec.org/hpg/omp2001. All results shown conform to Base Metrics rules, meaning that the benchmark codes were not modified. For better presentation of the graph, we have normalized all results with the lowest published result as of December 23, 2001 for each platform. The results for the Fujitsu PRIMEPOWER 2000 563 MHz system have been divided by the Fujitsu PRIMEPOWER 2000 18-processor result. The results for the SGI Origin 3800 500 MHZ R14 K system have been divided by the SGI Origin 3800 8-processor result. Finally, the results for the HP PA-8700 750 MHz system have been divided by the HP PA-8700 16-processor result.

The benchmarks WUPWISE, SWIM, and APSI show good scalability up to 128 processors (Figure 3). The scalability of APSI has a dip between 64 and 100 processors, and this is under investigation.

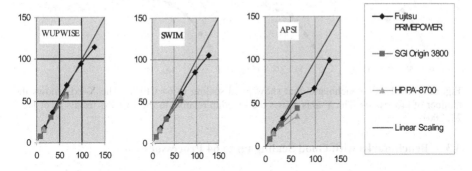

Fig. 3. OMPM2001 benchmarks that show good scalability. The X-axis shows the number of processors. The Y-axis shows performance relative to the lowest published result as of Dec. 23, 2001 for each platform.

3.2   Benchmarks with Superlinear Scaling

The benchmark APPLU shows superlinear scaling on HP PA-8700, on the SGI Origin 3800, and on the Fujitsu PRIMEPOWER 2000 due to a more efficient usage of the cache as more processors are used (Figure 4). HP's PA-8700 processor has 1.5MB primary data cache (and no secondary cache), and the R14000 in SGI's Origin 3800 has an 8MB off-chip unified secondary cache. The cache of the 8- processor Origin 3800 system (64MB unified L2 cache) is insufficient to fit the critical data, but the cache of the 16 processor system holds it. The cache of the 64 processor PA-8700 system (96MB L1 data cache) is large enough to hold the critical data. Preliminary results on the Fujitsu PRIMEPOWER 2000 system, which also has an 8MB off-chip secondary cache, show superlinear scaling when going from 8 processors with a 64MB L2 cache to 16 processors with a 128MB L2 cache, and to 22 processors with a 176MB L2 cache.

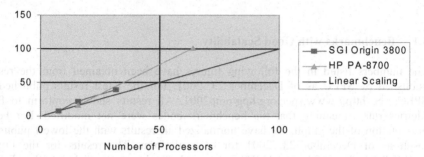

Fig. 4. OMPM2001 benchmark APPLU, which shows superlinear scalability. The Y-axis shows the speedup relative to the lowest published result as of Dec. 23, 2001 for each platform.

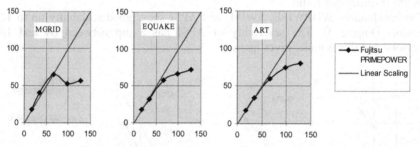

Fig. 5. OMPM2001 benchmarks that show good scaling to 64 CPUs. The X-axis shows the number of processors. The Y-axis shows the speedup relative to the 18-CPU results as of Dec. 23, 2001.

3.3   Benchmarks with Good Scaling up to 64 Processors

The benchmarks EQUAKE, MGRID, and ART show good scaling up to 64 processors, but poor scaling for larger number of processors (Figure 5). This has been shown on a Fujitsu PRIMEPOWER 2000 system, and the results are normalized by the 18-processor result. MGRID and EQUAKE are sparse matrix calculations, which do not scale well to large number of processors. We expect that the large data set of the OMPL suite, described in the next section, will have better scaling behavior to a large number of processors.

3.4   Benchmarks with Poor Scaling

The benchmarks GALGEL, FMA3D, and AMMP show poor scaling for all number of processors beyond 8 or 16 CPUs (Figure 6). GALGEL uses the QR algorithm, which has been known to be non-scalable.

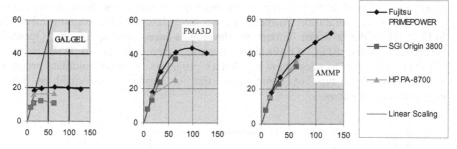

Fig. 6. OMPM2001 benchmarks that show poor scaling. The X-axis shows the number of processors. The Y-axis shows the speedup relative to the lowest published results as of Dec. 23, 2001 for each platform.

4   SPEC OMPL2001 Benchmark Development

As of this writing, the development of a new OpenMP benchmark suite, also referred to as SPEC OMPL2001, is underway. In contrast to the SPEC OMPM2001 suite, the target problem size (working set size and run time) is approximately 4x to 8x larger than SPEC OMPM2001. The entire suite should take no more than two days to complete on a 16-CPU 350MHz reference machine. On a reference system, we expect the maximum working set size would be around 6GB, which will require a support for a 64-bit address space[1]

There are still many unknowns in the final outcome of the SPEC OMPL2001 benchmark suite. We have set the following design goals for the new suite. SPEC OMPL2001 will exercise more code paths than SPEC OMPM2001, necessitating additional parallelization efforts. On some benchmark programs, I/O became a bottleneck in handling larger data sets. When I/O is an integral part of the application, parallel I/O would be called for. Alternatively, output can be trimmed down. ASCII input files are most portable and performance-neutral, but the overhead of converting large ASCII input files to floating-point binary can impact the execution time significantly. Furthermore, the C code will be made C++ friendly, so that C++ compilers can also be used for the benchmarks.

Figure 7 shows the scaling of the working set size and the execution time for a 32-processor system. For example, WUPWISE in SPEC OMPL2001 uses 3.5 times more memory than SPEC OMPM2001, and it takes three times longer to execute. This experiment is based on the latest SPEC OMPL2001 benchmark development kit, and thus the code and the data set are subject to changes before the final release of the benchmark.

5   Conclusion

In this paper we have analyzed the performance characteristics of published results of the SPEC OMPM2001 benchmark suite. We have found that many of the benchmark

---

[1] More memory would be needed if more CPUs were used.

programs scale well up to several tens of processors. We have also found a number of codes with poor scalability. Furthermore, we have described the ongoing effort by SPEC's High-Performance Group to develop a new release of the OpenMP benchmark suites, SPEC OMPL2001, featuring data sets up to 6GB in size. SPEC/HPG is open to adopt new benchmark programs. A good candidate program would represent a type of computation that is regularly performed on high-performance computers.

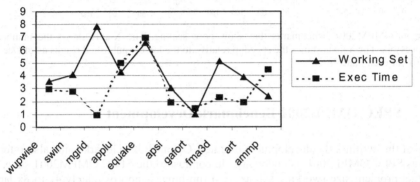

Fig. 7. Working set and execution time scaling of a preliminary version of the SPEC OMPL suite

5   Conclusion

In this paper we have analyzed the performance characteristics of published results of the SPEC OMPM2001 benchmark suite. We have found that many of the benchmark programs scale well up to several tens of processors. We have also found a number of codes with poor scalability. Furthermore, we have described the ongoing effort by SPEC's High-Performance Group to develop a new release of the OpenMP benchmark suites, SPEC OMPL2001, featuring data sets up to 6GB in size. SPEC/HPG is open to adopt new benchmark programs. A good candidate program would represent a type of computation that is regularly performed on high-performance computers.

Acknowledgement

The authors would like to thank all of those who developed the application programs and data sets used in the benchmark. Benchmark development would have been impossible without their hard work.

References

1. Vishal Aslot, Max Domeika, Rudolf Eigenmann, Greg Gaertner, Wesley B. Jones, and Bodo Parady, SPEComp: A New Benchmark Suite for Measuring Parallel Computer Performance, In Proc. Of WOMPAT 2001, Workshop on OpenMP Applications and Tools, Lecture Notes in Computer Science, vol. 2104, pages 1-10, July 2001.

2. Vishal Aslot and Rudolf Eigenmann, Performance Characteristics of the SPEC OMP2001 Benchmarks, in Proc. of the Third European Workshop on OpenMP (EWOMP'2001), Barcelona, Spain, September 2001.
3. Hidetoshi Iwashita, Eiji Yamanaka, Naoki Sueyasu, Matthijs van Waveren, and Ken Miura, The SPEC OMP2001 Benchmark on the Fujitsu PRIMEPOWER System, in Proc. of the Third European Workshop on OpenMP (EWOMP'2001), Barcelona, Spain, September 2001.
4. E. Barszcz, R. Fatoohi, V. Venkatkrishnan and S. Weeratunga, Solution of Regular Sparse Triangular Systems on Vector and Distributed-Memory Multiprocessors, Rept. No: RNR-93-007, NASA Ames Research Center, 1993.
5. Gelfgat A.Yu., Bar-Yoseph P.Z. and Solan A, Stability of confined swirling flow with and without vortex breakdown, Journal of Fluid Mechanics, vol. 311, pp.1-36, 1996.
6. Key, S. W. and C. C. Hoff, An Improved Constant Membrane and Bending Stress Shell Element for Explicit Transient Dynamics, Computer Methods in Applied Mechanics and Engineering, Vol. 124, pp 33-47, 1995.
7. M.J. Domeika, C.W. Roberson, E.W. Page, and G.A. Tagliarini, Adaptive Resonance Theory 2 Neural Network Approach To Star Field Recognition, in Applications and Science of Artificial Neural Networks II, Steven K. Rogers, Dennis W. Ruck, Editors, Proc. SPIE 2760, pp. 589-596(1996).
8. Hesheng Bao, Jacobo Bielak, Omar Ghattas, Loukas F. Kallivokas, David R. O'Hallaron, Jonathan R. Shewchuk, and Jifeng Xu, Large-scale Simulation of Elastic Wave Propagation in Heterogeneous Media on Parallel Computers, Computer Methods in Applied Mechanics and Engineering 152(1-2):85-102, 22 January 1998.
9. OpenMP Architecture Review Board, OpenMP Fortran Application Programming Interface Version 1.1, November 1999 (http://www.openmp.org/specs/mp-documents/fspec11.pdf)
10. OpenMP Architecture Review Board, OpenMP Fortran Application Programming Interface Version 2.0, November 2000 (http://www.openmp.org/specs/mp-documents/fspec11.pdf)

A Shared Memory Benchmark in OpenMP

Matthias S. Müller

High Performance Computing Center Stuttgart (HLRS),
Allmandring 30, 70550 Stuttgart, Germany
mueller@hlrs.de
http://www.hlrs.de/people/mueller

Abstract. The efficient use of the memory system is a key issue for the performance of many applications. A benchmark written with OpenMP is presented that measures several aspects of a shared memory system like bandwidth, memory latency and inter-thread latency. Special focus is on revealing and identifying bottlenecks and possible hierarchies in the main memory system.

1 Introduction

With the introduction of OpenMP the portability and ease of shared memory programming has been greatly improved. The concept is simple. All data can be accessed by every thread and the programmer can focus on the distribution of workload. Despite this simplicity many issues will influence the performance and several benchmarks have been designed to address them [4,6,1,3,5]. Since many scientific applications are memory bound the performance will strongly depend on an efficient use of the shared memory. Many issues will have an influence on the performance. Is it a flat SMP or hierarchical NUMA system? What are the different bandwidths and how is the memory distributed? Are there any bottlenecks I have to be aware of. Instead of looking at the memory system with complete applications [7] this benchmark tries to answer these or similar questions. Since HLRS has many scientific and industrial users with codes that are not cache friendly, the focus is on the main memory system. All tests are designed to run out of cache.

Since OpenMP provides a rather abstract interface to parallel programming that is not very hardware oriented, writing a shared memory benchmark is not trivial. Not only that it is undefined where the allocated memory is located physically and relative to the threads. The operating system is also free to schedule the threads to the available CPUs with its own policy. Both assignments may change during runtime. It is therefore unclear whether a portable benchmark written in OpenMP may reveal any information besides average values for a given architecture. The pragmatic approach taken here is, that if the operating system is smart enough to hide something from a good benchmark it should also not be relevant for a real application. As examples for different kind of shared memory systems the results of three different platforms are presented:

H. Zima et al. (Eds.): ISHPC 2002, LNCS 2327, pp. 380–389, 2002.

PC A dual processor Pentium IV pc with 2 Gigabyte of RDRAM memory. It serves as an example for a simple bus based system.

Hitachi SR8K The Hitachi SR8000 consists of nodes. Each node is a flat SMP system with eight CPUs.

HP N-Class This system also has eight CPUs, but the system is organized in cells. Each cell has local memory, the connection to remote memory is made via a fast interconnect.

2   Latencies

Since different threads communicate with each other by reading and writing shared memory, the latencies involved with these 'communications' are an important factor for the overall performance. Two different kind of latencies can be distinguished. First the latency to access the main memory, second the latency that occurs in the direct 'communication' between two threads.

2.1   Latency to Main Memory

With the rapidly increasing clock frequency the latency to the main memory measured in clock cycles is getting higher and higher. Therefore techniques have been developed to reduce or hide this latencies. Examples are caches, out of order execution or data prefetching. Nevertheless, as soon as these optimization techniques fail the cost of a memory access is dominated by the latency.

To measure the memory latency the core loop looks like this:

```
for(i=0; i<N; i++) {
    idx = index[idx];
}
```

The index field is initialized such that every access reads a different cacheline. The total memory amount that is referenced exceeds the cache size. The structure of the loop and the initialization of the index field should inhibit any optimization techniques. In Fig. 1 different latencies are shown. For the results labeled `serial` everything was measured in a serial region, `offset` denotes the owner of the indexfield that is used. A smart operating system could move the running thread to a CPU that is close to the memory or migrate pages. This is one of the reasons why for all other results all threads are measuring the latencies simultaneously. A comparison with the serial results should reveal several informations. If all observed latencies are significantly higher than in the serial case it indicates a limitation in the transaction system. If suddenly a two-level hierarchy is visible the optimization that acted in the serial case now fails.

The results of the HP N-Class are shown in Fig. 1. There is no visible hierarchy. When multiple threads are active the latency increases from about 250ns to 300ns. The same behavior is shown by the PC, where the latency increases from 170ns to 210ns. The latency of the Hitachi SR8K is around 390ns independent of the number of active threads.

382   Matthias S. Müller

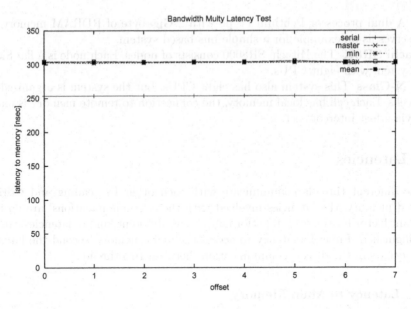

Fig. 1. Latency to Main Memory for HP N-Class. There is no visible hierarchy. When multiple threads are active the latency is $300\mu s$ instead of $250\mu s$.

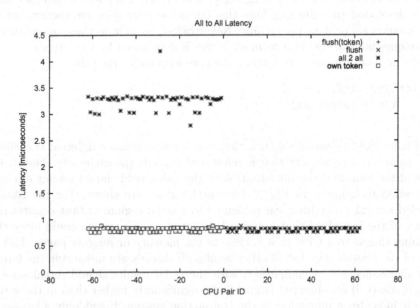

Fig. 2. Ping Pong Latencies for HP N-CLASS.

2.2   Interprocess Memory Latency

In a Shared Memory Programming Model threads communicate with each other by sharing variables. The question posed here is how long it takes before a modification made by one thread is visible to another thread. If the information exchange must be made by reading and writing from main memory the answer would be simple: two times the latency to main memory for the half trip time. One thread would write the memory, another would read it. In real life the situation is more complex. Since the two threads are not synchronized half a memory latency is added on average. The write latency could be higher than the read latency, because on some systems the whole cache line has to be read before it can be written. Altogether this results in a round trip latency between 4 and 7 memory read latencies as measured in the previous section. However, in a bus based system the second thread could fetch the cache line while it is written. Two CPUs might share the same cache or the two threads may run on the same CPU.

To measure the latency between two threads a simple ping-pong style benchmark is used.

```
rank=omp_get_thread_num();
if(rank == sender ){
  for(i=0;i<count ; i++){
    /*send:*/
    token=1;
    #pragma omp flush(token)
    /*receive:*/
    while(token != 0){
        #pragma omp flush(token)
    }
  }
}
if(rank == receiver){
  for(i=0; i<count ; i++){
    /*receive:*/
    while(token != 1){
        #pragma omp flush(token)
    }
    /*send:*/
    token=0;
        #pragma omp flush(token)
  }
}
```

The program is running with N threads. On all systems the number of threads was chosen equal to the number of CPUs. In order to observe possible hierarchies in the system the ping pong benchmarks is made between all

$N * (N - 1)$ pairs of threads. To identify a pair consisting of two threads with the thread number $n1$ and $n2$ the value $N * n1 + n2$ is assigned to the pair.

The code fragment above shows the original form of the benchmark. Several variants are used to examine more details. First the call to `omp flush(token)` is replaced by a call to `omp flush`. After this call a consistent view of the complete memory is guaranteed which makes it potentially more expensive. In the second modification there is not one single token, but the token used in the communication between two threads is always owned by one of the threads. Finally, in the all to all case $N/2$ pairs are communicating simultaneously. Each pair has its own token, which is owned by one of the threads.

In Fig. 2 the latencies of the HP N-Class are plotted vs. the pair ID. Negative thread IDs are used for the last two modifications. For all modifications a two level hierarchy is visible, for the all to all benchmark the latencies are significantly increased. The observed latencies are 3000ns/3300ns (10 read latencies) and 760ns/830ns (3 read latencies)for the other tests. The SR8K shows a inter-thread latency of 2000ns (5 read latencies) and the PC of 280ns. The PC is the only platform where the inter-thread latency is significantly lower than the predicted value of four to seven times the read latency. This indicates an active bus snooping. The all to all communication reveals again a bottleneck of the HP N-Class.

3   Bandwidth

To measure the memory bandwidth the access patterns from the well known stream benchmarks [2] are used. Reading from and writing to memory is measured with SUM and FILL respectively. COPY reads and writes at the same time, while DAXPY performs the operation $a(i) = a(i) + qb(i)$, with two loads and one store. The length of the vectors a and b are choosen such that all operations run out of cache. The unmodified stream 2 benchmark that measures memory performance for different vector lengths is used to select an appropriate vector length.

To achieve a high bandwidth a good memory subsystem and an appropriate connection of the CPU is required. In the shared memory case the requirements are more stringent. The memory system has to fullfil the concurrent requests of several CPUs. Any bottleneck will result in a loss of performance at some place. The purpose of the next two benchmarks is to find such bottlenecks.

3.1   Scale-up with Number of Threads

To avoid any overhead that might be introduced by OpenMP or inter-thread synchronization, each threads works on its own piece of memory. The memory is allocated and initialized in the parallel section by each thread. This should give optimal results if the memory is owned by the allocator or if a first touch algorithm is in place.

Each thread works independently, there is no OpenMP work-share directive. This should keep the likelihood of compiler performance bugs to a minimum.

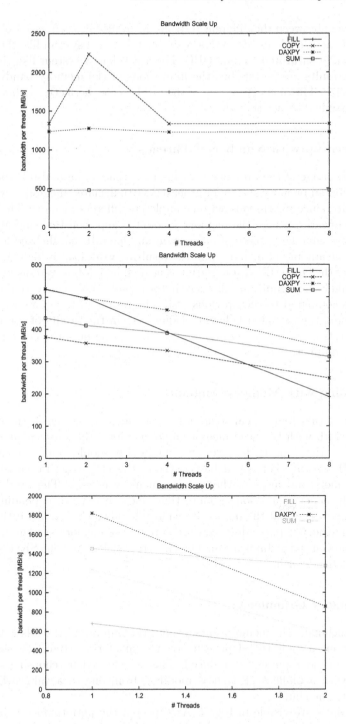

Fig. 3. Scale-up of Memory Bandwidth for Hitachi SR8K, HP N-Class and the PC (from top to bottom).

Only barriers between the different tests are introduced to synchronize the tests. The results are visible in Fig. 3. For a simple bus based system like the PC, the bus is already saturated by one CPU. For a cell based system like the HP N-Class the results are better, but the limitations of the memory bandwidth are clearly visible. The memory system of the SR8K scales perfectly, with a constant memory bandwidth per thread.

3.2   Speed-up with Number of Threads

The disadvantage of the previous benchmark is that it could also be performed by a distributed memory system or at least a virtual shared memory system with good results, since the threads act on completely different memory. Therefore we test a modification where the memory amount is kept constant and all threads work on the same large piece of memory in an OpenMP parallelized loop.

Since the amount of memory and the resulting work load is high enough, the overhead of the OpenMP parallelization is negligible. For the systems tested here, there should be no significant difference in the capability to scale with the number of threads compared to the previous test. The results of the systems presented here reflected the expected behavior. Differences in the results of speed-up and scale-up tests should be carefully analyzed not to be performance bugs in the OpenMP compiler.

3.3   Scaling with Memory Amount

Another test that may reveal structures in the memory system is the scaling of the bandwidth with the used memory amount. Since the start value is already much larger than the cache size, the bandwidth should remain constant for a flat SMP system. For a cell based system the cell structure could become visible as more and more of the systems memory is used. The total available bandwidth increases as soon as more than one cell is used, depending on the allocation strategy and the inter cell memory bandwidth. Finally, if the system is low on memory it puts additional pressure on the memory system. It may for example prevent page duplication. The PC, HP N-Class and the SR8K did show a constant bandwidth.

3.4   Memory Granularity

In this benchmark the memory amount and the number of threads is kept constant. To put an additional pressure on the memory system, the size of the requested memory pieces are reduced. This is achieved by choosing a `STATIC` schedule with a `CHUNKSIZE` that is modified from one to a value where each thread gets one chunk.

The results are visible in Fig. 4. As expected the performance is best for a large chunksize. If the chunksize is reduced, the bandwidth first remains high. This indicates that the parallel overhead remains constant due to the static

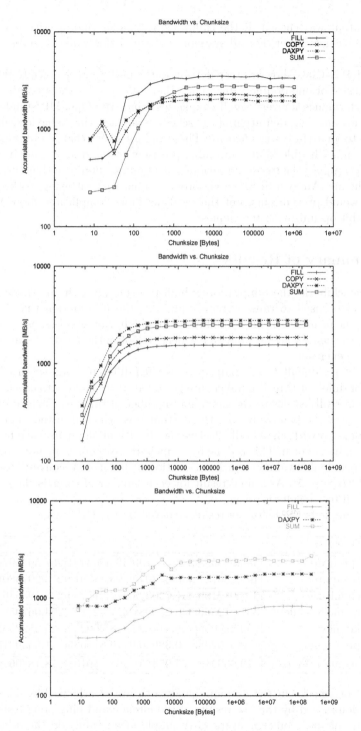

Fig. 4. Memory Bandwidth vs. Chunksize for Hitachi SR8K, HP N-Class and the PC.

schedule that was chosen. If the chunksize is smaller than a critical value the performance gets lower. For all systems the size of this value is consistent with the page size.

The HP N-Class shows a steep descent of the performance down to the smallest chunksize of one double value. The other two systems show an almost constant performance for very small chunksizes. Since the OpenMP standard does not provide an exact definition of the schedule policies, this behavior can not be deduced to an architectural feature. The standard states that 'a program should not rely on a schedule kind conforming precisely to the description given'. This gives the compiler the freedom to choose a minimum chunksize that corresponds to a cacheline. An own SPMD style programming without using work-share directives would prevent this, but the resulting more complicated loop structure may inhibit optimization techniques.

4 Summary of Results

For all benchmarks the temptation is high to come up with one single number that could be used to rank systems. However, the information that is lost by reducing the results to a single number is often high. Nevertheless, as an overview and to serve as a starting point for a deeper look into the results a summary would be beneficial.

Since it is difficult to compare systems with different numbers of CPUs the results for different thread numbers are provided. It also allows to check whether the system really scales to the maximum number of threads. Each result is the mean value of the bandwidth per thread that was achieved in the single tests.

To come up with an overall effective bandwidth all latencies have to be converted to a bandwidth. This is done by artificially assigning a memory amount of 8 bytes (one double value) to each transaction. A latency Δt converts to a bandwidth of $8Bytes/\Delta t$. As a final result for each number of threads the geometric mean of all test results is given.

An example is the following result achieved by the HP N-Class:

threads:	1	2	4	8
speed up :	355.953595	343.561199	325.763491	267.504597
scale up :	465.286751	440.674049	393.185014	273.207580
chunksize :	373.017109	341.720323	312.215239	233.193936
memory scale :	351.912566	338.211988	317.428108	264.067959
latency :	30.385659	28.412623	28.260233	25.982235
ping pong :	0.386203	0.386203	10.813685	1.953931
geometric mean :	79.639219	75.954131	125.350109	78.188061

The bottlenecks in the systems bandwidth are clearly visible. The overall value is dominated by the low results of the latency and ping pong tests due to the nature of the geometric average. A simple mean value, on the other hand, would be dominated by the larger number of bandwidth tests.

5   Conclusion and Outlook

The presented tests reveal a number of properties and bottlenecks of the system under investigation. The memory and inter-thread latency benchmarks did clearly reveal hierarchical systems. They also provide information about the costs of a memory access in a worst case scenario.

The scale-up test shows the scalability without interference with OpenMP overhead or compiler issues. In this sense the speed-up test is targeted to the same bottlenecks with a higher requirement for the compiler and run-time library. The variation of chunksizes shows the reaction of the memory system if the requested memory chunks are reduced. Although it already provides some insight, its exact interpretation is limited by the tolerant definition of the scheduling clause by the OpenMP standard. The scaling of the memory amount is targeted to systems where the operating system is not aware of a cell structure of the underlying hardware.

The benchmarking of further systems in the near future will show if these set of tests is sufficient to provide all relevant information about the shared memory system. Further tests are already in development, but in a too early stage to be presented here. A larger number of tests will also show whether the proposed benchmark summary is useful for a comparison of different systems.

Acknowledgements

I wish to thank Uwe Küster (HLRS) and Holger Berger (NEC) who provided the idea and basic code for the measurement of the memory latency.

References

1. J. M. Bull. Measuring synchronization and scheduling overheads in OpenMP. In *First European Workshop on OpenMP*, 1999.
2. John D. McCalpin. Memory bandwidth and machine balance in current high performance computers. *IEEE TCCA*, Dec. 1995. http://www.cs.virginia.edu/stream.
3. Matthias Müller. Some simple OpenMP optimization techniques. In R. Eigenmann and M.J. Voss, editors, *OpenMP Shared Memory Parallel Programming*, volume 2104 of *LNCS*, WOMPAT 2001, West Lafayette, IN, USA, 2001. Springer.
4. RWCP. Openmp version of NAS parallel benchmarks. http://pdplab.trc.rwcp.or.jp/Omni/benchmarks/NPB/index.html.
5. Mitsuhisa Sato, Kazuhiro Kusano, and Sigehisa Satoh. Openmp benchmark using PARKBENCH. In *Proceeding of Second European Workshop on OpenMP*, Edinburgh, Scotland, U.K., Sept. 2000.
6. SPEC. SPEComp 2001. http://www.spec.org.
7. Steven Cameron Woo, Moriyoshi Ohara, Evan Torrie, Jaswinder Pal Singh, and Anoop Gupta. The splash-2 programs: Characterization and methological considerations. In *Proceedings of the 22nd int. Symp. on Computer Architecture*, pages 24–36, June 1995.

Performance Evaluation of the Hitachi SR8000 Using OpenMP Benchmarks

Daisuke Takahashi, Mitsuhisa Sato, and Taisuke Boku

Center for Computational Physics,
Institute of Information Sciences and Electronics, University of Tsukuba
1-1-1 Tennodai, Tsukuba-shi, Ibaraki 305-8577, Japan
{daisuke,msato,taisuke}@is.tsukuba.ac.jp

Abstract. This paper reports the performance of a single node of the Hitachi SR8000 when using OpenMP benchmarks. Each processing node of the SR8000 is a shared-memory parallel computer composed of eight scalar processors with pseudo-vector processing feature. We have run the all of the SPEC OMP2001 Benchmarks and three benchmarks (LU, SP and BT) of the NAS Parallel Benchmarks on the SR8000. According to the results of this performance measurement, we found that the SR8000 has good scalability continuing up to 8 processors except for a few benchmark programs. The performance results demonstrate that the SR8000 achieves high performance especially for memory-intensive applications.

1 Introduction

OpenMP [1] has emerged as the standard programming model for shared-memory parallel programming. The OpenMP Application Program Interface (API) supports multi-platform shared-memory parallel programming in C/C++ and Fortran on all architectures, including Unix and Windows NT platforms.

Many computer hardware and software vendors have endorsed the OpenMP API and have released commercial compilers that can compile an OpenMP parallel program. Hitachi supports an OpenMP compiler [2] that generates native codes for the Hitachi SR8000.

The purpose of this paper is to evaluate the performance of OpenMP benchmark programs on a single node of the Hitachi SR8000. Although a large system of Hitachi SR8000 may contain several nodes which is connected by their high-speed network, we restrict ourselves to a single node in this paper. A single node of the SR8000 is considered as a shared-memory parallel computer composed of eight scalar processors with pseudo-vector processing feature [3,4,5,6,7]. The automatic parallelizing compiler is provided to perform parallelization for the shared-memory multiprocessors within each node of the SR8000.

We chose the SPEC OMP2001 Benchmarks [8] and the OpenMP version of NAS Parallel Benchmarks (NPB) [9] for performance evaluation.

SPEC OMP2001 is SPEC's first benchmark suite for evaluating performance of OpenMP applications. Aslot et al. [10] have presented this benchmark suite and described issues encountered in the creation of the OpenMP benchmarks. In

H. Zima et al. (Eds.): ISHPC 2002, LNCS 2327, pp. 390–400, 2002.

addition, some SPEC OMP2001 benchmark results have been presented [10,11,12]. Some performance characteristics of the SPEC OMP2001 Benchmarks have been also reported [10,12].

The NAS Parallel Benchmarks (NPB) [9] were designed to compare the performance of parallel computers and are widely recognized as a standard indicator of parallel computer performance. While the MPI version of NPB is available in public, its OpenMP version PBN (Programming Baseline for NPB) [13] is still restricted only for research. We parallelized NPB2.3-serial version by using OpenMP directives.

We have chose some benchmark programs from the NAS Parallel Benchmarks 2.3-serial version on the SR8000.

The rest of the paper is organized as follows. Section 2 describes the SPEC OMP2001 Benchmarks. Section 3 describes the NAS Parallel Benchmarks. Section 4 describes the Hitachi SR8000 architecture and its unique features. Section 5 gives the performance results. In section 6, we present some concluding remarks.

2 SPEC OMP2001 Benchmarks

The SPEC OMP2001 [8] benchmark suite consists of 11 large application programs that represent the type of software used in scientific technical computing.

All applications, except for *gafort*, are floating-point applications taken directly from the SPEC CPU2000 benchmark suite. Each application is either automatically or manually parallelized by inserting OpenMP directives to mark parallel regions of the code [10]. Each benchmark requires up to 2 GB of memory at runtime and runs for up to 10 hours on a modern single processor system.

Wupwise (Wuppertal Wilson Fermion Solver) is a program in the area of lattice gauge theory (quantum chromodynamics). Lattice gauge theory is a discretization of quantum chromodynamics, which is generally accepted to be the fundamental physical theory of strong interactions among quarks as the constituents of matter. Quark propagators are obtained by solving the inhomogeneous lattice-Dirac equation. The *wupwise* solves the inhomogeneous lattice-Dirac equation via the BiCGStab iterative method, which has become established as the method of choice. Its Fortran source code is 2200 lines long.

Swim is a weather prediction program. The model is based on the dynamics of finite-difference models of the shallow-water equations. Its Fortran source code is 400 lines long.

Mgrid is a very simple multigrid solver that computes a three-dimensional potential field. SPEC adapted it from the NAS Parallel Benchmarks, with modifications for portability and a different workload. Its Fortran source code is 500 lines long.

Applu is a solution of five coupled nonlinear PDE's, on a 3-dimensional logically structured grid, using an implicit pseudo-time marching scheme, based on two-factor approximate factorization of the sparse Jacobian matrix. Its Fortran source code is 4000 lines long.

Galgel is a particular case of the GAMM (Gesellschaft fuer Angewandte Mathematik und Mechanik) benchmark devoted to numerical analysis of oscillatory instability of convection in low-Prandtl-number fluids. Its Fortran source code is 15300 lines long.

Equake simulates the propagation of elastic waves in large, highly heterogeneous valleys, such as California's San Fernando Valley, or the Greater Los Angeles Basin. Computations are performed on an unstructured mesh that locally resolves wavelengths, using a finite element method. Its Fortran source code is 1500 lines long.

Apsi is a program to solve for mesoscale and synoptic variations in potential temperature. Its Fortran source code is 7500 lines long.

Gafort computes the global maximum fitness using a genetic algorithm. Its Fortran source code is 1500 lines long.

Fma3d is a finite element method computer program designed to simulate the inelastic, transient dynamic response of three-dimensional solids and structures subjected to impulsively or suddenly applied loads. Its Fortran source code is over 60000 lines long.

Art (Adaptive Resonance Theory) is neural network, which is used to recognize objects in a thermal image. Its C source code is 1300 lines long.

Ammp (Another Molecular Modeling Program) runs molecular dynamics (i.e. solves the ODE defined by Newton's equations for the motions of the atoms in the system) on a protein-inhibitor complex which is embedded in water. Its C source code is 13500 lines long.

3   The NAS Parallel Benchmarks

The NAS Parallel Benchmarks (NPB) were designed to compare the performance of parallel computers and are widely recognized as a standard indicator of computer performance. The NPB suite consists of five kernels and three simulated CFD applications derived from important classes of aerophysics applications. The five kernels mimic the computational cores of five numerical methods used by CFD applications. The simulated CFD applications reproduce much of the data movement and computation found in full CFD codes. Details of the benchmark specifications can be found in [9], and the MPI implementations of NPB are described in [14].

In this paper, we used three benchmarks (LU, SP and BT) from the serial version of NPB2.3. We parallelized the benchmarks using OpenMP directives.

The LU benchmark solves a finite difference discretization of 3-D compressive Navier-Stokes equations through a block-lower-triangular block-upper-triangular approximate factorization of the original difference scheme. The LU factored form is cast as a relaxation and is solved by SSOR.

SP (Scalar Pentadiagonal) equations and BT (Block Tridiagonal) equations form the basis of the 3-D CFD applications. Both are benchmark programs to solve partial differential equations, with SP based on the finite difference method

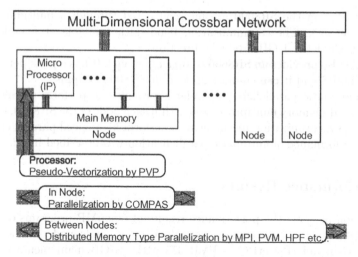

Fig. 1. Architecture of the SR8000 system

(Scalar ADI iteration), and BT based on the finite difference method (5×5 block ADI iteration).

4 Architecture of the Hitachi SR8000

The Hitachi SR8000 System is a distributed-memory parallel computer which consists of pseudo-vector SMP nodes. An SR8000 system consists of 4 to 512 nodes. Figure 1 shows an overview of the SR8000 system architecture.

The node processor of the SR8000 model F1 is a 2.67 ns PowerPC node with major enhancements made by Hitachi. For example, Hitachi added hardware barrier synchronization mechanism is provided for fast barrier operation. Each node contains eight RISC microprocessors which have the "Pseudo Vector Processing" (PVP) feature [3,4,5,6]. This allows data to be fetched from main memory, in a pipelined manner, without stalling the succeeding instructions. The result is that data is fed from memory into the arithmetic units as effectively as in a vector supercomputer. This feature was already available on the CP-PACS [3], and experiments have shown that this idea works well for vector processing.

The peak performance per processor, or IP, can be attained with two simultaneous multiply/add instructions, resulting in a speed of 1.5 GFLOPS on the SR8000 model F1. The processor has a first-level write-through 128 KB 4-way set associative on-chip data cache with 128-byte cache lines. The instruction cache is 2-way set associative 64 KB on-chip cache.

Eight processors are coupled to form one processing node all addressing a common part of the memory. Hitachi refers to this node configuration as COM-PAS, **Co**-operative **M**icro-**P**rosessors in single **A**ddress **S**pace. Peak performance of a node of the SR8000 model F1 is 12 GFLOPS and maximum memory capacity is 16 GB.

The nodes on the SR8000 are interconnected through a multidimensional crossbar network. The communication bandwidth available at a node of the SR8000 model F1 is 1 GB/s (single direction) × 2.

Hitachi also provides an SR8000 compact C model. It has a peak performance of 12 GFLOPS and it can connect up to four SR8000 systems.

The automatic parallelizing compiler is provided to perform parallelization for the shared-memory multiprocessors within each node of the SR8000, using the hardware's synchronization mechanism to perform high-speed parallel execution [2]. The performance of automatic parallelization was described in [2].

5   Performance Results

This section presents the performance results of OpenMP programs on the Hitachi SR8000. All measurements were done on a single node of the SR8000 compact C model (PowerPC + PVP 375 MHz, 8 GB main memory size, 12 GFLOPS peak performance, HI-UX/MPP 03-05).

5.1   SPEC OMP2001 Performance Results

The eight SPEC OMP2001 benchmark programs written in Fortran (*wupwise*, *swim*, *mgrid*, *applu*, *galgel*, *apsi*, *gafort* and *fma3d*) that we used were compiled with Hitachi's Optimizing Fortran 90 compiler V01-04-/B under the options "f90 -Oss -omp -model=C -64". These options instruct the compiler to set the optimization level to maximize the execution speed ("-Oss"), to specify whether to enable OpenMP directives ("-omp"), to generate an object for the compact C model of the SR8000 ("-model=C") and to use 64-bit addressing mode ("-64"), respectively.

For *galgel*, we used portability flag "-fixed=132" which specifies the 132 columns that can be written in one line. For *fma3d*, we also used portability flag "-conti199" is that up to 199 continuation lines can be written.

The three SPEC OMP2001 benchmark programs written in C (*equake*, *art* and *ammp*) that we used were compiled with Hitachi's Optimizing C compiler V01-04 under the options "cc -Os -omp -parallel=4 -model=C -64". These options instruct the compiler to specify various optimizing options including element parallelization and pseudo vectorization so that a user program can execute at the maximum speed ("-Os"), to specify whether to enable OpenMP directives ("-omp"), to set the parallelization level to level 4, to generate an object for the compact C model of the SR8000 ("-model=C") and to use 64-bit addressing mode ("-64"), respectively.

We ran the codes on a single node of the SR8000 using "ref" data sets. Table 1 shows the result of all SPEC OMP2001 Benchmarks (Medium size) up to 8 processors. The first column of a table indicates the benchmarks used. The second column gives the "Reference Time" in seconds. The next four columns contain the "Estimates Runtime" in seconds. Since the results shown in Table 1 have not yet been reviewed by SPEC, they are marked as "Estimates Runtime".

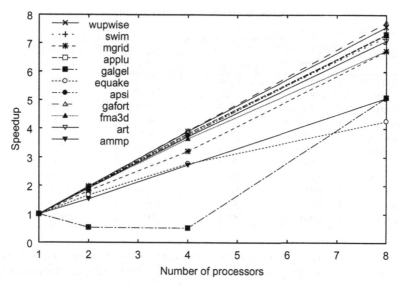

Fig. 2. Speedup of SPEC OMP2001 on the SR8000

Table 1. SPEC OMP2001 results on the SR8000

Benchmark	Reference Time	Estimates Runtime			
		1 CPU	2 CPUs	4 CPUs	8 CPUs
wupwise	6000	16336	8312	4193	2158
swim	6000	2400	1226	627	332
mgrid	7300	8458	4636	2645	1257
applu	4000	5414	2801	1418	740
galgel	5100	5309	10045	10453	1042
equake	2600	7924	4768	2863	1854
apsi	3400	6055	3148	1608	830
gafort	8700	26358	13353	6727	3408
fma3d	4600	14372	7550	3946	2129
art	6400	21844	11460	5874	3075
ammp	7000	55664	36678	20465	10950

Figure 2 shows the speedup of SPEC OMP2001 Benchmarks relative to the one-processor execution time on a single node of the SR8000. The figure shows that the speedup on 8 processors is about 4.3 to 7.7 for the benchmarks.

Mgrid and *fma3d* speedup between 6.5 and 7.0 on 8 processors, whereas *wupwise, swim, applu, apsi, gafort* and *art* speedup by a factor over 7.0 on 8 processors. *Wupwise* uses OpenMP versions of the LAPACK routines (`dznrm2.f` `zaxpy.f zcopy.f zdotc.f zscal.f`). Thus, *wupwise* achieved almost perfect scaling on the SR8000.

In *galgel*, the speedup is under 1.0 on 2 and 4 processors, as shown in Figure 2. Its speedup loss comes mainly from fork-join overhead. The key of the performance in *galgel* is a `PARALLEL DO` directive. *Galgel* includes about 90 `PARALLEL DO` blocks and all of them enclose only a few assignment statements without nested-DO loops [11]. Moreover, the fork-join overhead on 2 and 4 processor

Table 2. SPEC OMP2001 results

Benchmark	Reference Time	Estimates Runtime SR8000 (8 CPUs)	Base Runtime Origin 3800 (8 CPUs)	Base Runtime PRIMEPOWER2000 (18 CPUs)
wupwise	6000	2158	1553	1080
swim	6000	332	1655	1211
mgrid	7300	1257	1813	1322
applu	4000	740	1071	435
galgel	5100	1042	905	1170
equake	2600	1854	670	735
apsi	3400	830	1195	439
gafort	8700	3408	2363	1680
fma3d	4600	2129	1590	1210
art	6400	3075	1988	1685
ammp	7000	10950	2957	2174

executions is relatively high compared to on 8 processors execution due to implementation of Hitachi's OpenMP compiler. This is the reason why *galgel* shows poor speedup on 2 and 4 processors.

Similar to *galgel*, *equake* includes many **parallel for** directive which contains trivial computation. Thus, *equake* shows the lowest speedup among all of the SPEC OMP2001 Benchmarks.

Table 2 compares the results on the Hitachi SR8000 compact C model (PowerPC + PVP 375 MHz, 8 CPUs), SGI Origin 3800 (MIPS R14000 500 MHz, 8 CPUs) [8] and Fujitsu PRIMEPOWER2000 (SPARC64 GP 563 MHz, 18 CPUs) [8]. The first column of a table indicates the benchmarks used. The second column gives the "Reference Time" in seconds. The third column gives the "Estimates Runtime" in seconds. The next two columns contain the "Base Runtime" in seconds.

We found that the performance of the *swim* on the SR8000 is better than that of the swim on the Origin 3800 and PRIMEPOWER2000. This is mainly because the *swim* is a very memory-intensive program [12]. The SR8000 can hide memory latency effectively with software managed controlling of data movement by pseudo-vector processing feature.

On the other hand, the performances of the some benchmark programs (*wupwise, equake, gafort, fma3d, art* and *ammp*) are lower than those of both the Origin 3800 and PRIMEPOWER2000. In particular, with the *ammp*, the runtime of SR8000 (8 CPUs) is slower than the "Reference Time". *Ammp* includes many innermost loops that cannot be pseudo-vectorized due to data dependency. Although the Origin 3800 and PRIMEPOWER2000 have a second-level 8 MB off-chip cache, the processor of the SR8000 has no second-level cache. This is the reason why *ammp* shows poor performance on the SR8000.

5.2   NPB Performance Results

We measured the performance of a single node of the SR8000 by the NAS Parallel Benchmarks (NPB) [9].

Table 3. NPB2.3-serial (Class A) results on the SR8000

Benchmark	1 CPU		2 CPUs		4 CPUs		8 CPUs	
	Time	MFLOPS	Time	MFLOPS	Time	MFLOPS	Time	MFLOPS
LU	595.56	200.31	495.59	240.71	413.37	288.59	276.03	432.19
SP	291.02	292.11	155.15	547.93	85.69	992.03	40.59	2094.51
BT	584.33	287.99	303.13	555.15	168.39	999.37	102.75	1637.84

Table 4. NPB2.3-serial (Class B) results on the SR8000

Benchmark	1 CPU		2 CPUs		4 CPUs		8 CPUs	
	Time	MFLOPS	Time	MFLOPS	Time	MFLOPS	Time	MFLOPS
LU	2455.73	203.13	1851.08	269.48	1476.70	337.80	1150.83	433.45
SP	1196.93	296.60	619.81	572.77	316.29	1122.42	163.29	2174.09
BT	2579.87	272.18	1348.13	520.86	748.57	938.03	452.72	1551.03

The OpenMP versions of the NAS Parallel Benchmarks are derived from the serial versions of NAS Parallel Benchmarks 2.3 (NPB2.3-serial). We parallelized the benchmarks by inserting only the OMP directives. The code was compiled with Hitachi's Optimizing Fortran 90 compiler V01-04-/B. The compiler options used were "f90 -Oss -omp -model=C -i,EU,L -64". We ran the codes on a single node of the SR8000 using Class A and Class B data sets.

The results for LU, SP and BT are shown in the following. Tables 3 and 4 show the results of the performance of the NAS Parallel Benchmarks 2.3-serial version. The first column of tables indicate the benchmarks used. The next eight columns contain the elapsed time in seconds and the performance in MFLOPS. The performance of SP (Class B) is over 2100 MFLOPS, namely about 18% of the peak performance of a single node of the Hitachi SR8000 compact C model, as shown in Table 4.

Figures 3 and 4 show the speedup of the NAS Parallel Benchmarks relative to the one-processor execution time on a single node of the Hitachi SR8000. The figures show that the speedup on 8 processors is about 5.7 to 7.3 for two of the three benchmarks, LU being the exception. The reason why LU is slow to speed up is that it contains a wavefront-style loop; every loop in the loop nest has dependence across loop iterations, and the loop in this case is not parallelized. Although this loop cannot be parallelized easily in OpenMP, the automatic parallelization of Hitachi's compiler can parallelize such a wave front style loop [2].

6   Concluding Remarks

In this paper, we evaluated the SPEC OMP2001 Benchmarks and three benchmarks (LU, SP and BT) of the NAS Parallel Benchmarks on a single node of the Hitachi SR8000.

For the SPEC OMP2001 Benchmarks evaluated, we observed about 4.3 to 7.7 times speedup on 8 processors. On the other hand, for the NAS Parallel

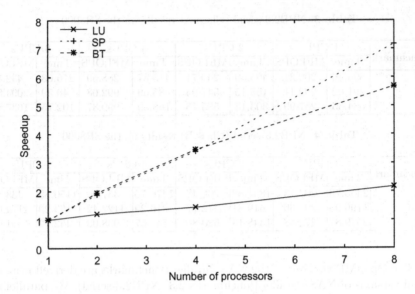

Fig. 3. Speedup of NPB2.3-serial (Class A) on the SR8000

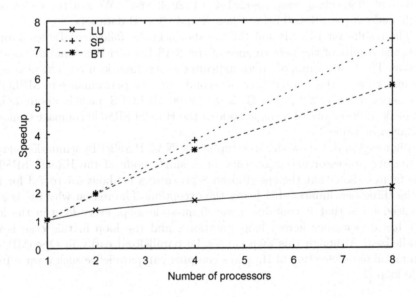

Fig. 4. Speedup of NPB2.3-serial (Class B) on the SR8000

Benchmarks evaluated, we observed about 5.7 to 7.3 times speedup, except for LU, on 8 processors.

A few benchmark programs include many innermost loops that cannot be pseudo-vectorized due to data dependency. Moreover, the processor of the

SR8000 has no second-level cache. This is the reason why these benchmark programs show poor performance on the SR8000.

According to the results of our performance measurement, we found that the SR8000 has good scalability continuing up to 8 processors except for a few benchmark programs.

The SR8000 can hide memory latency effectively with software managed controlling of data movement by pseudo-vector processing feature. The performance results demonstrate that the SR8000 achieves high performance especially for memory-intensive applications.

References

1. OpenMP: Simple, Portable, Scalable SMP Programming.
   (http://www.openmp.org)
2. Nishitani, Y., Negishi, K., Ohta, H., Nunohiro, E.: Implementation and evaluation of OpenMP for Hitachi SR8000. In: Proc. Third International Symposium on High Performance Computing (ISHPC 2000). Volume 1940 of Lecture Notes in Computer Science., Springer-Verlag (2000) 391–402
3. Nakazawa, K., Nakamura, H., Boku, T., Nakata, I., Yamashita, Y.: CP-PACS: A massively parallel processor at the University of Tsukuba. Parallel Computing **25** (1999) 1635–1661
4. Tamaki, Y., Sukegawa, N., Ito, M., Tanaka, Y., Fukagawa, M., Sumimoto, T., Ioki, N.: Node architecture and performance evaluation of the Hitachi Technical Server SR8000. In: Proc. 12th International Conference on Parallel and Distributed Computing Systems. (1999) 487–493
5. Shimada, K., Kawashimo, T., Hanawa, M., Yamagata, R., Kamada, E.: A superscalar RISC processor with 160 FPRs for large scale scientific processing. In: Proc. International Conference on Computer Design (ICCD'99). (1999) 279–280
6. Nishiyama, H., Motokawa, K., Kyushima, I., Kikuchi, S.: Pseudo-vectorizing compiler for the SR8000. In: Proc. 6th International Euro-Par Conference (Euro-Par 2000). Volume 1900 of Lecture Notes in Computer Science., Springer-Verlag (2000) 1023–1028
7. Brehm, M., Bader, R., Heller, H., Ebner, R.: Pseudovectorization, smp, and message passing on the Hitachi SR8000-F1. In: Proc. 6th International Euro-Par Conference (Euro-Par 2000). Volume 1900 of Lecture Notes in Computer Science., Springer-Verlag (2000) 1351–1361
8. Standard Performance Evaluation Corporation (SPEC): SPEComp 2001.
   (http://www.spec.org/hpg/omp)
9. Bailey, D., Barton, J., Lasinski, T., Simon, H.: The NAS parallel benchmarks. NAS Technical Report RNR-91-002, NASA Ames Research Center, Moffett Field, CA (1991)
10. Aslot, V., Domeika, M., Eigenmann, R., Gaertner, G., Jones, W.B., Parady, B.: SPEComp: A new benchmark suite for measuring parallel computer performance. In: Proc. International Workshop on OpenMP Applications and Tools (WOMPAT 2001). Volume 2104 of Lecture Notes in Computer Science., Springer-Verlag (2001) 1–10
11. Iwashita, H., Yamanaka, E., Sueyasu, N., van Waveren, M., Miura, K.: The SPEC OMP2001 benchmark on the Fujitsu PRIMEPOWER system. In: Proc. Third European Workshop on OpenMP (EWOMP 2001). (2001)

12. Aslot, V. Eigenmann, R.: Performance characteristics of the SPEC OMP2001 benchmarks. In: Proc. Third European Workshop on OpenMP (EWOMP 2001). (2001)
13. Jin, H., Frumkin, M., Yan, J.: The OpenMP implementation of NAS parallel benchmarks and its performance. NAS Technical Report NAS-99-011, NASA Ames Research Center, Moffett Field, CA (1999)
14. Bailey, D., Harris, T., Saphir, W., Van der Wijngaart, R., Woo, A., Yarrow, M.: The NAS parallel benchmarks 2.0. NAS Technical Report NAS-95-020, NASA Ames Research Center, Moffett Field, CA (1995)

Communication Bandwidth of Parallel Programming Models on Hybrid Architectures

Rolf Rabenseifner

High-Performance Computing-Center (HLRS), University of Stuttgart
Allmandring 30, D-70550 Stuttgart, Germany
rabenseifner@hlrs.de
www.hlrs.de/people/rabenseifner/

Abstract. Most HPC systems are clusters of shared memory nodes. Parallel programming must combine the distributed memory parallelization on the node inter-connect with the shared memory parallelization inside of each node. This paper introduces several programming models for hybrid systems. It focuses on programming methods that can achieve optimal inter-node communication bandwidth and on the hybrid MPI+OpenMP approach and its programming rules. The communication behavior is compared with the pure MPI programming paradigm and with RDMA and NUMA based programming models.

Keywords: OpenMP, MPI, Hybrid Parallel Programming, Threads and MPI, HPC.

1 Motivation

Today, most systems in high performance computing (HPC) are clusters of SMP (symmetric multi-processor) nodes, i.e., they are hybrid architectures, shared memory systems are inside of each node, and a distributed memory parallel (DMP) system is across the node boundaries. To achieve a minimal parallelization overhead, often a hybrid programming model is proposed, e.g., OpenMP [21] or automatic compiler based thread parallelization inside of each SMP node, and message passing (e.g., with MPI [16]) on the node interconnect. Another often used programming model is the flat and pure massively parallel processing (MPP) MPI model, where separate single-threaded MPI processes are running on each CPU. Using the hybrid programming model instead of the MPP-MPI model has the advantage that there is no message passing overhead inside of each SMP node, because the threads can access the data provided by other threads directly by accessing the shared memory instead of passing the data through a message.

The hybrid MPI+OpenMP programming model is already used in many applications, but often there is only a small benefit as, e.g., reported with the climate model calculations of one of the Gordon Bell Prize finalists at SC 2001 [14], or sometimes losses are reported compared to the MPP-MPI model, e.g., as shown with an discrete element modelling algorithm in [12].

H. Zima et al. (Eds.): ISHPC 2002, LNCS 2327, pp. 401–412, 2002.

One of the major drawbacks of the hybrid MPI-OpenMP programming model is based on a very simple usage of this hybrid approach: If the MPI routines are invoked only outside of parallel regions, all threads except the master thread are sleeping while the MPI routines are executed.

This paper will discuss this phenomenon and other hybrid MPI-OpenMP programming strategies. In Sect. 2, an overview on hybrid programming models is given. Sect. 3 shows different methods to combine MPI and OpenMP. Further rules on hybrid programming are discussed in Sect. 4. Pure MPI can also be used on hybrid architectures, as shown Sect. 5. Sect. 6 presents the results of a parallel communication benchmark. Sect. 7 compares the MPI based programming models with major shared and virtual shared memory models.

2   Programming Models on Hybrid Architectures

The available programming models depend on the type of cluster hardware. If the node interconnect allows cache-coherent or non-cache-coherent non-uniform memory access (ccNUMA and nccNUMA), i.e., if the memory access inside of each SMP node and across the cluster interconnect is implemented by the same instructions, then one can use programming models which need a shared memory access across the whole cluster. This includes OpenMP on the whole cluster, usage of nested parallelism inside of OpenMP, but also OpenMP with cluster extensions, that are primarily based on a first touch mechanism [11] or on data distribution extensions [15]. These cluster extensions may also benefit from the availability of software based shared virtual memory (SVM) [5,25,26]. At NASA/Ames, a hybrid approach was developed. The parallelization is organized in two levels: The upper level is process based, and in the lower level each process is multi-threaded with OpenMP. The processes are using a Fortran wrapper around the System V shared memory module *shm*, that allows to fork the processes, to initialize a shared memory segment, to associate portions of this segment with Cray pointer based array in each process, and to make a barrier synchronization over all processes. This system is named as Multi Level Parallelism (MLP) and it allows very flexible, dynamic and simple way of load balancing: At each start of a parallel region inside of each MLP process, the number of threads, i.e., the number of used CPUs, may be adapted [8]. Although MLP is a proprietary method of NASA/Ames, the programming style based on *shm* is non-proprietary.

If the node interconnect requires different methods for accessing local and cluster-wide memory, but if there are remote direct memory access (RDMA) methods available, i.e., if one node can access the memory of another node without interaction of a CPU on that node, then further programming methods are available: Such systems can be programmed with Co-Array Fortran [20] or Unified Parallel C (UPC) [7,9]. In Co-Array Fortran, the access to an array of another process or thread is done by using an additional trailing array subscript in square brackets addressing that process or thread. Both language extensions can also be used to program clusters of SMP nodes, because they neither add a message passing overhead nor the overhead of additional copies. A key issue for a more widespread usage of UPC and Co-Array Fortran is the availability of (portable) commpiling systems for a wide range of platforms with a clear development path to achieve an optimal performance, as it was presented for MPI by the early MPICH implementation [10]. Another approach to use the RDMA hardware is based on one-sided communication, e.g., in Cray's *shmem* library or in MPI-2 [17]. These library-based methods allow to store (fetch) data to (from) the memory of another process in a SPMD environment. The *shmem* library was ported by many vendors to their systems. All programming models available for RDMA-class node-interconnect are also usable on NUMA-class interconnects.

The third class of hardware supports neither NUMA access nor RDMA. Only pure message passing is available on the node-interconnect. Programming models designed for this class of hardware have the major advantage that they are applicable to all other already mentioned classes. This paper focuses on this type of hardware. The commonly accepted standard for message passing between the nodes is the Message Passing Interface (MPI) [16,17]. The major programming styles are pure MPI, i.e., the MPP model that uses each CPU for one MPI process, and hybrid models, e.g., MPI on the node-interconnect and OpenMP or automatic or semi-automatic compiler based thread-parallelization inside of each SMP node. Inside of each node mainly two different SMP parallelization strategies are used: (a) A coarse-grain SPMD-style parallelization similar to the work distribution betwen the processes in a message passing program is applied; this method allows a similar computational efficiency as with the pure MPI parallelization; the efficiency of the communication is a major factor in the omparance of this hybrid approach with the pure MPI solution. The present paper is focused on the communication aspects. (b) A fine-grained SMP parallelization is done in an incremental effort of parallelizing loops inside of the MPI processes. The efficiency of such hybrid solution depends on both, the efficiency of the computation (Amdahl's law must be considered on both levels of parallelization) and of the communication, as shown in [6] for the NAS parallel benchmarks. Different SMP parallelization strategies in the hybrid model are also studied in [27]. High-Performance Fortran (HPF) is also available on clusters of SMPs. In [3], HPF based on hybrid MPI+OpenMP is compared with pure MPI.

3   MPI and Thread-Based Parallelization

This model was already addressed by the MPI-2 Forum in Sect. 8.7 *MPI and Threads* in [17]. For hybrid programming, the MPI-1 routine MPI_Init() should be substituted by a call to MPI_Init_threads() which has the input argument named *required* to define which thread-support the application requests from the MPI library, and the output argument *provided* which is used by the MPI library to tell the application which thread-support is available. MPI libraries may support the following thread-categories (higher categories are supersets of all lower ones):

T0 – No thread-support, represented *provided=MPI_THREAD_SINGLE*.

T1a – The MPI process may be multi-threaded but only the master thread may call MPI routines **AND** only while the other threads do not exist, i.e., parallel threads created by a parallel region must be destroyed before an MPI routine is called. This class is not mentioned in the MPI standard and an MPI library supporting this class (and not more) must also return provided=MPI_THREAD_SINGLE because of the lack of this definition in the MPI-2 standard[1].

T1b – The definition T1a is relaxed in the sense, that more than one thread may exist during the call of MPI routines, but all threads except the master thread must sleep, i.e., must be blocked in some OpenMP synchronization. As in T1a, an MPI library supporting T1b but not more must also return provided=MPI_THREAD_SINGLE.

T2 – Only the master thread will make calls to MPI routines. The other threads may run other application code while the master thread calls an MPI routine. This is allowed if the MPI library returns a value greater or equal to MPI_THREAD_FUNNELED in *provided*.

T3 – Multiple threads may make MPI-calls, but only one thread may execute an MPI routine at a time. This requires *provided* ≥ MPI_THREAD_SERIALIZED.

T4 – Multiple threads may call MPI without any restrictions. This hybrid programming style is available when *provided* = MPI_THREAD_MULTIPLE was returned.

The constants are monotonic, i.e.,
MPI_THREAD_SINGLE ≤ MPI_THREAD_FUNNELED ≤ ... .

Usually, the application cannot distinguish whether an OpenMP parallelization needs T1 or T2 to allow calls to MPI routines outside of OpenMP parallel regions, because it is not defined, whether at the end of a parallel region the team of threads is sleeping or is destroyed. And usually, this category is chosen, when the MPI routines are called outside of parallel regions. Therefore, one should summarize the cases T1a and T1b to only one case:

T1 – The MPI process may be multi-threaded but only the master thread may call MPI routines **AND** only outside of parallel regions (in case of OpenMP) or outside of parallelized code (if automatic parallelization is used). We define here an additional constant THREAD_MASTERONLY with a value between MPI_THREAD_SINGLE and MPI_THREAD_FUNNELED.

4   Rules with Hybrid Programming

T1 is the most simple hybrid programming model with MPI and OpenMP, because MPI routines may be called only outside of parallel regions. The new cache coherence rules in OpenMP 2.0 guarantee, that the outcome of an MPI routine is visible to all threads in a subsequent parallel region[2], and that the outcome of all threads of a parallel region is visible to a subsequent MPI routine.

---

[1] This may be solved in the revision 2.1 of the MPI standard.
[2] There is still a lack in the draft from Nov. 2001 for the C language binding

T2 can be achieved by surrounding the call to the MPI routine with the OMP MASTER and OMP END MASTER directives inside of a parallel region. One must be very careful, because OMP MASTER does not imply an automatic barrier synchronization or an automatic cache flush either at the entry to or at the exit from the master section. If the application wants to send data computed in the previous parallel region or wants to receive data into a buffer that was also used in the previous parallel region (e.g., to use the data received in the previous iteration), then a barrier with implied cache flush is necessary prior to calling the MPI routine, i.e., prior to the master section. If the data or buffer is also used in the parallel region after the exit of the MPI routine and its master section, then also a barrier is necessary after the exit of the master section. If both barriers must be done, then while the master thread is executing the MPI routine, all other threads are sleeping, i.e., we are going back to the case T1b.

T3 can be achieved by using the OMP SINGLE directive, which has an implied barrier only at the exit (unless NOWAIT is specified). Here again, the same problems as with T2 must be taken into account.

These problems with T2 and T3 arise, because the communication needs must be funneled from all threads to one thread (an arbitrary thread in T3, and the master thread in T2). Only T4 allows a direct message passing from each thread in one node to each thread in another node.

Based on these reasons and because T1 is available on nearly all clusters, most hybrid and portable parallelization is using only the programming scheme described in T1. This paper will evaluate this hybrid model by comparing it with the non-hybrid model described in the next section.

5   MPP-MPI on Hybrid Architectures

Using a pure MPI model, the cluster must be viewed as a hybrid communication network with typically fast communication paths inside of each SMP node and slower paths between the nodes. It is important to implement a good mapping of the communication paths used by application to the hybrid communication network of the cluster. The MPI standard defines virtual topologies for this purpose, but the optimization algorithm isn't yet implemented in most MPI implementations. Therefore, in most cases, it is important to choose a good ranking in MPI_COMM_WORLD. E.g., on a Hitachi SR8000, the MPI library allows two different ranking schemes, round robin (ranks 0, N, 2*N, ... on node 0; ranks 1, N+1, 2*N+1, ... on node 1, ...; with N=number of nodes) and sequential (rank 0–7 on node 0, ranks 8–15 on node 1, ...), and the user has to decide which scheme may fit better to the communication needs of his application.

The MPP-MPI programming model implies additional message transfers due to the higher number of MPI processes and higher number of boundaries. Let us consider, for example, a 3-dimensional cartesian domain decomposition. Each domain may have to transfer boundary information to its neighbors in all six cartesian directions ($\uparrow\downarrow \rightleftarrows \swarrow\nearrow$). Bringing this model on a cluster with 8-way SMP nodes, on each node, we should execute the domains belonging to a $2\times2\times2$ cube.

Domain-to-domain communcation occurs as node-to-node (inter-node) communication and as intra-node communication between the domains inside of each cube. Hereby, each domain has 3 neighbors inside the cube and 3 neighbors outside, i.e., in the inter-node and the intra-node communication the amount of transferred bytes should be equivalent. If we compare this MPP-MPI model with a hybrid model, assuming that the domains (in the MPP-MPI model) in each 2×2×2 cube are combined to a super-domain in the hybrid model, then the amount of data transferred on the node-interconnect should be the same in both models. This implies, that in the MPP-MPI model, the total amount of transferred bytes (inter-node plus intra-node) will be twice the number of bytes in the hybrid model. This result is independent from the way, the inner-node parallelization is implemented in the hybrid model, i.e., whether it is done in a coarse grained domain decomposition style or as fine grained loop parallelism.

6   Benchmark Results

The following benchmark results will compare the communication behavior of the hybrid MPI+OpenMP model with the pure MPP-MPI model. Based on the domain decomposition scenario discussed in the last section, we compare the bandwidth of both models and the ratio of the total communication time presuming that in the MPP-MPI model, the total amount of transferred data is twice the amount in the hybrid model. The benchmark was done on a Hitachi SR8000 with 16 nodes from which 12 nodes are available for MPI parallel applications. Each node has 8 CPUs. The effective communication benchmark b_eff is used [13,22,23]. It accumulates the communication bandwidth values of the communication done by each MPI process. To determine the bandwidth of each process, the maximum time needed by all processes is used, i.e., this benchmark models an application behavior, where the node with the slowest communication controls the real execution time. To compare both models, we use the following metrics:

- b_eff – the accumulated bandwidth average for several ring and random patterns;
- 3-d-cyclic-Lmax – a 3-dimensional cyclic communication pattern with 6 neighbors for each MPI process; the bandwidth is measured with 8 MB messages.
- 3-d-cyclic-avg – same, but an average of 21 different message sizes.

For each metrics, the following rows are presented in Tab. 1:

- b_{hybrid}, the accumulated bandwidth b for the hybrid model measured with a 1-threaded MPI process on each node (12 MPI processes),
- and in parentheses the same bandwidth per node,
- b_{MPP}, the accumulated bandwidth for the MPP-MPI model (96 MPI processes with sequential ranking in MPI_COMM_WORLD),
- and in parentheses the same bandwidth per process,

Table 1. Comparing the hybrid and the MPP communication needs.

		b_eff	3-d-cyclic-Lmax	3-d-cyclic-avg
b_{hybrid}	[MB/s]	1535	5638	1604
(per node)	[MB/s]	(128)	(470)	(134)
b_{MPP}	[MB/s]	5299	18458	5000
(per process)	[MB/s]	(55)	(192)	(52)
b_{MPP}/b_{hybrid}	(measured)	3.45	3.27	3.12
s_{MPP}/s_{hybrid}	(assumed)	2	2	2
T_{hybrid}/T_{MPP}	(concluding)	1.73	1.64	1.56

- b_{MPP}/b_{hybrid}, the ratio of accumulated MPP bandwidth and accumulated hybrid bandwidth,
- T_{hybrid}/T_{MPP}, the ratio of execution times T, assuming that total size s of the transferred data in the MPP model is twice of the size in the hybrid model, i.e., $s_{MPP}/s_{hybrid} = 2$, as shown in Sect.5.

Note, that this comparison was done with no special optimized topology mapping in the MPP model. The result shows, that the MPP communication model is faster than the communication in the hybrid model. There are at least two reasons: (1) In the hybrid model, all communication was done by the master thread while the other threads were inactive; (2) One thread is not able to saturate the total inter-node bandwidth that is available for each node.

This communication behavior may be a major reason when an application is running faster in the MPP model than in the hybrid model.

7   Comparison

The comparison in this paper focuses on bandwidth aspects, i.e., how to achieve a major percentage of the physical inter-node network bandwidth with various parallel programming models.

7.1   Hybrid MPI+OpenMP versus Pure MPI

Although the benchmark results in the last section show a clear advantage of the MPP model, there are also advantages of the hybrid model. In the hybrid model there is no communication overhead inside of a node. The message size of the boundary information of one process may be larger (although the total amount of communication data is reduced). This reduces latency based overheads in the inter-node communication. The number of MPI processes is reduced. This may cause a better speedup based on Amdahl's law and may cause a faster convergence if, e.g., the parallel implementation of a multigrid numerics is only computed on a partial grid. To reduce the MPI overhead by communicating only through one thread, the MPI communication routines should be relieved by unnecessary local work, e.g., concatenation of data should be better done by copying the data to a scratch buffer with a thread-parallelized loop, instead of using derived MPI datatypes. MPI reduction operations can be split into the inter-node communication part and the local reduction part by using user-defined operations, but a local thread-based parallelization of these operations may cause problems because these threads are running while an MPI routine may communicate.

Hybrid programming is often done in two different ways: (a) the domain decomposition is used for the inter-node parallelization with MPI and also for the intra-node parallelization with OpenMP, i.e., in both cases, a coarse grained parallelization is used. (b) The intra-node parallelization is implemented as a fine grained parallelization, e.g., mainly as loop parallelization. The second case also allows automatic intra-node parallelization by the compiler, but Amdahl's law must be considered independently for both parallelizations.

If the other application threads do not sleep while the master thread is communicating with MPI then communication time T_{hybrid} in Tab. 1 counts only the eighth (a node has 8 CPUs on the SR8000) because only one instead of 1 (active) plus 7 (idling) CPUs is communicating. In this hybrid programming style, the factor T_{hybrid}/T_{MPP} must be reduced to the eighth, i.e. from about 1.6 to about 0.2. But in this case, the application must implement a load balancing algorithm to guarantee that the load on the communicating master thread is equal to the load on the other threads. This means that minimal transfer time can be achieved with the hybrid model, but at the costs of implementing an optimal load balancing between the thread(s) that computes and communicates and those threads that only compute.

7.2   MPI versus Remote Memory Access

Now, we compare the MPI based models with the NUMA or RDMA based models. To access data on another node with MPI, the data must be copied to a local memory location (so called halo or shadow) by message passing, before it can be loaded into the CPU. Usually all necessary data should be transferred in one large message instead of using several short messages. Then, the transfer speed is dominated by the asymptotic bandwidth of the network, e.g., as reported for 3-d-cyclic-Lmax in Tab. 1 per node (470 MB/s) or per process (192 MB/s). With NUMA or RDMA, the data can be loaded directly from the remote memory location into the CPU. This may imply short accesses, i.e., the access is latency bound. Although the NUMA or RDMA latency is usually 10 times shorter than the message passing latency, the total transfer speed may be worse. E.g., [8] reports on a ccNUMA system a latency of 0.33–1 μs, which implies a bandwidth of only 8–24 MB/s for a 8 byte data. This effect can be eliminated if the compiler has implemented a remote pre-fetching strategy as described in [18], but this method is still not used in all compilers.

The remote memory access can also be optimized by buffering or pipelining the data that must be transferred. This approach may be hard to automate, and current OpenMP compiler research already studies the bandwidth optimization on SMP clusters [24], but it can be easily implemented as an directive-based

Table 2. Memory copies from remote memory to local CPU register.

Access method	copies	remarks	bandwidth $b(message\ size)$
2-sided MPI	2	internal MPI buffer + application receive buffer + application receive buffer	$b_\infty/(1 + \frac{b_\infty T_{lat}}{size})$, e.g., $300\,\mathrm{MB/s} / (1 + \frac{300\,\mathrm{MB/s} \times 10\,\mu s}{10\,\mathrm{kB}})$ $= 232\,\mathrm{MB/s}$
1-sided MPI	1	application receive buffer	same formula, but probably better b_∞ and T_{lat}
UPC,	1	page based transfer	extremely poor
Co-Array Fortran, HPF,	0	**word based access**	8 byte / T_{lat}, e.g., 8 byte / $0.33\,\mu s = \mathbf{24\,MB/s}$
OpenMP with	0	latency hiding with pre-fetch	b_∞
cluster extensions	1	latency hiding with buffering	see 1-sided communication

optimization technique: The application thread can define the (remote) data it will use in the next simulation step and the compiled OpenMP code can pre-fetch the whole remote part of the data with a bandwidth-optimized transfer method. Table 2 summarizes this comparison.

7.3 Parallelization and Compilation

Major advantages of OpenMP based programming are that the application can be *incrementally parallelized* and that one still has a single source for serial and parallel compilation. On a cluster of SMPs, the major disadvantages are, that OpenMP has a flat memory model and that it does not know buffered transfers to reach the asymptotic network bandwidth. But, as already mentioned, these problems can be solved by tiny additional directives, like the proposed migration and memory-pinning directives in [11], and additional directives that allow a contiguous transfer of the whole boundary information between each simulation step. Those directives are optimization features that do not modify the basic OpenMp model, as this would be done with directives to define a full HPF-like user-directed data distribution (as in [11,15]). Another lack in the current OpenMP standard is the absence of a strategy of combining automatic parallelization with OpenMP parallelization, although this is implemented in a non-standardized way in nearly all OpenMP compilers. This problem can be solved, e.g., by adding directives to define scopes where the compiler is allowed to automatically parallelize the code, e.g., similar to the parallel region, one can define an *autoparallel* region. Usual rules for nested parallelism can apply, i.e., a compiler can define that it cannot handle nested parallelism.

An OpenMP-based parallel programming model for SMP-clusters should be usable for both, fine grained loop parallelization, and coarse grained domain decomposition. There should be a clear path from MPI to such an OpenMP cluster programming model with a performance that should not be worse than with pure MPI or hybrid MPI+OpenMP.

It is also important to have a good compilation strategy that allows the development of well optimizing compilers on any combination of processor, memory access, and network hardware. The MPI based approaches, especially the hybrid MPI+OpenMP approach, clearly separate remote from local memory access optimization. The remote access is optimized by the MPI library, and the local memory access must be improved by the compiler. Such separation is realized, e.g., in the NANOS project OpenMP compiler [2,19]. The separation of local and remote access optimization may be more essential than the chance of achieving a zero-latency by remote pre-fetching (Tab. 2) with direct compiler generated instructions for remote data access. Pre-fetching can also be done via macros or library calls in the input for the local (OpenMP) compiler.

8   Conclusion

For many parallel applications on hybrid systems, it is important to achieve a high communication bandwidth between the processes on the node-to-node inter-connect. On such architectures, the standard programming models of SMP or MPP systems do not longer fit well. The rules for hybrid MPI+OpenMP programming and the benchmark results in this paper show that a hybrid approach is not automatically the best solution if the communication is funnelled by the master thread and long message sizes can be used. The MPI based parallel programming models are still the major paradigm on HPC platforms. OpenMP with further optimization features for clusters of SMPs and bandwidth based data transfer on the node interconnect have a chance to achieve a similar performance together with an incremental parallelization approach, but only if the current SMP model is enhanced by features that allow an optimization of the total trafic, e.g., with an user-directed optimization of page migration. and by features for latency-hiding, e.g., by allowing a user-directed transfer of the total boundary all at once.

9   Future Work

In the future, we also want to examine hybrid programming models based on aspects of latency and latency-hiding, especially in combination with vectorizing codes. Optimization of hybrid MPI+OpenMP programming will be done with thread-parallel MPI techniques and compared with topology-optimized non-hybrid MPI parallelization. To evaluate cluster programming models, we want to compare proposed distributed OpenMP extensions [11,15] with fully thread-parallel MPI and hybrid MPI+OpenMP solutions varying the chunk sizes and cluster parameters to examine the influence of latency and bandwidth of local and remote memory accesses. The work should be a basis to study and to optimize MPI-based parallelization libraries on hybrid systems.

   If OpenMP with the described cluster extensions should be used as a basic programming concept, it is important, that an automatic analysis of the remote and local memory access (e.g., integrated in existing analysis tools, e.g., in VAMPIR [4]), and improved tools for detecting race-conditions are available [1].

Acknowledgments

The author would like to acknowledge his colleagues and all the people that supported these projects with suggestions and helpful discussions. He would especially like to thank Alice Koniges, David Eder and Matthias Brehm for productive discussions of the limits of hybrid programming, Bob Ciotti and Gabrielle Jost for the discussions on MLP, Gerrit Schulz for his work on the benchmarks, Gerhard Wellein for discussions on network congestion in the MPP model, and Thomas Bönisch, Matthias Müller, Uwe Küster, and John M. Levesque for discussions on OpenMP cluster extensions and vectorization.

References

1. Assure and AssureView, http://www.kai.com/parallel/kappro/assure/.
2. Eduard Ayguade, Marc Gonzalez, Jesus Labarta, Xavier Martorell, Nacho Navarro, and Jose Oliver, *NanosCompiler: A Research Platform for OpenMP Extensions*, in proceedings of the 1st European Workshop on OpenMP (EWOMP'99), Lund, Sweden, Sep. 1999.
3. Siegfried Benkner, Thomas Brandes, *High-Level Data Mapping for Clusters of SMPs*, in proceedings of the 6th International Workshop on High-Level Parallel Programming Models and Supportive Environments, HIPS 2001, San Francisco, USA, April 2001, Springer LNCS 2026, pp 1–15.
4. Holger Brunst, Wolfgang E. Nagel, and Hans-Christian Hoppe, *Group Based Performance Analysis for Multithreaded SMP Cluster Applications*, in proceedings of Euro-Par2001, R. Sakellariou, J. Keane, J. Gurd, L. Freeman (Eds.), Manchester, UK, August 28.–31., 2001, LNCS 2150, Springer, 2001, pp 148–153.
5. R. Berrendorf, M. Gerndt, W. E. Nagel and J. Prumerr, *SVM Fortran*, Technical Report IB-9322, KFA Julich, Germany, 1993, www.fz-juelich.de/zam/docs/printable/ib/ib-93/ib-9322.ps.
6. Frank Cappello and Daniel Etiemble, *MPI versus MPI+OpenMP on the IBM SP for the NAS benchmarks*, in Proc. Supercomputing'00, Dallas, TX, 2000. http://citeseer.nj.nec.com/cappello00mpi.html
7. William W. Carlson, Jesse M. Draper, David E. Culler, Kathy Yelick, Eugene Brooks, and Karen Warren, *Introduction to UPC and Language Specification*, CCS-TR-99-157, May 13, 1999, http://www.super.org/upc/, www.gwu.edu and http://projects.seas.gwu.edu/~hpcl/upcdev/upctr.pdf.
8. Robert B. Ciotti, James R. Taft, and Jens Petersohn, *Early Experiences with the 512 Processor Single System Image Origin2000*, proceedings of the 42nd International Cray User Group Conference, SUMMIT 2000, Noordwijk, The Netherlands, May 22–26, 2000, www.cug.org.
9. Tarek El-Ghazawi, and Sébastien Chauvin, *UPC Benchmarking Issues*, proceedings of the International Conference on Parallel Processing, 2001, pp 365–372, http://projects.seas.gwu.edu/~hpcl/upcdev/UPC_bench.pdf.
10. W. Gropp and E. Lusk and N. Doss and A. Skjellum, *A high-performance, portable implementation of the MPI message passing interface standard*, in Parallel Computing 22–6, Sep. 1996, pp 789–828.
11. Jonathan Harris, *Extending OpenMP for NUMA Architectures*, in proceedings of the Second European Workshop on OpenMP, EWOMP 2000, www.epcc.ed.ac.uk/ewomp2000/.

12. D. S. Henty, *Performance of hybrid message-passing and shared-memory parallelism for discrete element modeling*, in Proc. Supercomputing'00, Dallas, TX, 2000. http://citeseer.nj.nec.com/henty00performance.html

13. Alice E. Koniges, Rolf Rabenseifner, Karl Solchenbach, *Benchmark Design for Characterization of Balanced High-Performance Architectures*, in proceedings, 15th International Parallel and Distributed Processing Symposium (IPDPS'01), Workshop on Massively Parallel Processing, April 23-27, 2001, San Francisco, USA.

14. Richard D. Loft, Stephen J. Thomas, and John M. Dennis, *Terascale spectral element dynamical core for atmospheric general circulation models*, in proceedings, SC 2001, Nov. 2001, Denver, USA.

15. John Merlin, *Distributed OpenMP: Extensions to OpenMP for SMP Clusters*, in proceedings of the Second European Workshop on OpenMP, EWOMP 2000, www.epcc.ed.ac.uk/ewomp2000/.

16. Message Passing Interface Forum. *MPI: A Message-Passing Interface Standard*, Rel. 1.1, June 1995, www.mpi-forum.org.

17. Message Passing Interface Forum. *MPI-2: Extensions to the Message-Passing Interface*, July 1997, www.mpi-forum.org.

18. Matthias M. Müller, *Compiler-Generated Vector-based Prefetching on Architectures with Distributed Memory*, in High Performance Computing in Science and Engineering '01, W. Jer and E. Krause (eds), Springer, 2001.

19. The NANOS Project, Jesus Labarta, et al, http://research.ac.upc.es/hpc/nanos/.

20. R. W. Numrich, and J. K. Reid, *Co-Array Fortran for Parallel Programming*, ACM Fortran Forum, volume 17, no 2, 1998, pp 1–31, www.co-array.org and ftp://matisa.cc.rl.ac.uk/pub/reports/nrRAL98060.ps.gz.

21. OpenMP Group, www.openmp.org.

22. Rolf Rabenseifner and Alice E. Koniges, *Effective Communication and File-I/O Bandwidth Benchmarks*, in Recent Advances in Parallel Virtual Machine and Message Passing Interface, proceedings of the 8th European PVM/MPI Users' Group Meeting, Santorini/Thera, Greece, LNCS 2131, Y. Cotronis, J. Dongarra (Eds.), Springer, 2001, pp 24–35.

23. Rolf Rabenseifner, *Effective Bandwidth (b_eff) and I/O Bandwidth (b_eff_io) Benchmark*, www.hlrs.de/mpi/b_eff/ and www.hlrs.de/mpi/b_eff_io/.

24. Mitsuhisa Sato, Shigehisa Satoh, Kazuhiro Kusano and Yoshio Tanaka, *Design of OpenMP Compiler for an SMP Cluster*, in proceedings of the 1st European Workshop on OpenMP (EWOMP'99), Lund, Sweden, Sep. 1999, pp 32–39. http://citeseer.nj.nec.com/sato99design.html

25. Alex Scherer, Honghui Lu, Thomas Gross, Willy Zwaenepoel, *Transparent Adaptive Parallelism on NOWs using OpenMP*, in proceedings of the Seventh Conference on Principles and Practice of Parallel Programming (PPoPP '99), May 1999, pp 96–106.

26. Weisong Shi, Weiwu Hu, and Zhimin Tang, *Shared Virtual Memory: A Survey*, Technical report No. 980005, Center for High Performance Computing, Institute of Computing Technology, Chinese Academy of Sciences, 1998, www.ict.ac.cn/chpc/dsm/tr980005.ps.

27. Lorna Smith and Mark Bull, *Development of Mixed Mode MPI / OpenMP Applications*, in proceedings of Workshop on OpenMP Applications and Tools (WOMPAT 2000), San Diego, July 2000.

Performance Comparisons of Basic OpenMP Constructs*

Achal Prabhakar[1,2], Vladimir Getov[1,3], and Barbara Chapman[2]

[1] Parallel Architectures and Performance Team, CCS-3
Los Alamos National Laboratory, New Mexico, USA
[2] Department of Computer Science, University of Houston, Texas, USA
[3] School of Computer Science, University of Westminster, London, UK

Abstract. OpenMP has become the de-facto standard for shared memory parallel programming. The directive based nature of OpenMP allows incremental and portable developement of parallel application for a wide range of platforms. The fact that OpenMP is easy to use implies that a lot of details are hidden from the end user. Therefore, basic factors like the runtime system, compiler optimizations and other implementation specific issues can have a significant impact on the performance of an OpenMP application. Frequently, OpenMP constructs can have widely varying performance on different operating platforms and even with different compilers on the same machine. This makes it very important to have a comparative study of the low-level performance of individual OpenMP constructs. In this paper, we present an enhanced set of microbenchmarks for OpenMP derived from the EPCC benchmarks and based on the SKaMPI benchmarking framework. We describe the methodology of evaluation followed by details of some of the constructs and their performance measurement. Results from experiments conducted on the IBM SP3 and the SUN SunFire systems are presented for each construct.

1 Introduction

OpenMP is increasingly being thought of as the industry standard for parallel programming on shared memory systems. The directive based interface and incremental parallelism are two major features of OpenMP that add ease of use and convenience to parallelising existing and new applications. As in any other programming model, the overheads of supporting the facilities for parallel execution can have a considerable impact on the overall application's performance. If not used with caution, the overheads due to the OpenMP programming model itself can dominate the overall execution time of the application, voiding the parallelisation effort. Although much depends on the application's structure and algorithm, the low-level performance of basic constructs in OpenMP can have

---

* This work was partially supported by the U.S. Department of Energy through Los Alamos National Laboratory contract W-7405-ENG-36 and by the Los Alamos Computer Science Institute under grant LANL 03891-99-23.

H. Zima et al. (Eds.): ISHPC 2002, LNCS 2327, pp. 413–424, 2002.

a significant effect. The performance of OpenMP constructs will vary across different architectures and with different compilers on the same machine; the runtime system and the compiler are key factors in the performance of a parallel program. Often there is a significant difference in overheads incurred by similar constructs which suggest the use of one over the other. Thus information on performance penalties and scalability of OpenMP constructs is very useful, both for understanding the underlying implementation and architecture as well as for creating efficient parallel programs using OpenMP. In this paper, we describe a benchmark suite that is designed to evaluate the overheads incurred by OpenMP constructs and that adds some innovative measurements.

Benchmarking and evaluation of high performance computers and programming models is a well established field. In addition to producing benchmark codes, a number of individuals and groups in the past have come up with techniques and tools for making, analyzing and presenting benchmark results. Notably the GENESIS [5], ParkBench [6], EuroBen, NAS parallel benchmarks [1] and the EPCC low-level OpenMP benchmarks [3] have been widely used. The nature of measurements, the methodology and the design of benchmarks suites keeps changing with the advent of new machines and architectures. The goal of our project was to evaluate and compare the raw performance of the low-level facilities offered by OpenMP without going into application level details. In addition, we aimed to provide an efficient benchmark suite so that it can execute in the shortest time possible and still deliver fairly accurate results. Given the high cost of supercomputer time, we should not spend more of it than necessary on benchmarking. Also, we assumed that it is not possible to include every possible measurement and method, and thus the benchmark suite should be extensible and modular, allowing for the easy addition of new measurements and methodologies, possibly even measurements for future programming models. At the same time, no single programming model should be dependent on the availability of any other one, giving the user freedom to exclude any of them. For instance, if one is not interested in benchmarking MPI, it should be possible to compile and execute the suite on a platform that does not have MPI installed. The starting point for our OpenMP benchmarks was the popular suite of low-level benchmarks developed at EPCC to measure the overheads of OpenMP constructs.

The rest of the paper is organised as follows. Section 2 discusses the methodology and design of the benchmarks suite. Section 3 describes some of the important tests in the benchmarks suite. Space constraints do not permit us to cover all the contructs and the variants. Section 4 analyses the results from our experiments; it is followed by the conclusion in Section 5.

2   Evaluation Methodology

In general, the process of making a measurement, recording the results, collating them and writing to the output as well as accepting user input and constraints is similar for each measurement, regardless of the programming model used or the scope of the activity. The only difference in the whole process is the actual section of code that is executed and measured. As such it was obvious to consider the use of a framework that provides these basic facilities and allows a system for measurements routines to be "plugged" into the framework [7]. This way any new measurement will only require that the user writes the code that performs the operation to be measured and informs the framework of the new routine. This solves the problem of having an integrated and extensible suite.

SKaMPI, which stands for Special Karlsruhe MPI benchmarks [8,9], from the University of Karlsruhe is one such benchmarking suite for MPI. SKaMPI provides a framework specifically for building extensible benchmarking suites. The important features of SKaMPI, such as automatic parameter refinement, automatic control of error, ability to merge the output results from multiple runs, and an intuitive user input mechanism, make it very useful for our work. The benchmark suite described here is based on the SKaMPI framework, with extensions made in both CASC at LLNL [4] and CCS-3 at LANL.

Since OpenMP [2] is directive based, it is not possible to directly measure the overheads imposed by each OpenMP construct. The methodology followed is to execute and measure a piece of code without an OpenMP directive (T_{ref}) and then measure the same code with the OpenMP directive of interest(T_m). The reference measurement is then subtracted from the OpenMP-enabled measurement to get the overhead for the particular construct (T_o).

$T_m = T_{ref} + T_o$, thus $T_o = T_m$ - T_{ref}.

This is however not always necessary; for instance, the OpenMP `barrier` directive is an executable statement and can be measured without requiring a reference measurement. There are additional complications involved. As an example, consider the `lock-unlock` measurement. Should the measurement be the time taken by the master to perform the lock, or should it be the total time spent by all the threads? The decision in each such case is dependent on which method gives the most relevant information as regards the effect of the overheads on the total execution time of an application.

3   Basic Constructs and Evaluation Results

3.1   Lock - Unlock

The OpenMP `lock-unlock` primitive provide fine grained control over synchronization between threads. As an alternative to using `critical` regions, there are instances when the explicit `lock-unlock` is advantageous over using `critical` regions. While the performance difference between `critical` and `lock-unlock` depends a lot on the underlying runtime system, in general, the `lock-unlock` combination is expected to incur less overhead than `critical`. The conventional approach to measure the lock/unlock overhead is to have all the threads make repeated invocations of lock followed by unlock and averaging the result to get a fairly accurate estimate of the overhead. However, if the runtime system does not guarantee lock-fairness, it may be possible that the thread which is timing

416   Achal Prabhakar, Vladimir Getov, and Barbara Chapman

the loop can grab the lock in every iteration without giving a chance to other
threads. In such a case, the results would reflect the `lock-unlock` overhead on
a single thread, which is clearly not correct. In order to circumvent the fairness
problem, our approach is to ensure that in every iteration, each thread gets an
opportunity to grab the lock at-most once. This can be accomplished by making
sure that all threads wait after each iteration for other threads. This is achieved
by placing a OpenMP `barrier` after the unlock operation. The sequence of
events is illustrated in Fig 1. Although it is possible that the same thread can
grab the lock at the next iteration, the probability that the timing will be offset
by more than the time for a single thread lock acquire is very low.

The reference time T_{ref} is calculated by the code in Program 1. Program 2
shows the pseudo code for the actual `lock-unlock` time T_m measurement.

```
for ( 1 to N )
  do_some_work
#pragma omp barrier
end for
```

```
for ( 1 to N )
  omp_set_lock()
  do_some_work
  omp_unset_lock()
#pragma omp barrier
end for
```

Program 1. Lock-unlock reference time measurement.

Program 2. Lock-unlock measurement (no assumptions of lock-fairness)

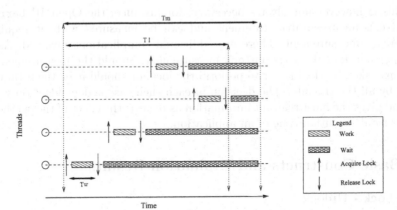

Fig. 1. Timing Diagram for the Lock/Unlock Measurement

It is subtracted from the reference time to offset the `barrier` cost and dummy
work cost. The measured quantity now is the sum of times that each thread
had spent in performing the `lock-unlock` operation, regardless of whether the
runtime system ensures lock fairness or not. This gives an accurate estimate of
the effect that the number of threads has on the lock performance.

$T_m = T_b + T_l + T_w$ or $T_m = T_{ref} + T_l$

where $T_{ref} = T_b + T_w$ Thus $T_l = T_m - T_{ref}$

3.2   Critical

The OpenMP `critical` directive is similar to the `lock-unlock` primitive: both guarantee exclusivity to a thread in a region. A thread waits at the beginning of a `critical` region until no other thread is executing the same `critical` region. In principal a `critical` region is equivalent to enclosing the structured block with calls to `omp_set_lock` and `omp_unset_lock`, however the performance difference is highly dependent on the underlying runtime implementation of both `critical` and `lock-unlock` used by the operating platform. In general, the `lock-unlock` primitives are expected to provide better performance than the `critical` directive, but the actual difference in performance and scalability depends on the implementation and might vary with the number of threads.

The measurement approach is similar to that for the `lock-unlock` primitives. A dummy work routine enclosed in a `critical` region is called repeatedly by all the threads in the team with the master thread responsible for timing the loop. The final result is averaged over the number of iterations to get an estimate of the overhead for a single `critical` region. The pseudo code for the measurement routine is shown in Program 3.

```
for ( 1 to N )
    #pragma omp critical
        do_some_work()
end for
```

Program 3. Pseudo code for measuring overhead of `Critical` directive assuming lock fairness.

Since the compiler decides how to translate a `critical` directive into more basic primitives, the behavior of the `critical` directive with respect to lock-fairness is implementation dependent. In case the runtime system does not guarantee that a thread can not acquire a mutex repeatedly, the code segment in program 3 will not yield the correct overhead. Like the `lock-unlock` measurement, the worst case scenario would be that the master thread, doing the timing, acquires the mutex for all iterations of the loop. The result would be the overhead for executing a `critical` region with only one thread.

Our solution is to force every thread to wait at the end of the `critical` region by having an OpenMP `barrier` directive. This guarantees that for any single iteration, all the threads would have executed the `critical` region at least once. The timing diagram for this measurement is similar to the `lock-unlock` (Fig. 1.). It should be noted that the reference measurement is done over the same number of threads as for the actual measurement. The result from the reference measurement for n threads is subtracted from the actual measurement for the same number of threads. Program 4 and 5 show the pseudo code for measuring the `critical` overhead and the reference time respectively.

```
for ( 1 to N )
#pragma omp critical
  do_some_work();
#pragma omp barrier
end for
```

```
for ( 1 to N )
  do_some_work()
#pragma omp barrier
end for
```

Program 4. Critical directive measurement (no assumption of lock fairness.)

Program 5. Reference measurement for critical directive.

3.3 Barrier

The barrier is a global synchronization directive. Each thread waits at the barrier until all the other threads have arrived at it. Once all threads have reached the barrier, each thread continues to execute the code after the barrier in parallel. There is an implied flush of shared variables for the barrier directive.

The barrier measurement is similar to the atomic directive measurement. The work routine is placed in the barrier scope and is executed repeatedly by all the threads (see Program 7). The measured time is for the execution of the n thread barrier plus the time taken by the work routine. It should be noted that since the barrier has an implied flush, its overhead is automatically included in the barrier overhead. The reference measurement calculates the execution time of the work routine for a single thread, which is then subtracted from the barrier measurement to get the overhead for the barrier directive (see Program 6).

```
for ( 1 to N )
  do_some_work()
end for
```

```
for ( 1 to N )
#pragma omp barrier
  do_some_work()
end for
```

Program 6. Barrier directive reference measurement.

Program 7. Barrier directive measurement.

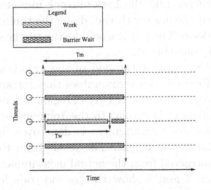

Fig. 2. OpenMP Single

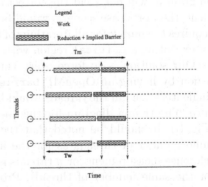

Fig. 3. OpenMP For-Reduction

3.4 Single

The OpenMP `single` directive is another one of the many synchronization directives available; it ensures that only a single thread will execute the block of code in the scope of the directive. The executing thread need not necessarily be the master thread. There is an implied `barrier` after the `single` directive. The approach for measuring the overhead of the `single` directive is illustrated in the timing diagram in Figure 2. The block of code containing the `single` directive and the work routine is executed for a number of times. The master thread performs the timing, which is averaged over the number of iterations performed for the loop (see pseudo code in Program 8). The reference time calculation is similar to that for the `barrier` directive (Prog. 6).

```
for ( 1 to N )
#pragma omp single
   do_some_work()
end for
```

Program 8. Single directive measurement.

```
for ( 1 to N )
#pragma omp for reduction(+:red_var)
  for( 1 to N * num_threads)
     red_var += do_some_work()
  end for
end for
```

Program 9. reduction clause with for directive measurement.

3.5 Reduction

A `reduction` clause is used in conjunction with work-sharing constructs to perform reduction on the shared variable according to the specified reduction operator. Since this clause is used in conjunction with the work sharing constructs, there is always an implied `barrier` unless the enclosing construct has a nowait clause, in which case the reduction computation is not guaranteed until a `barrier` is encountered. The measurement of the `reduction` clause is important as it is one of the most frequent operations in any parallel program. Also the use of the `reduction` clause, rather than a manual implementation of a reduction, is usually expected to provide opportunities to the compiler for optimization.

Since it is always used in conjunction with a work-sharing construct, the reference time measurement for this clause has to estimate the overhead of the enclosing construct. Consider the case of `reduction` with the OpenMP for directive. We provide tests for measuring the overhead of the omp `for reduction` as well as the overhead of just introducing the `reduction` clause in the omp for directive. Program 9 shows the pseudo code for measuring the overhead of the omp `for reduction` directive. In this case the reference time is simply the overhead of the loop and the work routine (Program 10). In order to measure the overhead of the `reduction` clause itself, the reference routine has to incorporate the omp `for` directive also, as shown in Program 11.

```
for( 1 to N )
  red_var += do_some_work()
end for
```

Program 10. Reduction clause reference measurement.

```
#pragma omp for
for( 1 to N * Num_threads )
    red_var += do_some_work()
end for
```

Program 11. Reduction clause without for directive reference measurement.

3.6 Parallel & Parallel-for

The `parallel` construct is the only way to start a parallel region in OpenMP. When a thread encounters a parallel region, it creates a new team of threads and itself becomes the master of the team. The code immediately following the `parallel` construct and within the dynamic scope of the directive is executed in parallel by all the threads in the team, including the thread that created the team. There is an implied `barrier` at the end of the `parallel` clause.Measurement of the `parallel` region is simpler than for the other directives. The measurement routine consists of a work routine within a `parallel` region which is repeatedly invoked and the final result is averaged by the master thread. The reference time is the execution time of the work routine for a single thread.

Parallel `for` is the compound parallel work share directive. It specifies a parallel region with only a single work sharing directive. When a thread encounters a `parallel-for` directive, it creates a new team of threads. The work is shared among the threads according to the schedule policy in effect. After completion of the work all the threads wait at the implied `barrier`.

Measuring the overhead of the `parallel for` directive involves repeatedly timing the work routine contained within the `parallel for` directive (Prog.12). The reference time is the time taken for the execution of the work routine for a single thread (Prog.13).

```
for ( 1 to N )
#pragma omp parallel for
  for ( 1 to N * Num_threads )
    do_some_work()
  end for
end for
```

Program 12. Parallel-for directive measurement.

```
for ( 1 to N )
  do_some_work()
end for
```

Program 13. Parallel-for directive reference measurement.

4 Presentation and Discussion of Results

The experiments were conducted on an IBM SP3 8-way SMP with 375 MHz Power3 processors, using the VisualAge C++ Professional / C for AIX Compiler, Version 5 and on a SUN SunFire 6800 24-way UltraSPARC-III, 750 MHz using the Sun WorkShop 6 update 2 C 5.3 2001/05/15 compiler. All the times mentioned in the results are in microseconds.

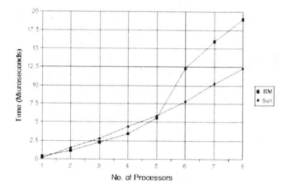

Fig. 4. Lock-unlock overhead on the IBM SP3 and SUN SunFire

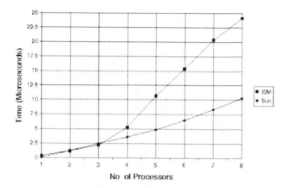

Fig. 5. Critical overhead on the IBM SP3 and SUN SunFire

On the IBM SP3, the `lock-unlock` overheads increase rapidly from 5 to 6 processors (Fig. 4). Strangely, on the SUN the overheads of the `critical` directive (Fig. 5) are lower than those for the `lock-unlock` primitives, even though there is an implied flush at the end of the `critical` region, suggesting that it will incur additional overheads. The difference can be attributed to better optimization in the implementation of the `critical` directive. On the IBM, the cost of `lock-unlock` is less than that of the `critical` directive, with the gap increasing with increasing number of processors, most probably due to increasing cost of the implied flush. The lower cost of `critical` on the SUN SunFire and the `lock-unlock` on the IBM SP3 suggest alternate implementations and / or difference in cost of the flush on the two systems. However, for both constructs the SUN machine shows better scalability than the IBM.

The `barrier` implementation on the SUN appears to be quite efficient and scales well with an increasing number of processors. The overhead of a `barrier` is very small, around 3 microseconds for 8 processors (Fig. 6). Markedly different is the IBM implementation which is costly and also scales poorly. The poor `barrier` performance on the IBM may be responsible for the low performance of all other `barrier` synchronized directives, such as the `parallel`, `parallel-for`,

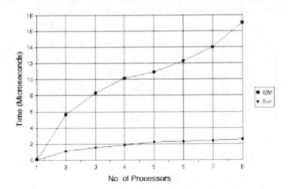

Fig. 6. Overhead of the `barrier` directive on the IBM SP3 and SUN SunFire

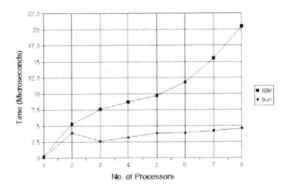

Fig. 7. Overhead of the `single` directive on the IBM SP3 and SUN SunFire

and `reduction` constructs. Once again, we notice a sharp increase for 2 and 8 processors on the IBM.

For fewer than 7 processors, the overhead of the `single` directive (Fig. 7) is lower than the overhead of the `barrier` on the IBM SP3, even though there is an implied `barrier` at the end of a `single` directive. However, with more than 7 processors the cost is higher than that of `barrier`, which is the expected behaviour. A reason for the discrepancy could be the overlap between the first phase (check-in) of the `barrier` with the computations involved with the `single` directive. On the SUN SunFire however, the cost of `single` is always greater than that of the `barrier` and closely follows the `barrier` curve, suggesting that most of the cost of the `single` directive is in the implied `barrier`. There is a sharp increase in the overhead of the `single` directive for 2 processors. We noticed this increase for a majority of runs performed and decided to keep it as being more statistically representative. However, this might have been due to disturbances when the experiments were conducted. In general, the cost of `single` on the SUN appears to be scalable and as efficient as the implied `barrier`. On the IBM though, the cost of `single` increases rapidly with an increasing number of processors, with a sharp increase from 1 to 2 and 6 to 8 processors.

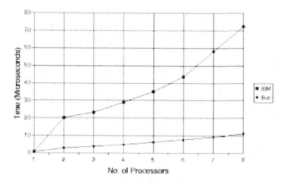

Fig. 8. Overhead of the `for-reduction` directive on the IBM SP3 and SUN SunFire

The `reduction` operation is extremely expensive on the IBM SP3, with the overhead being as high as 72 microseconds for 8 processors (Fig. 8). The poor scalability suggests an implementation that is based on atomic updates of the shared reduction variable. Also the costly `barrier` adds to the overheads of `reduction`. The SUN SunFire shows a much better performance, both in terms of overhead and scalability. The slope is very similar to that of the `barrier` overhead curve, pointing towards a tree based reduction algorithm that is combined with the execution of the `barrier`.

Overheads for the `parallel` directive on the IBM SP3 increase sharply for 2 processors, remain almost constant up to 5 processors and then increase rapidly for 7 and 8 processors (Fig. 9). Since there is an implied `barrier` at the end of the `parallel` region, subtracting the cost of the `barrier` from the overhead of the `parallel` region, gives us an estimate of the overhead due to `parallel` region creation and management. This simple calculation shows a large overhead for 2 processors which reduces up to 6 processors and rises again for 7 processors. On the SUN SunFire machine, a `parallel` region incurs lower overheads and they scale better than on the IBM SP3. However the overhead is always more than twice the cost of the `barrier`, suggesting that on the SUN there is a complete `barrier` at the start as well as the end of the `parallel` region. Unlike the IBM, subtracting 2 times the cost of the `barrier` from the `parallel` region overhead yields a more uniform curve.

5   Conclusions

The importance of benchmarks in analysing and tuning the performance of parallel programs cannot be overemphasized. We have presented an enhanced and exhaustive set of benchmarks for evaluating the performance of OpenMP constructs that used the work carried out at EPCC on OpenMP overheads measurement as a starting point. The implementation of the benchmarks is based on the extensible SKaMPI framework and allows great flexibility for tailoring the benchmark both externally and internally. We have shown the results for two

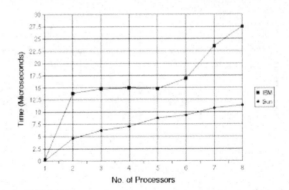

Fig. 9. Overhead of the `parallel` directive on the IBM SP3 and SUN SunFire

different machines, the IBM SP3 and the SUN SunFire 6800. Critical discussion of the results has been presented.

Acknowledgements

We convey our sincere thanks to the staff at the San Diego Supercomputing Center and at the University of Ulm in Germany for allowing use of their computing facilities and for the technical support while running the experiments presented in this paper.

References

1. D. H. Bailey, E. Barszcz, J. T. Barton, D. S. Browning, R. L. Carter, D. Dagum, R. A. Fatoohi, P. O. Frederickson, T. A. Lasinski, R. S. Schreiber, H. D. Simon, V. Venkatakrishnan, and S. K. Weeratunga. The NAS Parallel Benchmarks. *The International Journal of Supercomputer Applications*, 5(3):63–73, Fall 1991.
2. OpenMP Architecture Review Board. OpenMP Fortran Application Program Interface, Version 1.1, November 1999.
3. J. M. Bull. Measuring Synchronisation and Scheduling Overheads in OpenMP. In *EWOMP '99, Lund, Sep., 1999.*, 1999.
4. Bronis R. de Supinski and John May. Benchmarking Pthreads Performance. In *Proceedings of the 1999 International Conference on Parallel and Distributed Processing Techniques and Applications (PDPTA '99)*, June 1999.
5. A. J. G. Hey. The genesis distributed memory benchmarks. *Parallel Computing*, 17(10–11):1275–1283, 1991.
6. R. Hockney and M. Berry. Public international benchmarks for parallel computers report. Technical report, Parkbench Committee, 1994.
7. R. W. Hockney and V. S. Getov. Low-level benchmarking: Performance profiles. In *Proc. Euromicro Workshop on PDP,IEEE CS Press*, Jan. 1998.
8. R. Reussner, P. Sanders, L. Prechelt, and M. Mueller. SKaMPI: A detailed accurate MPI benchmark. In *Springer Lecture Notes in Computer Science.*, volume 1497, pages 52–59, 1998.
9. R.H. Reussner. Skalib: Skampi as a library - technical reference manual. Technical report, Department of Informatic, University of Karlsruhe, Germany, 1999.

SPMD OpenMP versus MPI on a IBM SMP for 3 Kernels of the NAS Benchmarks

Géraud Krawezik[1,2], Guillaume Alléon[2], and Franck Cappello[1]

[1] Université Paris-Sud, Laboratoire de Recherche en Informatique, Bâtiment 490,
91405, Orsay Cedex, France
{gk,fci}@lri.fr
[2] EADS – Corporate Research Center, 31707 Blagnac Cedex, France

Abstract. Shared Memory Multiprocessors are becoming more popular since they are used to deploy large parallel computers. The current trend is to enlarge the number of processors inside such multiprocessor nodes. However a lot of existing applications are using the message passing paradigm even when running on shared memory machines. This is due to three main factors: 1) the legacy of previous versions written for distributed memory computers, 2) the difficulty to obtain high performances with OpenMP when using loop level parallelization and 3) the complexity of writing multithreaded programs using a low level thread library. In this paper we demonstrate that OpenMP can provide better performance than MPI on SMP machines. We use a coarse grain parallelization approach, also known as the SPMD programming style with OpenMP. The performance evaluation considers the IBM SP3 NH2 and three kernels of the NAS benchmark: FT, CG and MG. We compare three implementations of them: the NAS 2.3 MPI, a fine grain (loop level) OpenMP version and our SPMD OpenMP version. A breakdown of the execution times provides an explanation of the performance results.

1 Introduction

1.1 Current Hardware and Programming Trends in HPC

In the past decades, parallel computers have been implemented using a large variety of architectures: vector, MPP, SMP, COMA, NUMA, Clusters. Currently three kind of them are used: vector, cluster and SMP. Large configurations are built from clustering vector or SMP nodes. This technology may involve a message passing network or a NUMA switch. The most popular machines, gathering several memory models (message passing + shared memory), corresponds to a trade-off between the manufacturing cost and the features required by customers. Basically, customers are asking for shared memory machines (more precisely SMP) as they are simpler to program. However, the manufacturing cost precludes to build very large SMP nodes. That is why NUMA switch are very promising as a solution to fulfill the requirements of the programmers while keeping manufacturing costs moderate.

H. Zima et al. (Eds.): ISHPC 2002, LNCS 2327, pp. 425–436, 2002.

Because many parallel machines of the last decade had a distributed memory architecture, a lot of applications have been ported for the message passing paradigm. Performance evaluations of large cluster of SMP demonstrates the good performance of the message passing implementations compared to more exotic model such as hybrid of message passing and shared memory [1] [2] [3] [4]. Many years of experience in message passing have lead to a fairly good maturity of this parallel programming model.

These two trends leave the programmer with a very difficult question: is it worthwhile to migrate an existing large application already programmed with the message passing paradigm to a shared memory version for SMP or NUMA parallel machines.

1.2   The Need for a Fair Comparison in Simple Conditions

Answering the programmer's question is very hard since it covers several issues: -how difficult is it to migrate an existing message passing code to a shared memory version? -what would be the performance of the resulting code? -would the performance vary for different shared memory machines.

The first issue considers the programming effort. There are several ways to transform an existing message passing program into a shared memory one. A first one is to use a set of compiler directives to implement a loop level parallelization of a sequential version of the message passing code. Loop-level parallelism (fine grain), is very appealing as it is close to automatic-parallelization: one profiles a code to detect the most expensive loops, and then use directives such as !$OMP DO to distribute their execution among several threads. It also enables incremental development, as one starts from his serial code, and then can parallelize one element at a time. The OpenMP standard provides a complete set of directives for this purpose. A much more complicated approach is multithreading a sequential code using a low level threads library such as the Pthreads. In this case the programmer must manage by himself their creation, destruction and coordination, as well as work distribution. An intermediate approach is to use OpenMP for SPMD programming. This paper considers this last approach. While it is very close to multithreading, it benefits of the portability of OpenMP and removes the complexity of low level thread management. The programmer must only concentrate on work parallization, array references distribution and loop nest optimization. We will show that it is very convenient if there is an existing message passing implementation of the application. Another advantage is that the programmer can benefit of the code maturity, especially for the loop nest optimizations.

The last two issues not only depend on the hardware but also on the operating system management of the shared resources (CPU, Memory, Disk). It is very difficult to compare message passing and shared memory features on a NUMA machine because the management of these resources can be different, especially for the virtual memory pages allocation/management. Thus, we believe that answering the programmer's question requires several steps. This paper addresses the very first which is checking that shared memory versions of well known and tuned message passing programs can provide better performance than the original implementation on simple operating conditions. So we will compare the performance of message passing and shared memory implementations of three kernels of the NAS 2.3 benchmark on a SMP machine. Using it removes the memory pages allocation/migration problem that can lead to non-reproducible behaviors. Conducting experiments without this problem should provide fair information about the reasons behind the better performance of one model over the others. Not all SMP machines can be used for this first step. The IBM SP machine with NH2 nodes has been chosen for the comparison because 1) it features several years of improvements for both the message passing and the shared memory environments and 2) the NH2 node uses a crossbar switch between the processors and the memory providing one of the highest memory bandwidth for a 16 way multiprocessor. A subsequent step toward the answer of the programmer question would be to add the memory pages allocation/migration problem. This part of the problem will not be addressed in this paper.

2   SPMD Programming Style with OpenMP

An important reason for the high performance of MPI programs on regular applications is the domain decomposition strategy used by the programmer. It provides a framework for expressing/exploiting the locality properties of the application at two levels. First, the main task is decomposed in coarse grain communicating subtasks, with a structure minimizing communications. Second, the programmer can investigate the locality properties of the subtask programs almost without considering the first decomposition level. Usual optimization techniques are used to reduce as much as possible the computing time of the subtasks. Years of improvements have lead to the existing MPI programs that are highly tuned at the two levels.

The popular programming technique with OpenMP is different. OpenMP provides a set of directives for task parallelism. However for regular applications, most of the programmers use the set of directives provided to express loop level parallelization. Programming using only this kind of parallelization essentially neglects the benefit of domain decomposition: reduction of data exchanges between the processors and high potential for local optimizations. Blocking may be achieved with loop level parallelization using loop rewriting or nested parallelism. However very few programmers actually intent to use these approaches because of their complexity or because nested parallelism is implemented by a serialization for most of the compilers. Moreover, there is no obligation for domain decomposition in a shared memory machine and especially for SMPs. All processors may access any data stored at any place, at any time. So one can write a shared memory program without considering that data exchange between processors might be a huge source of performance degradation even in a shared memory machine.

In this paper we will consider another strategy close to the one used to design high performance message passing programs on distributed memory architectures. The program is structured in such a way that 1) data exchanges are minimized, 2) exchanges use data sets as large as possible, 3) sets are continuous chunks of data 4) computations may use blocking and loop optimization, 5) synchronization is minimized. Because high performance MPI programs carefully follow these recommendations, we can fairly easily derive shared memory versions of the applications from them.

2.1   Previous Works

SPMD programming with OpenMP is not a novel idea. The OpenMP specification explicitly presents the coarse grain parallelization through the use of the Section directive.

Several researchers have compared the performance of different implementations of the same programs using several models, including pure MPI, fine grain OpenMP, coarse grain OpenMP, hybrid MPI + OpenMP.

The two programming styles of OpenMP (loop level parallelization and SPMD) are compared in [2] for a conjugate gradient code. The experiments are conducted on Compaq ES40 nodes. For 4 processors, the advantage of the SPMD version over the loop level parallelization is about 17% (we obtain an improvement of more that 40% with 4 processors for SPMD for the NAS CG).

In [5], the author compares SPMD OpenMP and MPI for Ocean Models. He pinpoints 1) the better communication performance of the shared memory implementations and 2) some difficulties for adapting MPI applications to SPMD OpenMP. He remarks that from his results we cannot conclude that the shared memory implementations provide better performance than the message passing one. We will demonstrate that for a SMP, SPMD OpenMP actually provides better performances than MPI for three NAS benchmark kernels.

SPMD OpenMP and MPI are compared for the game of life code and a master slave application in [4]. The authors argue that SPMD OpenMP can provide better performance than MPI because of 1) load imbalance, 2) computation grain too fine 3) memory limitation due to replication and 4) poor optimization or limited scaling of MPI implementation. They also report interesting results on a SGI cluster of multiprocessors about fine grain OpenMP reaching nearly the same performance as SPMD OpenMP for the game of life. Our conclusion about fine grain and SPMD OpenMP are the opposite: SPMD OpenMP always provides better performance than fine grain OpenMP.

The paper [6] presents another comparison of the two programming models for a seismic processing code on a SUN Enterprise. None of the models is showing significantly better performance than the other. The authors notice that their analysis of the executions show behavior differences of the programs, but mainly due to the compilation phase.

None of these experiments compare the performance of message passing and SPMD OpenMP programming on SMP multiprocessors considering well known and tuned kernels applications. Moreover, there are very few explanations about the fundamental reasons of the better performance of one model over the others.

2.2   High Performance SPMD OpenMP

Thus, the next parts of this article consider a method based on the transformation of existing MPI codes into SPMD OpenMP ones. The strategy to get better performance with SPMD OpenMP is based on our previous study presented in [1]. In this paper we have shown that the main reasons behind the low performance of hybrid programming with loop level parallelization are: 1) the limited number of parallelized loop nests and 2) the difficulty of optimizing loop nests with OpenMP. So the general strategy to reach high performance is to keep the computation part of the original MPI code. The performance improvement compared to this version will come from a lower communication cost for OpenMP due to an optimization of all data exchanges.

Like a message passing program, a SPMD OpenMP program will have some computation parts and some communication parts. When deriving it from an MPI one, we will adapt them independently. In this paper, there will be no attempt to merge the computation and data exchange parts in the SPMD OpenMP program, even though this could be an interesting issue.

Keeping MPI Computing Loops. Our goal in using the SPMD programming style with OpenMP is to benefit from the advantages provided by domain decomposition.Thus, we use the same decomposition strategy between OpenMP threads that is used between MPI processes in the original code. This implies an obvious translation of the initial part of the MPI code to OpenMP by replacing the different MPI calls (MPI_Comm_Size, MPI_Comm_Rank) by calls to the equivalent OpenMP library functions.

This allows to use the decomposition routines of the MPI program. We also tried to maintain the same structures for the arrays used during the computation. The few changes that appeared are due to unexpected behaviors of the compiler, or because the translation of MPI communications into shared memory exchanges required the use of additional data sets.

A naïve idea is to 'privatize' every global variable, as they are supposed to be in the message passing version. But we have noticed that using the !$OMP THREADPRIVATE directive may lead to excessive system function calls to get the thread id during the execution of subroutines. In the IBM XLF 7.1 compiler, they were included in various places and especially inside the core of the most expensive loop nests.

To avoid this problem, we made the main arrays 'shared', each processor identifying its own chunk of data by indexing the arrays with its thread id. Thus, in the main body of a program, an array A that was accessed by A(j) in MPI will be accessed as A(j, thread_ID) in SPMD OpenMP. As a local variable inside a subroutine is automatically made private, we do not need to change anything on this point.

Replacing MPI Communications by Shared Memory Operations. In the process of adapting the MPI communication into OpenMP data exchanges, we must consider point-to-point communications and global operations.

The first approach for replacing MPI point-to-point communication by shared memory operations is to use shared variables. Another approach is to remove explicit exchanges or communication by making all the data shared, and readjusting the array boundaries for calculation to their 'physical' values. We use

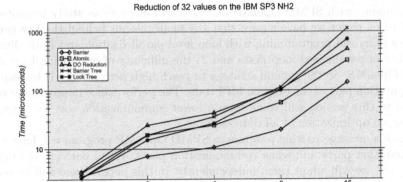

Fig. 1. Performance of five implementations of the reduction operation on an IBM NH2 SMP node.

a more MPI-like approach, which keeps the communication buffer, since some other studies ([7], [5]) have proved the advantages of using shadowing for the zones that will be exchanged between threads.

We have made the buffers shared between processors, the process of exchanging data being the following:

1. The threads fill the communication buffers
2. A barrier !$OMP BARRIER is used to synchronize the threads
3. The threads read the data from the communication buffers
4. Another !$OMP BARRIER if necessary

In the programs used for the performance evaluation, there are two kinds of global operations: AllReduce and AllToAll. Implementing them only with OpenMP functions or directives guaranties the code portability.

For the AllReduce, we have tested several implementations, using a DO directive with Reduction clause, OpenMP lock functions, ATOMIC, CRITICAL sections, shared memory with Barrier and variations of them such as building a binary tree of locks. Figure 1 presents their performance.

For example, the equivalent of a MPI_AllReduce(send, recv,..., MPI_Max, ...) call can be represented with the following OpenMP equivalent:

```
        double precision send(total_threads), recv
!$OMP PARALLEL PRIVATE(recv) SHARED(send)
        ... calculations done on send(thread_ID) ...
!$OMP BARRIER
        do i = 1, total_threads
          recv = max(recv, send(i))
        enddo
        ...
!$OMP END PARALLEL
```

Figure 1 shows that the shared memory version with barrier is the best for all the number of processors. In all the comparison results presented in the following section, we have used this version.

The AllToAll is implemented on the same way, using a shared array which is indexed using the threads id. In such a model, a call to MPI_AllToAll(sendbuf, recvbuf, ...) will be replaced by the following shared memory equivalent:

```
        double precision sendbuf(total_chunk, total_threads),
                        recvbuf(total_chunk)
!$OMP PARALLEL SHARED(sendbuf) PRIVATE(recvbuf)
        ... calculations done on sendbuf (*, thread_ID) ...
!$OMP BARRIER
        offset2 = me * (total_chunk / total_threads)
        do proc = 1, total_threads
            offset1 = (proc-1) * (total_chunk / total_threads)
            do i = 1, total_chunk / total_threads
                recvbuf (i + offset1) = sendbuf (i + offset2, proc)
            enddo
        enddo
        ...
!$OMP END PARALLEL
```

This is a naïve implementation of the AllToAll operation, and as in the case of the AllReduce, we are testing other ways of doing it, for example by using a permutation of the indexes of processors, so that the accesses to the data of one thread is done by one thread at a time.

3   Performance Evaluation

3.1   Hardware and Software Conditions

We conducted the experiments on an IBM SP3 NH2 SMP machine. The results are presented up to 16 processors. We have executed the same programs on a SGI O3800 to provide results with a larger number of processors, but the variations of the execution times for a same program lead us to drop the results although they confirm the presented result set. As explained in [7], performance evaluation on such machines is quite difficult.

NH2 nodes connect 16 Power3+ 375 Mhz processors to a 16 GB main memory using a crossbar switch featuring about 14.2 GB/s bandwidth. Each processor has a 64KB L1 data cache and a 8MB L2 cache. The compilation uses XLF 7.1 with the following flags: -qsmp=noauto:omp -O3 -bmaxdata:0x80000000 -qcache=auto -qarch=pwr3 -qtune=pwr3 -qmaxmem=-1 -qunroll -qnosave. The same set of flags are used for the MPI implementation (except the -qsmp option).

We consider 3 kernels of the NPB 2.3 with two data set sizes (A and B) and three implementations: MPI NPB 2.3, fine grain OpenMP and SPMD OpenMP. We chose these three kernels because in their MPI implementation, they feature various computation to communication ratio and communication patterns.

For class A, CG is a communication intensive benchmark using almost only reductions as communication patterns. Its computation to communication ratio decreases with the number of nodes. For class B, the communication becomes less significant compared to the computation. MG features stencil communications and FT implements transpositions with all-to-all global operations. The computation per communication ratio of FT remains constant for A and B dataset size.

The fine grain implementation of CG and MG derives from the SDSC hybrid version, from which we removed all MPI calls. For FT we applied the RWCP omp-C fine grain parallelization to the serial version of the NPB 2.3.

The SPMD OpenMP version is built from the MPI one, keeping the loop nests and replacing communications by shared memory global operations. In addition, for FT we have replaced the MPI 3-steps transposition by a 1-step algorithm.

For the two OpenMP implementations, we use the static scheduling option. Several other scheduling policies where tested (dynamic and guided), but the static option gave the best performance. For better performance, threads are binded to processors for the SPMD version. It did not provide any performance improvement for the OpenMP fine grain version.

We run the experiment under PPE 2.3 parallel programming environment, in dedicated mode.

3.2 MPI versus Fine Grain OpenMP versus SPMD OpenMP

Figure 2 presents the performance comparison of the three implementations of the three NAS kernels according to the number of processors on a IBM SP NH2 node.

The figure 2 shows for CG a significant improvement of the performance of SPMD OpenMP over both MPI (up to 14% better) and fine grain OpenMP (up to 138%). The picture is less impressive for CG class B. SPMD OpenMP stays comparable to MPI. The advantage over the fine grain version is only 17%. For MG, SPMD OpenMP provides a good improvement compared to MPI (up to 28% for class A and B). Its advantage over fine grain OpenMP is also substantial: up to 46% for class A and up to 56% for class B. FT also demonstrates that SPMD OpenMP can provide significant improvement compared to MPIfor various dataset sizes: up to 15% for class A and up to 27% for class B. The SPMD OpenMP version of FT also provides better performance than the fine grain version (up to +70% for Class A and up to 24% for Class B). It is interesting to notice that, for FT, fine grain OpenMP is slightly more efficient than MPI.

As a general result, we can consider that SPMD OpenMP is a good alternative to MPI in most of the cases and is always significantly better than a fine grain OpenMP implementation.

3.3 Breakdown of the Execution Time

To understand the reasons behind the results, we have decomposed the execution times between computation and communication for the three implementations.

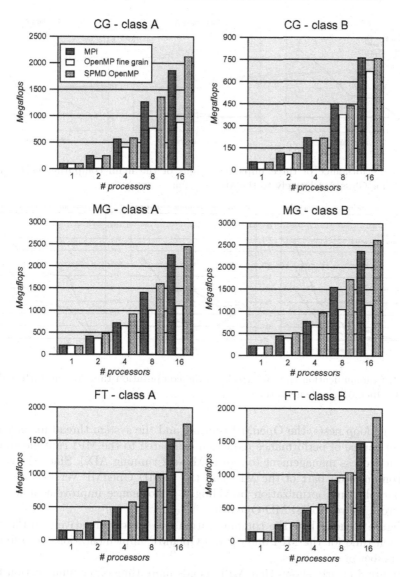

Fig. 2. Performance of three implementations of the NAS Benchmark on the IBM NH2 SMP node.

Figure 3 presents the time spent for all three approaches in the computation part of the program relatively to the MPI version.

This figure clearly shows that the fine grain version cannot compete with the other ones due to a bad utilization of the memory hierarchy. The only exception is MG, for which the communication due to the torus structure is the main cause of performance loss of OpenMP fine grain. Improving this part (like in the RWCP implementation) should solve this problem. The figure shows that with

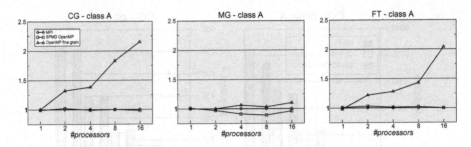

Fig. 3. Performance of three implementations of the main computing parts of the kernels for Class A relatively to the MPI version

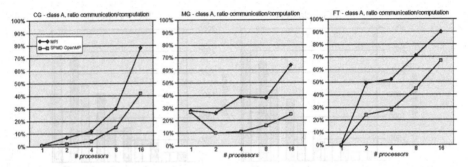

Fig. 4. Communication time relatively to the computation time for the MPI and the SPMD OpenMP implementations.

the same loop nests, the OpenMP runtime and the system thread management are not source of performance reduction compared to the MPI runtime and the system process management (on an IBM NH2 running AIX). Since the core of the computation part of the MPI and the SPMD OpenMP versions are quite comparable, any optimization providing a performance improvement for MPI may work as well for SPMD OpenMP.

Figure 4 shows how the communication times evolve relatively to the computation time for the MPI and SPMD OpenMP implementations according to the number of processors.

Figure 4 clearly shows that MPI spends more time on communication than SPMD OpenMP. This is obviously due to the faster implementation of global data operations for the latter. It is noticeable that the aggregation of optimized loop nests and optimized global operations actually leads to good performance for the SPMD OpenMP version. This is not obvious because one may expect to have some complex memory hierarchy effects such as unnecessary cache coherence operations between the two parts.

4   Conclusion and Future Works

High performance programming of shared memory machines is a difficult problem. This paper has provided two contributions: First, we have demonstrated that a SPMD parallelization of the NAS benchmark with OpenMP is more efficient than a loop level parallelization (up +138% of performance increase for CG Class A). One reason for that is the poor efficiency of the parallelized loop nests compared to the one of the SPMD version. This result on a SMP machine confirms some previous experiments on NUMA machines. However, while they focused on the problem of memory allocation and management, we demonstrate that even in the case where such problem is negligible, loop level parallization may not be able to achieve the performance of a well tuned MPI or SPMD OpenMP version. We hope that the current efforts of the community for improving the performance of loop level parallelization will contradict this result in the near future.

Second, the SPMD implementation reaches higher performance than MPI for most of applications and data sets conditions. The improvement increases for applications with low computation per communication ratio. In such cases, SPMD OpenMP can provide up to 28% performance improvement compared to MPI (MG class A). The reasons behind this result are 1) we basically keep the structure of the MPI application and especially the loop nests, 2) we can take advantage of the shared memory to provide global operations such as reduction and all-to-all with higher performance than their MPI counterparts.

Keeping the same loop nests as the one of the MPI version suggests that any improvement on the MPI version can be directly translated to the SPMD OpenMP version. Moreover, improving the computing part of a program will reduce the computation per communication ratio. This will further increase the advantage of SPMD OpenMP over MPI.

From these results, we can conclude that the SPMD OpenMP programming approach is a good alternative to MPI for SMP machines. We will pursue our tests on NUMA machines (IBM Regatta, SGI O3800) to check that this result is applicable to them.

The future works include: 1) designing and developing a high performance global operation library for SPMD OpenMP programming and 2) designing an hybrid MPI+OpenMP version of the NAS benchmarks based on the SPMD OpenMP approach.

Acknowledgements

The authors would like to thank Patrick Worley from ORNL for granting us access to the computers of this facility, and Alain Mango from CINES for his help on the proper usage of the IBM SP3. Franck Cappello would like to greatly thank Pr. Mitsuhisa Sato, Pr. Taisuke Boku, and Pr. Akira Ukowa of the Center For Computational Physics at the University of Tsukuba for having accepted him as a research fellow during the summer of 2001.

References

1. F. Cappello and D. Etiemble. Mpi versus mpi+openmp on ibm sp for the nas benchmarks. In *Supercomputing 2000: High-Performance Networking and Computing (SC2000)*, 2000. http://www.sc2000.org/proceedings/techpapr/index.htm.

2. P. Kloos amd F. Mathey and P. Blaise. Openmp and mpi programming with a cg algorithm. In *Proceedings of the Second European Workshop on OpenMP (EWOMP 2000), http://www.epcc.ed.ac.uk/ewomp2000/proceedings.html*, 2000.
3. A. Kneer. Industrial mixed openmp/mpi cfd application for practical use in free-surface flow calculations. In *International Workshop on OpenMP Applications and Tools, WOMPAT 2000, http://www.cs.uh.edu/wompat2000/Program.html*, 2000.
4. L. Smith and M. Bull. Development of mixed mode mpi/ openmp applications. In *International Workshop on OpenMP Applications and Tools, WOMPAT 2000, http://www.cs.uh.edu/wompat2000/Program.html*, 2000.
5. A. J. Wallcraft. Spmd openmp vs mpi for ocean models. In *First European Workshop on OpenMP - EWOMP'99, http://www.it.lth.se/ewomp99/programme.html*, 2000.
6. B. Armstrong, S. Wook Kim, and R. Eigenmann. Quantofying differences between openmp and mpi using a large-scale application suite. In *Proceedings of the 3rd International Symposium on High Performance Computing (ISHPC2000 - WOMPEI2000), http://research.ac.upc.es/wompei/program.html*, 2000.
7. B. Chapman, A. Patil, and A. Prabhakar. Performance oriented programming for numa architectures. In *LNCS 2104, International Workshop on OpenMP Applications and Tools, WOMPAT 2001, West Lafayette, IN, USA*, 2001.

Parallel Iterative Solvers for Unstructured Grids Using an OpenMP/MPI Hybrid Programming Model for the GeoFEM Platform on SMP Cluster Architectures

Kengo Nakajima[1] and Hiroshi Okuda[2]

[1] Research Organization for Information Science and Technology (RIST)
2-2-54 Naka-Meguro, Meguro-ku, Tokyo 153-0061, Japan
nakajima@tokyo.rist.or.jp
[2] Department of Quantum Engineering and System Sciences
The University of Tokyo
7-3-1 Hongo, Bunkyo-ku, Tokyo 113-8656, Japan
okuda@garlic.q.t.u-tokyo.ac.jp

Abstract. An efficient parallel iterative method for unstructured grids developed by the authors for SMP cluster architectures on the GeoFEM platform is presented. The method is based on a 3-level hybrid parallel programming model, including message passing for inter-SMP node communication, loop directives by OpenMP for intra-SMP node parallelization and vectorization for each processing element (PE). Simple 3D elastic linear problems with more than 8×10^8 DOF have been solved by 3×3 block ICCG(0) with additive Schwarz domain decomposition and PDJDS/CM-RCM reordering on 128 SMP nodes of Hitachi SR8000/MPP parallel computer, achieving performance of 335.2 GFLOPS. The PDJDS/CM-RCM reordering method provides excellent vector and parallel performance in SMP nodes.

1 Introduction

In recent years, shared memory symmetric multiprocessor (SMP) cluster architecture has become very popular for massively parallel computers. For example, all Accelerated Strategic Computing Initiative (ASCI) machines have adopted this type of architecture [1]. In 1997, the Science and Technology Agency of Japan (now, the Ministry of Education, Culture, Sports, Science and Technology, Japan) began a 5-year project to develop a new supercomputer, the Earth Simulator [2]. The goal is the development of both hardware and software for earth science simulations. The Earth Simulator has SMP cluster architecture and consists of 640 SMP nodes, where each SMP node consists of 8 vector processors. The present study was conducted as part of the research toward developing a parallel finite-element platform for solid earth simulation, named GeoFEM [3]. In this architecture, *loop directives + message passing* style *hybrid* programming model appears to be very effective when message passing such as MPI [4] is used in inter-SMP node communication, and when intra-SMP node parallelization is guided by loop directives such as OpenMP [5]. A significant amount of research on this issue has been conducted in recent 2 or 3 years [6, 7], but most studies

H. Zima et al. (Eds.): ISHPC 2002, LNCS 2327, pp. 437–448, 2002.

have focused on applications involving structured grids such as the NAS Parallel Benchmarks (NPB) [8], with very few examples for unstructured grids.

In this study, parallel iterative methods on unstructured grids for SMP cluster architecture have been developed for Hitachi SR8000 [9] parallel computers at the University of Tokyo [10]. A parallel programming model with the following 3-level hierarchy has been developed:

- Inter-SMP node   MPI
- Intra-SMP node   OpenMP for parallelization
- Individual PE   Compiler directives for (pseudo) vectorization [9]

The entire domain is partitioned into distributed local data sets [3], and each partition is assigned to one SMP node (Fig. 1). In order to achieve efficient parallel/vector computation for applications with unstructured grids, the following 3 issues are critical:

- Local operation and no global dependency
- Continuous memory access
- Sufficiently long loops

A special reordering technique proposed by Washio et. al. [11, 12] has been integrated with parallel iterative solvers with localized preconditioning developed in the Geo-FEM project [3] in order to attain local operation, no global dependency, continuous memory access and sufficiently long loops.

In the following part of this paper, we give an overview of GeoFEM's parallel iterative solvers, local data structure and reordering techniques for parallel and vector computation on SMP nodes, and present the results for an application to 3D solid mechanics.

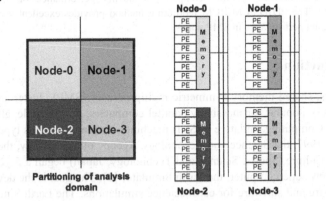

Fig. 1. Parallel FEM computation on SMP cluster architecture. Each partition corresponds to an SMP node.

2   Parallel Iterative Solvers in GeoFEM

The incomplete lower-upper (ILU)/Cholesky (IC) factorization method is one of the most popular preconditioning techniques for accelerating the convergence of Krylov iterative methods. Among ILU preconditioners, ILU(0), which does not allow fill-in beyond the original non-zero pattern, is the most commonly used. Backward/forward substitution (BFS) is repeated in each iteration. BFS requires global data dependency, and this type of operation is not suitable for parallel processing in which the locality is of utmost importance. Most preconditioned iterative processes are a combination of the following:

- Matrix-vector products
- Inner dot products
- DAXPY ($\alpha x + y$) operations [13] and vector scaling
- Preconditioning operations

The first 3 operations can be parallelized relatively easily [13]. In general, preconditioning operations (i.e., BFS) represent almost 50 % of the total computation if ILU(0) is implemented as the preconditioner. Therefore, a high degree of parallelization is essential for the BFS operation.

Localized ILU(0) is a *pseudo* ILU(0) preconditioner that is suitable for parallel processors. This method is not a *global* method, rather, it is a *local* method on each processor or domain. The ILU(0) operation is performed for each processor by zeroing out matrix components located outside the processor domain. This *localized* ILU(0) provides data locality on each processor and good parallelization because no inter-processor communications occur during ILU(0) operation.

However, localized ILU(0) is not as powerful as the global preconditioner. Generally, the convergence rate worsens as the number of processors and domains increases [14, 15]. At the critical end, if the number of processors is equal to the number of the degrees of freedom (DOF), this method performs identically to diagonal scaling. In order to stabilize localized ILU(0) preconditioning, additive Schwarz domain decomposition (ASDD) for overlapped regions [16] has been introduced.

3   Reordering Methods for Parallel/Vector Performance Using SMP Nodes

As shown in Fig. 1, the entire domain is partitioned into local data sets and each local data set corresponds to one SMP node.

3.1   Cyclic Multicolor – Reverse Cuthil McKee Reordering

In order to achieve efficient parallel/vector computation for applications with unstructured grids, the following 3 issues are critical:

- Local operations and no global dependency
- Continuous memory access
- Sufficiently long loops

For unstructured grids, in which data and memory access patterns are very irregular, the reordering technique is very effective for achieving highly parallel and vector performance. The popular reordering methods are hyperplane/reverse Cuthil-McKee (RCM) and multicoloring [17]. In both methods, elements located on the same hyperplane (or classified in the same color) are independent. Therefore, parallel operation is possible for the elements in the same hyperplane/color and the number of elements in the same hyperplane/color should be as large as possible in order to obtain high granularity for parallel computation or sufficiently large loop length for vectorization.

Hyperplane/RCM (Fig. 2(a)) reordering provides fast convergence of IC/ILU-preconditioned Krylov iterative solvers, yet with irregular hyperplane size. For example in Fig. 2(a), the 1st hyperplane is of size 1, while the 8th hyperplane is of size 8. In contrast, multicoloring provides a uniform element number in each color (Fig. 2(b)). However, it is widely known that the convergence of IC/ILU-preconditioned Krylov iterative solvers with multicolor ordering is rather slow. Convergence can be improved by increasing the number of colors, but this reduces the number of elements in each color.

The solution for this trade-off is cyclic multicoloring (CM) on hyperplane/RCM [11]. In this method, the hyperplanes are renumbered in a cyclic manner. Figure 2(c) shows an example of CM-RCM reordering. In this case, there are 4 colors; the 1st, 5th, 9th and 13th hyperplanes in Fig. 2(a) are classified into the 1st color. There are 16 elements in each color. In CM-RCM, the number of colors should be large enough to ensure that elements in the same color are independent.

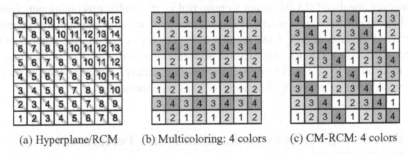

(a) Hyperplane/RCM   (b) Multicoloring: 4 colors   (c) CM-RCM: 4 colors

Fig. 2. Example of hyperplane/RCM, multicoloring and CM-RCM reordering for 2D geometry

3.2   DJDS Reordering

The compressed row storage (CRS) [13] matrix storage format is highly memory-efficient, however the innermost loop is relatively short due to matrix-vector operations as follows:

```
do i= 1, N
  do j= 1, NU(i)
    (operations)
    F(i)= F(i) + A(k1)*X(k2)
  enddo
enddo
```

The following loop exchange is then effective for obtaining a sufficiently long innermost loop:

```
do j= 1, NUmax
  do i= 1, N
    (operations)
    F(i)= F(i) + A(k1)*X(k2)
  enddo
enddo
```

Descending-order jagged diagonal storage (DJDS) [12] is suitable for this type of operation and involves permuting rows into an order of decreasing number of non-zeros, as in Fig. 3(a). As elements on the same hyperplane are independent, performing this permutation inside a hyperplane does not affect results. Thus, a 1D array of matrix coefficients with continuous memory access can be obtained, as shown in Fig. 3(b).

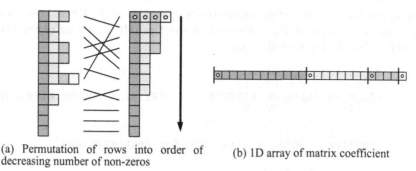

(a) Permutation of rows into order of decreasing number of non-zeros

(b) 1D array of matrix coefficient

Fig. 3. DJDS reordering for efficient vector/parallel processing

3.3   Distribution over SMP Nodes: Parallel DJDS Reordering

The 1D array of matrix coefficients with continuous memory access is suitable for both parallel and vector computing. The loops for this type of array are easily distributed to each PE in an SMP node via loop directives. In order to balance the computational load across PEs in the SMP node, the DJDS array should be reordered again in cyclic manner. The procedure for this reordering, called parallel DJDS (PDJDS) is described in Fig. 4:

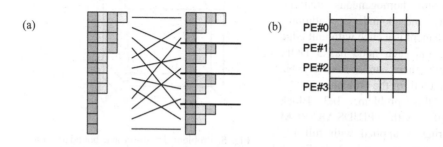

Fig. 4. PDJDS reordering for an SMP node: Example with 4 PEs per SMP node (a) Cyclic reordering (b) 1D array assigned to each PE after reordering and load-balancing.

3.4   Summary of Reordering Methods

The reordering procedures for increasing parallel/vector performance of the SMP cluster architecture described in this section are summarized as follows:

(1) RCM reordering on the original local matrix for independent sets.
(2) CM reordering to obtain loops whose length is sufficiently long and uniform.
(3) DJDS reordering for efficient vector processing, producing 1D arrays of coefficients with continuous memory access.
(4) Cyclic reordering for load-balancing among PEs on an SMP node.
(5) PDJDS/CM-RCM reordering is complete.

The typical loop structure of the matrix-vector operations for PDJDS /CM-RCM reordered matrices based on the pseudo-vector and OpenMP directives of the Hitachi SR8000 is described in the following:

```
do col= 1, COLORtot
  do j= 1, NUmax(col)
!$OMP PARALLEL DO PRIVATE (iS, iE, i) : OpenMP Directive
    do pe= 1, SMP_PE_tot
      iS= NstartU(col,j,pe)
      iE= NendU  (col,j,pe)
*VOPTION, INDEP  : Directive for Vectorization
      do i= iS, iE
        (operations)
      enddo
    enddo
  enddo
enddo
```

4   Examples

The proposed methods were applied to large-scale 3D solid mechanics example cases, as described in Fig. 5, which represent linear elastic problems with homogeneous material property and boundary conditions. Each element is a cube with unit edge length, and each node has 3 DOF, therefore there are $3 \times Nx \times Ny \times Nz$ DOF in total for the problem.

For this problem, 3×3 Block ICCG(0) with PDJDS/MC-RCM reordering is applied with full LU factorization for each 3×3 diagonal block. One ASDD operation is applied

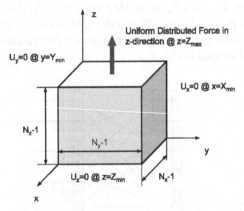

Fig. 5. Problem definition and boundary conditions for 3D solid mechanics example cases.

to each iteration. Vector performance was evaluated on an NEC SX-4 (JAERI/CCSE) and a Hitachi SR2201 (University of Tokyo), and SMP parallel performance was tested on the Hitachi SR8000. In each case, the number of colors was set to 99, corresponding to an average vector length of (total number of FEM nodes) / (number of PEs or SMP nodes × 99 × NPE(number of PEs on each SMP node)) where NPE = 1 for SX-4 and SR2201, and NPE=8 for SR8000.

4.1 Vector/Vector Parallel Performance

Before computation on the Hitachi SR8000, vector performance was evaluated on an NEC SX-4 and Hitachi SR2201. Parallelization for SMP nodes was not applied; matrices are therefore reordered by DJDS/CM-RCM. An example including 41^3 nodes (206,763 DOF) was solved using 1 processor of the NEC SX-4, achieving 969 MFLOPS performance for the linear solver component with peak performance of 2 GFLOPS (48.5% of peak performance) [3].

Evaluations for parallel computing were conducted on the Hitachi SR2201 at the University of Tokyo [10]. Pseudo vectorization can also be applied on the SR2201, and each PE performs as a vector processor. The largest case was 196,608,000 DOF on 1024 PEs. A performance of 68.7 GFLOPS was achieved. Total peak performance of the system was 300 GFLOPS with 1024 PEs [3]. The 68.7 GFLOPS performance then corresponds to 22.9% of the peak performance. The work ratio is higher than 90% if the problem size for 1 PE is sufficiently large, more than 24,000 DOF in this case [3].

4.2 SMP Parallel Performance

The following cases were tested on the Hitachi SR8000 at the University of Tokyo. 2 systems were used. One is Hitachi SR8000/128 (8 PEs, 8GFLOPS peak performance and 8GB memory for each node. 128 nodes (1024 PEs), 1.0TFLOPS peak performance and 1.0TB memory for total system). The other is Hitachi SR8000/MPP (8PEs, 14.4GFLOPS peak performance and 16GB memory for each node. 128 nodes (1024 PEs), 1.8TFLOPS peak performance and 2.0TB memory for total system):

(1) The increase in speed for fixed problem size (3×128^3=6,291,456 DOF) using between 1 and 8 SMP nodes (SR8000/MPP).
(2) Communication/synchronization overhead for intra-SMP node parallelization for various problem sizes using 1 SMP node (SR8000/MPP).
(3) Effect of matrix storage and reordering for various problem sizes using 1 SMP node (SR8000/128).
(4) Performance evaluation for various problem sizes using 1 to 128 SMP nodes (SR8000/MPP).

Figure 6 shows the results of (1). In this example, the size of the entire problem was fixed at 3×128^3 (6,291,456) DOF, and the increase in speed was evaluated for 1 to 8 SMP nodes. The number of iterations for convergence ($\varepsilon = 10^{-8}$) was 333 (1-node), 337 (2-nodes), 338 (4-nodes), and 341 (8-nodes), indicating that the number of iterations remains almost constant as the number of nodes increases. This is due to the ASDD. Speedup rate at 8 SMP nodes was 6.49, which corresponds to the 81.1% of the linear (ideal) speedup. Speedup effect for many nodes is worse than the ideal speedup due to smaller problem size per node.

444   Kengo Nakajima and Hiroshi Okuda

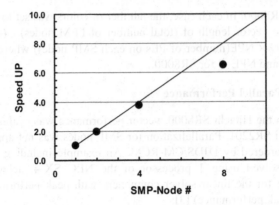

Fig. 6. Relationship between number of SMP nodes and the speedup on the Hitachi SR8000/MPP with DJDS/CM-RCM reordering. The total problem size is fixed at 3×128³ (6,291,456) DOF. Speedup rate for 8 SMP nodes is 6.49.

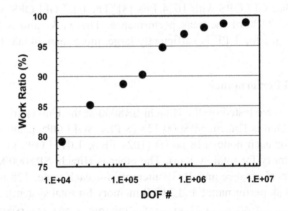

Fig. 7. Work ratio for various problem sizes on the Hitachi SR8000/MPP with 1 SMP node woth PDJDS/CM-RCM reordering. The work ratio is more than 90% if the problem size is 3×40³ (192,000) DOF (24,000 DOF/PE).

Figure 7 shows the results of (2). Communication/synchronization overhead occurs for parallel processing in each SMP node. The work ratio was measured for various problem sizes from 3×16³ (12,288) DOF to 3×128³ (6,291,456) DOF on 1 SMP node. Measurements were made using the XCLOCK system subroutine of the Hitachi compiler [9]. The results show that overhead is more than 30% for the smallest problem size, and less than 10% for the problem size of 3×40³ (192,000) DOF (24,000 DOF/PE) and less than 5% for 3×64³ (786,432) DOF (98,304 DOF/PE). According to these results, communication/synchronization overhead for intra-SMP node communication is almost negligible if the problem size is sufficiently large. Figure 8 shows the results of (3), demonstrating the effect of reordering. In this case, the following 3 cases were compared:

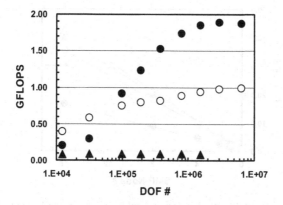

Fig. 8. Effect of coefficient matrix storage method and reordering for various problem sizes on the Hitachi SR8000/128 with 1 SMP node. The performance of the solver without reordering is very low due to synchronization overhead during forward/backward substitution for the IC factorization (BLACK Circles: PDJDS/CM-RCM, WHITE Circles: PDCRS/CM-RCM, BLACK Triangles: CRS no reordering).

- PDJDS/CM-RCM reordering
- Parallel descending-order compressed row storage (PDCRS) /CM-RCM reordering
- CRS without reordering

PDCRS/CM-RCM reordering is identical to PDJDS/CM-RCM except for the storage of matrices in a CRS manner after permutation of rows into the order of decreasing number of non-zeros, where the length of the innermost loop is shorter than that for PDJDS. The elapsed execution time was measured for various problem sizes from 3×16^3 (12,288) DOF to 3×128^3 (6,291,456) DOF on 1 SMP node. PDCRS is faster than PDJDS for smaller problems, but PDJDS outperforms PDCRS for larger problems as a result of pseudo vectorization. The effect of vector length is much more significant than that of comm./sync. overhead in SMP node shown in Fig.7. The cases without reordering exhibit very poor performance. Parallel computation is impossible for forward/backward substitution (FBS) in the IC factorization process even in the simple geometry examined in this study. This FBS process represents about 50% of the total computation time. If this process is not parallelized, the performance reaches less than 20% of that with reordering. The number of iterations for convergence is also larger for cases without reordering.

Figure 9 and 10 show the results of (4). The problem size is fixed for one SMP node and the number of nodes was varied between 1 and 128. The largest problem size was $128\times3\times128^3$ (805,306,368) DOF, for which the performance was about 335.2 GFLOPS, corresponding to 18.6% of the total peak performance of the 128 SMP nodes. Figure 9 and 10 show that the performance at small problem sizes per SMP node (3×32^3=98,304 DOF), was almost 50% of that for the larger problems. Figure 11 shows the results with various problem sizes on 128 SMP nodes. Parallel work ratio among SMP nodes by MPI is more than 95% if problem size is sufficiently large.

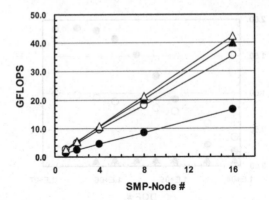

Fig. 9. Problem size and GFLOPS rate for various problem sizes on the Hitachi SR8000/MPP. Problem Size/PE is fixed. Largest case is 100,663,296 DOF on 16 SMP nodes (128 PEs). Maximum performance is 42.4 GFLOPS (Peak Performance = 230.4 GFLOPS). (BLACK Circles: $3 \times 32^3 = 98,304$ DOF/SMP node, WHITE Circles: 3×64^3, BLACK Triangles: 3×80^3, WHITE Triangles: 3×128^3).

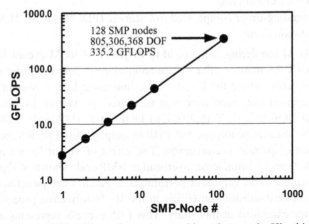

Fig. 10. Problem size and GFLOPS rate for various problem sizes on the Hitachi SR8000/MPP. Problem Size/PE is fixed as 6,291,456 DOF. Largest case is 805,306,368 DOF on 128 SMP nodes (1024 PEs). Maximum performance is 335.2 GFLOPS (Peak Performance = 1.8 TFLOPS)

5   Conclusion

In this study, an efficient parallel iterative method for unstructured grids was developed for SMP cluster architectures on the GeoFEM platform using *loop directives + message passing* type parallel programming model with the following 3 level hierarchy:

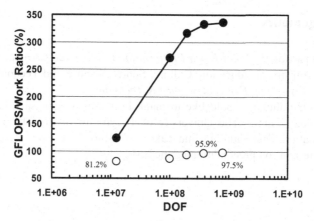

Fig. 11. Problem size and GFLOPS rate/iterations for convergence ($\varepsilon=10^{-8}$) in various problem sizes on 128 SMP nodes of the Hitachi SR8000/MPP. The largest case is 805,306,368 DOF on 128 SMP nodes (1024 PEs). Maximum performance was 335.2 GFLOPS (Peak Performance = 1.8 TFLOPS). (BLACK Circles: GFLOPS rate, WHITE Circles: Parallel work ratio (inter SMP nodes)). Parallel work ratio is more than 95 % if problem size is sufficiently large.

- Inter-SMP node   MPI
- Intra-SMP node   OpenMP for parallelization
- Individual PE   Compiler directives for (pseudo) vectorization

Simple 3D elastic linear problems with more than 8×10^8 DOF were solved by 3×3 block ICCG(0) with additive Schwarz domain decomposition and PDJDS/CM-RCM reordering on 128 SMP nodes of a Hitachi SR8000/MPP, achieving performance of 335.2 GFLOPS (18.6% of peak performance). PDJDS/CM-RCM reordering provides excellent vector and parallel performance in SMP nodes. Without reordering, parallel processing of forward/backward substitution in IC/ILU factorization was impossible due to global data dependency even in the simple examples in this study. Communication/synchronization overhead in a SMP node is less than 10% if the problem size is 3×40^3 (192,000) DOF (24,000 DOF/PE) or larger. The parallel work ratio for inter SMP node communication was found to be higher than 95% if the problem size for each node was sufficiently large. The proposed method was also tested on an NEC SX-4, achieving performance of 969 MFLOPS (48.5% of peak performance) for a problem with 2×10^5 DOF using a single processor and on a Hitachi SR2201 with 68.7 GFLOPS (22.9% of peak performance) for a problem with 1.97×10^8 DOF using 1024 processors.

In this study, an effective hybrid parallel programming model for SMP cluster architecture was developed. Future study includes:

- porting the methods to other SMP cluster hardware, and
- applying the methods to real-world problems with more complicated geometries.

Acknowledgements

This study is a part of *the Solid Earth Platform for Large-Scale Computation* project funded by the Ministry of Education, Culture, Sports, Science and Technology, Japan through *Special Promoting Funds of Science & Technology*.

Furthermore the authors would like to thank Professor Yasumasa Kanada (Information Technology Center, The University of Tokyo) for discussions on high performance computing, Drs. Shun Doi and Takumi Washio (NEC C&C Research Laboratory) for discussions on preconditioning methods.

References

1. Accelerated Strategic Computing Initiative (ASCI) Web Site: http://www.llnl.gov/asci/
2. Earth Simulator Research and Development Center Web Site: http://www.es.jamstec.go.jp/
3. GeoFEM Web Site: http://geofem.tokyo.rist.or.jp/
4. MPI Web Site: http://www.mpi.org/
5. OpenMP Web Site: http//www.openmp.org/
6. Falgout, R. and Jones, J.: "Multigrid on Massively Parallel Architectures", *Sixth European Multigrid Conference*, Ghent, Belgium, September 27-30, 1999.
7. Cappelo, F. and Etiemble, D.:"MPI versus MPI+OpenMP on the IBM SP for the NAS Benchmarks", *SC2000 Technical Paper*, Dallas, Texas, 2000.
8. NPB (NAS Parallel Benchmarks) Web Site: http://www.nas.nasa.gov/Research/Software/swdescription.html#NPB/
9. Hitachi SR8000 Web Site: http://www.hitachi.co.jp/Prod/comp/hpc/foruser/sr8000/
10. Information Technology Center, The University of Tokyo Web Site: http://www.cc.u-tokyo.ac.jp/
11. Washio, T., Maruyama, K., Osoda, T., Shimizu, F. and Doi, S.: "Blocking and reordering to achieve highly parallel robust ILU preconditioners", *RIKEN Symposium on Linear Algebra and its Applications*, The Institute of Physical and Chemical Research, 1999, pp.42-49.
12. Washio, T., Maruyama, K., Osoda, T., Shimizu, F. and Doi, S.: "Efficient implementations of block sparse matrix operations on shared memory vector machines", *SNA2000: The Fourth International Conference on Supercomputing in Nuclear Applications*, 2000.
13. Barrett, R., Bery, M., Chan, T.F., Donato, J., Dongarra, J.J., Eijkhout, V., Pozo, R., Romine, C. and van der Vorst, H.: *Templates for the Solution of Linear Systems: Building Blocks for Iterative Methods*, SIAM, 1994.
14. Nakajima, K., Okuda, H.: "Parallel Iterative Solvers with Localized ILU Preconditioning for Unstructured Grids on Workstation Clusters", *International Journal for Computational Fluid Dynamics* 12 (1999) pp.315-322
15. Garatani, K., Nakamura, H., Okuda, H., Yagawa, G.: "GeoFEM: High Performance Parallel FEM for Solid Earth", *HPCN Europe 1999*, Amsterdam, The Netherlands, *Lecture Notes in Computer Science* 1593 (1999) pp.133-140
16. Smith, B., Bjørstad, P. and Gropp, W.: *Domain Decomposition, Parallel Multilevel Methods for Elliptic Partial Differential Equations*, Cambridge Press, 1996.
17. Saad, Y.: *Iterative Methods for Sparse Linear Systems*, PWS Publishing Company, 1996.

A Parallel Computing Model
for the Acceleration of a Finite Element Software

Pierre de Montleau[1], Jose Maria Cela[2],
Serge Moto Mpong[1], and André Godinass[1]

[1] University of Liège, Département MSM, Bâtiment B52/3, 1 chemin des Chevreuils,
4000 Liège 1, Belgium
p.demontleau@ulg.ac.be
[2] Centro Europeo de Paralelismo de Barcelona (CEPBA), Universidad Politechnica
de Catalua, c/ Jordi Girona, Módulo D6, 08034 Barcelona, Spain
cela@ac.upc.es

Abstract. This paper presents the OpenMP parallelization of the Finite
Element code LAGAMINE. It is a non-linear great deformations code for
solid mechanics. The parallelization approach uses coarse grain: for the
element loop, to build the stiffness matrix and for the direct solver, to
compute the LU factorisation. Application is proposed to a stamping
simulation.

1 Introduction

The finite element code LAGAMINE is developed by the department MSM of
the University of Liège since 1982. It is a solid, non linear, great deformations
code that has been adapted to numerous finite elements and constitutive laws.
Its modular architecture has been efficient to allow improvements towards applications very far from the initial goal: rolling simulation, soil simulation. The
present researches on micro-macro modelling increase the requirement of CPU
time. Even with 5000 DOF, it is now a strong limitation because of the complexity of the constitutive laws. Obviously, large size simulations with 100000 Degrees
of Freedom cannot reasonably be performed. This is why we are concerned by
the acceleration of the code.

Because of the architecture of the available parallel computers, a Silicon
Graphics SG3800 with 64 processors and an Alpha Server ES40 from Compaq
with four processors, which are both shared memory computers, we decided to
use OpenMP protocol. Moreover, OpenMP appears easiest to use. Using MPI
protocol for example requires a priori an important effort for the partitioning of
the mesh and for the parallel associated resolution.

In the non linear process of LAGAMINE, more or less 98% of the CPU
time is consumed on one hand, in the element loop to integrate the constitutive
equations and to compute the tangent stiffness matrix, on the other hand to
solve the linear system; the balance between these two parts changing seriously
with respect to the kind of problem. Also, the first goal is to parralelize the

H. Zima et al. (Eds.): ISHPC 2002, LNCS 2327, pp. 449–456, 2002.
© Springer-Verlag Berlin Heidelberg 2002

element loop including OpenMP instructions in the program and thereafter the direct solver. The parallelization of a direct solver is a complex problem while the parallelization of an iterative one is generally quite simpler. However, in regards to the size of the actual problems (5 or 6000 Degrees of Freedom) and considering they won't be greater than 100000 DOF in a nearest future, we chose to keep a direct solver. Moreover, we had the opportunity to use the parallel one developed by J.M. Cela of the CEPBA department. In the following, we describe briefly the equations solved, then we explain the settled parallelization and finally we propose different examples of speedup performed, either for the element loop or for the solver. Unfortunately, speedups for the full parallel program are not available at the time where this paper is written.

This work has been partially carried out in the frame of an European Community (EC) project, in the "Centro Europeo de Paralelismo de Barcelona" (CEPBA). A very useful tool called PARAVER developed by this centre has been used. This software provides a graphic interface that shows the work of each thread and OpenMP events. Moreover, time spent in different areas of the program can be performed. This tools allows one to see very quickly the regions where time is consumed and the tasks balance between the different threads.

1.1   Description of the Code

We present here the very general equations solved by the program in the simplest case. Let consider a domain Ω. If $\underline{u}$ is the displacement, $\underline{v}$ the velocity and $\underline{\sigma}$ the Cauchy stress tensor, the equilibrium equations are

$$\nabla \underline{\sigma} = \underline{f} \qquad \text{on } \Omega \tag{1}$$

$$\underline{\sigma}\,\underline{n} = \underline{T} \qquad \text{on } \partial_T \Omega \tag{2}$$

$$\underline{u} = \underline{u}_d \qquad \text{on } \partial_u \Omega \tag{3}$$

where $\underline{f}$ are the body forces, $\underline{T}$ is the applied loading on the boundary $\partial_T \Omega$ of Ω and $\underline{u}_d$ the imposed displacement on the boundary $\partial_u \Omega$ of Ω. Let call $\underline{\dot{\varepsilon}}$ the strain rate, with $\underline{\dot{\varepsilon}} = 1/2(\nabla \underline{v} + \nabla \underline{v}^T)$. The constitutive equations are

$$\underline{\dot{\sigma}} = g\left(\underline{\sigma}, \underline{\dot{\varepsilon}}\right) \tag{4}$$

where g is a non linear function. For the computation of such a non-linear system, an incremental method is used. The final load (or the time) is split in steps and the load is applied step by step. At each step, assuming an initial value for the strain rate, the constitutive equations (4) are integrated. Following a Newton-Raphson iterative algorithm the assumed strain rate is corrected until the equilibrium (1) is attempt. The load is increased up to the final load.

In the finite element method, the domain Ω is discretized in elements and the unknown of the problem is the vector composed by the N displacements and temperature of the nodes. The Newton Raphston method is then written in its discrete form

$$\underline{F}_{ext} = \underline{F}_{int} + \underline{\underline{K}}\,\delta\underline{u} \tag{5}$$

where $\underline{\underline{K}}$ is the tangent matrix (computed by perturbation for example), $\underline{F}_{ext}$ is the N vector of the nodal equivalent external forces, $\underline{F}_{int}$ the N vector of the nodal equivalent internal forces computed from the integration law. In (5), $\delta\underline{u}$ is the displacement correction, computed at each iteration by solving the linear system (5) of dimension N. Convergence is attempt as far as $\underline{F}_{ext} = \underline{F}_{int}$ for the wanted accuracy. The overall structure of the code is described on the Figure 1 below.

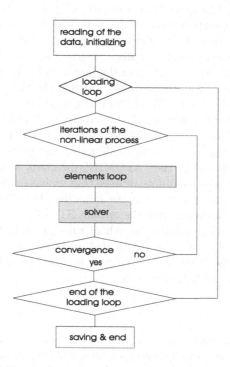

Fig. 1. Structure of the code LAGAMINE

The grey squares are the parallelized sections of the code where the most of time is spent.

2   Setting Parallelization

In view to parralelize using OpenMP, we have to change the sequential order of the program. That requires a dependency analysis and eventually modifications that preserve the semantic, but exhibit independent instructions. A priori, they are few modifications in the original program, except in the linear system resolution algorithm.

2.1   The Assembling Loop

The assembling loop is a loop over all the elements of the mesh. For each element, the associated law is called and the constitutive equations are integrated to compute the internal forces and the tangent matrix (see equation (5)). Local force vector and stiffness matrix of each element are also assembled in the global force and stiffness vector in this loop. The element loop is called at each iteration of the non linear process (see Fig. 1). The parallelization method is simply to build a PARALLEL DO loop. Each iteration or pack of iterations (one iteration is here the computation required for one element) are shared between the available processors. In fact, the most important difficulty was the analysis of the dependencies. Indeed, the loop calls a choice of one hundred elements, each of them calling specific laws. Therefore the number of variables is large. In this parallel zone, the status of all the variables is defined by the programmer. We use the SHARED, PRIVATE, and FIRSTPRIVATE attributes. The most important modifications have concerned variables as counters that lead clearly to sequential sequences or arrays that had both PRIVATE and SHARED components, so we had to create new variables. For the building of the global matrix, we use an OpenMP critical instruction, the "ATOMIC" directive, to force the system to authorize only one thread to update at once that global (shared) variable. Thus, those instructions are not executed at the same time by different threads. Using OpenMP, tasks can be shared with the options STATIC or DYNAMIC. For our problems, elements using different length of time are called. Thus we use the option DYNAMIC(N). Iterations are divided in the given size N packs. As soon as a thread finishes the iterations, a new pack is allocated to it. We obtained the best results for values of N from 1 to 5.

2.2   The Solver

The LU factorisation used to solve the linear system seems initially completely sequential. In fact, it is possible to reorder the equation numbering and following, to exhibit parallel sequences. In the frame of the European research project, we had the opportunity to get such a parallel linear solver, issued from the library CAESAR of the CEPBA department (see Lazaretta and al. 2001 [3]). This solver performs meanly the four steps described thereafter.

- The first point concerns the matrix storage. In the previous code the skyline storage was used. However, for sparse matrices, the Morse storage is generally preferred: only non zero elements of the matrix are stored. Thus, the necessary storage space can be hardly reduced. However the addressee of the matrix elements is more complex (indirect addressee).
- In the second step, the fill-in of the matrix is reduced as much as possible. That means that the equation numbering is modified in view to compute as less operations as possible during the factorisation. For this, we use the METIS software based upon partition graph of the matrix. See below an example of renumbering performed by METIS for a 12000 equations elastic problem.

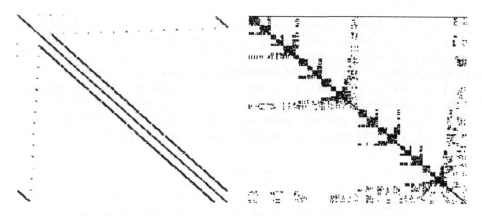

Fig. 2. Sparse matrix aspect before and after METIS renumbering

- The goal of the third point is to improve cache effects. It consists in a post renumbering that exactly maintains the fill-in of the matrix obtained by METIS, but that compresses as much as possible the nodes of the matrix. It may avoid miss cache effects during the factorisation where a work array of the size of the problem is used. Moreover, this post renumbering is a vital step in case of out of core, for disk access reasons. It is not the case for the size of the present problems performed.
- The symbolic factorisation is then performed. It gives the L and U matrix structure. The external loop of the LU algorithm is then parallelised with openMP. Thanks to Metis, a part of the iterations of this loop are independent and can be directely computed in parallel. However, data dependencies remain in this loop, and a synchronisation mechanism between threads is needed. Unfortunately openMP does not provide at the present a wait-signal mechanism. We have used an active wait mechanism. An array called "factorised" of length equal to the matrix dimension is used to mark if a row/column is factorised or not. Because a row/column is factorised by a single thread, it is guaranteed that only one thread is writing on an element of this array. When a thread needs to use the data of the factorised i-th row/column, it waits asking in a loop until "factorised(i)" is true. This synchronisation mechanism can introduce a strong overhead in the parallel execution. In order to minimise this overhead, the external loop of the factorisation is modified. We perform a domain decomposition of the factorisation elimination tree. In this way we obtain not connected sets of rows/columns with the same computational load. This sets can be factorised without any synchronisation between threads. All the synchronisations are concentrated at the end of the factorisation process. The time spent in the synchronisation mechanism depends on the matrix structure, i.e. the structure of the elimination tree. We have observed that for our matrices this time is 1% or 2% of the total factorisation time.

In most cases all the previous steps are performed one for all at the beginning of the process, because the matrix keeps the same structure during all the process. At each step or iteration, only the "numerical" factorisation applying the parallel process performed above and the triangular resolutions are performed.

The solver has been tested for a 6000 equations problem on an IBM machine with 128 processors for technical reasons. Results are described on the Figure 2 below. Note that the system is quite small, but results are already satisfactory. Better efficiency can reasonably be expected for larger size problems.

Table 1.

Number of CPU	1	2	4	8	12
Speedup	1.00	1.86	3.36	5.91	7.61
Efficiency (%)	100	93	84	73	63

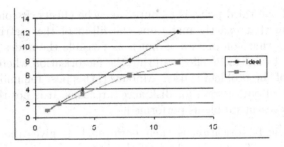

3   Applications

3.1   Deep Drawing with Evolution of the Structure

We present here the results obtained for the deep drawing simulation example, realised as part of the present PhD thesis of Laurent Duchne in the MSM Laboratory. The material is steel with a small elasticity limit. The mesh of this part has 4020 finite elements nodes and 1504 volume elements, for a total of 7000 degrees of freedom. The simulation is a three dimension mechanic computation, which consists in the deformation of the initial steel part with a punch until the wanted final shape. The material is characterised by the elasto-plastic Minty-law that includes recrystallization. The computation, after 10 days on only one processor of the Alpha Server, never succeed completely, generally because of hardware crash. On a Silicon Graphix SG3800 machine, with 8 processors, the process succeeded and needed 2 days of computation with a speed-up over than four.

More precisely, the speedup and the efficiency have been performed only for the element loop because the parallel solver was not available at this time. Any-

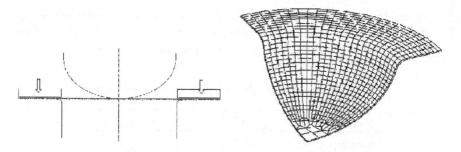

Fig. 3. Experimental set-up of deep drawing

way, the skyline solver is quite efficient in this very special case because of the structure shape.

Table 2. Speedup for the element loop, performed with SG3800

Number of CPU	1	2	4	8	16	32
Speedup	1.	2.02	4.06	8.06	15.36	29.31
Efficiency (%)	100	101	101	101	96	91

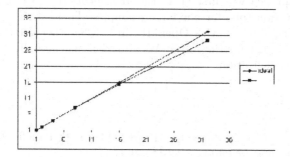

The computation length have been measured for the routine calling the loop. Results are shown on the Figure 2. A time computation error (more or less 5%) or cache effects can explain the speedup over the ideal one.

3.2 Upsetting of an Elastic Cube

The second example concerns the upsetting of a cube. The mesh of this cube has 4096 finite elements nodes and 3375 volume elements, for a total of 11500 degrees of freedom. The behaviour of the material is characterised by a linear elastic law. In this example the time spent in the element loop is quite negligible compared with the time spent in the solver, owing to the elastic law. The computation has been done on one processor, two processors, 4 processors and eight processors of a Silicon Graphix SG3800 computer. The efficiency for the element loop is described on the Table 3 below. The efficiency is less good than for the previous example because of the small length of time spent in each element

Table 3. Speedup for the upsetting of an elastic cube

Number of CPU	1	2	4	5
Speedup	1.	1.82	3.41	6.06
Efficiency (%)	100	91	85	75

The interesting part is here the solver. In this case the skyline method takes 1020 seconds CPU time on the SG3800. Sequentially, the sparse solver using METIS renumbering (see Figure 2 for the matrix shape) reduces the length of time to 158 seconds. Unfortunately, parallel benchmark are not available at the present time.

4   Conclusion

We have shown a model of parallelization of a finite element software simple to realize. Our approach was first of all the identification of areas that consume a lot of computation time for our large examples. Then, those zones has been parallelized either directly with OpenMP directives either getting the parallel library of the CEPBA for the solver. We consider that actually the code has a good level of parallelization.

References

1. Chergui, J., Lavalle P.F., Proux J-P., Roult C.: OpenMP Paralllisation multitches pour machines mmoire partage. IDRIS (2001)
2. Kaypis, G., Kumar, V.: METIS a software Package For Partitioning Unstructured Graphs, Partitioning Meshes and Computing Fill-Reducing Ordering of Sparse Matrices. University Of Minnesota, Dep. of Comp. Sc. (1998)
3. Larrazabal, G., Cela, J.M.: A Parallel Iterative Solver for the Schur Complement System, Proc. IEEE International Conference on Parallel Processing, ICPP (2001).

Towards OpenMP Execution on
Software Distributed Shared Memory Systems

Ayon Basumallik, Seung-Jai Min, and Rudolf Eigenmann*

School of Electrical and Computer Engineering
Purdue University, West Lafayette, IN 47907-1285
http://www.ece.purdue.edu/ParaMount

Abstract. In this paper, we examine some of the challenges present in
providing support for OpenMP applications on a Software Distributed
Shared Memory(DSM) based cluster system. We present detailed measurements of the performance characteristics of realistic OpenMP applications from the SPEC OMP2001 benchmarks. Based on these measurements, we discuss application and system characteristics that impede
the efficient execution of these programs on a Software DSM system. We
point out pitfalls of a naive translation approach from OpenMP into the
API provided by a Software DSM system, and we discuss a set of possible
program optimization techniques.

1 Introduction

The current OpenMP 2.0 API is designed to be used on shared-memory multi-
processors. In this paper we describe challenges and opportunities in providing
OpenMP support via a Software Distributed Shared Memory (DSM) system.
Implementing OpenMP via Software DSM [1] is one possible avenue for making
OpenMP amenable to distributed-memory systems, such as cluster architectures.

Alternative approaches to OpenMP implementation on distributed systems
include language extensions and architecture support. Several recent papers have
proposed language extensions. For example, in [2–4] the authors propose data
distribution directives similar to the ones designed for High-Performance Fortran (HPF). Other researchers have proposed page placement techniques to
map data to the most suitable processing nodes [2]. In [5] remote procedure
calls are used to employ distributed processing nodes. Another related approach
is to use explicit message passing for communication across distributed systems and OpenMP within shared-memory multiprocessor nodes [6]. Providing
architectural support for OpenMP essentially means building shared-memory
multiprocessors (SMPs). While this is not new, an important observation is
that increasingly large-scale SMPs continue to become commercially available.
For example, the largest machine on which SPEC OMP 2001 benchmarks have
been reported recently is a 128-processor Fujitsu PRIMEPOWER2000 system
(www.spec.org/hpg/omp2001/results/omp2001.html).

---

* This work was supported in part by the National Science Foundation under Grant
No. 9703180, 9975275, 9986020, and 9974976.

H. Zima et al. (Eds.): ISHPC 2002, LNCS 2327, pp. 457–468, 2002.

The present paper is meant to complement, rather than compete with, these related approaches. The goal is to measure quantitatively the challenges faced when implementing OpenMP on a Software DSM system. We have found that, while results from small test programs have shown promising performance, little information on the behavior of realistic OpenMP applications on Software DSM systems is available. This paper is meant to fill this void. We will pinpoint areas needing improvement and discuss opportunities for addressing the challenges. In particular, we will describe program transformation techniques that may reduce overheads incurred by a naive translation of OpenMP programs onto a Software DSM system.

In our measurements we use a commodity cluster architecture consisting of 16 PentiumII/Linux processors connected via standard Ethernet networks. We expect the scaling behavior of this architecture to be representative of that of common cluster systems with modest network connectivity. We have measured basic OpenMP low-level performance attributes via the kernel benchmarks introduced in [7]. These benchmarks are meant to characterize the performance behavior of OpenMP constructs, such as the time taken to fork and join a parallel region or to execute a barrier. We have also measured a small, highly parallel program, demonstrating upper bounds on the scalability of a Software DSM application. In order to understand the performance characteristics of realistic OpenMP applications, we have measured three of the SPEC OMP2001 applications. We have chosen the applications *wupwise, swim* and *equake*, which are among the most scalable of SPEC's OpenMP benchmarks. *Wupwise* and *swim* are Fortran codes whereas *equake* is written in C. In order to execute these OpenMP programs on a Software DSM system we have hand-translated them into the API provided by the Treadmarks system [8]. We will describe some of the key transformation steps.

The rest of the paper is organized as follows. In Section 2 we will describe the transformations applied to the OpenMP benchmarks in order to execute them on the Treadmarks Software DSM system. In Section 3 we present and discuss our measurements. In Section 4 we describe techniques for improving the performance of OpenMP/Software DSM programs. Section 5 concludes the paper.

2   Translating OpenMP Applications into Software DSM Programs

Common translation methods for loop-parallel languages employ a microtasking scheme. Here the *master processor* begins program execution as a sequential process and pre-forks *helper processes* on the participating processors, which *sleep* until needed. On encountering a parallel construct, the master wakes the helpers and sets up the requisite execution environment. Such microtasking schemes are typically used in shared-memory environments with low communication latencies and fully shared address-spaces.

Software distributed shared-memory systems usually exhibit significantly higher communication latencies and they do not support fully-shared address spaces. In the Treadmarks Software DSM system used in our work, shared-memory sections can be allocated on-request. However, all process-local address spaces are private – not visible to other processes. Typically, there are also constraints on the amount of shared-memory which can be allocated. This is an issue for applications with large data-sets. These properties question the benefit of a microtasking scheme because (1)the helper wakeup call performed by the master process would take significantly longer and (2) the data environment communicated to the helpers could only include addresses from the explicitly allocated shared address space. Therefore, in our work we have chosen an SPMD scheme, as is more common in applications for distributed systems. In an SPMD scheme all processes begin execution of the program in parallel. Sections that need to be executed only by one processor must be marked explicitly and executed conditionally on the process identification.

We translate OpenMP workshare constructs in the usual method, by modifying lower and upper bound of the loops according to the iteration space assigned for each participating process. Currently we support static scheduling only. All parallel constructs are placed between a pair of barrier synchronizations. The barriers perform the dual functions of synchronization and maintaining coherence of shared data. Our SPMD approach is different from that adopted in [1] and it resolves, to an extent, the issues described earlier. We do not need explicit wakeup calls. We may also substitute communication of certain data items by redundantly computing them and thus reduce the amount of memory that must be allocated as shared.

The translation of serial program sections now becomes non-trivial. The need to maintain correct control flow precludes the possibility of executing the serial section by only the master process. Thus, following the SPMD method, parts of the serial section are redundantly executed by all participating processes. However, correctness constraints dictate that shared-memory updates as well as certain operations like I/O in the serial section be done only by the master process. Each such portions is marked to be executed by the master only and is followed by a barrier to ensure shared-memory coherence.

3   Performance Measurements

In this section we describe and discuss some measurements carried out on the performance of OpenMP kernels, microbenchmarks and real-application benchmarks on our cluster. The programs were hand-translated using the transformations described in Section 2, without any further optimizations.

3.1   Speedup Bounds: Performance of an OpenMP Kernel

In order to test the basic capability of scaling parallel code, we measured the performance of a simple OpenMP kernel, which is a small, highly parallel program that calculates the value of π using the integral approximation $\int_0^1 \frac{4.0}{(1.0+x^2)}\, dx$. The OpenMP constructs used within the program include a single OMP DO

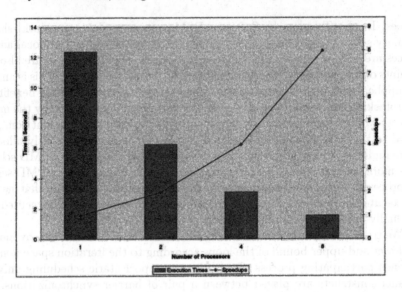

Fig. 1. Execution Time and Speedup of the *PI* program on 1,2,4 and 8 processors

with a REDUCTION clause. The execution times and speedups obtained(shown in Figure 1) provides an upper bound on the performance we can expect for our system. This kernel makes use of only 3 shared scalar variables.Thus the coherence actions at the barrier terminating the parallel region are minor. This point will assume significance when we compare the performance of this kernel to real applications.

3.2   Performance of the OpenMP Synchronization Microbenchmark

In order to understand the overheads incurred by the different OpenMP constructs on our system, we have used the kernel benchmarks introduced in [7]. In the present work we are primarily interested in the synchronization overheads. The performance of our system on the Synchronization Microbenchmark is shown in Figure 2.

The trends displayed in Figure 2 look similar to those for the NUMA machines (such as the SGI Origin 2000) enumerated in [7]. This is consistent with the fact that a NUMA machine is conceptually similar to a Software DSM system. However, for a Software DSM system, the overheads are now in the order of milliseconds as compared to microseconds overhead in NUMA SMPs. The dominant factor is the *barrier* overhead, since the barrier also maintains coherence of the shared-memory. These coherence overheads grow with the increase in shared memory activity within parallel constructs in a Software DSM. This fact is not captured here since the Synchronization Microbenchmark, like the *PI* kernel benchmark, uses a very small amount of shared memory. To quantitatively understand the behavior of OpenMP programs that utilize shared data substantially, we look at three of the most scalable applications in the SPEC OMP2001 benchmark suite.

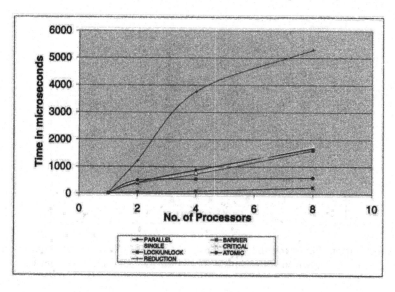

Fig. 2. OpenMP Synchronization Overheads as measured by the OpenMP Synchronization Microbenchmark. The overheads have been measured on a system of 1,2,4 and 8 processors.

3.3   Performance Evaluation of Real Application Benchmarks

The SPEC OMP2001 suite of benchmarks [9] consists of realistic OpenMP C and Fortran applications. We selected three of the more scalable benchmarks, namely *SWIM* and *WUPWISE* from the set of Fortran applications and *EQUAKE* from the set of C applications. The trends in our performance measurements was found to be consistent for all three applications. So, for brevity, we shall primarily limit our discussion of exact measurements to the *EQUAKE* benchmark.

Execution Time. Figure 3 shows the execution times for the *EQUAKE* benchmark run using the *train* dataset. The times shown here include the startup time. For each processor, the total elapsed time has been expressed as a sum of the user and the system times.

The figure shows a speedup in terms of user-times from one to eight processors. However, considering total elapsed time, the overall speedup is much less. For 8 processors, the performance degrades so that no speedup is achieved. This fact is consistent with the growing system time component as we go from serial to 8 processor execution. We verified that this system time component is not caused by system scheduling or paging activity. Instead, we attribute the system time to shared-memory coherence activities.

Detailed Measurements of Program Sections. The *equake* code has two small serial sections and several parallel regions, which cover more than 99% of the serial execution time. We first look at two parallel loops *main_2* and *main_3*

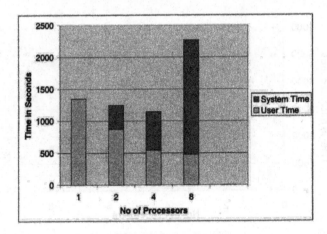

Fig. 3. Execution Times for the *equake* SPEC OpenMP Benchmark running on 1,2,4 and 8 processors, shown in terms of the measured user times and system times.

which show the desirable speedup behavior. The times spent within the parallel loop and the time spent at the barrier are shown separately. Loop *main_3* is one of the more important parallel loops where around 30% of the program time is spent. Figure 4 shows the behavior of these loops.

Loop *main_2* shows a speedup of about two and loop *main_3* shows almost linear speedup on 8 processors in terms of elapsed time. In both cases the loop time itself speeds up almost linearly, but the barrier time increases considerably. In fact, *main_2* has a considerable increase in barrier overheads from the serial to 8 processor execution. In *main_3*, the barrier overhead is within a more acceptable range. The reason for this is that the loop body of *main_2* is very small. It consists of a loop of 3 iterations containing a single assignment statement. Compared to that, *main_3* has a larger loop body which considerably amortizes the cost of the barrier.

We next look at three loops, *main_1 meminit_1* and *meminit_2* , which incur large barrier overheads. Figure 5 shows the behavior of these loops.

In all these three loops, loop times speed up linearly, but the barrier times increase to 15 to 30 times the loop times on 8 processors. So, for these loops, there is a substantial slowdown as we go from serial to 8-processor execution. These loops mainly contain shared memory writes. Especially *meminit_1* and *meminit_2* contain only shared memory initialization. So the barrier time becomes unacceptably more expensive than the loops themselves. As discussed previously, the barrier cost increases with the number of shared memory writes in the parallel region.

We next look at the performance of the serial sections (*serial_1* and *serial_2*) and the most time consuming parallel region (*smvp_0*) within the program. Figure 6 shows the behavior of these regions.

As previously discussed in Section 2, barriers have to be placed within a serial region so that shared memory writes by the master processor become visible

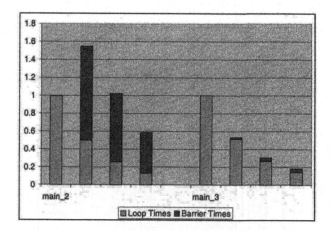

Fig. 4. Normalized Execution Times for the parallel loops, *main_2* and *main_3*, on 1,2,4 and 8 processors. The total execution time for each is expressed as a sum of the time spent within the loop and the time spent on the barrier at its end. Timings for each loop are normalized with respect to the serial version.

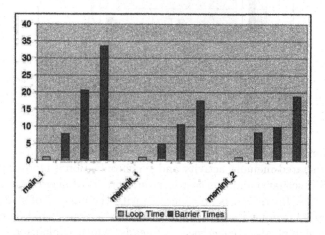

Fig. 5. Normalized Execution Times for the parallel loops *main_1 meminit_1* and *meminit_2* on 1,2,4 and 8 processors. The total execution time for each is expressed as a sum of the time spent within the loop and the time spent on the barrier at its end.

to the other processors. We see that this results in serial section *serial_1* experiencing a slowdown as we go from the serial to 8-processor execution. Section *serial_2* in unaffected since it does not have barriers in-between.

The parallel region *smvp_0* also suffers because of the same reason - it contains several parallel loops and after each of these, a barrier has to be placed. The parallel region *svmp_0* takes up more than 60% of the total execution time, hence its performance has a major impact on the overall application.

To summarize our measurements, we note that a naive transformation of realistic OpenMP applications for Software DSM execution does not give us the

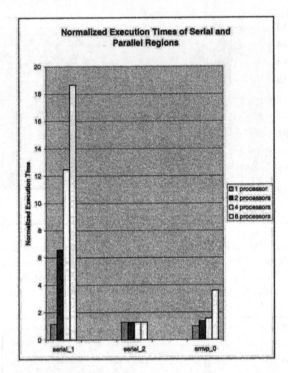

Fig. 6. Normalized Execution Times for the serial sections *serial_1*, *serial_2* and the parallel region *smvp_0* . For each region the times are normalized with respect to the serial version.

desired performance. A large part of this performance degradation is owing to the fact that shared-memory activity and synchronization is more expensive in a Software DSM scenario and this cost is several orders of magnitude higher than in SMP systems. Barrier costs now increase with the number of shared memory writes in parallel regions. Realistic applications use a large shared-memory space, and thus have high data coherence overheads, which explains the seemingly contradicting performance results of kernels and real-applications.

Our measurement motivate the need for optimizations when we try to run real OpenMP applications in a Software DSM scenario. The results discussed above indicate that these optimizations should be primarily aimed at optimizing shared-memory synchronization in programs, irrespective of the underlying Software DSM specifics. In our subsequent discussion, we present some of the possible optimizations.

4   Ongoing Work: Improving OpenMP Performance on Software DSM Systems

In this section we describe transformations that can optimize OpenMP programs on a Software DSM system. Many of these techniques have been discussed in other contexts. We expect their combined implementation in an OpenMP/ Software DSM compiler to have a significant impact. The realization of such a compiler is one objective of our ongoing project.

Data Prefetch and Forwarding: Prefetch is a fundamental technique for overcoming memory access latencies. Closely-related to prefetch is *data forwarding*, which is producer-initiated. Forwarding has the advantage that it is a one-way communication (producer forwards to all consumers) whereas prefetching is a two-way communication (consumers request data and producers respond). An important issue in prefetch/forwarding is to determine the optimal prefetch point. Prefetch techniques have been studied previously [10], albeit not in the context of OpenMP applications for Software DSM systems. We expect prefetch/forwarding to significantly lower the cost of OpenMP END PARALLEL region constructs, as it reduces the need for coherence actions at that point.

Barrier elimination: Two types of barrier eliminations will become important. It is well known that the usual barrier that separates two consecutive parallel loops can be eliminated if permitted by data dependences [11]. Similarly, within parallel regions containing consecutive parallel loops, it may be possible to eliminate the barrier separating the individual loops. Barriers in the serial section, as described in Section 2, can be eliminated in some situations, if the shared data written are not read subsequently within the same serial section.

Data privatization: Private data in a Software DSM system are not subject to costly coherence actions. An extreme of a Software DSM program can be viewed as a program that has only private, and no shared, data. The necessary data exchanges between processors are performed by copy operations via shared memory or possibly via explicit messages. Such a program is equivalent to a message-passing program. Many points are possible in-between this extreme and a program that has all OpenMP shared data placed in shared DSM address space. For example, shared data with read-only accesses in certain program sections can be made "private with copy-in" during these sections. Similarly, shared data that are exclusively accessed by the same processor can be privatized during such a program phase. For large phases, the benefit of this scheme is obvious. In programs with small read-only or exclusive phases, the copy-in and copy-out operations at phase boundaries become significant, moving closer the the "message-passing" extreme mentioned above.

Figure 7 demonstrates a simple form of this optimization, applied to the *equake* benchmark. The program reads data in a serial region from a file. In the original code, since all the processors use the data, a shared attribute is given. However, we have modified the program so that the input data, which are read-only after the initialization, become private to all processes. Figure 7 shows that even this simple optimization substantially reduces the execution times. With this new scheme, the resulting speedup on four processors is close to two.

Page placement: Software DSM systems may place memory pages on fixed home processors or the pages may migrate between processors. Fixed page placement

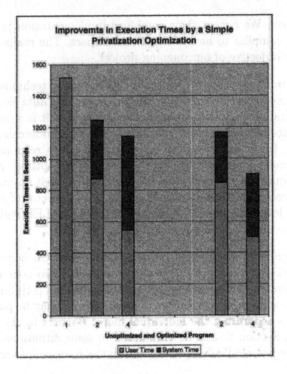

Fig. 7. Performance Improvements with a simple data privatization optimization

leads to high overhead if the chosen home is not the processor with the most frequent accesses to this page. Migrating pages can incur high overhead if the pages end up changing their home frequently. In both cases the compiler can help direct the page placing mechanism. It can estimate the number of accesses made to a page by all processors and choose the best home.

Automatic data distribution: Data distribution mechanisms have been well researched in the context of distributed memory multiprocessors and languages such as HPF. Many of these techniques are directly applicable to a Software DSM system. Automatic data distribution is easy in regular data structures and program patterns. Hence, a possible strategy is to apply data distribution with explicit messaging in regular program sections and rely on Software DSM mechanism for other data and program patterns. This idea has been pursued in [12], and we will combine our approaches in a collaborative effort. Planned enhancements include other data and loop transformation techniques for locality enhancement.

Adaptive optimization: A big impediment for all compiler optimizations is the fact that input data is not known at compile-time. Consequently, compilers must make *conservative assumptions* leading to suboptimal program performance. The potential performance degradation is especially high if the compiler's decision making chooses between transformation variants whose performance differs substantially. This is the case in Software DSM systems, which typically possess several tunable parameters. We are building on a prototype of a dynamically adaptive compilation system [13–15], called ADAPT, which can dynamically compile program variants, substitute them at runtime, and evaluate them in the executing application.

In ongoing work we are creating a compilation system that integrates the presented techniques. As we have described, several of these techniques have been proposed previously. However, no quantitative data of their value on large, realistic OpenMP applications on Software DSM systems is available. Evaluating these optimizations in the context of the SPEC OMP applications is an important objective of our current project.

Our work complements efforts to extend OpenMP with latency management constructs. While our primary emphasis is on the development and evaluation of compiler techniques, we will consider small extensions that may become critical in directing compiler optimizations.

5 Conclusions

In this paper we have studied the feasibility of executing OpenMP applications on a commodity cluster architecture via a distributed shared memory system. We have found that, although small OpenMP programs can exhibit near-ideal speedups, a naive translation of realistic OpenMP programs onto the API provided by the Software DSM system is not sufficient. We have found that coherence actions performed at synchronization points are major sources of overheads. We found that the efficient execution of realistic OpenMP applications on distributed processor architectures via a Software DSM system requires substantial program optimizations, one of which we demonstrated through a simple experiment. We are in the process of realizing such optimizations in a new optimizing compiler.

References

1. Y.C. Hu, H. Lu, A.L. Cox, and W. Zwaenepoel. OpenMP for Networks of SMPs. *Journal of Parallel and Distributed Computing*, 60(12):1512–1530, December 2000.
2. J. Bircsak, P. Craig, R. Crowell, Z. Cvetanovic, J. Harris, C. Nelson, and C. Offner. Extending OpenMP for NUMA Machines. In *Proc. of the IEEE/ACM Supercomputing'2000: High Performance Networking and Computing Conference (SC2000)*, November 2000.
3. V. Schuster and D. Miles. Distributed OpenMP, Extensions to OpenMP for SMP Clusters. In *Proc. of the Workshop on OpenMP Applications and Tools (WOMPAT2000)*, July 2000.
4. T.S. Abdelrahman and T.N. Wong. Compiler support for data distribution on NUMA multiprocessors. *Journal of Supercomputing*, 12(4):349–371, October 1998.

5. Mitsuhisa Sato, Motonari Hirano, Yoshio Tanaka, and Satoshi Sekiguchi. OmniRPC: A Grid RPC Facility for Cluster and Global Computing in OpenMP. In *Workshop on OpenMP Applications and Tools (WOMPAT2001)*, July 2001.
6. Lorna Smith and Mark Bull. Development of Mixed Mode MPI / OpenMP Applications. In *Proc. of the Workshop on OpenMP Applications and Tools (WOMPAT2000)*, July 2000.
7. J.M.Bull. Measuring Synchronization And Scheduling Overheads in OpenMP. In *Proc. of the European Workshop on OpenMP (EWOMP99)*, October 1999.
8. C. Amza, A.L. Cox, S. Dwarkadas, P. Keleher, H. Lu, R.Rajamony, W. Yu, and W. Zwaenepoel. TreadMarks: Shared Memory Computing on Networks of Workstations. *IEEE Computer*, 29(2):18–28, February 1996.
9. V.Aslot, M.Domeika, R.Eigenmann, G.Gaertner, W.B. Jones, and B.Parady. SPEComp: A New Benchmark Suite for Measuring Parallel Computer Performance. In *Proc. of the Workshop on OpenMP Applications and Tools (WOMPAT2001), Lecture Notes in Computer Science, 2104*, pages 1–10, July 2001.
10. E.H.Gornish, E.D.Granston, and A.V.Veidenbaum. Compiler-directed Data Prefetching in Multiprocessors with Memory Hierarchies . *Proceedings of ICS'90, Amsterdam, The Netherlands*, 1:342–353, June 1990.
11. C.W. Tseng. Compiler optimizations for eliminating barrier synchronization. In *Proc. of the 5th ACM Symposium on Principles and Practice of Parallel Programming (PPOPP'95)*, July 1995.
12. J.Zhu and J.Hoeflinger. Compiling for a Hybrid Programming Model Using the LMAD Representation. In *Proc. of the 14th annual workshop on Languages and Compilers for Parallel Computing (LCPC2001)*, August 2001.
13. M.J.Voss and R.Eigenmann. High-level adaptive program optimization with adapt. In *Proc. of the ACM Symposium on Principles and Practice of Parallel Programming (PPOPP'01)*, pages 93 – 102. ACM Press, June 2001.
14. M.J.Voss and R.Eigenmann. A framework for remote dynamic program optimization. In *Proc. of the ACM SIGPLAN Workshop on Dynamic and Adaptive Compilation and Optimization (Dynamo'00)*, January 1999.
15. M.J.Voss and R.Eigenmann. Dynamically adaptive parallel programs. In *International Symposium on High Performance Computing*, pages 109–120, Kyoto, Japan, May 1999.

Dual-Level Parallelism Exploitation with OpenMP in Coastal Ocean Circulation Modeling

Marc González[1], Eduard Ayguadé[1], Xavier Martorell[1],
Jesús Labarta[1], and Phu V. Luong[2]

[1] European Center for Parallelism of Barcelona (CEPBA)
Technical University of Catalunya (UPC), Barcelona, Spain
[2] University of Texas
Engineer Research and Development Center
Major Shared Resource Center
Vicksburg, MS 39180

Abstract. Two alternative dual-level parallel implementations of the Multi-block Grid Princeton Ocean Model (MGPOM) are compared in this paper. The first one combines the use of two programming paradigms: message passing with the Message Passing Interface (MPI) and shared memory with OpenMP (version called MPI-OpenMP); the second uses only OpenMP (version called OpenMP-Only). MGPOM is a multiblock grid code that enables the exploitation of two levels of parallelism.
The MPI-OpenMP implementation uses MPI to parallelize computations by assigning each grid block to a unique MPI process. Since not all grid blocks are of the same size, the workload between processes varies. OpenMP is used within each MPI process to improve load balance. The alternative OpenMP-Only implementation uses some extensions proposed to OpenMP that defines thread groups in order to efficiently exploit the available two levels of parallelism. These extensions are supported by a research OpenMP compiler named NanosCompiler.
Performance results of the two implementations from the MGPOM code on a 20-block grid for the Arabian Gulf simulation demonstrate the efficacy of the OpenMP-Only versions of the code. The simplicity of the OpenMP implementation as well as the possibility of using and simply defining policies to dynamically change the allocation of OpenMP threads to the two levels of parallelism is the main result of this study and suggests to consider this alternative for the parallelization of future applications.

Keywords: OpenMP and MPI implementations, multiple levels of parallelism, multiblock grid, coastal ocean circulation model.

1 Introduction

In recent years, OpenMP [1] has emerged as an industrial library for parallel programming in shared-memory computers. Parallel performance is achieved without significantly sacrificing execution time when it is ported across a range of shared-memory platforms. Moreover, its simplicity makes the conversion of a sequential

H. Zima et al. (Eds.): ISHPC 2002, LNCS 2327, pp. 469–478, 2002.

code to a parallel code for improving performance much easier and without major code modifications.

One of the key features which is not currently exploited by most current commercial and experimental systems is the use of OpenMP for multiple levels of parallelism. In general, OpenMP can be used to exploit the multi-level parallelism in most scientific and engineering numerical applications. This technique, however, has not been fully applied to any of these applications because they achieve satisfactory speed-ups when executed in mid-size parallel platforms or because most current systems supporting OpenMP (compilers and associated thread--level layer) sequentialize nested parallel constructs. This has originated the current practice of exploiting multiple levels of parallelism through a combination of different programming models and interfaces, MPI [2] coupled with OpenMP for example. In such cases, MPI is usually used for communication at the outer levels between subdomains or between block grids (multiblock grid), while OpenMP is used to parallelize the inner levels within each subdomain or block.

In addition to the possibility of nesting parallel constructs, OpenMP offers the possibility of controlling the number of threads usable at each level of parallelism (clause NUM_THREADS available in OpenMP v2.0). Some extensions have been proposed to OpenMP in order to allow a cleaner and effective control over the work distribution when dealing with multiple levels of parallelism through the definition of thread groups [4]. This proposal is the one used to express the parallelism in the MGPOM application and will be discussed in Section 2. Other proposals consist in offering work queues and an interface for inserting application tasks before execution- for example, the Illinois--Intel Multithreading library [5] or the WorkQueue mechanism [6] proposed by KAI, in which work can be created dynamically, even recursively, and put into queues. The proposal used in this paper is simpler and allows finer control over the allocation of threads to the multiple levels of parallelism in the application.

In this study, OpenMP is used for both outer as well as inner levels. Details of the extensions to OpenMP are discussed in Section 2. Its application to the MGPOM on a 20-block grid of the Arabian Gulf simulation is presented in Section 3. Parallel performance results are presented in Section 4. The conclusions of this study are in Section 5.

2   NanosCompiler Extensions

The NanosCompiler [3] and runtime library are serving as a research platform for proposing and evaluating extensions to the OpenMP language definition. One of these extensions consists of the dynamic creation of thread groups and the definition of the actual composition of groups at runtime. Groups can be created to exploit both loop- and task-level parallelism.

In the fork/join execution model defined by OpenMP, a program begins execution as a single process or thread. This thread executes sequentially until a PARALLEL construct is found. At this time, the thread creates a team of threads and it becomes its master thread. All threads execute the statements enclosed lexically within the parallel constructs. Work-sharing constructs (DO, SECTIONS and SINGLE) are provided to divide the execution of the enclosed code region among the members of a team. All threads are independent and may synchronize at the end of each work-sharing construct or at specific points (specified by the BARRIER directive). Exclusive execution mode is also possible through the definition of CRITICAL regions.

The SECTIONS directive is a non-iterative work-sharing construct which specifies the enclosed sections of code (each one delimited by a SECTION directive) are divided among threads in the team. Each section becomes a task which is executed once by a thread in the team. The DO work-sharing construct is used to divide the iterations of a loop into a set of independent tasks, each one executing a chunk of consecutive iterations. Finally, the SINGLE work-sharing construct informs that only one thread in the team is going to execute the work.

In this study, a *group of threads* is composed of a subset of the total number of threads available in the team to run a parallel construct. The threads participating in a parallel construct are identified following the active numeration inside the current team (from 0 to omp_get_num_threads()-1). In a parallel construct, the programmer may define the number of groups and the composition of each group. When a thread in the current team encounters a parallel construct defining groups, the thread creates a new team and it becomes its master thread. The new team is composed of as many threads as groups are defined; the rest of the threads are reserved to support the execution of nested parallel constructs. In other words, the groups definition establishes the threads that are involved in the execution of the parallel construct and the allocation strategy or scenario for the inner levels of parallelism that might be spawned. When a member of this new team encounters another parallel construct (nested to the one that caused the group definition), it creates a new team and deploys its parallelism to the threads that compose its group.

The GROUPS clause allows the user to specify thread groups. It can only appear in a PARALLEL construct or combined PARALLEL DO and PARALLEL SECTIONS constructs.

C$OMP PARALLEL [DO|SECTIONS] [GROUPS(gspec)]

Different formats for the groups specifier gspec are allowed [4]. In this paper we only comment on the two more relevant. For additional details concerning alternative formats as well as implementation issues, please refer to this publication.

GROUPS(ngroups,weight)

In this case, the user specifies the number of groups (ngroups) and an integer vector (weight) indicating the relative amount of computation that each group has to perform. Vector weight is allocated by the user in the application address space and it has to be computed from information available within the application itself (for instace iteration space, computational complexity or even information collected at runtime). The runtime library determines, from this information, the composition of the groups. The algorithm assigns all the available threads to the groups and ensures that each group at least receives one thread. The main body of the algorithm is shown in Figure 1 (using Fortran90 syntax).

The library generates two internal vectors (masters and howmany). In this algorithm, nthreads is the number of threads that are available to spawn the parallelism in the parallel construct containing the group definition.

```
    howmany(1:ngroups) = 1
    do while (sum(howmany(1:ngroups)) .lt. nthreads)
        pos = maxloc(weight(1:ngroups)/
                            howmany(1:ngroups))
        howmany(pos(1)) = howmany(pos(1)) + 1
    end do
    masters(1) = 0
    do i = 1, ngroups-1
        masters(i+1) = masters(i) + howmany(i)
    end do
```

Fig. 1. Skeleton of the algorithm used to compute the composition of groups.

The most general format allows the specification of three parameters in the group definition:

```
GROUPS(ngroups, masters, howmany)
```

The first argument (ngroups) specifies the number of groups to be defined and consequently the number of threads in the team that is going to execute the parallel construct. The second argument (masters) is an integer vector with the identifiers (using the active numeration in the current team) of the threads that will compose the new team. Finally, the third argument (howmany) is an integer vector whose elements indicate the number of threads that will compose each group. The vectors have to be allocated in the memory space of the application, and their content and correctness have to be guaranteed by the programmer. Notice that this format must be used when the default mapping explained before does not provide the expected performance.

3   Application of NanosCompiler Extensions to MGPOM

MGPOM is a standard Fortran77 multiblock grid code. A parallel version of MGPOM uses MPI asynchronous sends and receives to exchange data between adjacent blocks at the interfaces. OpenMP has been used as a second level of parallelization within each MPI process to improve the load balance in the simulation of the Arabian Gulf [7]. This area is extended from 48 East to 58 East in longitude and from 23.5 North to 30.5 North in latitude (left part of Figure 2). The computational 20-block grid shown on the right part of Figure 2 (also used in [7]) is used in this study.

OpenMP with NanosCompiler Extensions is used to parallelize the serial version of the MGPOM code at the outer levels (block to block) as well as at the inner levels (within each block). The number of threads used to exploit the inner level of parallelism depends on the size of each grid block. Figure 3 shows the use of OpenMP directives and GROUPS construct implemented into the main program and a subroutine of the serial MGPOM code version. This figure shows a version in which the runtime library, using the default algorithm described in Section 2, determines the composition of the groups. The GROUPS clause has as input arguments the number of blocks (maxb) and a vector with the number of grid points in each block (work).

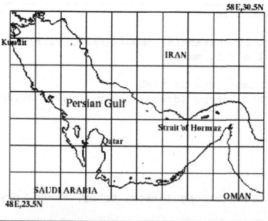

Fig. 2. The Persian Gulf coastline (left). The 20-block grid with blocks of different size (right).

Notice that the GROUPS clause is the only non-standard use of OpenMP. The exploitation of multiple levels of parallelism is achieved through the nesting of PARALLEL DO constructs.

Table 1 shows the number of grid points in each block for the 20-block grid. This is the composition of the work vector. The runtime library supporting the code generated by the NanosCompiler would generate the allocation of threads to groups shown in Tables 2 and 3 assuming 20 and 30 processors, respectively. Who and Howmany are the two internal vectors generated by the library with the master of each group (who) and the number of threads to be used in each group (howmany). With 20 processors, the critical path is determined by the largest block (i.e. block number 8), yields a theoretical speedup of 12.5. With 30 processors, the critical path is determined by the block with the largest ratio size/howmany (i.e. block number 17), with a theoretical speed-up of 19.4.

```
              PROGRAM main
              ...
C$OMP PARALLEL DO PRIVATE(n) GROUPS(maxb, work)
              DO 414 n = 1, maxb
                 ...
                 IF (mode .ne. 2) THEN
                    CALL baropg (drhox, drhoy, drx2d, dry2d, n)
                    ...
                 ENDIF
414           CONTINUE
              ...
              END

              SUBROUTINE baropg (drhox, drhoy, drx2d, dry2d, nb)
              ...
C$OMP PARALLEL DO PRIVATE (i,j,k)
              DO 200 j = 1, jm
                 DO 200 k = 1 kb
                    DO 200 I = 1, im
                       rho(i,j,k,nb) = rho(i,j,k,nb) - rmean(i,j,k,nb)
200           CONTINUE
              ...
              RETURN
              END
              ...
```

Fig. 3. Excerpt of the OpenMP implementation.

Table 1. Number of grid points in each block for the twenty-block case.

Block	1	2	3	4	5	6	7
Size	1443	1710	1677	2150	700	989	2597
Block	8	9	10	11	12	13	14
Size	2862	2142	1836	1428	1881	1862	2058
Block	15	16	17	18	19	20	-
Size	1470	1280	1848	2318	999	2623	-

Table 2. Default allocation of threads to groups with 20 processors.

Block	1	2	3	4	5	6	7	8	9	10
Howmany	1	1	1	1	1	1	1	1	1	1
Who	0	1	2	3	4	5	6	7	8	9
Block	11	12	13	14	15	16	17	18	19	20
Howmany	1	1	1	1	1	1	1	1	1	1
Who	10	11	12	13	14	15	16	17	18	19

According to the allocation of threads to groups shown in Table 3, the average work per thread has a large variance (from 1848 in block 17 to 700 in block 5). This variance results in a noticeable load imbalance. To reduce this load imbalance, several blocks could be gathered into a cluster such that the work distribution is equally divided among groups. To achieve this, the user can define its own composition of groups, as shown in Figure 4.

Table 3. Default allocation of threads to groups with 30 processors.

Block	1	2	3	4	5	6	7	8	9	10
Howmany	1	1	1	2	1	1	2	2	2	2
Who	0	1	2	3	5	6	7	9	11	13
Block	11	12	13	14	15	16	17	18	19	20
Howmany	1	2	2	2	1	1	1	2	1	2
Who	15	16	18	20	22	23	24	25	27	28

```
       PROGRAM main
         . . .
       CALL compute_groups(work, maxb, who, howmany)
         . . .
C$OMP PARALLEL DO PRIVATE(n) GROUPS(maxb, who, howmany)
       DO 414 n = 1, maxb
         . . .
          IF (mode .ne. 2) THEN
             CALL baropg (drhox, drhoy, drx2d, dry2d, n)
             . . .
          ENDIF
414       CONTINUE
         . . .
       END
```

Fig. 4. Excerpt of the OpenMP implementation with user-defined groups.

User function compute_groups decides how many OpenMP threads are devoted to the execution of each block and which OpenMP thread is going to be the master in the exploitation of the inner level of parallelism inside each block. This function has as input arguments the number of blocks (maxb) and a vector with the number of grid points in the block (work). It returns two vectors with the master of each group (who) and the number of threads to be used in each group (howmany). This is the information that is later used in the GROUPS clause.

Table 4 shows the allocation of the threads to groups as well as the identities of blocks in a cluster. A cluster of two blocks with 20 processors is shown in this case. In this case, the work imbalance is noticeably reduced.

4   Performance Results

In this section we evaluate the behaviour of two parallel versions of MGPOM. The MPI-OpenMP version exploits two levels of parallelism by combining the two programming paradigms: MPI to exploit the inter-block parallelism and OpenMP to exploit the intra-block parallelism. The OpenMP-Only version exploits the two-levels of parallelism using OpenMP and the extensions offered by the NanosCompiler and supporting OpenMP runtime system. For the compilation of the multilevel OpenMP version we use the NanosCompiler to translate from extended OpenMP Fortran77 to plain Fortran77 with calls to the supporting runtime library NthLib. We use the native f77 compiler to generate code for an SGI Origin2000 system [8]. The flags are set to -mips4 -64 -O2. The experiments have been performed on system with 64 R10k processors, running at 250 MHz with 4 Mb of secondary cache each.

Table 4. Allocation of threads to groups with 20 processors after clustering.

Cluster	1	2	3	4	5
Blocks	1, 9	2, 13	3, 12	4, 11	5, 8
Howmany	2	2	2	2	2
Who	0	2	4	6	8
Cluster	6	7	8	9	10
Blocks	6, 20	7, 19	10, 17	14, 15	16, 18
Howmany	2	2	2	2	2
Who	10	12	14	16	18

Table 5. Speed-up with respect to the sequential execution for the multi-level OpenMP and the mixed MPI/OpenMP versions.

	OpenMP-Only	MPI-OpenMP
10	9.6	
20	15.3	14.7
30	18.7	
40	23.9	22.2
50	26.7	
60	27.8	23.0

Table 5 shows the speed-up achieved by the two versions of the program. The MPI-OpenMP version uses the same number of threads per MPI process. For this reason performance numbers are available for only 20, 40 and 60 processors.

In order to understand these results, we have instrumented the application with the –P option of the NanosCompiler and generated a trace suitable for being analyzed with Paraver [9]. Each iteration of the application time step loop can be divided in 5 different parts: P1, P2, P3, P4 and P5. Basically, parts P1 and P4 perform independent computations over the blocks. However, parts P2, P3 and P5 are responsible for updating the boundaries of the blocks; these parts require communication in the MPI version and are the source of cache coherence overheads in the OpenMP version. Table 6 shows the average speed-up of each of these individual parts for the OpenMP-Only version.

Notice that parts P1 and P4 scale much better than the rest of the parts. The parallelization of parts P2, P3 and P5 (update of the boundaries) perform accesses to memory with no data locality with respect to the access pattern performed in parts P1 and P4. This causes a large number of secondary cache misses and cache invalidations, which add overhead to the overall parallel execution time. In addition to that, parts P2, P3 and P5 include a large number of small loops whose parallelization is not efficient, due to the size of the iteration space traversed and due to the barriers that have to be performed to enforce the correct update of the boundary elements.

Table 6. Speed-up (relative to the execution with 10 processors) of each part of the time step loop for the OpenMP-Only version.

	P1	P2	P3	P4	P5
10	1	1	1	1	1
20	1.8	1.2	1.3	1.8	1.2
30	2.2	1.6	1.8	2.3	1.4
40	3.2	1.7	1.9	3.5	1.5
50	4.1	2.0	2	4.5	1.7
60	4.9	2.1	2.1	5.3	1.8

In order to conclude this section, Table 7 shows the relative performance (w.r.t the execution with 20 processors of the OpenMP-only version in Table 5) for three OpenMP versions of the application: "no_groups" exploits two levels of parallelism but does not define groups (similar to using the NUM_THREADS clause), "groups" performs a homogeneous distribution of the available threads among the groups, and "weighted" performs a weighted distribution of the available threads among the groups. Notice that the performance of both "groups" and "weighted" versions is greater than "no_groups". Also, the "weighted" performs better due to load unbalance that exists among blocks. In summary, additional information is needed in the NUM_THREADS clause to boost the performance of applications with multiple levels of parallelism.

Table 7. Speed-up with respect to OpenMP-only with 20 processors for three different OpenMP versions.

	no_groups	groups	weigthed
20	0.44	1	1
40	0.33	1.5	1.7
60	0.26	1.51	1.9

5   Conclusions

The main purpose of this paper has been to examine the performance achievable when exploiting nested parallelism in two equivalent parallel versions of a coastal ocean circulation modeling application. One of the versions relies on mixing two programming paradigms to exploit the parallelism: MPI and OpenMP. The other version uses only OpenMP (using the possibility offered by the language definition of nesting parallel constructs and some extensions offered by the research OpenMP NanosCompiler).

The paper summarizes the extensions to OpenMP proposed to efficiently exploit nested parallelism. The extensions are mainly based on the definition of thread groups. The composition of the groups can be dynamically decided by the supporting OpenMP runtime system (using a predefined allocation strategy based on the amount of work to be performed by each group) or by the user using his/her own algorithm for deciding the allocation. The paper presents results in which the default allocation is overridden by the user in order to better balance the work distribution by overlapping the execution of several grid blocks on the same group of threads.

Regarding the parallelization of MGPOM, the main conclusion from this study is that OpenMP alone is able to achieve a similar (or even better) performance than the one that could be achieved mixing two programming paradigms (such as MPI and OpenMP). The use of a single programming paradigm makes the development, tuning and maintenance process of parallel applications simpler. Close analyses of the results show that the scalability of the application is limited by parts of the application that update the boundaries of the blocks. The parallelization with OpenMP performs a static scheduling of loop iterations that degrade the locality of the memory hierarchy. In addition to that, the parallelization with OpenMP forces a large number of synchronizations that add extra overheads to the parallel execution time. We are currently investigating alternative parallelization strategies for these parts that may improve the efficiency of the parallel execution.

Acknowledgments

This work has been supported by the Spanish Ministry of Science and Technology and the European Union FEDER program under contract TIC2001-0995-C02-01, and by the European Center for Parallelism of Barcelona (CEPBA).

References

1. OpenMP Organization. OpenMP Fortran Application Interface, v. 2.0, www.openmp.org, June 2000.
2. M. Snir, S. Otto, S. Huss-Lederman, D. Walker and J. Dongarra. MPI – The Complete Reference: Volume 1, the MPI Core. MIT Press, Cambridge, 1998.
3. E. Ayguadé, M. Gonzalez, J. Labarta, X. Martorell, N. Navarro, J. Oliver, NanosCompiler: A Research Infrastructure for OpenMP Extensions. 1st European Workshop on OpenMP (EWOMP'99), Lund (Sweden). September/October 1999.
4. M. Gonzalez, J. Oliver, X. Martorell, E. Ayguade, J. Labarta and N. Navarro. OpenMP Extensions for Thread Groups and Their Runtime Support. In Workshop on Languages and Compilers for Parallel Computing, August 2000.
5. M. Girkar, M. R. Haghighat, P. Grey, H. Saito, N. Stavrakos and C.D. Polychronopoulos. Illinois-Intel Multithreading Library: Multithreading Support for Intel Architecture--based Multiprocessor Systems. Intel Technology Journal, Q1 issue, February 1998.
6. S. Shah, G. Haab, P. Petersen and J. Throop. Flexible Control Structures for Parallelism in OpenMP. In 1st European Workshop on OpenMP, Lund (Sweden), September 1999.
7. P. Luong, C.P. Breshears and L.N. Ly, Application of Multiblock Grid and Dual-Level Parallelism in Coastal Ocean Circulation Modeling. Journal of Applied Mathematical Modelling, submitted for publication.
8. Silicon Graphics Computer Systems SGI. Origin 200 and Origin 2000. Technical Report, 1996.
9. European Center for Parallelism of Barcelona. Paraver and Instrumentation Packages Reference Manual. http://www.cepba.upc.es/paraver.

Static Coarse Grain Task Scheduling
with Cache Optimization Using OpenMP

Hirofumi Nakano[1], Kazuhisa Ishizaka[2], Motoki Obata[2],
Keiji Kimura[2], and Hironori Kasahara[2]

[1] Waseda University,
3-4-1 Ohkubo, Shinjuku-ku, Tokyo, 169-8555, Japan
hnakano@oscar.elec.waseda.ac.jp
[2] Waseda University & Advanced Parallelizing Compiler Project
{ishizaka,obata,kimura,kasahara}@oscar.elec.waseda.ac.jp

Abstract. Effective use of cache memory is getting more important
with increasing gap between the processor speed and memory access
speed. Also, use of multigrain parallelism is getting more important to
improve effective performance beyond the limitation of loop iteration
level parallelism. Considering these factors, this paper proposes a coarse
grain task static scheduling scheme considering cache optimization. The
proposed scheme schedules coarse grain tasks to threads so that shared
data among coarse grain tasks can be passed via cache after task and data
decomposition considering cache size at compile time. It is implemented
on OSCAR Fortran multigrain parallelizing compiler and evaluated on
Sun Ultra80 four-processor SMP workstation, using Swim and Tomcatv
from the SPEC fp 95. As the results, the proposed scheme gives us 4.56
times speedup for Swim and 2.37 times on 4 processors for Tomcatv
respectively against the Sun Forte HPC 6 loop parallelizing compiler.

1   Introduction

With increasing gap between processor and memory access speeds, locality optimization for cache is getting more important to improve effective performance
of multiprocessor system.

Also, it is getting difficult to improve performance of multiprocessor system
dramatically using traditional loop parallel processing with maturity of loop
parallelization techniques. To overcome the difficulty and to get scalable performance improvement, exploitation of multigrain parallelism, which hierarchically uses coarse grain parallelism among loops and subroutines, loop parallelism
among loop iterations and (near) fine grain parallelism among instructions or
statements [1,2,3], is needed.

As to cache optimization by compilers, there has been made various studies, such as, affine partitioning [4,5,6] which unifies multiple loop restructures,
a vertical execution of tasks after loop decomposition [7], a cache optimization
among coarse grain tasks for a single processor [8] and for shared memory multiprocessors with dynamic task scheduling [9].

H. Zima et al. (Eds.): ISHPC 2002, LNCS 2327, pp. 479–489, 2002.
© Springer-Verlag Berlin Heidelberg 2002

This paper describes a static coarse grain task scheduling with cache optimization [2,3,9] based on data localization method [10,11]. This scheme is implemented on OSCAR multigrain parallelizing compiler [12]. OSCAR compiler generates OpenMP Fortran whose coarse grain tasks are statically scheduled to parallel threads with cache optimization from a sequential Fortran program.

In Section 2, coarse grain task parallel processing is described. Section 3 proposes a static coarse grain task scheduling with cache optimization using OpenMP. Also, the effectiveness of the proposed schemes is evaluated on Sun Ultra80 four-processor SMP workstation using Swim and Tomcatv from the SPEC fp 95 benchmark suite in Section 4. Finally, concluding remarks are described in Section 5.

2   Coarse Grain Task Parallel Processing

This section describes coarse grain task parallel processing, which is a part of multigrain parallel processing. Coarse grain task parallel processing uses parallelism among three kinds of macro-tasks, or coarse grain tasks, namely, block of pseudo assignment statements (BPA), repetition block (RB) and subroutine block (SB). The compiler decomposes a source program into macro-tasks. Also, it generates macro-tasks hierarchically inside a sequential repetition block and a subroutine block.

Coarse grain task parallelization by OSCAR compiler is performed in the following steps.

1. Decomposition of a source program into macro-tasks.
2. Analysis of data and control flows among macro-tasks and generation of Macro Flow Graph (MFG) representing data and control flows.
3. Analysis of Earliest Execution Condition (EEC) based on data and control dependence analysis that represents the condition, on which macro-task may start its execution earliest, and generation of Macro Task Graph (MTG) that represents the EEC.
4. Scheduling macro-tasks to processors or processor groups.
   When a macro-task graph has no conditional dependencies, macro-tasks are statically scheduled to processors or processor clusters at a compiler time and parallelized code is generated for each processor according to the scheduling results. When macro-task graph contains control dependencies, compiler generates dynamic scheduling routine to assign macro-tasks to processors or processor clusters at a run time and embeds the dynamic scheduling routine to the generated parallelized code with macro-task code in order to cope with runtime uncertainties.

In the following, the details of the above steps are described

2.1   Generation of Macro-Tasks [13]

The compiler first generates macro-tasks, namely, block of pseudo assignment statements (a basic block or a block merging several basic blocks), repetition

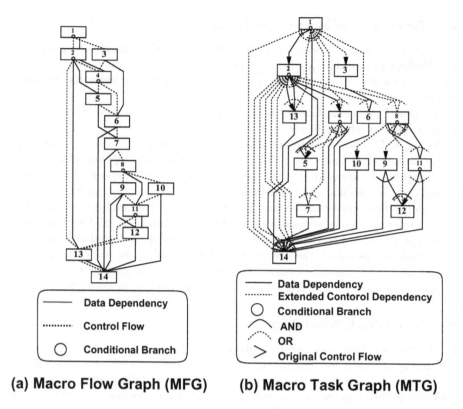

(a) Macro Flow Graph (MFG)   (b) Macro Task Graph (MTG)

Fig. 1. Macro Flow Graph and Macro Task Graph

blocks and subroutine blocks from a source program. Furthermore, compiler hierarchically decomposes the body of sequential repetition block and a subroutine block.

If a repetition block (RB) is a parallelizable loop, it is divided into different partial loops by loop iteration direction taking into consideration the number of processors, cache size and so on. These partial loops are defined as different macro-tasks to be executed in parallel.

2.2 Generation of Macro Flow Graph

After the generation of macro-tasks, the data and control flows among macro-tasks for each layer are analyzed hierarchically, and represented by macro flow graph (MFG) as shown in Fig.1(a).

In the Fig. 1(a), nodes represent macro-tasks, solid edges represent data dependencies among macro-tasks and dotted edges represent control flow. A small circle inside a node represents a conditional branch inside a macro-task. Though arrows of edges are omitted in the macro flow graph, it is assumed that the directions are downward.

2.3 Generation of Macro Task Graph

To extract parallelism among macro-tasks from macro flow graph, compiler analyses Earliest Executable Condition of each macro-task. Earliest Executable Condition represents the conditions on which macro-task may begin its execution earliest.

Earliest Executable Condition of macro-task is represented in macro task graph (MTG) as shown in Fig. 1(b).

In the MTG, nodes represent macro-tasks. A small circle inside nodes represents conditional branches. Solid edges represent data dependencies. Dotted edges represent extended control dependencies. Extended control dependency means ordinary normal control dependency and the condition on which a data dependence predecessor macro-task is not executed. Solid and dotted arcs connecting solid and dotted edges have two different meanings. A solid arc represents that edges connected by the arc are in AND relationship. A dotted arc represents that edges connected by the arc are in OR relationship. In macro task graph, though arrows of edges are omitted assuming downward, an edge having arrow represents original control flow edges, or branch direction in macro flow graph.

2.4 Macro-Task Scheduling

In the coarse grain task parallel processing, static scheduling and dynamic scheduling are used for assignment of macro-tasks to processors or processor clusters which are defined by compiler. A suitable scheduling scheme is selected considering the shape of macro task graph and target machine parameters such as the synchronization overhead, data transfer overhead and so on.

If a macro task graph has only data dependencies and is deterministic, static scheduling is selected. In the static scheduling, assignment of macro-tasks to processors or processor clusters is determined by a scheduler at compile time. Static scheduling is useful since it allows us to minimize data transfer and synchronization overhead without run-time scheduling overhead.

If a macro task graph has control dependencies, dynamic scheduling is selected to cope with runtime uncertainties like conditional branches. Scheduling routine for dynamic scheduling are generated and embedded into a parallelized program with macro-task code by the compiler to eliminate the overhead for runtime thread scheduling by OS.

3 Static Coarse Grain Task Scheduling with Cache Optimization

In this section, the static scheduling algorithm considering cache optimization for coarse grain task parallel processing is described. In this paper, macro-tasks are assigned to processors, or threads, because the SMP machine used for this performance evaluation has only four processors.

The overview of the proposed algorithm is shown below. When a macro-task MT_i is executed on a processor, the data defined or referred in MT_i is likely to be on cache of the processor when the macro-task finishes. If a succeeding macro-task MT_j that shares a lot of data with MT_i is assigned to the same processor immediately after MT_i, a large portion of the shared data can be transferred from MT_i to MT_j through a cache.

3.1   Macro-Task Decomposition

In the case where the amount of data defined or referred in a macro-task MT_i is much larger than cache size, even if a macro-task MT_j which shares a large amount of data with MT_i is assigned immediately after MT_i, a large part of shared data would be already replaced, and couldn't be transferred through a cache to MT_j. Therefore, in this case macro-tasks and data should be decomposed into smaller macro-tasks with data fitting to cache.

To reduce data transfer overhead among macro-tasks assigned to different processors and to transfer the shared data through a cache between macro-tasks assigned to the same processor, a loop aligned decomposition [10,11] considering both amount of shared data and parallelism among macro-tasks is useful.

The loop aligned decomposition can be applied to arbitrary macro-task graph, in which RBs like doall loops and reduction loops are connected by data dependence edges.

3.2   Static Scheduling Algorithm DT-Gain/CP/MISF_DLG

This section proposes DT-gain/CP/MISF_DLG algorithm (Data Transfer Gain/ Critical Path/Most Immediate Successors First considering Data Localization Group). This algorithm schedules macro-tasks to processors so that tasks inside a Data Localization Group (DLG), which is a group of tasks generated by Loop Aligned Decomposition sharing the large data, are assigned to the same processor and a task outside DLG or an entrance task of a DLG are assigned to a processor having the largest Data Gain, or the most shared data to be accessed by the task. If there are multiple combinations of a ready task and a processor having the same Data Gain, a combination with a task having the largest path length to the exit node on the MTG (Macro Task Graph) is chosen. Furthermore, if there are multiple such combinations, a combination with a task having the largest number of immediate successors is chosen.

The details of the algorithm are follows.

Step 1: Calculate the largest path length, or CP, from each task node to the exit node on a target Macro Task Graph (MTG).

Step 2: Find ready tasks of which the all preceding tasks finish their execution or preceding task does not exist.

Step 3: If there is a ready task belonging to a Data Localization Group of which one or more preceding tasks inside the same DLG are already assigned to a processor, assign the ready task to the same processor as the preceding tasks.

Repeat Step 3 until such ready tasks does not exist

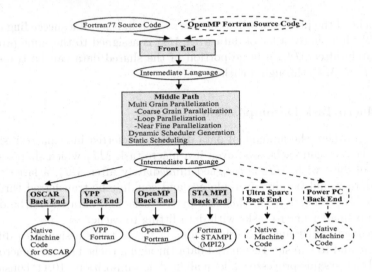

Fig. 2. A Composition of OSCAR Fortran Multigrain Parallelizing Compiler

Step 4: Calculate Data Gain for every combination of ready tasks out side
DLG or a ready task that is an entrance node of a DLG and idle processors.
Here, Data Gain is an amount of shared data existing on each processor to
be accessed by the ready task and means the data transfer amount to be
reduced if the ready task is assigned to the processor.

Assign a ready task to a processor in the combination with the largest Data
Gain. If there are combinations with the same largest data gain, choose
a combination including a ready task having the largest CP as the above
combination. Furthermore, if there are combinations having the largest CP,
choose a combination including a ready task having the most immediate
succeeding tasks.

If there exist tasks that have not been assigned, go to Step 2. Otherwise
finish scheduling.

4   Performance Evaluation

This section describes performance evaluation. Performance is evaluated on Sun
Ultra80 four-processor SMP workstation using Swim and Tomcatv from the
SPEC fp 95.

4.1   OSCAR Fortran Multigrain Parallelizing Compiler

OSCAR Fortran multigrain parallelizing compiler, on which the proposed
scheduling scheme, consists of a front end, a middle path and multiple back
ends is as shown in Fig. 2.

A front end reads in Fortran77 and OpenMP and generates intermediate
language.

A middle path analyses control flow and data dependence, restructures program, generates macro-tasks and exploits parallelism. It statically schedules coarse grain tasks considering cache optimization and generates parallelized intermediate code.

In OSCAR Fortran compiler, variety of back ends as shown in Fig. 2 are provided for different target machines like OSCAR multiprocessor system [13], UltraSparc processors, Fujitsu VPP vector supercomputers, heterogeneous cluster computing system with STAMPI and SMP machines with OpenMP. They generate assembly codes or parallelized Fortran codes with library calls or directives for each target machines.

The proposed static scheduling scheme for coarse grain tasks considering cache optimization is implemented on the middle path. Our coarse grain task parallel processing using OpenMP [3,9] uses "one time single level thread generation method" which forks and joins threads only once at the beginning and the end of execution respectively to reduce thread generation overhead. However, this method realizes hierarchical coarse grain task parallel processing using OpenMP, regardless single level thread generation, by generating different codes for each thread [9]. The generated OpenMP Fortran is compiled by a native compiler for target machine and executed. In other words, OSCAR Fortran multigrain parallelizing compiler is used as a preprocessor that translates Fortran77 into parallelized OpenMP Fortran.

4.2 Evaluation Environment

This subsection describes multiprocessor workstation Sun Ultra80, its compiler and benchmark programs used for the evaluation.

The specification of Sun Ultra80 four-processor SMP workstation and Forte loop automatic parallelizing compiler are shown in Table 1. Also, the used compile option for Forte compiler is shown in Table 2. In Table 2, Forte means compile options used in Forte compiler for a single processor and for automatic parallelization. Also OSCAR means compile options when OpenMP codes generated by OSCAR compiler are compiled by Forte compiler.

As application programs for the evaluation, Swim and Tomcatv from the SPEC fp 95 benchmark suite are used. Ref data set is used as input data.

In this evaluation, some parts of program sources are restructured by hand because the current version of OSCAR compiler has concentrated on developing original schemes and has not implemented some traditional program restructuring techniques. For example, in Swim, three subroutines, CALC1, CALC2 and CALC3, consume large execution time. These subroutines share a large amount of data. For these three subroutines, inline expansion and flexible cloning [14] are manually applied. Also, in Tomcatv, an array size is input from an input file at runtime. However, in this paper, array size of ref data set is described as a parameter to use static scheduling. Also, loop interchange and array subscript exchange are manually applied for Tomcatv.

Performance evaluation results for the restructured Swim and Tomcatv are shown in Fig. 3, 4, respectively. These figures show speedup against sequential

Table 1. The Specification of Sun Ultra80

Vender	Sun Microsystems
CPU	450MHz UltraSPARC-II
	4 processors SMP
L1 Instruction Cache	16Kbyte
	Pseudo 2-Way Set Associative
	Line size: 32byte
L1 Data Cache	16Kbyte, Direct-Map
	Line size: 32byte
	(Two 16byte sub-blocks)
	Write-through
	Non-allocating
L2 Unified Cache	4Mbyte, Direct-Map
	Line size: 64byte
	Write-back, Allocating
Main Memory	1024Mbyte
OS	Solaris8
Compiler	Forte[tm]
	HPC 6 update 1

Table 2. Compile Option

	Single processor	Multiprocessor
Forte		-fast -parallel
		-reduction
		-stackvar
	-fast	
OSCAR		-fast
		-explicitpar
		-mp=openmp

execution time by Forte compiler for a single processor. Also, each execution time was measured five times and the fastest time was as the plotted result.

In Fig. 3, the sequential execution time of Swim compiled by Forte was 99.7 seconds. The speedups (execution times) of automatic loop parallelization by Forte compiler were 1.51 times (66.1 seconds) for 2PEs, 1.60 times (62.5 seconds) for 3PEs and 1.66 time (60.2 seconds) for 4PEs. When OSCAR compiler was used as a preprocessor of Forte compiler, the speedups (execution times) were 1.23 times (81.3 seconds) for 1PE, 2.08 times (47.9 seconds) for 2PEs, 3.59 times (27.8 seconds) for 3PEs and 7.55 times (13.2 seconds) for 4PEs. The speedup by the proposed scheme was super linear by successful cache optimization.

In Fig. 4, the sequential execution time of Tomcatv compiled by Forte was 107.8 seconds. The speedups (execution times) of automatic loop parallelization by Forte compiler were 1.28 times (84.3 seconds) for 2PEs, 1.36 times (79.3 seconds) for 3PEs and 1.37 time (78.6 seconds) for 4PEs. When OSCAR compiler was used as a preprocessor of Forte compiler, the speedups (execution times)

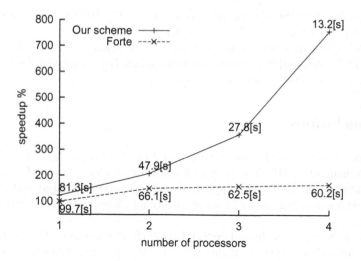

Fig. 3. Speedup of Swim

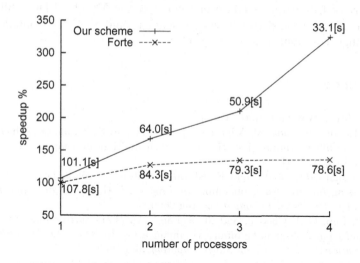

Fig. 4. Speedup of Tomcatv

were 1.07 times (101.1 seconds) for 1PE, 1.68 times (64.0 seconds) for 2PEs, 2.12 times (50.9 seconds) for 3PEs and 3.26 times (33.1 seconds) for 4PEs.

At these evaluations, L2 cache misses in Swim compiled by Forte compiler were about 3.4×10^8 for 1PE, 3.2×10^8 for 2PEs, 3.1×10^8 for 3PEs and 3.0×10^8 for 4PEs. On the contrary, L2 cache misses of OSCAR compiler were about 2.0×10^8 for 1PE, 1.8×10^8 for 2PEs, 9.3×10^7 for 3PEs and 4.5×10^7 for 4PEs. Also, though L2 cache misses in Tomcatv compiled by Forte compiler were about 2.7×10^8 for 1PE, 3.8×10^8 for 2PEs, 3.9×10^8 for 3PEs and 4.0×10^8 for 4PEs,

488   Hirofumi Nakano et al.

L2 cache misses of OSCAR compiler were about 2.0×10^8 for 1PE, 2.4×10^8 for 2PEs, 1.3×10^8 for 3PEs and 9.3×10^7 for 4PES.

These results show the proposed static coarse grain task scheduling scheme realizes 4.56 times and 2.37 times speedup against Forte compiler for Swim and Tomcatv by the improvement of L2 cache hit.

5 Conclusions

This paper has proposed a static coarse grain task scheduling with cache optimization using OpenMP. The proposed scheme is implemented on OSCAR Fortran multigrain parallelizing compiler, sequential Fortran is input and OpenMP Fortran whose coarse grain tasks are statically scheduled with cache optimization is output.

Its performance is evaluated on Sun Ultra80 four-processor SMP workstation, using Swim and Tomcatv from the SPEC fp 95. The results of evaluation show us that speedups of Swim and Tomcatv for 4 processors were 7.55 times and 3.26 times respectively against sequential execution time of them compiled by Forte.

A part of this research has been supported by METI/NEDO Millennium Project IT21 "Advanced Parallelizing Compiler" and STARC "Compiler cooperative single chip multiprocessor" project.

References

1. APC. http://www.apc.waseda.ac.jp/.
2. M. Okamoto, K. Aida, M. Miyazawa, H. Honda, and H. Kasahara. A hierarchical macro-dataflow computation scheme for oscar multi-grain compiler. *Trans. IPSJ*, 35(4):513–521, 1994.
3. H. Kasahara, M. Obata, and K. Ishizaka. Automatic coarse grain task parallel processing on smp using openmp. In *Proc. 12th Workshop on Languages and Compilers for Parallel Computing*, Aug 2000.
4. A. W. Lim, G. I. Cheong, and M. S. Lam. An affine partitioning algorithm to maximize parallelism and minimize communication. In *Proc. 13th ACM SIGARCH International Conference on Supercomputing*, Jun 1999.
5. A. W. Lim, S. Liao, and M. S. Lam. Blocking and array contraction across arbitrarily nested loops using affine partitioning. In *Proc. of the Eighth ACM SIGPLAN Symposium on Principles and Practice of Parallel Programming*, Jun 2001.
6. A. W. Lim and M. S. Lam. Cache optimizations with affine partitioning. In *Proc. of the Tenth SIAM Conference on Parallel Processing for Scientific Computing*, Mar 2001.
7. S. Vajracharya, S. Karmesin, P. Beckman, J. Crotinger, A. Malony, S. Shende, R. Oldehoeft, and S. Smith. Smarts: exploiting temporal locality and parallelism through vertical execution. In *Proc. of the 1999 international conference on Supercomputing*, Jun 1999.
8. D. Inaishi, K. Kimura, K. Fujimoto, W. Ogata, M. Okamoto, and H. Kasahara. A cache optimization with earliest executable condition analysis. In *Technical report of IPSJ*, Aug 1998.

9. K. Ishizaka, M. Obata, and H. Kasahara. Coarse grain task parallel processing with cache optimization on shared memory multiprocessor. In *Proc. 14th Workshop on Languages and Compilers for Parallel Computing*, Aug 2001.
10. A. Yoshida, K. Koshizuka, M. Okamoto, and H. Kasahara. A data-localization scheme among loops for each layer in hierarchical coarse grain parallel processing. *Trans. IPSJ*, 40(5):2054–2063, 1999.
11. A. Yoshida, S. Yagi, and H. Kasahara. A data-localization scheme for macrotask-graph with data dependencies on smp. In *Technical report of IPSJ, 2001-ARC-141*, Jan 2001.
12. H. Kasahara. *Parallel Processing Technology*. CORONA PUBLISHING CO., LTD., 1991.
13. H. Kasahara, H. Honda, A. Mogi, A. Ogura, K. Fujiwara, and S. Narita. A multi-grain parallelizing compilation scheme for oscar. *Proc. 4th Workshop on Languages and Compilers for Parallel Computing*, Aug 1991.
14. K. Yoshii, G. Matsui, M. Obata, S. Kumazawa, and H. Kasahara. An analysis-time procedure inlining scheme for multi-grain automatic parallelizing compilation. In *Technical report of IPSJ, ARC/HPC*, Mar 2000.

High Performance Fortran –
History, Status and Future

Hans P. Zima

University of Vienna, Austria

Abstract. High Performance Fortran (HPF) is a data-parallel language that was designed to provide users with a high-level interface for programming scientific applications, while delegating to the compiler the task of generating an explicitly parallel message-passing program. This lecture will outline the original motivation for the development of HPF and provide an overview of various versions of the language. The focus will be on a study of the expressivity of the HPF approach in the context of the requirements posed by "real" applications, and the need to achieve target code performance close to that of hand-written MPI programs. We will identify a set of important language elements for the efficient solution of a range of advanced applications in science and engineering. Dealing with such problems does not only need the flexible data and work distribution features included in the HPF-2 Standard, but also requires additional capabilities such as the explicit control of some aspects of communication. The lecture concludes with an outlook to future research and development in HPF-related languages and compilers.

H. Zima et al. (Eds.): ISHPC 2002, LNCS 2327, p. 490, 2002.
© Springer-Verlag Berlin Heidelberg 2002

Performance Evaluation for Japanese HPF Compilers with Special Benchmark Suite

Hitoshi Sakagami and Shingo Furubayashi

Computer Engineering, Himeji Institute of Technology, 2167 Shosha, Himeji, Japan
{sakagami,shingo}@comp.eng.himeji-tech.ac.jp

Abstract. The lack of performance portability has been disheartening scientific application users to develop portable programs written in HPF. As the users would like to run the same source code on different parallel machines as fast as possible, we have investigated the performance portability for Japanese HPF compilers (NEC and Fujitsu) with a special benchmark suite. We got good performance in most cases with DISTRIBUTE and INDEPENDENT directives on NEC SX-5, but Fujitsu VPP800 required to explicitly force no communication inside parallel loops with additional LOCAL directives. It was also found that manual optimizations for communication with HPF/JA extensions were very useful to tune parallel performance.

1   Introduction

HPF was defined as a standard language for data parallel programming, but many users were discouraged to develop parallel applications because they could only get poor performance due to immature compilers in the early stage. This situation prevents the spread of parallel computing into the computational science community, even parallel programming is mostly restricted to the computer science community.

JAHPF (Japan Association for HPF) [1], which has started in 1996, is an informal coalition of HPC users and computer vendors in Japan to promote the HPF language. JAHPF has defined a set of HPF extensions as HPF/JA [2] to enhance capability of HPF, and is now expanded into a new organization, HPF Promoting Consortium (HPFPC) [3]. Recently Japanese computer vendors have released HPF compilers for their vector-parallel supercomputers as a part of HPFPC's activities. We found that their HPF compilers could achieve good performance for real-world scientific applications [4].

The users want to run the same source code on different machines as fast as possible. In this paper, we have checked performance portability between Japanese HPF compilers that were implemented on NEC SX-5 and Fujitsu VPP800, using a special benchmark suite. NEC SX-5 consists of interconnected nodes. A node contains multiple, up to 16, vector processors (10GFLOPS) that share the memory within the node. Thus, an inter-node communication (16GB/s per pair of nodes) is much slower than an intra-node transfer (64GB/s per processor) based on the shared memory architecture [5]. Fujitsu VPP800 also consists of multiple vector processors (8GFLOPS), but all processors are uniformly connected with crossbar switches unlike NEC SX-5 and the communication speeds (3.2GB/s per pair of processors) between

H. Zima et al. (Eds.): ISHPC 2002, LNCS 2327, pp. 491–502, 2002.

any pairs of processors are constant [6]. As NEC SX-5 and Fujitsu VPP800 are vector-parallel computers, we focused on vector-parallel execution. Thus we not only inserted DISTRIBUTE directives but also other necessary HPF directives to execute target loops with a vector-parallel mode. It is important to carefully and manually tune communications especially for a large number of processors because of the Amdahl's law. HPF/JA has defined manual SHADOW and REFLECT directives to provide this feature. A trial to improve parallel performance with these HPF/JA extensions has been also carried out. The compiler version and options that we have used are summarized in Table 1.

Table 1. The summary of compiler version and options

NEC SX-5	HPF/SX V2 hpf driver:Rev.1.2	-O
Fujitsu VPP800	UXP/V HPF V20L20 L00121	-Wh,-Lt,-Owl -Wv,-m3

2   Benchmark Suite

A special benchmark suite, called the TNO HPF benchmark suite [7], has been designed to test performance portability between compilers. It consists of a set of HPF programs that test various aspects of efficient parallel code generation. The benchmark suite consists of a number of template programs that can generate test HPF programs with different alignments, distributions, assignments, and directives. This gives an insight in the effectiveness of the various algorithms used by the compiler. The lines prefixed with "!Dn!" in the template files specify the distribution and alignment of the used array, where n is a digit for variations. The lines prefixed with "!An!" are the different assignment and those with "!In!" specify the different directives. A simple script file is used to generate many test-files from single template file for any possible combinations of various test spaces. If the prefix D ranges from 1 to 6 and I from 1 to 2, twelve test-files from D1-I1 to D6-I2 are generated. To make D3-I2 test file, for example, the script deletes all lines which contain "!Dn!" excluding "!D3!" or "!Im!" excluding "!I2!", and removes "!*!" style comments from each line that includes "!D3!" or "!I2!" to activate HPF directives. These test-files are then executed and timed. They have run the TNO benchmark suite on three compilers: the PREPARE prototype compiler, the PGI-HPF compiler, and the GMD Adaptor HPF compiler. Thus we can compare Japanese compilers with above three.

We applied the following template files in this paper.

- init1D.tmpl: initialization of a one-dimensional array.
- copy1D.tmpl: copying between two one-dimensional arrays.
- stencil1D.tmpl: several stencil assignments.
- SCR1D.tmpl: assignment where sender computes rule applies.
- semireg1D.tmpl: several semi-regular assignments.

The target loops are coded with FORALL statements in the original TNO benchmark suite, but most actual scientific applications are still written in Fortran77 and loops are coded with DO statements. In addition, we are focusing on vector-parallel execution and expecting automatic loop vectorization. Japanese compilers have mature experience in vectorizing DO loops and therefore we converted FORALL statements to DO statements. To achieve best performance, we substituted a

NUMBER_OF_PROCESSORS function for a static parameter, which tells an exact number of processors at run time to compilers, in PROCESSORS directives to avoid ambiguous number of processors in parallel execution. We also enlarged a length of target loops and a number of iteration for accurate measurement, and we omitted block-like and cyclic-like distributions from the original TNO benchmark suite.

2.1   Initializing an Array

We use a simple one-dimensional array assignment that does not need communication to measure the overhead that is introduced by local enumeration. An additional DO loop is placed around the target DO loop to keep the total measured time far away from the system clock resolution. Our init1D.tmpl template is shown in Fig. 1.

```
      program init1D
      integer, parameter :: N=100000,imaxa=10000
      integer            :: i,iter,l,u,s
      real, dimension(N) :: a
      real*8 :: hpfja_gettod,start_time,end_time
!HPF$ PROCESSORS P(NUMBER_OF_PROCESSORS())
!D1,2!!HPF$ TEMPLATE tmpl(N)
!D3,4!!HPF$ TEMPLATE tmpl(N+2)
!D5,6!!HPF$ TEMPLATE tmpl(3*N+7)
!D1,3,5!!HPF$ DISTRIBUTE tmpl(BLOCK)   onto P
!D2,4,6!!HPF$ DISTRIBUTE tmpl(CYCLIC) onto P
!D1,2!!HPF$ ALIGN (i) with tmpl(  i  ) :: a
!D3,4!!HPF$ ALIGN (i) with tmpl(  i+2) :: a
!D5,6!!HPF$ ALIGN (i) with tmpl(3*i+7) :: a
!I1!!no HPF directive
      start_time = hpfja_gettod()
      do iter = 1, imaxa
        call values (N,iter,l,u,s)
!I2!!HPF$ INDEPENDENT
        do i = 1, u, s
          a(i) = 1.0
        end do
      end do
      end_time = hpfja_gettod()
      write(*,*) end_time - start_time
      stop
      end program
      subroutine values (n,iter,l,u,s)
      integer :: n,l,u,s,iter
      l = iter
      l = 1; u = n; s = 1
      return
      end subroutine
```

Fig. 1. Our init1D.tmpl template

In case of NEC SX-5, the target DO loop is vectorized but not parallelized with only the BLOCK DISTRIBUTE directive. As the HPF compiler on NEC SX-5 converts a HPF source into a Fortran source as a preprocessor, we can check how the compiler works by looking into the precompiled Fortran code. We found that the target loop was translated into a serial code with an execution guard, which was a most naive implementation. For a CYCLIC distribution, the loop is not even vectorized. Adding an INDEPENDENT directive, the loops with both distributions are vector-parallelized and result in good achievement regardless of three alignments, but execution errors occurred at a number of processors is 1 and 4 with alignment (3*i+7).

The target loop is neither vectorized nor parallelized with only BLOCK or CYCLIC DISTRIBUTE directives on Fujitsu VPP800. The loop with a BLOCK distribution is vector-parallelized with an additional INDEPENDENT directive, but it does not indicate improvement of parallel performance and even inferior performance for alignment (3*i+7). The INDEPENDENT directive also make the loop with a CYCLIC distribution vector-parallelize, but it shows low performance for all alignments even no communication is needed for execution. An initialization of the CYCLIC DO loop seems to waste time tremendously.

The measured results in seconds with the DISTRIBUTE and INDEPENDENT directives are given in Table 2 for both machines. "-----" indicates that it takes more than 60 seconds and shows poor performance, and "error" means that an error occurs and an execution is aborted.

Table 2. The execution time [sec] for init1D.tmpl

(a) NEC SX-5

proc#	tmpl(i) BLOCK	CYCLIC	tmpl(i+2) BLOCK	CYCLIC	tmpl(3*i+7) BLOCK	CYCLIC
1	0.4177	0.4208	0.4329	0.4359	0.4328	error
2	0.2148	0.2178	0.2309	0.2333	0.2300	0.2323
4	0.1133	0.1163	0.1290	0.1321	0.1292	error
8	0.0625	0.0655	0.0778	0.0804	0.0774	0.0805

(b) Fujitsu VPP800

proc#	tmpl(i) BLOCK	CYCLIC	tmpl(i+2) BLOCK	CYCLIC	tmpl(3*i+7) BLOCK	CYCLIC
1	0.9387	-----	0.9399	-----	1.1849	-----
2	0.8667	-----	0.8669	-----	0.9604	-----
4	0.8463	-----	0.8387	-----	0.8838	-----
8	0.8487	-----	0.8495	-----	0.9382	-----

2.2   Copying an Array

We have defined a simple test in which array B is assigned to array A for different distributions of A and B as shown in Fig. 2 to measure the communication overhead.

In case of NEC SX-5, the target DO loop with any combination of array distributions is vector-parallelized with the INDEPENDENT directive, but decent parallel performance is obtained only if both array A and B have the same distributions. With an additional ON HOME(b(i)) directive, the parallel performance for different distributions of array A and B is drastically improved in spite of necessity of communications. According to the converted Fortran code, the target loop is treated as an assignment from array B to a temporary array instead of array A. As there is no reference code to array A in the program, the compiler does not generate communication codes which reflect contents of the temporary array to its original array A. Thus the results are outward and more tricky coding will be needed to measure the communication overhead correctly.

```
              program copy1D
              integer, parameter :: N=100000,imaxa=5000
              integer            :: i,iter,l,u,s
              real               :: a(N),b(N)
              real*8 :: hpfja_gettod,start_time,end_time
       !HPF$ PROCESSORS P(NUMBER_OF_PROCESSORS())
       !D1,2!!HPF$ DISTRIBUTE a(BLOCK)  onto P
       !D3,4!!HPF$ DISTRIBUTE a(CYCLIC) onto P
       !D1,3!!HPF$ DISTRIBUTE b(BLOCK)  onto P
       !D2,4!!HPF$ DISTRIBUTE b(CYCLIC) onto P
       !I1!!no HPF directive
              start_time = hpfja_gettod()
              do iter = 1, imaxa
                 call values (N,iter,l,u,s)
       !I2,3!!HPF$ INDEPENDENT
                 do i = l, u ,s
       !I3!!HPF$ ON HOME(b(i))
                    a(i) = b(i)
                 end do
              end do
              end_time = hpfja_gettod()
              write(*,*) end_time - start_time
              stop
              end program
```

Fig. 2. Our copy1D.tmpl template

The target loop is vector-parallelized with the INDEPENDENT directive only if the distribution of array A is same as that of array B on Fujitsu VPP800. Even the communication is not needed and the loop is vector-parallelized in the cyclic-cyclic case, it shows low performance. We could not find any changes with the ON HOME(b(i)) directive unlike NEC SX-5. Finally we fully force no communication with a handwritten LOCAL directive for the cyclic-cyclic distribution, but it still shows low performance. The target DO loop is still CYCLIC DO and it wastes the same amount of time as discussed in Section 2.1.

All results are shown in Table 3. "no vec" in the table means that the loop is parallelized but not vectorized and we did not measure such execution mode.

Table 3. The execution time [sec] for copy1D.tmpl

| | | NEC SX-5 | | | | Fujitsu VPP800 | |
| | | INDEPENDENT | | ON HOME(b(i)) | | INDEPENDENT | |
	proc#	b(BL)	b(CY)	b(BL)	b(CY)	b(BL)	b(CY)
a(BLOCK)	1	0.3332	1.7260	0.3125	0.4114	0.5939	no vec
	2	0.1590	-----	0.1586	0.2704	0.4804	no vec
	4	0.0830	-----	0.0825	0.1965	0.4574	no vec
	8	0.0486	-----	0.0445	0.1557	0.4642	no vec
a(CYCLIC)	1	1.7540	0.3365	0.3922	0.3126	no vec	-----
	2	-----	0.1669	0.2485	0.1608	no vec	-----
	4	-----	0.0933	0.1738	0.0926	no vec	-----
	8	-----	0.0479	0.1353	0.0463	no vec	-----

2.3   Stencil Assignments

Stencil operations are an important scheme in many applications and are subject to special optimizations especially with BLOCK distributions. The two assignments *A1* and *A2* are shown in Fig. 3 and we omit CYCLIC distributions from the original template.

```
        program stencil1D
        integer, parameter :: N=100000,imaxa=5000
        integer            :: i,iter,l,u,s
        real               :: a(N),b(N)
        real*8 :: hpfja_gettod,start_time,end_time
!HPF$ PROCESSORS P(NUMBER_OF_PROCESSORS())
!HPF$ DISTRIBUTE (BLOCK) onto P :: a,b
!I1!!no HPF directive
!A1!!I3!!HPF$ SHADOW b(0:1)
!A2!!I3!!HPF$ SHADOW b(0:10)
        start_time = hpfja_gettod()
        do iter = 1, imaxa
            call values(N,iter,l,u,s)
!I3!!HPFJ REFLECT b
!I2,3!!HPF$ INDEPENDENT
            do i = l, u, s
!I3!!HPF$ ON HOME(a(i)), LOCAL
!A1!          a(i) = b(i) + b(i+1)
!A2!          a(i) = b(i) + b(i+10)
            end do
        end do
        end_time = hpfja_gettod()
        write(*,*) end_time - start_time
        stop
        end program
```

Fig. 3. Our stencil1D.tmpl template

The target loop is not vectorized without the INDEPENDENT directive and vector-parallelized with it for both assignments on NEC SX-5. The HPF compiler on NEC SX-5 implements an automatic shadow optimization if stencil offsets are less than a default shadow width, namely four. The *A1* assignment allows this stencil optimization, because the stencil offset does not exceed the default shadow width. If the stencil offset exceeds the default shadow width, the compiler has to fall back to copying the remote values into a temporary variable. Thus the *A1* assignment shows better absolute performance than *A2*. To assess this automatic optimization, we interrupt the optimization by applying the *-Moverlap=size:0* compiler option and the performance for *A1* is found to be degraded to the same level of *A2*. As we can not get parallel speedups even with the automatic optimization, we should try a manual optimization with SHADOW, REFLECT and LOCAL directives. The manual optimization with the *I3* template shows best absolute and parallel performance for both assignments. Explicit REFLECT communications are more efficient than shift communications which are automatically generated by the compiler.

In case of Fujitsu VPP800, communication codes are generated within the parallel loop and the target DO loop is not vectorized unless we force no communication with the *I3* template. We can get better parallel performance than that of NEC SX-5.

All results are summarized in Table 4.

Table 4. The execution time [sec] for stencil1D.tmpl

(a) NEC SX-5

proc#	INDEPENDENT		INDEP. w/o auto opt.		SHAD.,REF.,LOCAL	
	b(i+1)	b(i+10)	b(i+1)	b(i+10)	b(i+1)	b(i+10)
1	0.6821	1.7850	1.9393	1.7857	0.6920	0.7050
2	0.6442	1.6085	1.6957	1.6118	0.6607	0.6571
4	0.6691	1.5468	1.5866	1.5607	0.6640	0.6611
8	0.6723	1.4987	1.5184	1.5037	0.6704	0.7008

(b) Fujitsu VPP800

proc#	INDEPENDENT				SHAD.,REF.,LOCAL	
	b(i+1)	b(i+10)			b(i+1)	b(i+10)
1	no vec	no vec			0.7907	0.7845
2	no vec	no vec			0.6092	0.6077
4	no vec	no vec			0.5471	0.5445
8	no vec	no vec			0.5452	0.5443

2.4   Sender Computes Rule

Normally, compilers use the Owner Computes Rule to determine on which processor the RHS of an assignment is to be evaluated. Sometimes, however, it can be advantageous to deviate from that rule and evaluate the RHS (or part of it) on the sender side. This is called the Sender Computes Rule and the benchmark template to estimate performance of this rule is given in Fig. 4. We omit CYCLIC distributions from the original template and add another variables to enhance advantage of the Sender Computes Rule.

The target loop is not vectorized without the INDEPENDENT directive. The NEC HPF compiler uses the Owner Computes Rule and vectorizes the loop with the INDEPENDENT directive, resulting into four communication assignments and poor performance. We can improve the performance to manually switch to the Sender Computes Rule with an ON HOME(b(i+10)) directive, only introducing one communication assignment. But the improvement seems to be too much because at least one communication is needed. The value is calculated on the sender side and is only assigned into a temporary array without communications in the converted Fortran code. Thus the results are also wrong just like in Section 2.2 and a mature code will be required. Still using the Owner Computes Rule, we can also improve the performance with SHADOW, REFLECT and LOCAL directives, namely the *I4* template. Even four variables must be reflected in the *I4* template, this template shows better performance than the *I2* template due to efficient communications of the explicit REFLECT directive. But large degradation in the parallel performance can be seen if the number of processors is more than four.

```
        program SCR1D
        integer, parameter :: N=100000,imaxa=5000
        integer            :: i,iter,l,u,s
        real               :: a(N),b(N),c(N),d(N),e(N)
        real*8 :: hpfja_gettod,start_time,end_time
!HPF$ PROCESSORS P(NUMBER_OF_PROCESSORS())
!HPF$ DISTRIBUTE (BLOCK) onto P :: a,b,c,d,e
!I1!!no HPF directive
!I4!!HPF$ SHADOW (0:10) :: b,c,d,e
        start_time = hpfja_gettod()
        do iter = 1, imaxa
            call values (N,iter,l,u,s)
!I4!!HPFJ REFLECT b,c,d,e
!I2,3,4!!HPF$ INDEPENDENT
            do i = 1, u, s
!I3!!HPF$ ON HOME(b(i+10))
!I4!!HPF$ ON HOME(a(i)), LOCAL
            a(i) = b(i+10)+c(i+10)+d(i+10)+e(i+10)
            end do
        end do
        end_time = hpfja_gettod()
        write(*,*) end_time - start_time
        stop
        end program
```

Fig. 4. Our SCR1D.tmpl template

The Fujitsu HPF compiler generates four communications within the target loop with the INDEPENDENT directive using the Owner Computes Rule. The ON HOME(b(i+10)) directive can switch to the Sender Computes Rule, but one communication is still inside the parallel loop and it is not vectorized. Thus we must add SHADOW, REFLECT and LOCAL directives to achieve the vector-parallel execution. We can get no degradation even with eight processors and better parallel performance than that of NEC SX-5.

All results are summarized in Table 5.

Table 5. The execution time [sec] for SCR1D.tmpl

proc#	NEC SX-5			Fujitsu VPP800
	INDEP.	HOME(b(i+10))	SHAD.,REF.,LOC.	SHAD.,REF.,LOC.
1	5.9302	0.7571	2.6345	1.0752
2	5.5627	0.4324	2.8569	0.7665
4	5.5410	0.2739	4.2012	0.8050
8	5.3991	0.1891	4.4310	0.6519

2.5   Semi-regular Assignments

Semi-regular gather and scatter assignments are used to test the efficiency of irregular communication patterns as shown in Fig. 5. The assignments *A1* and *A3* have an indirection array that is chosen such that no actual communication is needed, but *A2* and *A4* templates require real communications. The normal optimization scheme for parallel loops that have indirection in the array subscript expressions is to use an inspector/executor scheme [8]. In the inspector phase the communication pattern is measured by traversing the parallel loop and in the execution phase this communication pattern (often called communication schedule) is used to communicate the remote array elements. Normally this procedure is followed each time the parallel loop is invoked. If however, the schedule does not change between succeeding loop invocations, a smart compiler could reuse the schedule, either through analysis or special compiler directives. HPF/JA has defined this special feature as an INDEX_REUSE directive, but both compilers do not support this directive at this time and we must leave this optimization to the compilers.

```
            program semireg1D
            integer, parameter :: N=100000,imaxa=5000
            integer            :: i,iter,l,u,s,ind(N)
            real               :: a(N),b(N)
            real*8 :: hpfja_gettod,start_time,end_time
!HPF$ PROCESSORS P(NUMBER_OF_PROCESSORS())
!HPF$ DISTRIBUTE (BLOCK) onto P :: a,b
!I1!!no HPF directive
!I5!!HPF$ SHADOW b(*)
            do i = 1, N
!A1,3!   ind(i) = i
!A2,4!   ind(i) = N - i + 1
            end do
            start_time = hpfja_gettod()
            do iter = 1, imaxa
                call values(N,iter,l,u,s)
!I5!!HPFJ REFLECT b
!I2,3,4,5!!HPF$ INDEPENDENT
                do i = l, u, s
!A1,2!!I3!!HPF$ ON HOME(a(i))
!A3,4!!I4!!HPF$ ON HOME(b(i))
!A1,2!!I5!!HPF$ ON HOME(a(i)), LOCAL
!A1,2!       a(i) = b(ind(i))
!A3,4!       a(ind(i)) = b(i)
                end do
            end do
            end_time = hpfja_gettod()
            write(*,*) end_time - start_time
            stop
            end program
```

Fig. 5. Our semireg1D.tmpl template

The NEC HPF compiler can vector-parallelize the target loop for all assignments, but all result in fearful performance. We can specify gather communication with an ON HOME(a(i)) directive to the gather assignment, namely the *I3* template, but no improvement is observed. Adding an ON HOME(b(i)) directive to the scatter assignment, namely the *I4* template, explicit scatter communication is indicated to the compiler and drastic progress is found. As the assignment *A3* causes no communication indeed but *A4* needs actual communication, there is few difference in performance and this result may be incorrect. Looking into the converted Fortran code, the target loop is treated as an assignment from array *B* to a temporary array instead of array *A*. As array *A* is never referred in this program, the compiler does not generate codes to communicate from the temporary array to its original array A. Thus the results are also wrong just like in Section 2.2 and 2.4, and more sophisticated coding will be essential to test the efficiency of irregular communication patterns correctly. It is noted that the compiler on NEC SX-5 does not support the FULL SHADOW directive at this time and we could not run the $I5$ template.

In case of Fujitsu VPP800, the target loop is never vectorized except with the $I5$ template in which FULL SHADOW, REFLECT and LOCAL directives are introduced. Only a local part of the whole declared array area is usually allocated onto each processor, but the whole declared array area is allocated onto each processor with FULL SHADOW directives. Therefore memory usage is very inefficient when FULL SHADOW directives are used for a large array, but the conversion from global subscript to local one at reference of the array is not necessary and a program can be executed at high speed. In spite of patterns of indirection in the array subscript expressions, the communication cost of reflecting full shadow variables is same and we can get same performance. According to the overhead of full shadow reflections, only inferior parallel performance is achieved. We can not vector-parallelize the scatter communication in the same manner with HPF/JA extensions because there is no method to inversely reflect shadow contents to its original.

All results are shown in Table 6.

Table 6. The execution time [sec] for semireg1D.tmpl

	NEC SX-5 ON HOME(b(i)) A(ind(i))=B(i)		Fujitsu VPP800 SHAD.,REF.,ON HOME(a(i)),LOC. A(i)=B(ind(i))	
proc#	ind(i)=i	ind(i)=n-i+1	ind(i)=i	ind(i)=n-i+1
1	0.3806	0.3806	0.7840	0.7844
2	0.2366	0.2367	1.2387	1.2410
4	0.1635	0.1635	2.1427	2.1313
8	0.1226	0.1228	3.3161	3.2933

2.6   Modification of Benchmark Suite

Both HPF compilers inform that the target loop is parallelized and/or vectorized, but no information about usage of a temporary array is notified. In most cases, the HPF compiler on NEC SX-5 automatically introduces a temporary array to improve performance, and generates no communication code which copy values from the temporary array to the original array if no reference to the original array exists. Thus the TNO HPF benchmark suite is too simple to measure parallel performance correctly in such cases, and benchmark programs should be modified with more careful coding. To achieve this, reference codes for an assigned array are added as shown in Fig. 6. As the assigned array is used as an actual argument of a subroutine call statement, the compiler thinks that the array has a chance to be referred within the subroutine and values of the assigned array must be definite before calling the subroutine. Therefore communication codes are generated by the compiler and we can fairly measure the performance.

With the modified copy1D.tmpl template, we found that the ON HOME(b(i)) directive could not improve the parallel performance for different distributions of array A and B at all. If both array A and B have the same distributions, the target loop is executed with the vector-parallel mode without communication and this speed is far and away faster than communication speed of copying the array for different distributions.

```
...
do iter = 1, imaxa
   call values (N,iter,l,u,s)
   do i = 1, u, s
              = ... b(i) ...
      a(i) = ...
   end do
   call refer (a,N)
end do
...
subroutine refer (ref,n)
integer :: n
real    :: ref(n)
return
end subroutine
```

Fig. 6. Reference codes for an assigned array are introduced into the benchmark suite to measure parallel performance correctly

The modified SCR1D.tmpl template measured reasonable performance with the ON HOME(b(i+10)) directive, because four communication assignments are reduced into one due to the Sender Computes Rule as shown in Table 7. The performance using the Sender Computes Rule is better than that using SHADOW, REFLECT and LOCAL directives, even Fujitsu VPP800 shows best performance with those directives.

Table 7. The execution time [sec] for the modified SCR1D.tmpl template on NEC SX 5

proc#	before modification		after modification
	INDEP.	HOME(b(i+10))	HOME(b(i+10))
1	5.9302	0.7571	2.1140
2	5.5627	0.4324	1.8318
4	5.5410	0.2739	1.6558
8	5.3991	0.1891	1.7170

We can not get decent performance with the modified semireg1D.tmpl template for all cases due to the necessity of semi-regular communication. Although the HPF compiler on Fujitsu VPP800 can vector-parallelize the semi-regular loop with FULL SHADOW, REFLECT and LOCAL directives, an INDEX_REUSE HPF/JA extension that reuses the communication schedule would be very useful to get better parallel performance for semi-regular assignments.

3   Summary

Insufficient portability of performance has been discouraging scientific application users to develop portable programs in HPF and preventing the spread of parallel computing into the computational science community.

We have evaluated the performance portability of Japanese HPF compilers with a special set of programs, called the TNO benchmark suite. We could get good performance in most cases with DISTRIBUTE and INDEPENDENT directives on NEC SX-5, but we had to explicitly force no communication inside parallel loops with additional LOCAL directives and to insert necessary communications outside the loops with SHADOW and REFLECT directives on Fujitsu VPP800. We also found that manual optimizations for communication with HPF/JA extensions were very useful to tune parallel performance.

It was noted that the computation grain in this benchmark suite was very small and this fact was disadvantageous for Fujitsu VPP800 that was based on a distributed memory architecture, but up to 16 processors of NEC SX-5 could share memory and faster communications were performed in our benchmarks.

The HPF situation for computational science users has been drastically improved by Japanese HPF compilers. More cooperation and effort, however, will be still needed to keep progress going to a sufficient performance goal.

Acknowledgments

We would like to acknowledge Cybermedia Center, Osaka University for supporting NEC SX-5 environments, Data Processing Center, Kyoto University for Fujitsu VPP800.

This work was partly supported by the Project n0130 of Data Processing Center, Kyoto University.

References

1. http://www.hpfpc.org/jahpf/.
2. http://www.hpfpc.org/jahpf/spec/hpfja-v10-eng.pdf.
3. http://www.hpfpc.org/.
4. Sakagami, H. and Mizuno, T.: Compatibility Comparison and Performance Evaluation for Japanese HPF Compilers using Scientific Applications, Concurrency and Computation: Practice and Experience, to be published.
5. http://www.sw.nec.co.jp/hpc/sx-e/.
6. http://primepower.fujitsu.com/hpc/en/.
7. Denissen, W. and Sips, H. J.: Finding Performance Bugs with the TNO HPF Benchmark Suite. Concurrency and Computation: Practice and Experience, to be published.
8. Saltz, J., Crowley, K., Mirchandaney, R. and Berryman, H.: J. of Parallel and Distributed Computing, Vol. 8, pp. 303-312 (1990).

Evaluation of the HPF/JA Extensions on Fujitsu VPP Using the NAS Parallel Benchmarks

Kae Asaoka[1], Akio Hirano[1], Yasuo Okabe[2], and Masanori Kanazawa[1]

[1] Data Processing Center, Kyoto University
[2] Graduate School of Informatics, Kyoto University

Abstract. We have ported 5 codes in APR's HPF implementation of NAS Parallel benchmark, so that it is conformable to HPF 2.0 and HPF/JA specification. The porting is done with the full usage of HPF 2.0 and HPF/JA new features, while the base Fortran codes remain almost unmodified. Then we have measured the performance of these benchmark codes on Fujitsu VPP800 by HPF/VPP compiler. We have achieved fairly good acceleration ratio for all codes except MG, and better absolute performance than NPB 2.3 parallel implementation with MPI for EP and BT.

1 Introduction

5 years have passed since the HPF 2.0 specification was released [6], but deployment of HPF in scientific computing field makes not so much progress as expected. Some reasons are pointed out. The HPF language level is too high, too abstract to specify detailed directions on data transfer. Many language features are left in approved extensions, and thus the language is not necessarily stable; there is difficulty in writing codes with perfect portability. On the other hand, it is very much costly to develop really high-performance HPF compilers which supports all features including approved extensions.

To cope with these difficulties, JAHPF (Japan Association for High Performance Fortran) started a coalition of the 3 Japanese supercomputer manufacturers and more than 20 experienced HPC programmers, and published the HPF/JA extension specification [8]. Many new features in HPF/JA are imported from either of the compilers of the three vendors [7,13,15], and full-set (HPF 2.0 and HPF/JA) compilers are now available on supercomputers by the vendors [10,9].

In this paper, we argue on the effectiveness of the HPF 2.0 and HPF/JA extensions. To evaluate this on NAS Parallel Benchmark (NPB) [1] codes, we have adopted APR's implementation [16], a well-known open-source HPF implementation of the NPB and we have added not a few HPF and HPF/JA directives. Then we have done performance measurement of the ported codes, compiled by Fujitsu's HPF/VPP compiler [4] on the Fujitsu VPP800 system at Kyoto University, a genuine distributed-memory parallel vector supercomputer.

The rest of the paper includes, brief overview of the HPF 2.0 and HPF/JA specifications, our porting guideline of APR's implementation of NPB, how the

H. Zima et al. (Eds.): ISHPC 2002, LNCS 2327, pp. 503–514, 2002.

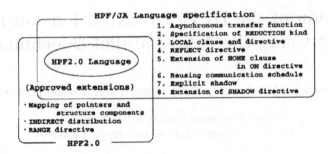

Fig. 1. Inclusion of the specifications

codes EP, BT, SP, FT and MG are originally designed, how they can be parallelized via HPF and how they have been tuned, and the results of benchmarking on VPP and evaluation of the effectiveness of HPF 2.0 and HPF/JA features.

2   HPF and HPF/JA Extensions

The design goal of HPF is the realization of semi-automatic parallelizing compilers for distributed parallel computers. Once a programmer specifies the distribution of array data onto processors by HPF directives, the compiler partitions the computation, allocates it onto processors by means of the "owner-computes" rule, and adds the necessary data transfer and synchronization.

The language specification of HPF has been discussed in HPFF (High Performance Fortran Forum) [17] since 1992. HPFF released the first specification HPF 1.0 in 1993. In 1997 the language was repartitioned into two components, the HPF 2.0 language and the set of Approved Extensions [6]. Soon after that JAHPF (Japanese Association for High Performance Fortran) [19] started its activities and released the HPF/JA 1.0 extension specification [8]. The HPF/JA extensions are classified into the following two major purpose. One is to enlarge the application area, and the other is to compensate for incompetence of current compilers.

The syntax of HPF 2.0 is that of Fortran95 plus HPF directives which are recognized as just comments in Fortran syntax. An HPF directive has the directive origin like !hpf$. An HPF/JA directive has the directive origin !hpfj, which distinguishes itself from HPF 2.0 directives.

Inclusion relation among these specifications is shown in Fig. 1. HPF/JA includes all of the HPF 2.0 language specification, most of the HPF 2.0 approved extensions and some native extensions. Some features in approved extensions are explicitly excluded so that vendors can develop HPF compilers with less difficulty.

3   Porting the Benchmark Codes

In this work, we use Applied Parallel Research Inc. (APR)'s earlier implementation of NPB in HPF [16] as the base codes.

This is known as a free and open-source HPF code and has been evaluated on many platforms [12]. This includes five benchmark codes for NPB, EP, MG, FT, SP and BT. Most of the codes are written in ordinary Fortran77 and not many HPF directives are used[1]. In contrast, we add as many HPF and HPF/JA directives in order to get the maximum performance, while we change the original FORTRAN codes as little as possible.

Fujitsu's HPF/VPP compiler recognizes DO loops and FORALL loops that have explicitly distributed arrays specified by either **processors**, **distribute** or **align** directives as potentially parallelizable loops. HPF/VPP can automatically detect these loops and can determine whether each of these is parallelizable or not, even when the programmer does not specify its parallelizability via any **independent** directive explicitly.

However, when the loop has an inner DO loop, the control variable of the inner loop must be declared as a *private* variable. Otherwise unnecessary synchronization occur repeatedly, that heavily decrease the performance.

Thus we have decided that we insert **independent** directives to all parallelizable loops and add **new** clause with all private variables explicitly. We also specify the partition of computation by **on** directives explicitly, and add both **resident** directives and **local** HPF/JA directives for all processor-local variables.

3.1   EP

EP Algorithm. EP (Embarrassingly Parallel) is a benchmark code for random number generation used in Monte Carlo method. Brief overview of the code is shown in Fig. 2 (a).

As shown in the figure, DO loop 150 is the kernel loop. In the loop, **randlc**, a function for computing random seeds, and **vrandlc**, a subroutine for generating a pseudorandom number sequence, are called, then in DO loop 140 a pair of random numbers is generated. The number of pairs that fall in each domain of ten domains is enumerated by counter variables $q0$, $q1$,... and $q9$. This process is repeated **nn** times.

All process except the enumeration on the domains can be computed fully independent on DO loop 150, and hence, in parallel.

Use of Templates. In HPF, inserting a **independent** directive just before the DO loop 150 does not necessarily means the loop is computed *in parallel*. This is because HPF partitions computation along with some distributed array in the owner-computes manner. Here in this code no distributed array occurs.

So we add a **template** directive that declares a template named **dummy**, as shown in Fig. 2 (b). The size is **nn**, the same as the number of repetition of DO loop 150, and the mapping of template is specified as block distribution. The enumeration on the ten regions is a typical summation among all processors; this can be parallelized by specifying the counter variables $q0$, $q1$,..., $q9$ in a **reduction** clause.

---

[1] MG and FT contain APR directives, which are extensions beyond the HPF language, for APR's Forge xHPF compiler.

```
                                    !hpf$ processors p(4)
                                    !hpf$ template dummy(nn)
                                          ! declares the template
                                    !hpf$ distribute dummy(block) onto p
                                          ! specifies the processor mapping
                                          :
                                    !hpf$ independent,
                                    !hpf$& new(k,i,t1,t2,t3,x,l,...),
   dimension x(2*nk)                !hpf$& reduction(q0,q1,...,q9)
   do 140 k=1,nn                          do 140 k=1,nn
                                             ! the parallelized loop
     t3 = randlc(t1,t2)             !hpf$ on home(dummy(k)) begin
       ! generates a seed of random numbers (t1)    :
     call vranlc(2*nk,t1,a,x)            t3 = randlc(t1,t2)
       ! generates a random number sequence (x)     call vranlc(2*nk,t1,a,x)
          :                                  :
     do 150 i=1,nk                        do 150 i=1,nk
          :                                     :
       ...! generates a pair of random numbers     l=...
       l=... ! computes its domain            if (l .eq. 0) then
       if (l .eq. 0) then                        q0=q0+1.0d0
          q0=q0+1.0d0                         else if (l .eq. 1)then
       else if (l .eq. 1)then                   q1=q1+1.0d0
          q1=q1+1.0d0                              :
             :                                    q9=q9+1.0d0
          q9=q9+1.0d0                           end if
          end if  ! counts up         150   end do
150  end do                          !hpf$ end on
140 end do                            140 end do
```

(a) program overview   (b) parallelized version

Fig. 2. Parallelizing EP

Extrinsic Language Binding. HPF 2.0 include the extrinsic language subprogram binding features defined in Fortran95 specifications. This is declared by
a extrinsic clause. HPF/VPP supports HPF_GLOBAL, this is the default when
no declaration is, HPF_LOCAL and FORTRAN_LOCAL bindings. An HPF_LOCAL
subprogram is called on each processor independently, and when a distributed
array is bound as a parameter only the partitioned part is passed.

A FORTRAN_LOCAL subprogram is also called on each processor independently but this works quite the same as ordinary extrinsic subprograms compiled
in Fortran compilers.

The subprograms **randlc** and **vrandlc** can be executed fully independently
among processors. Thus we add extrinsic(fortran_local) clauses.

We have also replaced the code for pseudorandom number sequence with the
vectorizable code imported from NPB2.3-serial[2].

3.2   BT

BT Algorithm and Data Mapping. BT (Block Tridiagonal simulated CFD
application) benchmark solves multiple, independent systems of non diagonally
dominant, block tridiagonal equations with a (5×5) block size. This is based on
the 3-dimensional ADI method.

We adopt the block distribution on the outermost index as the distribution
of the 3-dimensional array. Our VPP800 system has not so many processors,

---

[2] This change accelerates the performance 13.6 times on VPP800 (#PE is 4) with
HPF/VPP.

```
                                        !hpf$ processors p(4)
                                        !hpf$ distribute (*,block)
                                        !hpf$&        onto p :: x,u
                                        !hpf$ shadow u(0,2:2)
      !hpf$ processors p(4)
      !hpf$ distribute (*,block)        !hpfj reflect u
      !hpf$&        onto p :: x,u       !hpf$ independent,new(i,j)
              :                       p     do 100 j = 3, n-2
      !hpf$ independent,new(i,j)      p !hpf$ on home(x(:,j)) begin
 p        do 100 j = 3, n-2           p !hpfj local begin
 p    !hpf$ on home(x(:,j)) begin     v p     do 200 i = 1,n
 p          do 200 i = 1,n            v p       x(i,j)=x(i,j)+u(i,j-2)
 p 5          x(i,j)=x(i,j)+u(i,j-2)  v p   &        +u(i,j-1)+u(i,j)
 p    &          +u(i,j-1)+u(i,j)     v p   &        +u(i,j+1)+u(i,j+2)
 p    &          +u(i,j+1)+u(i,j+2)     p 200   end do
 p    200   end do                    p !hpfj end local
 p    !hpf$ end on                    p !hpf$ end on
 p    100 end do                      p 100 end do
```

(a) naive code   (b) using **shadow** and **reflect** directives

Fig. 3. Access to data on the adjacent processors

but each has a vector unit; the maximum performance is achieved when multiply nested loops are parallelized at the outermost loop and vectorized in the innermost loop.

Access to the Data on the Adjacent Processors. Fig. 3 (a) shows an overview of a naive HPF implementation of the kernel of the BT benchmark. Here the arrays are simplified into 2-dimensional ones just for brief description.

In the code in Fig. 3 (a), the arrays x and u are distributed blockwise at the second dimension, and parallelized at DO loop 100. Here access to the array x is always *local* (and hence *resident*), but access to the array u is not necessarily local since the indices of the second dimension are $j-1$, $j-2$, $j+1$ and $j+2$, not j; the elements should be mapped onto the adjacent processors at the boundary of the partition.

HPF 2.0 defines the **shadow** directives for this kind of loops. The directive specifies that the processor has a region (shadow region) on its local memory so that it can keep copy of the data on neighbor processors.

Fig. 3 (b) shows a code that utilizes the **shadow** directives. Here the distributed array u is declared as it has shadow region, two elements upward and downward each, on the second dimension. The **reflect** directive used here is defined in HPF/JA extension. The array u is only referred, is not defined, in the parallelized loop. Specifying **reflect** just before the loop, the access to u is assured to be *local* and so we can insert **local** directive in the loop.

This also makes the inner loop vectorized. The actual performance of the loop 100 in these two codes (a) and (b) of Fig. 3 is, 24.8*msec* for (a) and 0.113*msec* for (b), when we measured these on VPP800 (# PE is 4) with n=64.

Transposition of Distributed Array Data. The DO loop in Fig. 4 (a) shows the code for transposition between 4-dimensional arrays in the APR implementation of the BT benchmark.

```
     !hpf$ processors p(4)
     !hpf$ distribute u(*,*,*,block) onto p
     !hpf$ align with u :: ut,r,rt
              :
     !hpf$ independent,new(k,j,i,m)
  p        do k =1,nz
  p   !hpf$ on home(u(:,:,:,k)) begin
  p          do j = 1,ny
  p            do i = 1,nx
  p              do m = 1,5
  p 1              ut(m,i,k,j) = u(m,i,j,k)
  p 1              rt(m,i,k,j) = r(m,i,j,k)
  p              end do
  p            end do
  p          end do
  p   !hpf$ end on
  p          end do
```

```
     !hpfj asyncid id
            :
     !hpfj asynchronous(id), nobuffer begin
  1      forall(m=1:5,i=1:nx,j=1:ny,k=1:nz)
     &        ut(m,i,k,j) = u(m,i,j,k)
  1      forall(m=1:5,i=1:nx,j=1:ny,k=1:nz)
     &        rt(m,i,k,j) = r(m,i,j,k)
     !hpfj end asynchronous
     !hpfj asyncwait(id)
```

(a) transposition by a DO loop

(b) transposition by asynchronous FORALL loops

Fig. 4. Transposition of 4-dimensional arrays

In the 3-dimensional ADI method, 3-D space is scanned along each of three axes in turn. Multiply nested loops that has dependency along one axis but can be executed independently along other two axes occurs one by one.

When the scan is done along the distribution of the array, the dependency blocks parallelization. In the APR's implementation[3], a temporal array is declared as a *shadow* of the original array. Before entering the multiple loop the all of the array data is copied and in the loop the shadow is referred instead of the original. After exiting the loop the shadow is written back to the original.

This copy process is actually transposition of an 4-dimensional array and the code for it is like Fig. 4. The code in Fig. 4 (a) the quadruple nested DO loop copies arrays **u** and **r** onto **ut** and **rt** with the transposition of the 3rd index and the 4th index.

HPF/VPP compiles the code like (a) into elementwise data transfer in the parallelized loop; this runs correctly but is very slow. HPF/VPP supports the asynchronous data-transfer extension in HPF/JA specification and this kind of bulk copy among arrays are done by this efficiently. A code for transposition using asynchronous data-transfer mode is shown in Fig. 4 (b).

The FORALL statement in the **asynchronous** block in (b) is not a executed statement but just specifies the data transfer pattern. The **nobuffer** clause means that no buffering is needed during the copy.

Measured performance of the codes (a) and (b) with **nx=ny=nz**=64 on VPP800 (# PE is 4) is, 6,533msec for (a) and 0.345msec for (b).

3.3   SP

SP (Scalar Pentadiagonal simulated CFD application) is a benchmark code that solves multiple, independent systems of non diagonally dominant, scalar, pentadiagonal equations. SP and BT are similar in many respects, but there is a fundamental difference with respect to the communication to computation ratio.

---

[3] originally written by Sisira Weeratunga, NASA

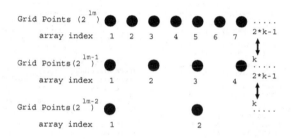

Fig. 5. Alignment among grid arrays in MG

3.4   FT

FT (3-D FFT PDE) is a benchmark code for 3-dimensional Fast Fourier Transform. The data is stored in a 3-dimensional array and the array is distributed blockwise at the outermost dimension. FT does 1-dimensional FFT for each axis serially. 1-dimensional FFT has dependency in the axis, but no dependency in the other two axes.

The original APR implementation *redistributes*[4] the arrays before and after FFT along distributed axis. However, HPF/VPP has a syntactic restriction that dynamic arrays (that can be redistributed) must have full shadows, which causes a the shortage of memory. Instead we introduce temporary arrays and transpose the arrays, as the code in BT and SP described in Sec. 3.2

The code for 1-D FFT is a subroutine, which can be called fully independently among processors. Thus we have specified the subroutine as FORTRAN_LOCAL by `extrinsic` clause.

3.5   MG

MG (Multigrid) is a benchmark that solves 3-D Poisson PED. The number of grid points at each dimension becomes twice at each iteration step, from 2^1 to 2^{l_m}. Namely as many arrays as the number of steps (l_m) are needed. The value of l_m is 8 on CLASS A and CLASS B benchmarks and 9 on CLASS C.

In the original APR code, only one 1-dimensional array is declared and it is passed to subroutines as a 3-dimensional array. The array is dynamically redistributed into block distribution along the outermost dimension by APR'S native directive. Since HPF/VPP cannot parallelize this kind of parameter passing, we have modified the code as l_m 3-dimensional arrays are declared distinctively. Each array is distributed in block distribution along the outermost dimension, so that the corresponding grid points (see Fig. 5) are surely on the same processor. This can be written with `template` and align directives in HPF 2.0 (Fig. 6).

In the original code, one subroutine is used for all steps. The subroutine is passed an array with different size at each call. This can be written in HPF as a *prescriptive* subroutine call. However, HPF/VPP seems not to be able to determine an element of a distributed array that is passed in prescriptive manner is

---

[4] using APR native directive **PARTITION**

```
integer, parameter :: MM9 = 2**LM+2        ! 512 + 2
integer, parameter :: MM8 = (MM9-2)/2 + 2  ! 256 + 2
integer, parameter :: MM7 = (MM8-2)/2 + 2  ! 128 + 2
        :
integer, parameter :: MM2 = (MM3-2)/2 + 2  ! 4 + 2
integer, parameter :: MM1 = (MM2-2)/2 + 2  ! 2 + 2

integer, parameter :: KM1 = MM1            ! 4
integer, parameter :: KM2 = 2*KM1-1        ! 7
        :
integer, parameter :: KM8 = 2*KM7-1        ! 385
integer, parameter :: KM9 = 2*KM8-1        ! 769

real(8), dimension(MM9,MM9,MM9) :: v
real(8), dimension(MM9,MM9,MM9) :: r , u
real(8), dimension(MM8,MM8,MM8) :: r8 , u8
real(8), dimension(MM7,MM7,MM7) :: r7 , u7
        :
real(8), dimension(MM2,MM2,MM2) :: r2 , u2
real(8), dimension(MM1,MM1,MM1) :: r1 , u1
```

```
!hpf$ processors proc(40)
!hpf$ template tmp0(MM9,MM9,KM9)
!hpf$ distribute (*,*,block) onto proc     :: tmp0
!hpf$ align (i,j,k) with tmp0(i,j,k)       :: v
!hpf$ align (i,j,k) with tmp0(i,j,k)       :: r,u
!
!hpf$ template tmp1(KM9)
!hpf$ distribute (block)       onto proc   :: tmp1
!hpf$ align (i,j,k) with tmp1(k*2-1)       :: r8,u8
!hpf$ align (i,j,k) with tmp1(k*4-3)       :: r7,u7
        :
!hpf$ align (i,j,k) with tmp1(k*128-127) :: r2,u2
!hpf$ align (i,j,k) with tmp1(k*256-255) :: r1,u1

!hpf$ shadow (0,0,1:1) :: r,u
!hpf$ shadow (0,0,1:1) :: r8,u8
!hpf$ shadow (0,0,1:1) :: r7,u7
        :
!hpf$ shadow (0,0,1:1) :: r2,u2
!hpf$ shadow (0,0,1:1) :: r1,u1
```

Fig. 6. Codes for distribution and alignment of the arrays in MG

resident on the processor or not, and the code in the subroutine is not vectorized at all. To avoid this situation, we dare to modify the code as it has a subroutines for each distributed array and the parameter is passed in *descriptive* manner.

In the MG code, a processor refers the data on the neighbor processors. Thus we have declared the arrays with shadow area, one element on the upper boundary and one on the lower boundary by shadow directive, and have specified reflect and local directives appropriately, just as described in Section 3.2. We have also utilized the directives asynchronous data transfer described in Section 3.2 in exchanging lowermost and second-uppermost (or uppermost and second-lowermost) subarrays of each dimension of 3-D arrays.

4   Performance Evaluation

To evaluate the performance and parallelism of the ported benchmark codes, we have done a performance measurement on Fujitsu VPP800/63 system at Kyoto University Data Processing Center. The compiler is Fujitsu UXP/V Fortran V20L20 L01091 and Fujitsu UXP/V HPF V20L20 L01041. During the benchmarking we use a subroutine **gettod** and **fjhpf_gettod** that tell the elapse time in micro seconds.

First we will show the results of NPB CLASS C benchmark, with 1, 2, 4, 8, 16 and 32 processors (PEs; processing elements). The results are shown in Table 1. Here the upper rows show the elapse time in seconds, and the acceleration ratio when the elapse time of the code compiled by the ordinary vectorizing Fortran compiler and executed on 1 PE is 1 is given in parentheses. FT cannot be executed on 1 PE because of the shortage of memory, the acceleration ratio is computed with the elapse time of the code compiled HPF/VPP and executed on 2 PEs as 2 instead. As is shown in the table and Fig. 7, we have obtained fairly good scalability on EP, FT, BT and SP. The performance of MG looks worse than the results of other benchmarks, but the acceleration ratio here we have gotten is much better than the result of MG with HPF in previous works [2]. We may say that, if a base code is a well designed Fortran program with enough consideration of parallel execution, we can get enough good performance with HPF by specifing HPF/JA directives carefully.

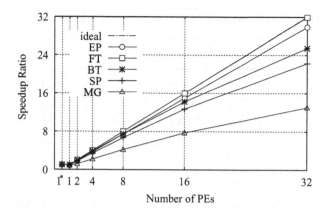

Fig. 7. Acceleration ratio

Table 1. Elapsed time (Seconds) and acceleration ratio (CLASS=C)

Benchmark	1PE*	1PE	2PE	4PE	8PE	16PE	32PE
EP	443.96	473.04	237.21	118.76	59.43	29.69	14.86
	(1.00)	(0.93)	(1.87)	(3.73)	(7.47)	(14.95)	(29.87)
FT	—	—	3684.18	1843.09	921.28	460.64	230.52
			(2.00)	(3.99)	(7.99)	(15.99)	(31.96)
BT	17337.00	17608.00	8807.90	4515.00	2319.40	1222.00	681.48
	(1.00)	(0.98)	(1.96)	(3.83)	(7.47)	(14.18)	(25.44)
SP	9155.30	10260.00	5160.40	2647.30	1365.90	721.59	411.30
	(1.00)	(0.89)	(1.77)	(3.45)	(6.70)	(12.68)	(22.25)
MG	87.61	100.23	76.00	39.31	20.71	11.38	6.72
	(1.00)	(0.87)	(1.15)	(2.22)	(4.23)	(7.69)	(13.03)

(* Compiled by ordinary Fortran compiler)   (Seconds)

Next we have evaluated the effectiveness of HPF/JA extension, in comparison to execution of codes without HPF/JA proper extensions, namely code only with directives in the HPF 2.0 and the approved extension. We have also measured the performance of the same code compiled by Adaptor [18], GMD's open-source free full-set HPF 2.0 compiler, and executed with UXP/V MPI V20L20 L00011. The elapse time is measured by the **system_clock** service subroutine. The results are shown in Table 2[5].

The result shows that HPF/JA directives works very effectively on HPF/ VPP. In other words, specifying HPF/JA directives fully is almost mandatory to get good performance. EP and FT codes perform well on Adaptor; almost the same performance as that on HPF/VPP. The performance of the Adaptor compiler would very likely profit a lot on other benchmark codes from including the HPF/JA features. MG code is not parallelized at all by Adaptor, so we do not list the result on the table.

---

[5] In our code for EP, no HPF/JA native directive is used.

Table 2. Effectiveness of JA extensions (CLASS=A,4PE)

Benchmark	HPF/VPP	HPF/VPP (Without JA)	Adaptor
EP	$\Longrightarrow$	7.32	6.95
FT	42.30	605.78	42.52
BT	269.21	(*1)	599.36
SP	174.35	(*1)	535.18
MG	1.36	(*1)	—

(*1) More than 1hour   (Seconds)

Finally, we have measured the performance of our code with HPF/VPP and the NPB2.3 parallel code with UXP/V MPI on VPP800. The result in CLASS C is shown in Table 3 (a). Our codes for EP and BT get better absolute performance than NPB2.3 with MPI, although the algorithms used are different and thus we cannot naively compare NPB2.3 and our HPF implementation.

Comparison of the performance and vectorization ratio among APR's original code compiled by (non-parallel) vectorizing Fortran compiler, our ported code compiled by Fortran compiler and HPF/VPP, NPB 2.3-serial code and NPB 2.3 parallel code are shown in Table 3 (b).

5   Concluding Remarks

We have ported 5 codes in APR's earlier HPF implementation of NAS Parallel benchmark, so that it is conformable to HPF 2.0 and HPF/JA specification, with the full usage of HPF 2.0 and HPF/JA new features but with little modification in the base Fortran codes. Then we have measured the performance on Fujitsu VPP800 by HPF/VPP compiler. We have obtained good acceleration ratio for all codes except MG, and better absolute performance than NPB 2.3 parallel implementation with MPI for EP and BT.

There has been a known work by Hitachi that have implemented all 8 benchmarks in HPF and have evaluated [11]. They have written fully new code for the best performance. In contrast, we have posed ourselves a porting guideline, in that we modify the base Fortran code by APR as little as possible; just adding HPF 2.0 and JA extension directives. Similar approach was adopted in the work by a group in NAS [3], where BT, SP, LU, FT, CG, and MG are implemented in HPF using the NPB2.3-serial as base code, although they do not utilize the HPF/JA extension.

Fortunately the base codes we have used here have enough intrinsic parallelism and we can achieve considerably good performance on VPP even by such simple porting. However, as shown in Sec. 3.2, only a slight difference of coding makes, indeed leads to more than ten thousand times performance degradation. It should be also noted that we have been to modify the base Fortran code by the implementation restriction of the compiler, as noted in Sec. 3.4

Table 3. Comparison to NPB2.3

(a) Absolute performance
(CLASS=C)

(b) Vectorization ratio (CLASS=A)

Benchmark (#PE)	HPF /VPP	NPB 2.3
EP (40)	11.96 (718)	15.83 (542)
FT (32)	230.52 (1,720)	44.96 (8,817)
BT (36)	571.72 (5,013)	588.09 (4,874)
SP (36)	348.46 (4,162)	61.22 (23,688)
MG (32)	6.72 (23,170)	4.40 (35,388)

(Seconds, (MFLOPS))

Benchmark	Original Fortran	Ported code Fortran	HPF/4PE	NPB2.3-serial (1PE)	NPB2.3 (4PE)
EP	403.43 (1.3%)	27.71 (88.8%)	7.32 (85.8%)	39.48 (58.2%)	9.89 (57.6%)
FT	167.02 (13.7%)	154.43 (13.3%)	42.30 (11.9%)	27.21 (71.8%)	6.97 (67.5%)
BT	1002.73 (29.1%)	992.28 (29.4%)	269.21 (20.8%)	1187.48 (47.2%)	304.76 (47.1%)
SP	696.59 (10.5%)	642.92 (11.5%)	174.35 (10.0%)	165.75 (31.7%)	24.98 (88.8%)
MG	2.61 (52.3%)	2.43 (58.4%)	1.36 (35.0%)	3.38 (29.4%)	0.92 (29.5%)

(Seconds, (ratio))

We have implemented all benchmarks among 5 published ones by APR, but porting other 3 NPB remain as future works. There has some projects implementing standard HPF benchmark codes, like HPFbench [5]. We are planning to do evaluation of HPF/JA extensions on those benchmarks also.

References

1. Bailey, D., Barszcz, E., Barton, J., Browning, D., Carter, R., Dagum, L., Fatoohi, R., Fineberg, S., Frederickson, P., Lasinski, T., Schreiber, R., Simon, H., Venkatakrishnan, V.,Weeratunga, S.: THE NAS PARALLEL BENCHMARKS, RNR Technical Report RNR-94-007, NASA Ames Research Center (1994)
2. Chamberlain, B.L, Deitz, S.J, Snyder, L.: A Comparative Study of the NAS MG Benchmark across Parallel Languages and Architectures, Proc. 2000 ACM/IEEE Supercomputing Conference on High Performance Networking and Computing (SC2000) (2000)
3. Frumkin, M., Jin, H., Yah, J.,: Implementation of NAS parallel benchmarks in high performance fortran, Technical Report NAS-98-009, NASA Ames Research Center (1998)
4. Fujitsu: UXP/V HPF users' manual for V20L20 L01041, J2U5-0450-01 (2001)
5. Hu, Y.C., Jin, G., Johnson, S.L., Kehagias, D., Shalaby, N.: HPF Bench: A High Performance Fortran Benchmark Suite, ACM Transactions on Mathematical Software (2000)
6. High Performance Fortran Forum: High Performance Fortran Language Specification, version 2.0, http://dacnet.rice.edu/Depts/CRPC/HPFF/versions/hpf2/hpf-v20/ (1997)
7. Iwashita, H., Sueyasu, N., Kamiya, S., van Waveren, M.: A comparison of HPF and VPP Fortran: How it has been used in the Design and Implementation of HPF/JA Extensions, HUG2000 (2000)

8. Japan Association for High Performance Fortran: HPF/JA Language Specification, version 1.0, http://www.hpfpc.org/jahpf/ (1999)
9. Mizuno,T., Sakagami, H., Murai, H.: Capability Comparison between Japanese HPF compilers using Scientific Applications, HUG2000 (2000)
10. Murai, H., Araki, T., Hayashi, Y., Suehiro, K.: Implementation and Evaluation of an HPF Compiler for Vector Parallel Machines, HPF/SX V2, HUG2000 (2000)
11. Ohta, H., Nishitani, Y., Kobayashi, A., Nunohir, E.: Parallelization of the NAS Parallel Benchmarks by an HPF Compiler "Parallel FORTRAN", IPSJ JOURNAL**38**, 9 (1997) 1830–1839
12. Saini, A., Bailey, D.H.: NAS Parallel Benchmark (Version 1.0) Results 11-96, ReportNAS-96-018, NASA Ames Research Center (1996)
13. Sato, M, Ohta, H., Nunohiro, E.: Development and Evaluations of HPF Translator "Parallel FORTRAN", IPSJ MAGAZINE **38**, 2 (1997)
14. Seo, Y., Iwashita, H., Ohta, H., Sakagami, H., Takahashi S.: HPF/JA: HPF Extensions for Real-World Parallel Applications, HUG98 (1998)
15. Kamachi,T., Kusano, K., Suehiro, K., Seo, Y., Tamura, M., Sakon, S.: Design and Implementation of an HPF Compiler and Its Performance Results, IPSJ JOURNAL **37**, 7 (1996)
16. ftp://ftp.infomall.org/tenants/apri/Benchmarks/
17. http://www.crpc.rice.edu/HPFF/
18. http://www.gmd.de/SCAI/lab/adaptor/
19. http://www.hpfpc.org/jahpf/
20. http://www.nas.nasa.gov/Software/NPB/

Three-Dimensional Electromagnetic Particle-in-Cell Code Using High Performance Fortran on PC Cluster

DongSheng Cai[1], Yaoting Li[1], Ken-ichi Nishikawa[2],
Chiejie Xiao[1], and Xiaoyan Yan[1]

[1] Institute of Information Sciences and Electronics,
The University of Tsukuba, Ibaraki 305-8573, Japan
{cai,ytli,cjxiao,yxy}@is.tsukuba.ac.jp
[2] Department of Physics and Astronomy, Rutgers University,
136 Frelinghuysen Road, Piscataway, New Jersey 08854-8019, USA
kenichi@physics.rutgers.edu

Abstract. A three-dimensional full electromagnetic particle-in-cell (PIC) code, TRISTAN (Tridimensional Stanford) code, has been parallelized using High Performance Fortran (HPF) as a RPM (Real Parallel Machine). In the simulation, the simulation domains are decomposed in one-dimension, and both the particle and field data located in each domain that we call the sub-domain are distributed on each processors. Both the particle and field data on a sub-domain is needed by the neighbor sub-domains and thus communications between the sub-domains are inevitable. Our simulation results using HPF exhibits the promising applicability of the HPF communications to a large scale scientific computing such as 3D particle simulations.

1 Introduction

This paper reports on parallelization of Tridimensional Stanford (TRISTAN) code [1] that is a three-dimensional electromagnetic full particle code developed at Stanford University on a two-way PentiumPro PC cluster that consists of 10 distributed SMPs using High Performance Fortran.

In our parallel program, the simulation domain is decomposed into the sub-domains as shown in Fig. 1. The Particle-In-Cell (PIC) computation in TRISTAN to be performed on a certain sub-domain or on a certain processor where the sub-domain is distributed will typically require the data from their neighbor processors to proceed the whole PIC simulations. Here we distribute the field arrays and the particles over processors as indicated in Fig. 1. Thus the data must be transferred between processors in each time step so as to allow PIC simulation to proceed in time. These inter-processor communications in each time step need to be programmed in HPF constructs.

The amount of inter-processor communications needed for a parallel program basically depends on the algorithms and the scales of the physical problem sizes

H. Zima et al. (Eds.): ISHPC 2002, LNCS 2327, pp. 515–525, 2002.
© Springer-Verlag Berlin Heidelberg 2002

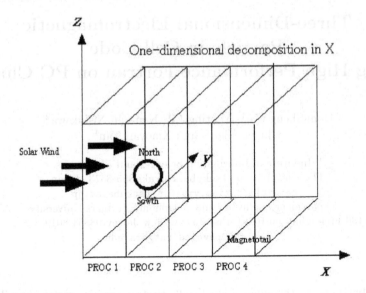

Fig. 1. Coordinate of the simulation domains and domain decomposition in x.

adopted in the simulations. In PIC simulations, they are the way decompose the simulation domains, the sizes of the sub-domain boundaries, and the number of the particles in a cell, respectively.

The pgHPF [2] compiler of Portland Group Inc. aims to realize the standard High Performance Fortran specification and can be installed on a number of parallel machines. Executables produced by the pgHPF compilers are unconstrained, and can be executed on any compatible IA-32 processor-based system regardless of whether the pgHPF compilers are installed on that system or not. From the HPF programmer's point of view, the differences between versions of the pgHPF runtime library have little effect on program developments.

In parallel programming models, usually, the SPMD models using MPI (Message Passing Interface) or PVM (Parallel Virtual Machine) are one of the most poplar model. The biggest HPF advantage is its programming styles. Once the simulation domains are decomposed and the data are distributed to each sub-domains or over processors using simple HPF compiler directives, other HPF programming styles are very similar to those in usual Fortrans. Of course, the biggest problems here is the performance issues comparing with those using MPI or PVM.

Actually, pgHPF is based on a RPM (PGI Proprietary Communications - Real Parallel Machine) protocol. This transport mechanism was developed by PGI to model the behavior of PVM among a homogeneous group of hosts on a network. It offers both greater programming efficiency and performance than PVM with fewer requirements. In this paper, to archive a similar high parallel performance comparing with that of MPI or PVM using HPF in a full electromagnetic PIC simulation, some careful optimizations of inter-processor communications are proposed.

Our code is same as the TRISTAN code except the parallelization part, which utilizes charge-conserving formulas and radiating boundary conditions [1]. It is written in HPF so that that the code can be run on any parallel computers with the HPF compilers.

The parallelization part of our HPF TRISTAN code is similar to Liewer et al. [3] and Decyk [4]. We separate the communication parts from computation parts, and use both the particle manager and the field manager to localize the inter-processor communications [4].

The basic controlling equations of the plasmas are:

$$m_i \frac{dv_i}{dt} = q_i \cdot (E + v_i \times B) \tag{1}$$

$$\frac{\partial B}{\partial t} = - \bigtriangledown \times E \tag{2}$$

$$\frac{\partial E}{\partial t} = c^2 \bigtriangledown \times B - \frac{1}{\varepsilon_0} J \tag{3}$$

$$J = n \sum (q_i v_i) \tag{4}$$

The coordinate and one-dimensional domain decomposition using in the simulation domain is shown in Fig. 1. For parallel benchmarking purposes, we perform the real simulations of solar wind-magnetosphere interactions using the code. For the simulation of solar wind-magnetosphere interactions, the following boundary conditions were used for the particles [1]: (1) Fresh particles representing the incoming solar wind (unmagnetized in our test run) are continuously injected across the yz plane at $x = x_{min}$ with a thermal velocity plus a bulk velocity in the $+x$ direction; (2) thermal solar particle flux is also injected across the sides of our rectangular computation domain; (3) escaping particles are arrested in a buffer zone, redistributed there more uniformly by making the zone conducting in order to simulate their escape to infinity, and finally written off. We use a simple model for the ionosphere where both electrons and ions are reflected by the Earth dipole magnetic field. The effects of the Earth rotation are not included. Since the solar-winds and the Earth dipole magnetic field are included, the load-imbalance due to this asymmetry are expected in this HPF TRISTAN code.

2   Arrays in Original TRISTAN Code

The motivation of TRISTAN, a fully three-dimensional (3D) electromagnetic (EM) particle-in-cell(PIC) code written by Oscar Buneman and other collaborators in Stanford University, is to develop a general particle-in-cell code for space plasma simulations [1]. Here we only discuss the data structure and the data distribution over processors on the HPF TRISTAN code, and do not discuss the details of the numerical algorithms and the plasma physics in TRISTAN in the present report. For the physics of the PIC code, please refer to, for examples, [5] and [6].

The data structure of TRISTAN code consists of two primitive data types. The first one is the particle data as follows: $x(mp), y(mp), z(mp), u(mp), v(mp),$ $w(mp)$, where $mp=$ total number of particles, the positions and velocities of ions and electrons are recorded at $x(1:mh), y(1:mh), z(1:mh), u(1:mh), v(1:mh), w(1:mh)$, and $x(mh+1:mp), y(mh+1:mp), z(mh+1:mp), u(mh+1:mp), v(mh+1:mp), w(mh+1:mp)$, respectively, where $mh = mp/2$. The second one is the grided field data expressed as the triple-indexed arrays of EM (ElectroMagnetic) fields as follows:

$$ex(i,j,k), ey(i,j,k), ez(i,j,k),$$

and

$$bx(i,j,k), by(i,j,k), bz(i,j,k).$$

The original TRISTAN code uses "COMMON" block clause to save and transfer fields data between subroutines in the **MOVER**(push particles) and **DEPOSIT** (deposit current data to the field grids) subroutine calls. Meanwhile in the subroutines that processes the surfaces and edges of the grid data, the filed data are transferred by dummy arrays in the original code. In both of these subroutines, the field arrays are treated as single-indexed. On the other hand, triple-indexed field arrays are employed in the field solver subroutines. In the code, single-indexed arrays are converted automatically to the triple-indexed arrays when they passed over two subroutines.

Converting a serial Fortran program to a HPF program, we have to stress two points that are very important for rewriting TRISTAN in HPF: **1** the "COMMON" statement is restricted as suggested by pgHPF user guide and there they indicated 'We strongly recommended that programmers writing new F90 code use features like "MODULE" ... to avoid the use of "COMMON"...'[7][8], in case of data overlapping, and substituted it by "MODULE" block; and **2** to control the communications, all the arrays are treated as fixed indexes throughout the whole program. We control the communication parts using both the field and particle managers [4].

3   Field Data Domain Decompositions

The field data are decomposed over sub-domains of that number is equal to the number of the processors as indicated in Fig. 1. In processing the current deposition that is so-called the scatter part of the computations, to avoid large transients or variations of currents TRISTAN uses a 'smoother' that has 27 different weights, smoothing the current deposition. In **DEPOSIT** subroutine the smoothing is performed as follows:

ey(i+smx+1,j+smy,k+smz+1, Np)=ey(i+smx+1,j+smy,k+smz+1, Np)+...,

where $smx = -1:1, smy = -1:1, smz = -1:1$. Therefore, the current deposition of one particle will be related to three grids in each dimension, where one of them are at the backward grid and two of them at the forward grids in each dimension.

In the "MODULE" block, the field arrays are written in HPF directives as follows:

$$\textbf{REAL}, \textbf{DIMENSION}(nx, j, k, Np) :: ex, ey, ez$$
$$\textbf{REAL}, \textbf{DIMENSION}(nx, j, k, Np) :: bx, by, bz$$

where Np=the number of processor, $nx = i/Np + 3$ (here assuming i/Np is not necessarily equal to be integer exactly) keeping one guard cell in the left (backward) and two in right (forward) side of the sub-domains in the domain-decomposition direction (i. e., in the solar-magnetotail direction). Here the indices i, j and k corresponds to the numbers of field grids in x, y and z directions, respectively. Using the HPF directive "DISTRIBUTE", we, respectively, map the sub-domains to each processor on a distributed memory parallel computer:

$$\textbf{DISTRIBUTE}(*,*,*,BLOCK) \ ONTO \ Np :: ex,ey,ez$$
$$\textbf{DISTRIBUTE}(*,*,*,BLOCK) \ ONTO \ Np :: bx,by,bz$$

In order to separate the communication parts from the computation parts, each sub-domain keeps extra cells, the so-called guard or ghost cells, that store the field data information in the first and the last two grids of that sub-domain in the decomposition direction. Figure 1 illustrates this concept of the data mapping over the sub-domains or processors. Here the communications are required after updating the field data every time step. In the field manager [4], the data send to the neighbor processors are packed in the working arrays: $Cex(1, j, k, Np)$, $Cey(1, j, k, Np)$, and $Cez(1, j, k, Np)$, before they are send to the neighbor sub-domains. Thus the field data communications are performed by the HPF $CSHIFT$ construct after the data are packed in the working arrays. The followings are the related parts of the HPF programs in the field manager[4]:

```
Cex(1,:,:,:)=ex(2,:,:,:)

Cex=CSHIFT(Cex,+1,4)

ex(nx-1,:,:,:)=Cex(1,:,:,:)
```

. . .

4   Particle Data Domain Decompositions

The particle data can be written in HPF directives as follows:

$$\textbf{REAL}, \textbf{DIMENSION}(m, Np) :: x_e, y_e, z_e, x_i, y_i, z_i$$
$$\textbf{REAL}, \textbf{DIMENSION}(m, Np) :: u_e, v_e, w_e, u_i, v_i, w_i$$

where the subscripts i and e, respectively, stand for ion and electron, the number m is the array size in each sub-domain. To ensure that the enough space are

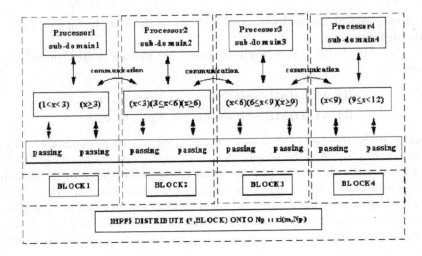

Fig. 2. Diagram of particle array decompositions and communications, with processor number $Np = 4$, grid number in decomposition direction=12.

reserved to store the particle data due to the load-imbalance, m must be 10-30 % larger than the average number of particles. The number Np is the number of processors, and is the index used in the HPF "DISTRIBUTE" directive. As the particles move in time in the simulations, the physical position of some particles may cross the sub-domain boundaries, and move to the neighbor sub-domains. When a particle moves from one sub-domain to another, the data of the particle left the sub-domain must be sent to the appropriate neighbor processor every time step. Before updating and sending the particle data, we have to sort the particles that should be send to another sub-domain, and pack them in the working arrays: $CRi(:, Np)$, $CLi(:, Np)$, $CRe(:, Np)$, and $CLe(:, Np)$. The number of the ions and electrons sent in right and left are denoted by the arrays $ionspsR(Np)$, $ionspsL(Np)$, $lecspsR(Np)$, and $lecspsL(Np)$, respectively. In our HPF TRISTAN, we send both the packed arrays and their particles number arrays to the neighbor sub-domains as follows:

```
CRi=CSHIFT(CRi, -1,2)

ionspsR=CSHIFT(ionspsR, -1)
```

. . .

Figure 2 shows the example of the particle data distributions and communications. After both the particle numbers and the packed working arrays are sent and received by each appropriate processors, the received particles are sorted and put into the appropriate part of the particle arrays in that sub-domain. The communications and sorting of these particles are performed in the particle manager [4].

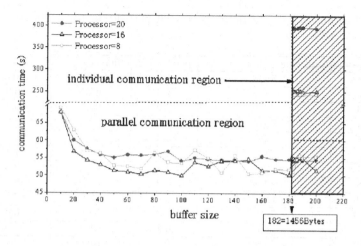

Fig. 3. Buffer sizes and CSHIFT communication times in HPF.

5 Dual PentiumPro PC Cluster

Our dual PentiumPro PC cluster consists of 16 PCs and each PC have dual $200MHz$ PentinumPros with 128MB EDD DIMM memories. The PCs in the PC cluster system are hooked through 100 Base-T ethernet with 100 Base-T switching Hub. Redhat Linux version 4.1 is used as their operating systems. The pgHPF compiler version 1.7 is installed for HPF computations.

6 Programming Comments on HPF Communications in PC Cluster

One of the most difficult HPF programming in our HPF TRISTAN code is the communication programming, especially, the determination of the buffer sizes which is used to pack the data to send to the neighbor processors. Of course, we can define a buffer size large enough to send the particle or grid data to neighbor processors at one time. However, as shown in Fig. 3, our experience shows that when the buffer sizes become larger than some critical values, in this case 1456 bytes in our PC cluster system, the communication suddenly becomes unstable, and the communication times suddenly jump up to 5 to 8 times larger than those less than the critical value 1456 bytes. As indicated in the figure, the communication times when the buffer size go beyond 1456 bytes are not uniquely determined and rather undeterministic. In order to avoid the sudden communication slow-down, we have to carefully to choose the buffer size. We have to split the particles or grids data into smaller pieces of buffers, pack the smaller data, and send the data to the neighbor processors one by one. Thus we can avoid the large slow-down of the simulations in this system due to the unstable HPF communications. In our HPF TRISTAN code, the buffer sizes can be varied and can be set without modifying the program. We can first evaluate

Table 1. Benchmark resluts with time step=100, particle number =1200,000, and grid number =185 × 65 × 65.

Processors	total time×Proc. No. (s)	speed up S_p	efficiency $\varepsilon(\%)$	$\varepsilon_{eff-grid}(\%)$
1	4836	1.0	100.	100
2	7412	1.3	65.2	96.9
3	7249	2.0	66.7	95.4
4	7075	2.7	68.4	93.9
5	7287	3.3	66.3	92.5
6	7171	4.0	67.4	91.1
7	7236	4.7	66.8	89.8
8	7499	5.2	64.5	88.5
9	7937	5.5	60.9	87.2
10	7627	6.3	63.4	86.0
11	7843	6.8	61.7	84.9
12	7824	7.4	61.8	83.7
13	8019	7.8	60.3	82.6
14	8063	8.4	60.0	81.5
15	8588	8.4	58.3	80.4
16	8263	9.4	58.5	79.4
17	8453	9.7	57.2	78.4
18	8469	10.3	57.1	77.4
19	8617	10.7	56.1	76.4
20	8533	11.3	56.7	75.5

the best buffer size and run the simulations. The best buffer size can be chosen as indicated in Fig. 3. For examples, in the figure, the best buffer sizes can be chosen between 640 to 1400 bytes.

The reason for performance degradation in communications with this longer packets than 1465 bytes are not yet investigated in detail. One possibility of this degradation is due to MTU of the ethernet. MTU is the Maximum Transmission Unit that IP is allowed to use for a particular interface. If your MTU is set too big, in this case beyond ~1456 bytes your packets must be fragmented, or broken up, by a switching hub along the path to the other PCs. This may result in a drastic decrease in throughput. However, we have not identified the source of this communication degradation. We would like to leave this investigation to our future research.

7 HPF TRISTAN Code Results

In Table 1, the parameter $\varepsilon_{eff-grid}$ is defined as:

$$\varepsilon_{eff-grid} = \frac{\text{total grid no. in decomposition direction - total guard cell number}}{\text{total grid no. in decomposition direction}}.$$

Table 1 shows the total times multiplied by the number of processors, speedups and parallel efficiency vs the number of processors. The total computation time of

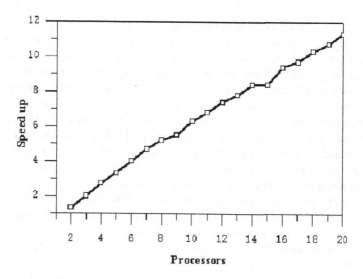

Fig. 4. Speed up vs processor number.

single processor are measured by the original version of TRISTAN code compiled by *pgf*77 compiler with the optimization level **-O2** option. Figure 4 shows the speed up vs processor number. Note that in the table we show the total CPU time multiplied by the processor number.

With fixing the problem size and increasing the processor number, the grid number in one sub-domain in decomposition direction is reduced gradually. For example, 60 extra ghost grid cells in total must be added to each sub-domains in decomposition direction or in x for 20 processors. It is about 25 percents of the total grid number in decomposition direction in this case. Thus the communication overhead become to be insignificant comparing with the total PIC computation time as we increase the number of the processors. The increase of the communication overhead reduces the parallel efficiency in the table. If the communication overhead is insignificant, it is very hard to improve the parallel efficiency of the code without varying the problem size. However, even the most advanced parallel computer nowadays, it is not so easy to increase the problem sizes as we increase the the number of the processors due to the large data sizes we have to store in each simulation runs. Thus the optimal parallel efficiency of the scalable relation between the problem sizes and the number of processors are difficult to measure in our simulation. However, Fig. 4 shows the high linearity of our HPF TRISTAN code and the code scales well. In addition, with the HPF compiler overhead and the load-imbalance overhead due to the Earth dipole field, the parallel efficiency around 60-65 % is affordable in this type of large scale simulations.

8   Concluding Remarks

In the present paper, we have successfully parallelized the three-dimensional full electromagnetic and full particle code using HPF. The code is originally the same as the TRISTAN code and the code is for the space plasma simulations. As shown in Fig. 4 and Table 1, fixing the problem size, our HPF TRISTAN code has a high linearity and scales well. However, our HPF code introduces about 70% overhead and the reason for this overhead is not yet investigated. We have also parallelized the three-dimensional skeleton-PIC code introduced by V. K. Decyk [4] in the same parallel algorithm [3] [4] using HPF. The HPF three-dimensional skeleton-PIC code introduces about 20% overheads[9]. One possibility to explain the larger overheads in our HPF TRISTAN code over the HPF skeleton-PIC code is that the more complicated data structures in HPF TRISTAN than those in the skeleton-PIC code. Our PCs in the cluster have no enough memory and this may degrade the performance of the PCs. Another possibility is the load-imbalance originated in the TRISTAN code as we discussed previously. Our TRISTAN code has the Earth dipole filed which generates and simulates the Earth magnetosphere in one of sub-domains, and this may cause a large load-imbalance. We would like to leave the detailed investigation to our future work.

The parallelization algorithm we used in our code is basically the same as [3][4]. We separate the communication parts from the computation parts. Thus the code can easily be converted to MPI or PVM code by replacing the HPF "CSHIFT" constructs to appropriate message passing interfaces. Our experiences show that the utilization of HPF "FORALL" or "DO INDEPENDENT" constructs in the data-parallel manner without separating the communication parts from the computation parts results in almost no gain of speedups or very poor speedups.

We have also compared the HPF skeleton-PIC code with the MPI or PVM skeleton-PIC code. The HPF code degradation of the total CPU time over the MPI or PVM code is only 10-15 % [9] in this case. Thus we expect that we should be able to enjoy the easier HPF programming with a very small performance degradation even in the more complicate codes like the TRISTAN codes.

Acknowledgment

The authors thanks Professor Viktor K. Decyk for his help using the skeleton PIC codes. The authors also thanks referees and Dr. Seo for invaluable comments regarding to the present paper.

References

1. O. Buneman, TRISTAN. In: Matsumoto H., and Omura, Y. (eds.): Computer Space Plasma Physics: Simulation Techniques and Software. Terra Scientific, Tokyo, (1993) 67-84
2. http://www.pgroup.com, (1998)
3. Liewer, P. C. and V. K. Decyk: A General Concurrent Algorithm for Plasma Particle-in-Cell Simulation Codes, J. Comput. Phys. 85, (1985) 302-322
4. Decyk, V. K.: Skeleton PIC codes for parallel computers, Comput. Physcs. Comm. 87, (1995) 87-94

5. C.K. Birdsall and A.B.Langdon, Plasma Physics via Computer Simulation, McGraw-Hill, New York, (1985)
6. D.W. Walker, Particle-in-cell Plasma Simulation codes on the Connection Machine, Computing Systems in Engineering, **1** (1991) 307-319
7. C.H. Koelbel, et al., The High Performance Fortran Handbook, The MIT press,(1994)
8. Ian Foster, Designing and Building Parallel Programs, Addison-Wesley, (1995)
9. Cai, D., Q. M. Lu, and Y. T. Li, Scalability in Particle-in-Cell code using both PVM and OpenMP on PC Cluster, *Proceedings of 3rd Workshop on Advanced Parallel Processing Technologies* (1999) 69-73

Towards a Lightweight HPF Compiler

Hidetoshi Iwashita[1], Kohichiro Hotta[1], Sachio Kamiya[1],
and Matthijs van Waveren[2]

[1] Strategy and Technology Division, Software Group
Fujitsu Ltd.
140 Miyamoto, Numazu-shi, Shizuoka 410-0396, Japan
{iwashita.hideto,hotta,kamiya.sachio}@jp.fujitsu.com
[2] Fujitsu European Centre for Information Technology Ltd.
Hayes Park Central, Hayes End Road
Hayes UB4 8FE, UK
waveren@fecit.co.uk

Abstract. The UXP/V HPF compiler, that has been developed for the VPP series vector-parallel supercomputers, extracts the highest performance from the hardware. However, it is getting difficult for developers to concentrate on a specific hardware. This paper describes a method of developing an HPF compiler for multiple platforms without losing performance. Advantage is taken of existing technology. The code generator and runtime system of VPP Fortran are reused for high-end computers; MPI is employed for general distributed environments, such as a PC cluster. Following a performance estimation on different systems, we discuss effectiveness of the method and open issues.

Keywords: HPF, compiler, distributed parallel computing, MPI, VPP Fortran

1 Introduction

The progress of the most recent computer hardware is remarkable. Only several years ago, vector computers provided the high performance needed to tackle HPC problems. But now, multiprocessor systems with scalar CPUs, which are becoming cheaper and more rapid each year, are assuming this position. Even the latest communication equipment, such as InfiniBand, Gigabit Ether, IEEE1394, and USB2.0 is starting to catch up in speed with the special hardware networks which support distributed parallel processing. Moreover, it is not only the speed of such change but diversity that is the latest tendency. There is a variety of SMP, SMP cluster, and cc-NUMA architectures with various cash construction and memory hierarchies on the market with the objective of using multiple CPUs simultaneously and effectively.

We have developed *UXP/V HPF*, the HPF compiler for VPP800 and VPP5000 [1] series vector-parallel computers [2]. This compiler offers valuable results, which include the world record of performance in HPF applications [3]. This is due to the runtime system and to the compiler being expert enough in the characteristic of VPP hardware that it can pull out the maximum performance. However the hardware lifecycles are decreasing, and there are greater variations in the hardware, making it

H. Zima et al. (Eds.): ISHPC 2002, LNCS 2327, pp. 526–538, 2002.

difficult to develop software specialized only in a specific hardware. Therefore, we need to rethink the development method, and reconstruct the compiler into layered modules in order to support multiple platforms. We also need technical breakthroughs in order to avoid performance falls due to simplistic generalizations.

As CPUs become cheaper and networks become quicker, the importance of parallel computing in a distributed environment is expected to increase. The most popular application interface to describe distributed parallel computing seems to be MPI [4] [5], which can be called a de-facto standard and includes all we need to do for parallel computing. However, even if MPI is useful for computer scientists and professional programmers, most HPC users seem to find it hard to use it to write real world applications. We don't think naked MPI will be the best answer to write parallel programs. We expect that MPI will be important not for programmers but for systems such as HPF compilers and parallelization support tools.

The Grid [6] is a recent remarkable technology as a platform of distributed environment. Many people expect it to increase in importance in the future.

OpenMP [7] is a language designed for a pure SMP environment and it does not have features to handle data locality. Therefore, it is not suitable for application in a distributed environment if no extensions are introduced. SGI and Compaq [8] have developed vendor-specific OpenMP language extensions in order to support data locality on their cc-NUMA architectures. The OpenMP Architecture Review Board (ARB) is discussing whether language extensions for distributed memory model are needed in OpenMP Version 3.0.

The SCore [9] technology contains a software distributed shared memory (SDSM) layer called SCASH, which works on distributed memory and which offers application software a view of shared memory. However, the OpenMP compiler for the SCore environment requires some language extensions to OpenMP to specify data locality in order to get a high performance [10].

VPP Fortran is an original data parallel language of Fujitsu and it is supported on all VPP series computers. Similar to the current HPF, it uses put/get communication that is supported on the VPP series, by their strong data transfer and hardware barrier facilities. Because the VPP Fortran language specification requires one-sided communication, it is difficult to adopt send/recv communication instead of put/get communication. We have the experience of implementing VPP Fortran on AP1000 and AP3000 distributed memory computers with the put/get communication method with little hardware support [11].

We wish to show in this paper that our HPF compiler can support multiple platforms without losing the performance. The structure of the HPF language processor is introduced in Section 2 and it is applied to VPP series vector-parallel computers in Section 3. In Section 4, development issues of the language processor on multiple platforms that are not restricted to the VPP series are discussed. Section 5 estimates the validity, and section 6 is the conclusion.

2   UXP/V HPF Compiler

UXP/V HPF system contains an HPF translator, which converts an HPF program into Fortran code, and a Fortran vector compiler. This section introduces the FLOPS compiler platform and some important passes in it, which constitute the HPF translator.

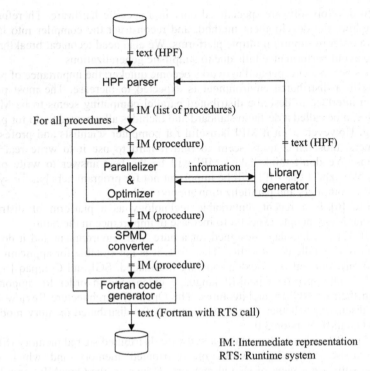

IM: Intermediate representation
RTS: Runtime system

Fig. 1. Configuration of the FLOPS HPF compiler

2.1 FLOPS Compiler

FLOPS (Fujitsu Labs' Optimizing and Parallelizing System) is the framework of source-to-source compilers whose main targets are distributed memory machines [12]. It is written in C and yacc (GNU Bison V1.27 for HPF). It has been used as a basis for parallel compiler products on AP3000 and VPP series computers since its research prototype was developed on the AP1000 scalar parallel computer [13]. The FLOPS framework defines the *intermediate representation (IM)* used in FLOPS compilers and provides access and utility functions onto the IM.

The configuration of the HPF version FLOPS compiler is shown in **Fig. 1**. IM is formed as C structures in product versions and can be input and output as a text of S-expression style in the research system. For all procedures (subroutines and functions) in the HPF source file, IM code is generated, parallelized, optimized, and finally output as a Fortran code.

2.2 SPMD Converter

While HPF program code represents single thread execution and a global name space, *SPMD (Single Program/Multiple Data) code* represents execution and data for each processor. **Fig. 2** shows an example of the function of the SPMD converter.

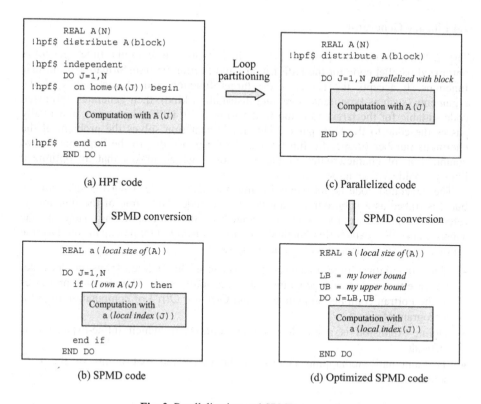

Fig. 2. Parallelization and SPMD conversion

2.3 Parallelizer and Optimizer

The conversion of (a) to (b) in **Fig. 2** implements simplistically the ON HOME directive. Good performance cannot be expected for the following reasons:

- Each iteration of the loop will have an extra cost of condition and branch. Because the conditional expression depends on the loop variable, it is hard for the Fortran compiler to optimize it.
- The loop will not scale to the number of processors because the loop bounds are still global.

Though it might be possible to convert (b) to (d) as an optimization of the SPMD code, our *Parallelizer* finds loop partitioning corresponding to the ON directive in the source code (a) and generates an intermediate code (c). Note that it is not always possible to find a suitable loop partitioning even if the ON directive is specified, since the generalized clause home (A (f (J))) requires computation of the inverse function of f and the relation of ON and DO constructs may not be simple.

Other functions of the *Parallelizer*, such as the generation of ON directives and the searching of loop independencies, makes automatic loop parallelization available in simple cases.

The *Optimizer* reduces the number of interprocessor communications and runtime system calls and arranges blocks of communications into asynchronous data transfers.

2.4 Library Generator

The *Library Generator* [14] & [15], which works at compile time, generates optimized parallel code for the HPF Library and Fortran 95 transformational intrinsic functions. It accepts the name of the target function and characteristics of the arguments (such as type, rank, size, and distribution kind), then generates optimized code suitable for the target function in the form of an HPF subprogram, and finally passes the code to the HPF parser. This implementation solves the problem of the enormous number of specific functions in HPF Library due to the large number of combination of characteristics of the arguments, which makes that an optimized library would have an impossible size.

The generated code is not expanded into the source code using inline-expansion but it is linked as a subprogram with the source code. The name of the function as referred to in the source code is changed to the corresponding name in the subprogram. We called this method *online-expansion*. Online-expansion has the following merits compared with inline-expansion:

- The compilation time increases less because the online-expansion does not increase the size of each program unit in the source code. Optimization processes of the Fortran compiler sometimes spend $O(n^2)$ or $O(n^3)$ of computation time for a program of size n.
- Online-expansion is available even in contexts in which inline-expansion is difficult.
- The automatic remapping facility at the entry and exit points of the HPF subprogram is useful.

3   Application to VPP Series

This section describes features of the VPP5000 hardware and how the current HPF system utilizes them.

3.1 Distributed Memory Machine VPP5000

The VPP5000 is the latest generation of Fujitsu vector-parallel supercomputers [1]. It is a distributed memory machine with up to 512 processor elements (PE) connected with a high-speed *crossbar network*. The throughput of the crossbar is 1.6 Gbyte/s for both input and output of each PE. Each PE has a vector unit of 9.6 Gflop/s, a scalar unit with VLIW RISC architecture, and up to 16 Gbyte of 45 ns SDRAM memory.

The *Data Transfer Unit (DTU)*, which the VPP series computers have in each PE, can directly access local memory and communicate with other DTUs. The DTU enables hardware put/get communication through the crossbar network without interrupting any CPUs. The DTU recognizes remote data through the virtual global address. In order to write data into remote memory (i.e., the put communication), the DTU reads data from local memory with the specified access pattern and sends the data to the remote DTU. The remote DTU then writes the data into its local memory using the given global address and stride pattern. In order to read remote data into local memory (i.e., the get communication), the DTU sends a request packet to the remote DTU, asking him to send back the specified data as a put communication.

The DTU can handle packets with a two-dimensional stride, which support at least two lower dimensions of a Fortran 90 array section. For example, the array section `A(i1:i2:i3,j1:j2:j3,1:n)` can be sent and received with only n packets (unless the packet size exceeds the limit) of zero-copy communication without interrupting any CPU.

3.2 Runtime System

The parallel runtime system (RTS), which is called from the generated code of the HPF translator, was made solely for the VPP800 and VPP5000 series. Parallel RTS has two main features, parallel execution management and interprocessor communication. From our experience with VPP Fortran [16], we applied the put/get communication method in RTS in order to extract the highest performance of the VPP800/5000.

By taking advantage of the rich features of DTU and crossbar network topology, RTS achieved almost the same throughput as the hardware peak performance of 1.6 GB/s/PE for big and regular data transfer, such as the transpose of a block-distributed array. RTS can send not only contiguous data, but can also combine some Fortran 90 array sections into one packet for sending and receiving with the DTU.

3.3 Characteristics of Communication on VPP

While large array data can be treated well on the VPP, applications that include many irregular accesses of small data tend to be inefficient. The cause of low performance is a relatively high latency, several 10s of microseconds in total with software and hardware overhead. Therefore, if the size of each packet is not much greater than 16KB (=10 [microsecond] x 1.6 [GB/second]), the latency dominates the throughput speed.

Unlike send/recv communications that imply loose synchronization, the put/get communication method often requires barrier synchronization in order to confirm if the remote storage can be referred and overridden. In order to support such frequent synchronizations, the VPP has a high-speed hardware barrier facility.

4   Multi-platform Development

This section discusses the possibility of developing compilers for multiple platforms starting from the current HPF compiler on VPP series.

4.1 Development for High-End Computers

Fujitsu has designed a data parallel language VPP Fortran and provides a compiler for it [16]. Even though two program codes written in VPP Fortran and in HPF/JA [17] reach almost the same peak performance, the RTS of HPF is 20% larger than the RTS of VPP Fortran. The difference is caused by the variety of data mapping in HPF (e.g., block cyclic mapping, indirect mapping, replication of partially distributed array, alignment with stride, scalar template, and alignment between non-distributed arrays) and by the dynamic management (e.g. automatic remapping at subprogram entry, redistribution and realignment while keeping the linkage of alignment, etc.). The heaviness of HPF RTS increases the cost of initialization at the entry points of subprograms. The cost becomes more important in the case of smaller applications.

Assuming that the VPP Fortran compiler will exist for the future high-end computers, we are considering to replace the SPMD converter and the succeeding passes of the HPF compiler and the RTS of HPF with those of VPP Fortran, as shown in **Fig. 3**. The following issues must be solved:

- The IM converter in **Fig. 3** is needed. -- It would convert IM generated by the parallelizer and the optimizer into a form that is acceptable to the VPP Fortran compiler. Both IMs of HPF and VPP Fortran are basically compatible with the exception of some small differences.
- Some features of HPF that should be supported in RTS are not supported in the RTS of VPP Fortran. -- They will be supported in the parallelizer and optimizer of the compiler as much as possible so that the enhancement of RTS will be minimized. For example, most redistribution can be solved at compile time with flow analysis and program conversion.

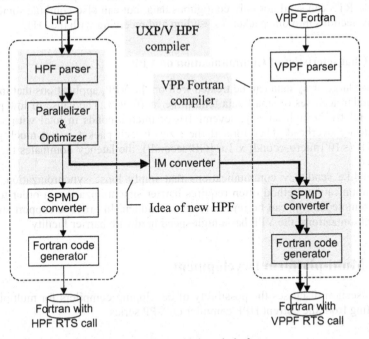

Fig. 3. HPF compiler on high-end platform

Especially the second issue requires more research. We have had a good experience with a highly tuned HPF/JA program, which can be shown to work well on the RTS of VPP Fortran. This is described in Section 5. Because the VPP Fortran RTS is lighter than the HPF RTS, the resulting system will be lightweight.

4.2 Development for a General Distributed Environment

Here we discuss how the UXP/V HPF compiler can be adapted to a general distributed environment that consists of multiple CPUs connected by a communication network, including PC clusters. Instead of RTS, we are trying *MPI* [4] as a communication layer. MPI is used on many distributed platforms. Implementation using MPI must be one of the following:

- Development of an RTS that calls MPI. The interface of the RTS will be changed from the current RTS.
- Changing the code generation so that the code contains direct MPI calls instead of RTS calls.

In both cases, most of the current passes including HPF parser, parallelizer, optimizer, and library generator can be used unmodified, but the SPMD converter must be modified. The advantage of the latter case is the portability of the compiler system because the RTS does not need to be recompiled for the different platforms. In the latter case, however, it is unclear if the runtime environment such as IDs of the active processors can be managed in the generated code without resorting to RTS. An online expansion technique similar to the library generator might be used.

While the put/get communication is available in MPI-2 [5], we adopt at first send/recv as the primitive of communication since it is more popular and is already evaluated on many platforms. Put/get communication is expected to be a good alternative in some cases.

The efficiency of communication is an open issue, since we do not rely on special communication hardware such as those of VPP. Communication aggregation will be a key technique in future development.

5   Performance Estimation

5.1 Performance Estimation for High-End Computers

This section estimates how the replacement of SPMD converter and RTS affects the peak performance of an HPF program for the high-end implementation. We have tuned the NAS Parallel BT benchmark code [18] on VPP5000 with HPF/JA language extensions and vector optimization [2]. In order to compare with this result, we estimated the performance of the generated code of the new compiler for high-end machines shown in **Fig. 3**. Instead of using the SPMD converter, we inspected the output code of the current HPF compiler and made the VPP Fortran compiler generate almost the same code, using VPP Fortran programming. This work is in effect an emulation of the ideal function of the SPMD converter. The conversion of HPF to VPP Fortran is described in **Table 1**.

The result of the comparison is shown in **Fig. 4**. For all data size classes S, A, B, and C, the new compiler was estimated to give a higher performance than the current compiler. The difference tends to become larger in smaller data size. To our impression, this is because the current HPF has the following expensive portions:

- Initialization at the entry points of user subprograms corresponding to dummy arguments.
- Handling of a variety of data mapping in RTS.

As a conclusion of this performance estimation, we confirmed that the intermediate code, which includes main features of HPF, can be translated into code that calls RTS of VPP Fortran. Because RTS of VPP Fortran is lighter than RTS of HPF, we expect that the resulting code will have a higher performance. Though they are not used in this tuned benchmark code, HPF has important features that VPP Fortran does not support, such as redistribution. Such features must be carefully implemented in order to keep the generated code and RTS lightweight. If the new compiler supports them with little help of RTS, it achieves almost the same performance as the VPP Fortran compiler on the same high-end computer.

Table 1. Conversion from HPF to VPP Fortran

HPF language items	Corresponding VPP Fortran language items
Data mapping with PROCESSORS, DISTRIBUTE, TEMPLATE, ALIGN, SHADOW, and SEQUENCE	Corresponding combination of PROCESSOR, INDEX PARTITION, GLOBAL, and LOCAL
INDEPENDENT directive	SPREAD DO directive with the loop decomposition generated by the HPF compiler
ON HOME construct and RESIDENT for execution of single processor	SPREAD REGION construct
ASYNCHRONOUS construct (in HFP/JA extension)	EQUIVALENCE (of local variable to global variable) and SPREAD MOVE construct
Asynchronous REFLECT directive (in HFP/JA extension)	OVERLAPFIX directive
Access of sequential and unmapped variables	Specified as LOCAL and careful manual maintenance of data consistency

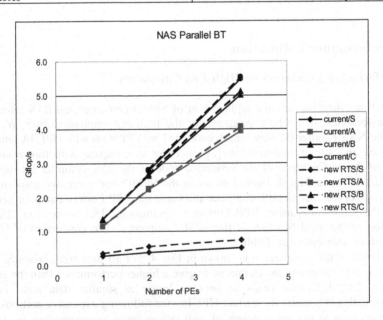

Fig. 4. Estimation of HPF compilers for high-end machines

5.2 Estimation of HPF Calling MPI

This section estimates the characteristics of the performance of the new HPF compiler that employs MPI as a communication primitive. Using the medium size SPEC OMPM2001 SWIM benchmark [19], we made the following two executable codes and evaluated them on the VPP5000.

HPF on RTS	An HPF program compiled with the current HPF compiler, which employs effective RTS developed only for VPP800 and VPP5000
HPF on MPI	An SPMD program with MPI, written manually with the purpose of estimating the performance of code that the new compiler will generate

Fig. 5 shows the result. The new HPF on MPI was estimated to give better performance than the current HPF on RTS. This estimation does not guarantee that the new compiler gives high performance for all application programs, but offers us very good prospects.

Using the analyzer tool, we measured the cost distribution. **Fig. 6** displays the total cost for each procedure of the benchmark program using all employed processors. Since the costs were measured on the basis of elapsed time, it means that there is a good load distribution and little overhead in the subprogram with the result that its cost is not much greater than the one of serial execution. In the MPI version, the column comm shows the total costs of MPI communication and the calling of MPI; in RTS version, the communication cost is included in each subprogram. We conclude from the performance estimation the following:

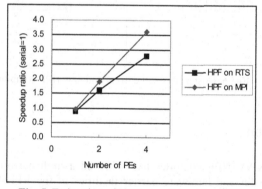

Fig. 5. Estimation of HPF compilers using MPI

- The RTS version has a large cost in the main program, and this may break the scalability of the parallel execution. The cost (elapsed time) includes allocation to the virtual global memory, broadcasting the global addresses, and interprocessor fork operation with large initialized data. Such initialization costs can be ignored in the typical long-term jobs that appear on the VPP. However, if we take short-term execution without a high-speed network into account, it should be improved.

- In contrast, the MPI version shows perfect load balancing in the main program, in which all processors do not have to do extra work (except the communication cost summed up into comm). This is because the implementation does not use virtual global memory or global variables. The initialization cost of the MPI environment is not visible in this measurement on VPP5000 but it may appear in measurements on other distributed memory environments.
- The communication cost of the MPI version increases more than twice between two and four processors. We use in this performance estimation only MPI_SEND, MPI_RECEIVE, and a collective communication MPI_SENDRECV for the purpose of interprocessor communication. So, more improvement might be possible if we use other functions such as MPI_REDUCE and non-blocking communication and if we take account of communication scheduling.
- We wrote an MPI version of the program, in which MPI functions are not directly called from the source code but called from a shell, which itself is called from the source code. The total cost of shell routines was trivial (only 0.07 second) compared to the whole cost of the program.

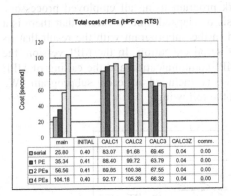

Total cost of PEs (HPF on RTS)

	main	INITIAL	CALC1	CALC2	CALC3	CALC3Z	comm.
serial	25.80	0.40	83.07	91.68	69.45	0.04	0.00
1 PE	35.34	0.41	88.40	99.72	63.79	0.04	0.00
2 PEs	56.56	0.41	89.85	100.38	67.55	0.04	0.00
4 PEs	104.18	0.40	92.17	105.28	66.32	0.04	0.00

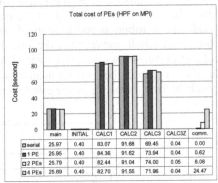

Total cost of PEs (HPF on MPI)

	main	INITIAL	CALC1	CALC2	CALC3	CALC3Z	comm.
serial	25.97	0.40	83.07	91.68	69.45	0.04	0.00
1 PE	25.95	0.40	84.36	91.62	73.94	0.04	0.62
2 PEs	25.79	0.40	82.44	91.04	74.00	0.05	8.08
4 PEs	25.89	0.40	82.70	91.55	71.96	0.04	24.47

Fig. 6. Cost analysis of current and new HPF compilers

6 Conclusion

The current UXP/V HPF compiler has achieved a high performance by using the strong data transfer and hardware barrier facilities of the VPP800/5000 effectively. The performance relies on high-level user tuning and a sufficient size of the problem. In order to develop compilers for a general distributed environment, we cannot rely on these facilities, high-level user tuning and large problem sizes. We have discussed how to develop compilers for multiple platforms starting from the current HPF compiler on VPP series. The keyword is *lightweight*. Technically, it is necessary to make the following items lightweight:

- Initialization cost at the beginning of the program and entry point of the subprograms, and
- Runtime information managed by RTS and the variety of data mapping handled by RTS.

In addition, the following lightness is also required:

- Reducing the user's (especially beginner's) load, in order to get a reasonable performance without perfect user tuning, and
- Reducing the redundancy of the development, in order to quickly support many platforms.

We have discussed performance estimations of the lightweight system on both high-end computers and on general distributed systems.
We consider common techniques such as MPI, Grid, and SCore as communication primitives. Instead of using put/get communication, loose synchronization, a feature of send/recv communication, will reduce the barrier synchronization and create opportunities for pipelined parallelism.

References

1. VPP5000 Series. http://primepower.fujitsu.com/hpc/en/vpp5000e/index.html
2. Hidetoshi Iwashita, Naoki Sueyasu, Sachio Kamiya, and Matthijs van Waveren. VPP Fortran and the Design of HPF/JA Extensions, *Concurrency: Practice and Experience.* To be published.
3. Tatsuki Ogino. Global MHD Simulation Code for the Earth's Magnetosphere Using HPF/JA, *Concurrency: Practice and Experience.* To be published.
4. Marc Snir, Steve Otto, Steven Huss-Lederman, David Walker, and Jack Dongarra. *MPI - The Complete Reference Volume 1, The MPI Core.* MIT Press, Cambridge, Massachusetts, 1999.
5. Message Passing Interface Forum. http://www.mpi-forum.org/
6. Global Grid Forum. http://www.gridforum.org/
7. OpenMP Architecture Review Board. http://www.openmp.org/
8. John Bircsak, Peter Craig, RaeLyn Crowell, Jonathan Harris, C. Alexander Nelson, and Carl D. Offner. Extending OpenMP for NUMA Architectures. In *WOMPAT2000*, San Diego, CA, July 2000.
9. Real World Computing Partnership. SCore Cluster System Software. http://pdswww.rwcp.or.jp/score/dist/score/html/index.html
10. Mitsuhisa Sato, Hiroshi Harada, Atsushi Hasegawa, and Yutaka Ishikawa. Cluster-enabled OpenMP: an OpenMP compiler for Software Distributed Memory System SCASH. In Proceedings of *JSPP2001*, pp.15-22, 2001. (In Japanese)
11. Tatsuya Shindo, Hidetoshi Iwashita, Tsunehisa Doi, and Junichi Hagiwara. An implementation and evaluation of a VPP Fortran compiler for AP1000. In *IPSJ SIGNotes High Performance Computing*, No.048-002, 1993.
12. Tatsuya Shindo, Hidetoshi Iwashita, Tsunehisa Doi, Junichi Hagiwara, and Shaun Kaneshiro. HPF Compiler for the AP1000. In *1995 International Conference on Supercomputing.* pp. 190-194, 1995.
13. AP3000 Homepage. http://primepower.fujitsu.com/hpc/en/ap3000-e/index.html
14. Matthijs van Waveren, Cliff Addison, Peter Harrison, David Orange, Norman Brown, and Hidetoshi Iwashita. Code Generator for the HPF Library and Fortran 95 Transformational Functions, *Concurrency: Practice and Experience.* To be published.
15. Matthijs van Waveren, Cliff Addison, Peter Harrison, David Orange, and Norman Brown. Code Generator for HPF Library on the Fujitsu VPP5000. *Fujitsu Scientific and Technical Journal,* vol. 35, no. 2, pp. 274-279, 1999. http://magazine.fujitsu.com/us/vol35-2/paper17.pdf

16. Hidetoshi Iwashita, Shin Okada, Makoto Nakanishi, Tatsuya Shindo, and Hiroshi Nagakura. VPP Fortran and Parallel Programming on the VPP500 Supercomputer, In poster session proceedings of *1994 International Symposium on Parallel Architectures, Algorithms and Networks (ISPAN'94)*. pp.165-172, 1994.
17. Japan Association for High Performance Fortran (JAHPF). *HPF/JA Language Specification Version 1.0*, November 1999. http://www.hpfpc.org/jahpf/spec/hpfja-v10-eng.pdf
18. NAS Parallel Benchmarks. http://www.nas.nasa.gov/Software/NPB/
19. SPEC OMP2001 Benchmark Suite. http://www.spec.org/hpg/omp2001/

Parallel I/O Support for HPF
on Computational Grids⋆

Peter Brezany[1], Jonghyun Lee[2], and Marianne Winslett[2]

[1] Institute for Software Science
University of Vienna, Liechtensteinstrasse 22, A-1090 Vienna, Austria
brezany@par.univie.ac.at
[2] Department of Computer Science, University of Illinois,
Urbana, IL 61801, USA
{jlee17,winslett}@cs.uiuc.edu

Abstract. Recently several projects have started to implement large-scale high-performance computing on "computational grids" composed of heterogeneous and geographically distributed systems of computers, networks, and storage devices that collectively act as a single "virtual supercomuter". One of the great challenges for this environment is to provide appropriate high-level programming models. High Performance Fortran (HPF) is a language of choice for development of data parallel components of Grid applications. Another challenge is to provide efficient access to data that is distributed across local and remote Grid resources. In this paper, constructs to specify parallel input and output (I/O) operations on multidimensional arrays on the Grid in the context of HPF are proposed. The paper also presents implementation concepts that are based on the HPF compiler VFC, the parallel I/O runtime system Panda, Internet, and Grid technologies. Preliminary experimental performance results are discussed in the context of a real application example.

1 Introduction

Many applications' computational requirements exceed the resources available at even the largest supercomputers. For these applications, Grid technology [14] promises access to increased computational capacity through the simultaneous, coordinated use of geographically separated large-scale computers linked by networks. Through this technology, the "size" of an application can be extended beyond the resources available at a partial location - indeed, potentially beyond the capacity of any one computing site anywhere [27].

The current main Grid research can be divided into two sub-domains: *Computational Grids* and *Data Grids*. Whereas a Computational Grid can be considered as a natural extension of a cluster computer system where large computing

⋆ The work described in this paper is being carried out as part of the research project "Aurora" supported by the Austrian Research Foundation, and was also supported by NASA under grant NAGW 4244 and by the US Department of Energy under grants B341494 to the Center for Simulation of Advanced Rockets and to the Center for Programming Models for Scalable Parallel Computing.

H. Zima et al. (Eds.): ISHPC 2002, LNCS 2327, pp. 539–550, 2002.

tasks are performed using distributed computing resources, a Data Grid [9,33] deals with the efficient management, placement and replication of large amounts of data on the Grid. However, once the data are in place, computational tasks can be run on the Grid using the provided data.

The fundamental challenge is to make Grid systems widely available and easy to use for a wide range of applications. This is crucial for accelerating their transition into fully operational environments.

Although first implementations of Grid infrastructures, such as Globus [13], support mainly the execution of message-based applications, it is foreseen that future Grid applications will require much more advanced programming models, e.g. based on parallel components specified in high-level programming languages and mechanisms for coordination of execution of these components [12]. Among such Grid applications, multi-physics applications are good examples. They are made of several high-performance simulation codes coupled together to simulate several physics behaviors. These codes are generally independently developed. Within a parallel code, the programmer may use a parallel language like HPF [20], Co-Array Fortran [32], HPC++[21], HPJava [8], OpenMP [35], etc. The coupling of the simulation codes could be carried out through the use of a Remote Method Invocation mechanism (Java RMI or CORBA) or an appropriate coordination language (e.g. the OpusJava approach [25]).

Parallel Grid application components are often data intensive - they typically require access to large (terabytes) remote datasets. Several scenarios may be identified when a component accesses a dataset:

(a) *Traditional I/O operations.* The component reads/writes a dataset produced/ consumed by a parallel component on a remote Grid site. This I/O operation is inherent in the given application algorithm; it can be optimized but not avoided. Another example of this kind of I/O is access to a dataset produced by a remote sensor or another instrument attached to the Grid.

(b) *Checkpoint.* For long-running production runs, it is desirable to save the state of certain arrays periodically (checkpoint) in order to resume (restart) from a previous state in case of a system failure, or when the job has moved to another Grid site by the Grid scheduler. A read operation on a checkpoint reads the entire checkpoint. The distinction between a checkpoint and an ordinary write operation is that a checkpoint operation should be performed as a single atomic action. The output of checkpoint operations usually needs to be moved to a safe remote place or several safe remote places (for reliability reasons), so that it will be availably for restart, if needed.

(c) *Snapshot.* The component outputs snapshots of certain arrays repeatedly, without overwriting previously output snapshots. The snapshots will be read and analyzed by visualization tools, typically on a different platform (e.g., a workstation), located at a different Grid site.

This paper describes an approach to providing HPF Grid codes with high-performance access to local and remote Grid datasets at a high abstraction level, using concepts that are familiar to any HPF programmer. This allows HPF users to specify the Grid I/O operations with minimal effort. We also discuss our prototype implementation of the new HPF constructs, in the context of an HPF compiler called the Vienna Fortran Compiler (VFC) [2], the Panda parallel I/O library [36], Internet and Grid technologies. Panda performs I/O operations on distributed multidimensional arrays, which are stored on parallel disks with a specified data organization. We refer to a collection of one or more such arrays, managed by Panda, as a scientific data repository (SDR) [7].

Although the concepts presented here are proposed in the context of HPF, they can be easily integrated into any other data-parallel language.

The following parts of the paper are organized as follows. Section 2 introduces the extensions to HPF to allow HPF programmers to access the functionality of a parallel I/O library. The implementation of the VFC/Panda interface and preliminary performance results are briefly discussed in Section 3. Related work is described in Section 4. We briefly present our conclusions in Section 5.

2   Language Support

We now present the new HPF language constructs and the underlying machine model. To indicate the relation of these language constructs to the concept of a scientific data repository (SDR), they are introduced by the prefix "!SDR$".

2.1   Data Distribution

The Abstract Grid Model. Fig. 1 shows an example grid architecture containing a set of Grid nodes connected by a wide-area network WN. Each Grid node contains a set of cluster or supercomputer nodes and IN, an interconnection network connecting the nodes. Each node is either a (1) computation node - has a processor and internal memory; (2) I/O node- has in addition a set of disks, and is responsible for controlling I/O activities only; or (3) combination node - same as an I/O node, except the processor fulfills both compute and I/O tasks.

Array Distribution Across I/O Nodes. The HPF concepts related to distribution of data onto computational nodes may be extended to describe the mapping of data onto I/O nodes. In the extensions proposed, I/O nodes are explicitly introduced by the declaration of an *I/O processor array*. For example, in relation to Fig. 1, the following declaration introduces a 1D I/O processor array *IOP*:

!SDR$ IOPROCESSORS :: IOP(2)

The array distribution onto the logical array of I/O processors is specified by a new directive IODISTRIBUTE, which in relation to Fig. 1 has, e.g. for a 2 D array *A*, the following form:

!SDR$ IODISTRIBUTE (BLOCK, *) ONTO IOP :: A

Mapping of logical I/O processors to physical processors and logical disks onto physical disks will depend on the particular I/O library.

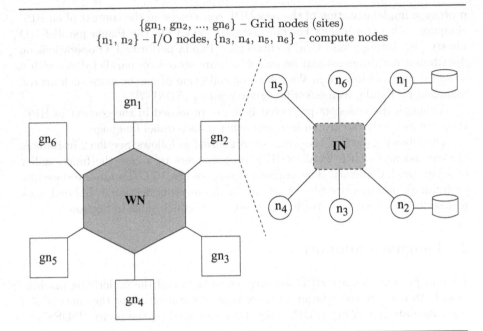

{gn$_1$, gn$_2$, ..., gn$_6$} – Grid nodes (sites)
{n$_1$, n$_2$} – I/O nodes, {n$_3$, n$_4$, n$_5$, n$_6$} – compute nodes

Fig. 1. The abstract Grid model

Data Replication Across Grid Nodes. Although network latencies will decrease in the future as network technology improves, there is still a performance difference between accessing data locally (on the same machine) or remotely over a network. By providing a copy of a data item close to a client application, access times can be reduced. In general, managing copies of data items is regarded as *replication*. Any unit of digital storage can be replicated: a single bit, a few bytes, arrays, objects or even an entire file. Replication can also help in load balancing and can improve reliability.

As opposed to "conventional" data items that exist only once in a data store, replicated data items require a particular naming convention. A set of identical replicas is identified by a logical name and each individual replica is identified by a physical name. Let us assume a physical file called *file1.DB* which is stored at site *gn1* in the directory */data*. Now a replica of *file1.DB* is created at site *gn3* in a similar directory. Thus, the following two physical names exist: *gn1/data/file1.DB* and *gn3/data/file1.DB*. The research aim of Data Grid projects addressing replication (e.g. [39]) is for a client application not to need to know about all physical names and thus about physical locations of a replicated item. For the client application it is sufficient to know the logical name, for instance *file1.DB*, and have an additional data structure called a *replica catalog* that maps the logical name to a set of physical names. In the rest of this section, we show how local and remote repositories are opened and closed, and how our model deals with replicas.

INTEGER :: localRepos, remoteRepos1, remoteRepos2, listLength
CHARACTER (LENGTH=40), DIMENSION (20) :: locationList
CHARACTER (LENGTH=40) :: location, replicaCatalogURL

. . .

```
D1:   !SDR$ OPEN ('MyRepository.panda', localRepos)
D2:   !SDR$ SEARCH_CATALOG ('RemoteRepository.panda', replicaCatalogURL,&
      !SDR$                        locationList, listLength)
D3:   !SDR$ SELECT_LOCATION ( locationList, listLength, location)
D4:   !SDR$ OPEN (location, remoteRepos)
         ... operations on a local repository denoted by localRepos and
            a remote repository denoted by remoteRepos ...
D5:   !SDR$ CLOSE (localRepos)
D6:   !SDR$ CLOSE (remoteRepos)
```

Fig. 2. Opening and closing local and remote repositories

The directives !SDR$ OPEN() and !SDR$ CLOSE() specify opening and closing a repository, respectively. The !SDR$ OPEN() directive must be given the name of the repository to open. It returns a repository handle in the second argument. All handles have type INTEGER. We assume that the appropriate library routine to use with a particular local or remote repository is apparent from the physical name of the repository. If the repository name is specified by a string variable identifier, it is analyzed at run time; otherwise, the analysis can be performed at compile time. In the examples below, we use the naming conventions of the replica management subsystem of the Globus Toolkit[1] [17].

Fig. 2 illustrates the use of the !SDR$ OPEN() and !SDR$ CLOSE() directives, access to the replica catalog, and the selection of an appropriate replica. The construct in line *D1* specifies the opening of the local repository *MyRepository.panda*, which is then denoted by the handle *localRepos* till this repository is closed in line *D5*. *RemoteRepository.panda* is a logical name of a repository which is stored (replicated) on one or more Grid nodes. The mapping between this logical name and physical storage locations is realized by a replica catalog. The string variable *replicaCatalogURL* stores the URL for such a replica catalog. The !SDR$ SEARCH_CATALOG() directive (line *D2*) accepts this variable and the logical repository name as input and returns a list of corresponding physical instances *locationList* and information about the length of this list *listLength* as output. The !SDR$ SELECT_LOCATION() directive (line *D3*) performs replica selection (its URL address is assigned to the output argument *location*); this might be based on interaction with the user, or on performance estimates provided by tools such as Network Weather Service [42], which provides network

---

[1] Other naming conventions have already been proposed by Data Grid projects and the Global Grid Forum [16] to describe some types of Grid datasets. For example, document [37] proposes a URL-like syntax [23] for creating logical and physical file names.

```
!SDR$ IODEF   transfer_type  ::   transfer_object_name
!SDR$    REPOSITORIES ::  LOCAL (localRepos),  REMOTE (remoteReposList)
!SDR$    INVOLVED :: array_list
!SDR$    IODISTRIBUTE ( dist_spec_1 ) ONTO  IO_proc_array ::  array_1
         ...
!SDR$    IODISTRIBUTE ( dist_spec_n ) ONTO  IO_proc_array ::  array_n
!SDR$ END IODEF
```

Fig. 3. Specification of a data transfer object

performance information, and the Grid Information Service [11], which provides storage system performance estimates. Finally the physical instance of the remote repository denoted by *location* is opened by !SDR$ OPEN() (line *D4*).

2.2   I/O Operations

Following [6], we propose HPF extensions to allow users to read and write a set of arrays, checkpoint a set of arrays, read arrays from a previously taken checkpoint, and read or write a snapshot of a set of arrays. All the proposed extensions allow applications to access both local and remote repositories.

The new HPF constructs allow users to define a *data transfer object*, specified by a block of directives as introduced in Fig. 3.

In this specification, the directive INVOLVED introduces a list of arrays which will be written or read to or from the repository, and the IODISTRIBUTE directives specify I/O distributions of these arrays. The way the arrays are to be transferred is determined by a keyword which stands in place of *transfer_type*. Due to the three I/O scenarios discussed in Section 1, the following transfer types can be specified in this context: TRADITIONAL, CHECKPOINT, and SNAPSHOT.

When the HPF application does not specify any I/O data distribution, the underlying library should select an appropriate distribution.

The clauses LOCAL and REMOTE specified in the REPOSITORIES directive introduce a local repository and a list of remote repositories, respectively, to/from the arrays are to be transferred. At least one repository must be given.

The example in Fig. 4 shows the specification of the operations to read and write checkpoints and write snapshots. In Fig. 4, the checkpoints are written into both local and remote repositories; the READ operation preferably reads a checkpoint from the local repository specified in the associated data transfer object, if available, else from the remote repository stored at the safe site. The WRITE snapshot operation writes snapshots into the remote repository.

3   Implementation and Preliminary Performance Results

Fig. 5 shows the architecture of the prototype implementation of our I/O support. In the figure, Panda is a parallel I/O library that supports the most common

```
PARAMETER :: M = ..., N = ...
REAL, DIMENSION (N, N) :: A, B, C, D
INTEGER :: localRepos, remoteRepos, remoteRepos2
!SDR$   IOPROCESSORS  :: IOP(M, M)

        ! declare transfer object of checkpoint type
            ... localRepos and remoteRepos are handles for the local
                and remote repositories, respectively, opened in Fig. 2 ...
!SDR$   IODEF   CHECKPOINT  :: Simulation01
!SDR$         REPOSITORIES  :: LOCAL (localRepos), REMOTE (remoteRepos)
!SDR$         INVOLVED :: A, B
!SDR$         IODISTRIBUTE (BLOCK, BLOCK ) ONTO IOP :: B
!SDR$   END IODEF
        ! write checkpoint
!SDR$   WRITE (Simulation01)

        ...
        ! read checkpoint (restart)
!SDR$   READ (Simulation01)

!SDR$   OPEN ('Snapshots.panda', remoteRepos2)
        ! declare transfer object of snapshot type
!SDR$   IODEF  SNAPSHOT  :: TemperatureAndPressure
!SDR$         REPOSITORIES :: REMOTE (remoteRepos2)
!SDR$         INVOLVED :: T, P
!SDR$         IODISTRIBUTE (CYCLIC, BLOCK) ONTO IOP :: T
!SDR$         IODISTRIBUTE (BLOCK, BLOCK) ONTO IOP :: P
!SDR$   END IODEF
        ! write new snapshot
!SDR$   WRITE (TemperatureAndPressure)
        ! close repositories
!SDR$   CLOSE (localRepos)
!SDR$   CLOSE (remoteRepos)
!SDR$   CLOSE (remoteRepos2)
```

Fig. 4. Specification of read and write operations for checkpoints and snapshots

I/O access patterns that occur in typical scientific applications. From a logical point of view, Panda operates on a parallel data store that can be considered to be a scientific data repository. Panda data servers can run on dedicated I/O nodes or combination nodes, and Panda clients run on compute nodes. The application on a particular node calls the local Panda client when it wishes to read or write data. During each I/O request, Panda servers read or write arrays from or to disks and move data to and from clients using message passing facilities such as the MPI library. When using Panda, each I/O chunk resulting from the distribution chosen for disk will be buffered and sent to (or read from) a file that one I/O node is in charge of, and that I/O node is responsible for reading, writing, gathering, and scattering the I/O chunk. Panda interacts with the underlying file system to perform I/O operations on a specific machine.

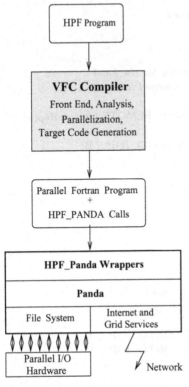

Fig. 5. I/O processing scheme

The experiments described here are taken from the first prototype implementation of Panda remote data accesses [26], implemented using the HTTP data transfer protocol and CGI application programs. (Later TCP-based implementations showed similar performance.) The bars in Fig. 6 represent the migration performance of a real scientific simulation, called WARP3D [24]. The simulation ran on the IBM SP at the Argonne National Laboratory using 4 compute processors; it generated 10 snapshots, 26.1 MB total, which were migrated to a workstation at the University of Illinois.

The first three bars show the performance of WARP3D with no (snapshot) I/O, with local I/O using WARP3D's native I/O approach, and with local I/O using Panda with 1 dedicated I/O processor. The first two bars show that the native I/O takes 4.5 seconds on average. In these configurations, the snapshot data had to be manually migrated for the remote visualization. Manual migration takes about 40 seconds, plus α extra time to set up, remove the FTP connection and make transfer requests. Thus the turnaround time for WARP3D using native I/O and manual data migration is $373 + 40 + \alpha = 413 + \alpha$ seconds.

The last bar shows the performance of WARP3D when using Panda for its local I/O and automatic data migration (each snapshot is first staged to the local repository and then immediately migrated to the remote repository) directly from WARP3D. It takes about only 5.5 seconds longer than the previous bar. This visible overhead is less than 2% of total run time, and would be even lower if we used multiple I/O nodes, where we can expect better aggregate transfer bandwidth. In our future work, we would like to integrate the Globus Toolkit Grid services with this setup.

As explained in [6], the VFC compiler translates an HPF code into the Fortran 90 [28] code containing calls to the runtime libraries. The !SDR$ directives are analyzed and translated into Fortran 90 data structures and sequences of procedure calls to the wrapper library, which provides a set of Fortran 90 subroutines that enable users to call Panda operations from a Fortran 90 program. The wrapper library performs all the transformations and conversions that are

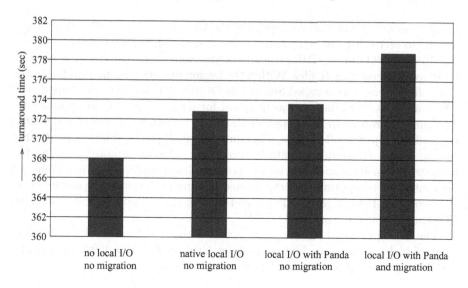

Fig. 6. The performance of WARP3D

necessary for coupling the C++ world of Panda to the Fortran 90 environment of VFC.

The current implementation provides support for arrays of simple and derived types. The system can accept a wide range of derived types, namely, those which consist of components having INTEGER, REAL and DOUBLE PRECISION types, both scalars and arrays, and of derived types that recursively match this criterion.

Paper [6] presents performance results of this HPF/Panda interface implementation achieved by benchmarks accessing local files on a Beowulf cluster system. In our experiments, we observed a performance decrease of 1 to 3% compared to the performance of the corresponding benchmark versions using the native Panda interface.

4   Related Work

In recent years, parallel I/O subsystems have become the focus of much research, leading to the design of parallel I/O hardware and matching system software. So far, the major software research efforts have led along two lines, i.e., parallel file system research (e.g., PIOFS [1], PIOUS [29], PPFS [22], and Galley [31]) and parallel I/O library research (e.g., PASSION [10], DRA [30], ROMIO [40], and Panda [36]). The development of HPF has almost fully focused on providing support for compute intensive parts of scientific and engineering applications and only minimally addressed their I/O aspects ([4,5,6,38]).

In the past four years, an increased attention has also been devoted to parallel I/O on Computational Grids. Global Access to Secondary Storage (GASS) [3] is a data movement and access service of the Globus toolkit. GASS allows applications to access files at remote sites using standard I/O with a few exceptions. The Remote I/O library (RIO) [15] allows programs to use the standard MPI-IO interface to access remote files. Within the Legion project [19], an object-oriented global file access system called Smart File Objects [41] has been developed. The I/O framework Armada [34] was designed for data-intensive Grid applications. The authors propose to access remote datasets through a network of distributed application objects, called ships. They provide small pieces of functionality that can be moved closer to the data by executing ships on remote hosts. So far, no system implementation has been reported. Grid Datafarm [18] is a Petascale data-intensive computing project initiated in Japan. The challenge involves construction of a Peta- to Exascale parallel file system exploiting local storage on PCs spread over the world-wide Grid.

The work presented in this paper provides the first framework for development of high-performance I/O support for Grid applications implemented in a high-level parallel programming language.

5   Conclusions

We have proposed a set of Grid-oriented HPF language constructs, based on the concept of a *data transfer object* and data replication mechanisms provided by underlying Grid services, for specifying parallel I/O operations on distributed multidimensional arrays in local and remote scientific data repositories. We designed the HPF extensions to be appropriate for use with a variety of parallel I/O libraries, and to be easy to master for HPF programmers with no special knowledge of I/O. We implemented the extensions in the VFC compiler, which translates the constructs into calls to the Panda parallel I/O library.

One interesting aspect of our proposed I/O constructs for HPF is that they offer great extensibility for the future. For example, their scope can be expanded to specify complex I/O access patterns. Further, an HPF compiler can analyze the code and offer hints about upcoming I/O operations to the underlying I/O library, to allow data to be prefetched and other preparations to be made. We plan to examine these possibilities in our future work.

References

1. Bassow, F.: IBM AIX Parallel I/O File System: Installation, Administration, and Use. IBM, Doc. SH34-6065-00 (1995)
2. Benkner, S., Neuhold, Ch., Egger, M., Sanjari, K., Velkov, B.: VFC - The Vienna HPF+ Compiler. In: Proceedings of the International Conference on Compilers for Parallel Computers, Linkoping (1998)
3. Bester, J., Foster, I., Kesselman, C., Tedesco, J., Tuecke, S.: GASS: A Data Movement and Access Service for Wide Area Computing Systems. In: Proceedings of the 6th Workshop on I/O in Parallel and Distributed Systems (1999)

4. Bordawekar, R.R., Choudhary, A.N.: Language and Compiler Support for Parallel I/O. In: Proceedings of the Working Conference on Programming Environments for Massively Parallel Distributed Systems, Switzerland (1994)
5. Brezany, P., Gerndt, M., Mehrotra, P., Zima, H.: Concurrent File Operations in a High Performance FORTRAN. In: Proceedings of Supercomputing '92 (1992) 230–237
6. Brezany, P., Czerwinski, P., Winslett, M.: A Generic Interface for Parallel Access to Large Data Sets from HPF Applications. Future Generation Computer Systems, 17 (2001) 977-985
7. Brezany, P., Winslett, M.: Advanced Data Repository Support for Java Scientific Programming. In: HPCN Europe 1999, Lecture Notes in Computer Science, Vol. 1593, Springer-Verlag, Berlin Heidelberg New York (1999) 1127–1136
8. Carpenter, B., Fox, G.: HPJava: Data Parallel Extensions to Java. In: Proceedings of the ACM Workshop on Java for High-Performance Network Computing, Palo Alto (1998)
9. Chervenak, A., Foster, I., Kesselman, C., Salisbury, C., Tuecke, S.: The Data Grid: Towards an Architecture for the Distributed Management and Analysis of Large Scientific Datasets. Journal of Network and Computer Applications (2001)
10. Choudhary, A. et. al.: PASSION: Parallel and Scalable Software for Input-Output. Technical Report SCCS-636, ECE Department, Syracuse University (1994)
11. Czajkowski, K., Fitzgerald, S., Foster, I., Kesselman, C.: Grid Information Services for Distributed Resource Sharing. In: Proceeding of the International Symposium on High Performance Distributed Computing (2001)
12. Denis, A., Perez, C., Priol, T.: Towards High Performance CORBA and MPI Middlewares for Grid Computing. In: Lee, C.A. (ed.) Grid Computing - GRID 2001, 2nd Int. Workshop, Denver, Lecture Notes in Computer Science, Vol. 2242, Springer Verlag, Berlin Heidelberg New York (2001)
13. Foster, I., Kesselman, C.: Globus: A Metacomputing Infrastructure Toolkit. International Journal on Supercomputer Applications, **2** (1997) 115-128
14. Foster, I., Kesselman, C., Tuecke, S.: The Anatomy of the Grid: Enabling Scalable Virtual Organizations. International Journal on Supercomp. Applications (2001)
15. Foster, I, Kohr, D., Krishnaiyer, R., Mogill, J.: Remote I/O: Fast Access to Distant Storage. In: Proceedings of the 5th Workshop on I/O in Parallel and Distributed Systems (1997) 14–25
16. Global Grid Forum. http://www.globalgridforum.org
17. Globus Project. Globus Toolkit 2.0 Beta Release. http://www.globus.org/gt2/
18. Grid Datafarm for Petascale Data Intensive Computing. http://datafarm.apgrid.org/
19. Grimshaw, A., Wulf, W., French, J., Weaver, A., Reynolds, P.: Legion: The Next Step toward a Nationwide Virtual Computer. Technical Report CS-94-21, Department of Computer Science, University of Virginia (1994)
20. High Performance Fortran Forum: High Performance Fortran. Version 2.0 (1997)
21. HPC++. High-Performance C++. http://www.extreme.indiana.edu/hpc++/
22. Huber, J., Elford, C.L., Reed, D.A., Chien, A.A., Bhune, S.S.: PPFS: A High-Performance Portable Parallel File System. In: Proceedings of the 9th ICS Conference, Barcelona (1995)
23. IETF: Rfc 2396. ftp://ftp.isi.edu/in-notes/rfc2396.txt
24. Koppenhoefer, K., Gullerud, A., Ruggieri, C., Dodds, R. Jr.: WARP3D: Dynamic Nonlinear Analysis of Solids Using a Preconditioned Conjugate Gradient Software Architecture. Structural Research Series 596, University of Illinois (1994)

550   Peter Brezany, Jonghyun Lee, and Marianne Winslett

25. Laure, E.: Distributed High Performance Computing with OpusJava. In: Proceedings of the ParCo99 Conference, Delft (1999)
26. Lee, J.: Web-based Data Migration for High-performance Scientific Codes. MS thesis, Department of Computer Science, University of Illinois at Urbana-Champaign (1999)
27. Messina, P.: Distributed Supercomputing Applications. In: Foster, I. Kesselman, C. (eds.), The Grid. Blueprint for a New Computing Infrastructure. Morgan Kaufmann (1999)
28. Metcalf, M., Reid, C.: Fortran 90/95 Explained. Oxford University Press (1996)
29. Moyer, S.A., Sunderam, V.S.: PIOUS: A Scalable Parallel I/O System for Distributed Computing Environment. In: Proceedings of the Scalable High-Performance Computing Conference (1994)
30. Nieplocha, J., Foster, I.: Disk Resident Arrays: An Array-Oriented I/O Library for Out-Of-Core Computations. In: Proceedings of the 6th Symposium on the Frontiers of Massively Parallel Computation, IEEE Computer Society Press, October (1996) 196–204
31. Nieuwejaar, M., Kotz, D.: The Galley Parallel File System. Parallel Computing, North-Holland (Elsevier Scientific) **23** (1997) 447–476
32. Numrich, R.W., Reid, J.K.: Co-Array Fortran for Parallel Programming. ACM Fortran Forum (1998)
33. Oldfield, R.: Summary of Existing and Developing Data Grids. White paper for the Remote Data Access group of the Global Grid Forum 1, Amsterdam (2001)
34. Oldfield, R., Kotz, D.: Armada: A Parallel File System for Computational Grids. In: Proceedings of the International Symposium on Cluster Computing and the Grid, Brisbane, Australia (2001) 194–201
35. OpenMP Consortium: OpenMP Fortran API, version 1.0 (1997)
36. Seamons, K.E., Winslett, M.: Multidimensional Array I/O in Panda 1.0. Journal of Supercomputing **10** (1995) 191–211
37. Segal, B.: Datagrid - Data Management. Deliverable DataGrid-D2.2 (2001)
38. Snir, M.: Proposal for I/O. Posted to HPFF I/O Forum (1992)
39. Stockinger, H.: Database Replication in World-Wide Distributed Data Grids. PhD Thesis, Institute of Computer Science and Business Informatics, University of Vienna, Austria (2001)
40. Thakur, R., Gropp, W., Lusk, E.: On Implementing MPI-IO Portably and with High Performance. In: Proceedings of the 6th Workshop on I/O in Parallel and Distributed Systems, ACM Press, (1999) 23–32
41. Weissman, J.: Smart File Objects: A Remote File Access Paradigm. In: Proceedings of the 6th Workshop on I/O in Parallel and Distributed Systems (1999)
42. Wolski, R., Spring, N., Hayes, J.: The Network Weather Service: A Distributed Resource Performance Forecasting Service for Metacomputing. Journal of Future Generation Computing Systems **15** (1999) 757-768

Optimization of HPF Programs with Dynamic Recompilation Technique

Takuya Araki, Hitoshi Murai, Tsunehiko Kamachi, and Yoshiki Seo

Parallel and Distributed Systems NEC Laboratory
Real World Computing Partnership

Abstract. Optimizing compilers perform various optimizations in order to exploit the best performance from computer systems. However, some kinds of optimizations cannot be applied if values of variables or system parameters are not known at compilation time. To solve this problem, we designed and implemented a system which collects such information at run time, and dynamically recompiles part of the program based on it. In our system, recompilation and management of runtime information are carried out on processors other than those which execute user programs. Therefore, recompilation cost does not affect the program execution time, unlike other similar systems. The evaluation result shows that quite high speedup can be attained with this method.

1 Introduction

Optimizing compilers perform various optimizations in order to exploit the best performance from computer systems. However, some kinds of optimizations cannot be applied if values of variables or system parameters are not known at compilation time.

For example, strength reduction, parallelization, and loop interchange are not allowed if values of variables which are used to examine correctness and effectiveness of the optimization are not known at compilation time. In addition, system parameters such as communication speed, cache size, and memory size may affect selections of optimization methods. To solve this problem, we designed and implemented a system which collects such information at run time, and dynamically recompiles part of the program based on it.

We used HPF (High Performance Fortran) in this system. By using HPF, various optimizations like parallelization and communication optimization can be achieved with dynamic recompilation.

We evaluated the system not only on homogeneous parallel computers, but also on a heterogeneous parallel system which is composed of a parallel computer Cenju-4 and a Linux-based PC cluster. Because system parameters may dynamically change on heterogeneous systems, dynamic recompilation would play an important role on such systems. Though current evaluation does not utilize such system parameters, we confirmed feasibility and effectiveness of the technique by the evaluation.

H. Zima et al. (Eds.): ISHPC 2002, LNCS 2327, pp. 551–562, 2002.

The reminder of this paper is structured as follows. Comparison with related work is discussed in section 2. Section 3 describes what kind of runtime optimization is targeted in this paper. Section 4 describes the dynamic recompilation system we implemented in this study. Section 5 shows evaluation result of the system. We summarize and conclude this paper in the last section.

2   Related Work

There are other optimization methods which use runtime information. One of them is multiple version loops [2]. It generates several versions of loops and selects one of them according to runtime information. This method does not need a runtime system unlike our system. However, in order to get optimal performance, this method should make a lot of versions of loops. This may cause explosion of code size.

There are many other systems which generate optimized code at runtime according to runtime information. Various JAVA JIT compilers are famous (one of recent researches in this field is shown in [1]). However, most of them utilize only frequency information of method calls, and decide whether it is effective to compile and optimize methods with the JIT compiler.

There are other systems which utilize other runtime information like values of variables, such as 'C [9], DyC [4], and Tempo [6]. But these systems target only sequential systems; thus, code generation and computation are done at the same processor. Since code generation time is added to total execution time, it should be fast; hence complex optimization methods cannot be applied. In addition, low-level code manipulation is needed in order to reduce code generation time. This makes these systems quite complex. On the other hand, our system performs recompilation on processors other than those which execute user programs. Thus, the effect of recompilation time is small.

Voss and Eigenmann proposed a system similar to ours [10]. Their system is interesting in that it can also be used to search optimal compilation flag at run time. Unlike our system, their system does not support distributed memory parallel systems, though it supports shared memory parallel systems.

3   Optimization with Dynamic Recompilation

Targets of dynamic recompilation are routines executed many times, like those in time expansion loops. Runtime information is collected in early iterations, and target routines are recompiled with the information. In later iterations, optimized routines are executed.

Currently, values of scalar variables and available memory size are used as runtime information. Following subsections describe what kind of optimizations can be applied using such information.

3.1   Values of Scalar Variables

If values of scalar variables do not change during execution (such variables are called "runtime constant"), specialization to the values is possible. Take the following programs for example:

```
do i = 1, 100        !HPF$ distribute a(block)    do j = 1, n
   a(i) = b(i) * c        do i = 1, 50                do i = 1, m
enddo                        a(i*2) = a(i*2-k)            ...a(i,j)...
                          enddo                       enddo
                                                   enddo
```

In the left program, if the value of c is known to be always 0, a(i) = b(i) * c can be replaced to a(i) = 0. This optimization eliminates multiplication and load of b(i).

In the middle program, this loop cannot be parallelized because the value of k is unknown. In this case, the distributed array a should be broadcast to all the processors in order to execute the loop sequentially. But if the value of k is odd, this loop can be parallelized, and only size k shift communication is needed.

In the right program, loop lengths of the two loops are given as variables. On vector super computers, it is desirable to make vector length long. Because innermost loops are vectorized, if loop length of the outer loop is longer than that of the inner loop, these loops are interchanged (if the dependencies permit).

But in this case, compilers cannot determine which loop length is long. In such a case, replacing n and m by constant values at runtime makes proper loop interchange possible.

Other possible optimizations using runtime constant include constant propagation/folding, conditional branch elimination, other loop transformations like loop fusion, loop collapse, and so on.

3.2 Available Memory Size

The HPF compiler used in this study supports two kinds of memory allocation scheme: "shrunk mode" and "noshrunk mode". In the shrunk mode, only fractions of arrays are allocated on each processor when the arrays are distributed. For example, if real a(100) is distributed onto 10 processors, 10 elements of a are allocated on each processor in the shrunk mode. In this mode, address translation is necessary. For example, if there is an access to a(15) in the original program, it should be translated to the access to the 5th element of a on the 2nd processor. This process is optimized as far as possible, but it is difficult to totally eliminate this cost.

On the other hand, all processors allocate the whole array (i.e. 100 elements of a) in the noshrunk mode. In the noshrunk mode, address translation is not necessary, but more memory is required. The noshrunk mode is referred to as "full shadow" in HPF/JA language specification [5].

If computers can be used exclusively, it is desired to allocate as many arrays as possible in the noshrunk mode in order to eliminate address translation cost. However, available memory size depends on target systems and input data. Thus, it is impossible to decide which arrays should be allocated in the noshrunk mode at compilation time. On the other hand, if available memory size can be known, it is possible to allocate optimal number of arrays in the noshrunk mode[1].

---

[1] In some cases, the shrunk mode is advantageous from the point of view of memory continuity and cache effect. However, we will not consider this effect here in order to simplify the discussion.

4 Dynamic Recompilation System

4.1 Directives

We introduced directives to specify which part of programs should be recompiled and how to optimize it with runtime information. We employed this strategy because it enables explicit control of dynamic recompilation, and the program still can be used as a valid HPF program.

In the current implementation, the unit of recompilation is a subroutine. There are two kinds of directives. One is for specifying target subroutines to be recompiled at runtime, which is specified at caller sites. The other is for specifying optimization methods, which is specified at callee sites.

As for caller site directives, take the following code for example:

```
      do iter = 1, 100
!DYN$ recompile,trigger(iter .eq. 10)
         call foo(a,b,c)
      enddo
```

Lines which start with !DYN$ are directives for dynamic recompilation. In this example, the directive specifies that the subroutine foo is the target of recompilation, and recompilation is triggered when the variable iter is equal to 10. Arbitrary logical expressions can be written in the trigger clause.

At callee sites, directives specify kind of optimization and its parameters. Take the following programs for example:

```
      subroutine foo(a,b,c)          subroutine foo(a)
      real a(100),b(100),c,d         real a(100), b(100)
      common /bar/ d           !HPF$ distribute (block) :: a,b
!DYN$ runtime_constant(c) begin !DYN$ candidates_of_fullshadow(b,a)
      do i = 1, 100                      ... a(i) ...
         a(i) = b(i) * c * d       end
      enddo
!DYN$ end runtime_constant
      end
```

Firstly, directives which specify optimization using runtime constant information are inserted in the left example. This program means that variable c enclosed by the directives is replaced by a constant value at recompilation time. This directive can be nested, and more than one variable can be specified in the directive.

Secondly, a directive which specifies optimization using available memory size information is inserted in the right example. This directive is inserted in the declaration part. In this case, array b and a are candidates of the noshrunk mode allocation in this order; b has higher priority than a. Note that all arrays are allocated in the shrunk mode before dynamic recompilation (except those explicitly specified to be allocated in the noshrunk mode). At runtime, available memory size and size of b and a are obtained as runtime information. When recompiling the subroutine, candidate arrays are allocated in the noshrunk mode in the specified order until available memory is used up.

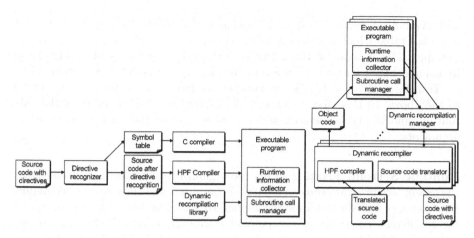

Fig. 1. Compilation System   **Fig. 2.** Runtime System

4.2   System Organization

In this section, we describe the organization of the system. It is composed of two parts: a (pre-execution) compilation system and a runtime system.

Compilation System. The compilation system generates executable programs according to directives. In the current implementation, the directive recognizer is implemented as a preprocessor of an HPF compiler. Figure 1 shows the organization of the compilation system.

The directive recognizer works as follows. Consider the first example given in Sect. 4.1. In this example, a directive is specified at a caller site. When the directive recognizer finds !DYN$ recompile,..., it translates the call to the subroutine foo into a library routine call "dcaller". Arguments of the library routine dcaller are: the original subroutine foo, a modified version of subroutine foo which collects runtime information during execution (after this, this subroutine is referred to as foo_info), a variable which is for holding a pointer to a recompiled subroutine, the logical value of the trigger clause, recompilation state which is used in dcaller, arguments of foo, and so on.

Next, take the second program given in Sect. 4.1 for example. In this case, directives !DYN$ runtime_constant are specified at a callee site. When the directive recognizer finds the directives, it generates a definition of a new subroutine foo_info. In this subroutine, a call to a library routine send_info(c,id_of_c) is inserted where originally !DYN$ ... begin was there. This library routine sends the value of the variable c and its id. The original subroutine foo without directives is also created.

In addition, if there are common blocks or subroutine calls in target subroutines, it is necessary to resolve the references to these symbols when the recompiled object code is loaded; these references should be linked with the corresponding addresses of the running program. To enable this, if there are such symbol references in target subroutines, the directive recognizer generates a symbol table which contains symbol names and its addresses in pairs. In the example, since a common block bar is used, address of the symbol (&bar) and its name ("bar") is stored to the table to look up the address from its name.

Transformation for !DYN$ candidates_of_fullshadow directives is similar to the above. In this case, send_info is inserted at the beginning of the subroutine, and available memory size and size of candidate arrays are given as arguments.

Runtime System. Figure 2 shows the organization of runtime system. Here, the executable program runs on a parallel computer, and the dynamic recompilation manager and the dynamic recompiler are executed on a different computer (or a different processor of the parallel computer) concurrently. As a result, overhead related to dynamic recompilation is not added to total execution time.

This approach reduces number of processors for computation. However, performance loss caused by this can be ignored if number of processors is large enough. For example, when there are 100 processors, the performance loss is only 1%.

The subroutine call manager (which corresponds to dcaller) works according to the current value of recompilation state as follows: (it corresponds to the argument of dcaller, and its initial value is "before recompilation")

before recompilation Check the logical value of the trigger clause.
  1. If it is false, execute the subroutine containing the runtime information collector (which corresponds to foo_info) and exit.
  2. If it is true, execute the subroutine containing the runtime information collector, and send recompilation request to the dynamic recompilation manager. Then change the state into "during recompilation", and exit.
during recompilation Ask the dynamic recompilation manager if recompilation finished.
  1. If recompilation successfully finished, change the state into "recompilation success". Then load recompiled object and execute it, and exit.
  2. If recompilation failed, change the state into "recompilation fail", execute the original subroutine and exit.
  3. If recompilation is not finished yet, execute the original subroutine and exit.
recompilation success Execute the previously loaded recompiled subroutine and exit.
recompilation fail Execute the original subroutine and exit.

This strategy makes it possible to recompile target subroutines in parallel with program execution.

The dynamic recompilation manager receives runtime information from the runtime information collector(which corresponds to send_info). When recompilation is requested, the dynamic recompiler is invoked and runtime information is passed to it.

The dynamic recompiler is composed of a source code translator and an HPF compiler. The source code translator modifies the original source code according to the runtime information. When the directive is !DYN$ runtime_constant, values of scalar variables are sent as runtime information. The most frequently occurred value is used for optimizing the source code. Here, the value of the variable may change after recompilation, thus "if statement" is inserted to check the value[2]. For example, the second example given in Sect. 4.1 is translated into the left program below if the most frequent value of c is 0 in the runtime information:

```
if(c .eq. 0) then                   subroutine foo(a)
  do i = 1, 100                      real a(100), b(100)
    a(i) = b(i) * 0 * d      !HPF$ distribute (block) :: a,b
  enddo                      !HPX$ noshrunk :: new_a,b
else                               copy_to_noshrunk(new_a,a)
  do i = 1, 100                    ... new_a(i) ...
    a(i) = b(i) * c * d            copy_to_shrunk(a,new_a)
  enddo                            end
endif
```

The backend compiler optimizes a(i) = b(i) * 0 * d into a(i) = 0.

When the directive is !DYN$ candidates_of_fullshadow, available memory size and size of candidate arrays are sent as runtime information. It is decided based on the information whether candidate arrays should be allocated in the noshrunk mode. Here, program translation varies according to the allocation class of candidate arrays. If candidate arrays are local arrays, they are simply allocated in the noshrunk mode. If they are arguments or variables declared in common blocks, new arrays are allocated in the noshrunk mode; values of original arrays are copied to the new arrays and computation is done on them. After the execution of the subroutine, values of the new arrays are copied back to the original arrays. The third example given in Sect. 4.1 is translated into the right program above if array b and a are both allocated in the noshrunk mode.

!HPX$ noshrunk is extension of our compiler which means specified arrays are allocated in the noshrunk mode. And copy_to_noshrunk and copy_to_shrunk copies values of shrunk(noshrunk) arrays to noshrunk(shrunk) arrays.

Object code is created by the HPF compiler after such transformation. The subroutine call manager loads the object code. Object code loader is implemented using GNU BFD library [3][3].

---

[2] This "if statement" can be removed by specifying another type of directive: !DYN$ runtime_constant_noguard. In this case, programmers are responsible for the fact that the value of the variable does not change.

[3] It would be possible to use dynamic link mechanism provided by operating systems (dlopen and dlsym routines). However, we did not use them because the platforms we used in this study do not support dynamic link mechanism. In addition, COFF format object files are directly manipulated on SX-4 system, because BFD library was not available on SX-4 system.

Object code loader works as follows: first, the loader reads each section (text, data, bss, etc) from the object file. Then, it checks the relocation entry, where information of unsolved symbol reference exists. If there are any unsolved references, the symbol table created at compilation time is looked up to obtain the address of the symbol. Then, reference to the symbol in the object file is changed to point to the address.

5 Evaluation

In this section, we describe evaluation results measured on a parallel computer Cenju-4 [7], a vector super computer SX-4 [8], and a heterogeneous parallel system which is composed of the Cenju-4 and a Linux-based PC cluster.

5.1 Cenju-4

Evaluation Condition. Executable programs run on PEs of the Cenju-4, and the dynamic compilation manager and the dynamic recompiler are executed on the front-end workstation of the Cenju-4. The Cenju-4 has 32 PEs connected via an interconnection network. Each PE consists of a 200MHz VR10000 RISC processor.

Three programs below are used for evaluation:

mul This program multiplies an array by a scalar value. Strength reduction is applied after recompilation. This corresponds to the first example of Sect. 3.1. The target subroutine is called 300 times.
cpy This program copies values within an array. This program cannot be parallelized at compilation time, but can be parallelized after recompilation. This corresponds to the second example of Sect. 3.1. The target subroutine is called 1000 times.
sp This is a program of NAS Parallel Benchmarks 1.0. The size of program we used is CLASS A. This program is used for evaluating optimization using available memory size information. Subroutines `xisweep`, `etasweep`, `ztasweep`, and `rhs` are recompiled. All local arrays and argument arrays accessed frequently are specified as candidates of noshrunk mode allocation. These subroutines are called 400 times.

In all programs, recompilation is triggered at the first iteration.

Evaluation Results. Table 1 shows the evaluation results of the programs. In this table, "Original" and "Recomp." show execution time (sec) of original programs and programs with recompilation. "Load Timing" means the iteration count when all recompiled subroutines are loaded. "Ideal" indicates execution time (sec) of ideal programs; in the ideal programs of `mul` and `cpy`, target variables of optimization are replaced by constants, and in that of `sp`, all arrays are allocated in the noshrunk mode. "Speedup" shows speedup ratio of the programs with recompilation compared to the original programs. "Ratio to Ideal"

Table 1. Evaluation on Cenju-4

#PE	mul						cpy						sp					
	1	2	4	8	16	32	1	2	4	8	16	32	1	2	4	8	16	32
Original	248	123	62.2	31.0	15.5	7.8	29.7	85.1	123	167	211	272	8821	4584	2438	1282	701	432
Recomp.	88.9	49.1	22.9	12.0	7.3	4.5	29.5	15.0	6.8	5.5	3.3	3.4	5476	2928	1586	861	484	333
Load Timing	3	6	9	15	26	38	45	18	12	9	8	7	6	9	15	28	48	74
Ideal	87.2	43.6	21.7	10.9	5.3	2.2	29.3	13.8	5.8	2.4	1.1	0.4	2521	1439	807	441	252	185
Speedup	2.8	2.5	2.7	2.6	2.1	1.7	1.0	5.7	18.1	30.0	63.9	80.0	1.6	1.6	1.5	1.5	1.4	1.3
Ratio to Ideal	0.98	0.89	0.95	0.91	0.73	0.49	0.99	0.92	0.85	0.44	0.33	0.12	0.46	0.49	0.51	0.51	0.52	0.56

shows relative speed of the programs with recompilation compared to the ideal programs.

Speedup ratio of `mul` is from 1.7 to 2.8. Ratio to the ideal program is from 0.98 to 0.49; it decreases as number of processors increases. This is because chance to execute optimized code is decreased; when many processors are used, execution speed increases and more computation is done during recompilation.

Speedup ratio of `cpy` is quite high: 80 times speedup at 32 processors. This is because there is expensive communication in the original program caused by a loop which cannot be parallelized, and the communication time increases as the number of processors increases. On the other hand, this loop is parallelized in the recompiled subroutine, and execution time decreases as the number of processors increases. Ratio to the ideal program shows similar tendency as `mul`.

In `sp`, all candidate arrays were allocated in the noshrunk mode, because available memory size was large enough. This program shows from 1.3 to 1.6 times speedup. Ratio to the ideal program is relatively low: about 0.5. This is because not all the arrays are candidates of noshrunk mode allocation, and there is copy overhead when argument arrays are allocated in the noshrunk mode.

5.2   SX-4

Evaluation Condition. Executable programs run on the SX-4, and the dynamic recompilation manager and the dynamic recompiler are executed on a PC. The vector length of the SX-4 system used for the evaluation is 256.

We used a program called "shallow" for evaluation. It is developed by APR. Inc. and solves a two-dimensional shallow water equations.

For evaluation, the program is modified so that problem size can be changed at execution time by reading it from a file. The problem size used for evaluation is 64 × 4096, and iteration number of time expansion loop was set to 1000. In addition, this program was modified not to distribute arrays; the program was executed on only one processor.

By setting problem size to 64 × 4096, most loops become 2-level nested loops whose inner loops iterate 64 times and outer loops iterate 4096 times. With dynamic recompilation, these iteration counts are treated as runtime constant. It is expected that this enables loop interchange and makes vector length long.

Recompiled subroutines were `calc1`, `calc2` and `calc3`, and recompilation is triggered at the first iteration.

560   Takuya Araki et al.

Table 2. Evaluation on SX-4

Original	35.4
Recomp.	20.8
Load Timing	99
Ideal	18.9
Speedup	1.7
Ratio to Ideal	0.91

Table 3. Evaluation on a heterogeneous system

System	PC cluster	Cenju-4	Hetero (1)	Hetero (2)
Original	33.5	48.0	19.5	24.3
Recomp.	21.1	18.8	13.9	10.8
Ideal	18.5	15.4	9.9	7.4
Speedup	1.58	2.55	1.40	2.25
Ratio to Ideal	0.88	0.82	0.71	0.69

Evaluation Results. Table 2 shows the evaluation results. Meaning of each table is the same as the previous evaluation. In the "ideal" program, problem size is set by `parameter` statements.

With dynamic recompilation, 1.7 times speedup was attained. It seemed that this is because vector length is made longer by loop interchange. To confirm this, we measured average vector length.

On SX-4, it is possible to profile programs by setting environment variables at execution time. We evaluated average vector length with this method. Average vector length of the original program was 64.1, that of the program with dynamic recompilation was 198.7, and that of the ideal program was 255.9. This proves that vector length was made longer by dynamic recompilation and it made the program faster.

In this evaluation, arrays are not distributed. This is because current implementation of the dynamic recompilation system replaces variables with constants on the HPF program level; therefore, even if beginnings and ends of loops are replaced by constants, they are modified to variables again in the parallelized fortran programs, which prevents expected optimization. To solve this problem is one of our future works.

5.3   Heterogeneous System

Evaluation Condition. The system we evaluated is composed of a Linux-based PC cluster and Cenju-4. The CPUs of the PC cluster are Pentium III 866MHz. They are connected via Gigabit Ethernet. The PC cluster and the Cenju-4 are connected via 100Base-T Ethernet.

Four processors of the PC cluster and four processors of the Cenju-4 are used for the evaluation. These processors are connected flatly by an MPI library which is implemented in our laboratory.

The dynamic recompilation manager is executed on another Linux-based PC. It also invokes the dynamic recompiler of the PC cluster. The dynamic recompiler of the Cenju-4 runs on the front-end workstation of the Cenju-4.

We used the previous program "mul" for this evaluation(the problem size is different), which only utilizes runtime constant information. Utilizing system parameters particular to heterogeneous systems is part of our future works.

Recompilation is triggered at the first iteration.

Evaluation Result. Table 3 shows the evaluation results. Meaning of each table is the same as the previous evaluations.

To compare the performance of the PC cluster with that of the Cenju-4, we also evaluated the program separately on the PC cluster and on the Cenju-4. Arrays are distributed with "BLOCK" distribution unlike the evaluation on the heterogeneous environment. On both the PC cluster and the Cenju-4, four processors (and four processes) are used for the evaluation.

Our HPF compiler accepts "GEN_BLOCK" distribution; this is a kind of block distribution which allows arbitrary size of block width. Amount of computation is proportional to the assigned data size because of owner computes rule. Thus, considering load balancing, it is desirable to distribute arrays using GEN_BLOCK so that assigned data size is proportional to the performance of each processor.

However, performance ratio of the PC cluster to Cenju-4 varies by conditions; Table 3 shows that the original program runs 1.4 times faster on the PC cluster than on the Cenju-4, but the ideal program runs 1.2 times faster on the Cenju-4 than on the PC cluster.

In order to evaluate the influence of load distribution, we evaluated the system using two distributions: "Hetero(1)" is based on the performance ratio of the original program, and "Hetero(2)" is based on that of the ideal program(block widths are not changed during program execution).

The evaluation results show that the dynamic recompilation technique is also effective on the heterogeneous system.

Comparing the performance of the program with recompilation, "Hetero(2)" provides better performance than "Hetero(1)". This is because triggering recompilation at the first iteration made recompiled routine run longer period than the pre-recompiled routine; since distribution of "Hetero(2)" is based on the performance of the ideal program, it is suitable for the recompiled routine. If recompilation is triggered at a later iteration, "Hetero(1)" can provides better performance than "Hetero(2)".

To always get the optimal performance, changing the block widths during program execution is needed. However, this is not easy because it cannot be known by programs when recompiled routines are loaded, and deciding appropriate block widths is difficult. To solve this problem is one of our future works.

6   Conclusions

In this study, we proposed, implemented, and evaluated a system which collects runtime information and recompiles part of the program according to the information. We used values of scalar variables and available memory size as runtime information. Evaluation results show that large speedup can be attained with dynamic recompilation. It clearly demonstrates effectiveness of our approach.

However, there remain many other programs which cannot be optimized with presented runtime optimization methods. Enhancing the system to support other runtime information and optimization methods to enlarge the applicable area is one of our future works.

We evaluated the system with four programs on three platforms. It is also included in future works to evaluate the system with more programs and on more platforms.

References

1. M. Arnold, S. Fink, D. Grove, M. Hind, and P. F. Sweeney. Adaptive optimization in the Jalapeño JVM. In *OOPSLA '00*, pages 47–65, 2000.
2. M. Byler, J. Davies, C. Huson, B. Leasure, and M. Wolfe. Multiple version loops. In *International Conference on Parallel Processing*, pages 312–318, 1987.
3. Free Software Foundation. LIB BFD, the binary file descriptor library, 1999. http://www.gnu.org/manual/bfd-2.9.1/bfd.html.
4. B. Grant, M. Mock, C. Chambers, and S. J. Eggers. An evaluation of staged runtime optimizations in DyC. In *Programming Language Design and Implementation*, pages 293–304, 1999.
5. Japan Association for High Performance Fortran. HPF/JA language specification, 1999. http://www.tokyo.rist.or.jp/jahpf/index-e.html.
6. R. Marlet. Tempo specializer - a partial evaluator for C. http://www.irisa.fr/compose/tempo.
7. T. Nakata, Y. Kanoh, K. Tatsukawa, S. Yanagida, N. Nishi, and H. Takayama. Architecture and the software environment of parallel computer Cenju-4. *NEC RESEARCH & DEVELOPMENT*, 39(4):385–390, 1998.
8. N. Nishi, S. Habata, M. Inoue, H. Matsumoto, and T. Kondo. SX-4 architecture for scalable parallel vector processing. In *International Symposium on Parallel and Distributed Supercomputing*, pages 45–50, 1995.
9. M. Poletto, W. C. Hsieh, D. R. Engler, and M. F. Kaashoek. 'C and tcc: A language and compiler for dynamic code generation. *ACM Transactions on Programming Languages and Systems*, 21(2):324–369, 1999.
10. M. J. Voss and R. Eigenmann. High-level adaptive program optimization with ADAPT. In *PPoPP*, pages 93–102, 2001.

Author Index

Lecture Notes in Computer Science

For information about Vols. 1–2251
please contact your bookseller or Springer-Verlag